Statistical Mechanics

T0202916

Franz Schwabl

Statistical Mechanics

Translated by William Brewer

Second Edition

With 202 Figures, 26 Tables,
and 195 Problems

 Springer

Professor Dr. Franz Schwabl
Physik-Department
Technische Universität München
James-Franck-Strasse
85747 Garching, Germany
E-mail: schwabl@ph.tum.de

Translator:
Professor William Brewer, PhD
Fachbereich Physik
Freie Universität Berlin
Arnimallee 14
14195 Berlin, Germany
E-mail: brewer@physik.fu-berlin.de

Title of the original German edition: *Statistische Mechanik*
(Springer-Lehrbuch) 3rd ed. ISBN 3-540-31095-9
© Springer-Verlag Berlin Heidelberg 2006

ISBN 978-3-642-06887-4   e-ISBN 978-3-540-36217-3

Springer is a part of Springer Science+Business Media

springer.com

© Springer-Verlag Berlin Heidelberg 2006
Softcover reprint of the hardcover 2nd edition 2006

Cover design: eStudio Calamar S. L., F. Steinen-Broo, Pau/Girona, Spain

A theory is all the more impressive the simpler its premises, the greater the variety of phenomena it describes, and the broader its area of application. This is the reason for the profound impression made on me by classical thermodynamics. It is the only general physical theory of which I am convinced that, within its regime of applicability, it will never be overturned (this is for the special attention of the skeptics in principle).

Albert Einstein

To my daughter Birgitta

Preface to the Second Edition

In this new edition, supplements, additional explanations and cross references have been added in numerous places, including additional problems and revised formulations of the problems. Figures have been redrawn and the layout improved. In all these additions I have pursued the goal of not changing the compact character of the book. I wish to thank Prof. W. Brewer for integrating these changes into his competent translation of the first edition. I am grateful to all the colleagues and students who have made suggestions to improve the book as well as to the publisher, Dr. Thorsten Schneider and Mrs. J. Lenz for their excellent cooperation.

Munich, December 2005   *F. Schwabl*

Preface to the Second Edition

Preface to the First Edition

This book deals with statistical mechanics. Its goal is to give a deductive presentation of the statistical mechanics of equilibrium systems based on a single hypothesis – the form of the microcanonical density matrix – as well as to treat the most important aspects of non-equilibrium phenomena. Beyond the fundamentals, the attempt is made here to demonstrate the breadth and variety of the applications of statistical mechanics. Modern areas such as renormalization group theory, percolation, stochastic equations of motion and their applications in critical dynamics are treated. A compact presentation was preferred wherever possible; it however requires no additional aids except for a knowledge of quantum mechanics. The material is made as understandable as possible by the inclusion of all the mathematical steps and a complete and detailed presentation of all intermediate calculations. At the end of each chapter, a series of problems is provided. Subsections which can be skipped over in a first reading are marked with an asterisk; subsidiary calculations and remarks which are not essential for comprehension of the material are shown in small print. Where it seems helpful, literature citations are given; these are by no means complete, but should be seen as an incentive to further reading. A list of relevant textbooks is given at the end of each of the more advanced chapters.

In the first chapter, the fundamental concepts of probability theory and the properties of distribution functions and density matrices are presented. In Chapter 2, the microcanonical ensemble and, building upon it, basic quantities such as entropy, pressure and temperature are introduced. Following this, the density matrices for the canonical and the grand canonical ensemble are derived. The third chapter is devoted to thermodynamics. Here, the usual material (thermodynamic potentials, the laws of thermodynamics, cyclic processes, etc.) are treated, with special attention given to the theory of phase transitions, to mixtures and to border areas related to physical chemistry. Chapter 4 deals with the statistical mechanics of ideal quantum systems, including the Bose–Einstein condensation, the radiation field, and superfluids. In Chapter 5, real gases and liquids are treated (internal degrees of freedom, the van der Waals equation, mixtures). Chapter 6 is devoted to the subject of magnetism, including magnetic phase transitions. Furthermore, related phenomena such as the elasticity of rubber are presented. Chapter 7 deals with the theory of phase transitions and critical phenomena; following a general overview, the fundamentals of renormalization group theory are given. In addition, the Ginzburg–Landau theory is introduced, and percolation is discussed (as a topic related to critical phenomena). The remaining three chapters deal with non-equilibrium processes: Brownian motion, the Langevin and Fokker–Planck equations and their applications as well as the theory of the Boltzmann equation and from it, the H-Theorem and hydrodynamic equations. In the final chapter, dealing with the topic of irreversiblility, fundamental considerations of how it occurs and of the transition to equilibrium are developed. In appendices, among other topics the Third Law and a derivation of the classical distribution function starting from quantum statistics are presented, along with the microscopic derivation of the hydrodynamic equations.

The book is recommended for students of physics and related areas from the 5th or 6th semester on. Parts of it may also be of use to teachers. It is suggested that students at first skip over the sections marked with asterisks or shown in small print, and thereby concentrate their attention on the essential core material.

This book evolved out of lecture courses given numerous times by the author at the Johannes Kepler Universität in Linz (Austria) and at the Technische Universität in Munich (Germany). Many coworkers have contributed to the production and correction of the manuscript: I. Wefers, E. Jörg-Müller, M. Hummel, A. Vilfan, J. Wilhelm, K. Schenk, S. Clar, P. Maier, B. Kaufmann, M. Bulenda, H. Schinz, and A. Wonhas. W. Gasser read the whole manuscript several times and made suggestions for corrections. Advice and suggestions from my former coworkers E. Frey and U. C. Täuber were likewise quite valuable. I wish to thank Prof. W. D. Brewer for his faithful translation of the text. I would like to express my sincere gratitude to all of them, along with those of my other associates who offered valuable assistance, as well as to Dr. H. J. Kölsch, representing the Springer-Verlag.

Munich, October 2002   *F. Schwabl*

Table of Contents

1. Basic Principles

1.1 Introduction

Statistical mechanics deals with the physical properties of systems which consist of a large number of particles, i.e. many-body systems, and it is based on the microscopic laws of nature. Examples of such many-body systems are gases, liquids, solids in their various forms (crystalline, amorphous), liquid crystals, biological systems, stellar matter, the radiation field, etc. Among their physical properties which are of interest are equilibrium properties (specific heat, thermal expansion, modulus of elasticity, magnetic susceptibility, etc.) and transport properties (thermal conductivity, electrical conductivity, etc.).

Long before it was provided with a solid basis by statistical mechanics, thermodynamics had been developed; it yields general relations between the macroscopic parameters of a system. The First Law of Thermodynamics was formulated by Robert Mayer in 1842. It states that the energy content of a body consists of the sum of the work performed on it and the heat which is put into it:

$$dE = \delta Q + \delta W . \qquad (1.1.1)$$

The fact that heat is a form of energy, or more precisely, that energy can be transferred to a body in the form of heat, was tested experimentally by Joule in the years 1843–1849 (experiments with friction).

The Second Law was formulated by Clausius and by Lord Kelvin (W. Thomson[1]) in 1850. It is based on the fact that a particular state of a thermodynamic system can be reached through different ways of dividing up the energy transferred to it into work and heat, i.e. heat is not a "state variable" (a state variable is a physical quantity which is determined by the state of the system; this concept will be given a mathematically precise definition later). The essential new information in the Second Law was that there exists a state variable S, the entropy, which for reversible changes is related to the quantity of heat transferred by the equation

---

[1] Born W. Thomson; the additional name was assumed later in connection with his knighthood, granted in recognition of his scientific achievements.

$$\delta Q = TdS \, , \tag{1.1.2}$$

while for irreversible processes, $\delta Q < TdS$ holds. The Second Law is identical with the statement that a perpetual motion machine of the second kind is impossible to construct (this would be a periodically operating machine which performs work by only extracting heat from a single heat bath).

The atomistic basis of thermodynamics was first recognized in the kinetic theory of dilute gases. The velocity distribution derived by Maxwell (1831–1879) permits the derivation of the caloric and thermal equation of state of ideal gases. Boltzmann (1844–1906) wrote the basic transport equation which bears his name in the year 1874. From it, he derived the entropy increase (H theorem) on approaching equilibrium. Furthermore, Boltzmann realized that the entropy depends on the number of states $W(E, V, \ldots)$ which are compatible with the macroscopic values of the energy E, the volume $V, \ldots$ as given by the relation

$$S \propto \log W(E, V, \ldots) \, . \tag{1.1.3}$$

It is notable that the atomistic foundations of the theory of gases were laid at a time when the atomic structure of matter had not yet been demonstrated experimentally; it was even regarded with considerable scepticism by well-known physicists such as E. Mach (1828–1916), who favored continuum theories.

The description of macroscopic systems in terms of statistical ensembles was justified by Boltzmann on the basis of the ergodic hypothesis. Fundamental contributions to thermodynamics and to the statistical theory of macroscopic systems were made by J. Gibbs (1839–1903) in the years 1870–1900.

Only after the formulation of quantum mechanics (1925) did the correct theory for the atomic regime become available. To distinguish it from classical statistical mechanics, the statistical mechanics based on the quantum theory is called quantum statistics. Many phenomena such as the electronic properties of solids, superconductivity, superfluidity, or magnetism can be explained only by applying quantum statistics.

Even today, statistical mechanics still belongs among the most active areas of theoretical physics: the theory of phase transitions, the theory of liquids, disordered solids, polymers, membranes, biological systems, granular matter, surfaces, interfaces, the theory of irreversible processes, systems far from equilibrium, nonlinear processes, structure formation in open systems, biological processes, and at present still magnetism and superconductivity are fields of active interest.

Following these remarks about the problems treated in statistical mechanics and its historical development, we now indicate some characteristic problems which play a role in the theory of macroscopic systems. Conventional macroscopic systems such as gases, liquids and solids at room temperature consist of 10^{19}–10^{23} particles per cm^3. The number of quantum-mechanical eigenstates naturally increases as the number of particles. As we shall see

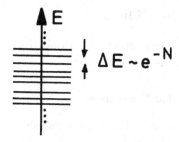

Fig. 1.1. Spacing of the energy levels for a large number of particles N.

later, the separation of the energy levels is of the order of e^{-N}, i.e. the energy levels are so densely spaced that even the smallest perturbation can transfer the system from one state to another one which has practically the same energy.

Should we now set ourselves the goal of calculating the motion of the $3N$ coordinates in classical physics, or the time dependence of the wavefunctions in quantum mechanics, in order to compute temporal averages from them? Both programs would be impossible to carry out and are furthermore unnecessary. One can solve neither Newton's equations nor the Schrödinger equation for 10^{19}–10^{23} particles. And even if we had the solutions, we would not know all the coordinates and velocities or all the quantum numbers required to determine the initial values. Furthermore, the detailed time development plays no role for the macroscopic properties which are of interest. In addition, even the weakest interaction (external perturbation), which would always be present even with the best possible isolation of the system from its environment, would lead to a change in the microscopic state without affecting the macroscopic properties. For the following discussion, we need to define two concepts.

The microstate: it is defined by the wavefunction of the system in quantum mechanics, or by all the coordinates and momenta of the system in classical physics.

The macrostate: this is characterized by a few macroscopic quantities (energy, volume, ...).

From the preceding considerations it follows that the state of a macroscopic system must be described statistically. The fact that the system passes through a distribution of microstates during a measurement requires that we characterize the macrostate by giving the probabilities for the occurrence of particular microstates. The collection of all the microstates which represent a macrostate, weighted by their frequency of occurrence, is referred to as a *statistical ensemble*.

Although the state of a macroscopic system is characterized by a statistical ensemble, the predictions of macroscopic quantities are precise. Their mean values and mean square deviations are both proportional to the number of particles N. The relative fluctuations, i.e. the ratio of fluctuations to mean values, tend towards zero in the thermodynamic limit (see (1.2.21c)).

1.2 A Brief Excursion into Probability Theory

At this point, we wish to collect a few basic mathematical definitions from probability theory, in order to derive the central limit theorem.[2]

1.2.1 Probability Density and Characteristic Functions

We first have to consider the meaning of the concept of a *random variable*. This refers to a quantity X which takes on values x depending upon the elements e of a "set of events" E. In each individual observation, the value of X is uncertain; instead, one knows only the probability for the occurrence of one of the possible results (events) from the set E. For example, in the case of an ideal die, the random variable is the number of spots, which can take on values between 1 and 6; each of these events has the probability $1/6$. If we had precise knowledge of the initial position of the die and the forces acting on it during the throw, we could calculate the result from classical mechanics. Lacking such detailed information, we can make only the probability statement given above. Let $e \in E$ be an event from the set E and P_e be its corresponding probability; then for a large number of attempts, N, the number of times N_e that the event e occurs is related to P_e by $\lim_{N\to\infty} \frac{N_e}{N} = P_e$.

Let X be a *random variable*. If the values x which X can assume are continuously distributed, we define the *probability density* of the random variable to be $w(x)$. This means that $w(x)dx$ is the probability that X assumes a value in the interval $[x, x+dx]$. The total probability must be one, i.e. $w(x)$ is normalized to one:

$$\int_{-\infty}^{+\infty} dx\, w(x) = 1 \ . \tag{1.2.1}$$

Definition 1: The mean value of X is defined by

$$\langle X \rangle = \int_{-\infty}^{+\infty} dx\, w(x)\, x \ . \tag{1.2.2}$$

Now let $F(X)$ be a function of the random variable X; one then calls $F(X)$ a random function. Its mean value is defined corresponding to (1.2.2) by[3]

$$\langle F(X) \rangle = \int dx\, w(x) F(x) \ . \tag{1.2.2'}$$

The powers of X have a particular importance: their mean values will be used to introduce the *moments* of the probability density.

---

[2] See e.g.: W. Feller, *An Introduction to Probability Theory and its Applications*, Vol. I (Wiley, New York 1968).

[3] In the case that the limits of integration are not given, the integral is to be taken from $-\infty$ to $+\infty$. An analogous simplified notation will also be used for integrals over several variables.

Definition 2: The *nth moment* of the probability density $w(x)$ is defined as

$$\mu_n = \langle X^n \rangle . \tag{1.2.3}$$

(The first moment of $w(x)$ is simply the mean value of X.)
Definition 3: The mean square deviation (or variance) is defined by

$$(\Delta x)^2 = \langle X^2 \rangle - \langle X \rangle^2 = \langle (X - \langle X \rangle)^2 \rangle . \tag{1.2.4}$$

Its square root is called the root-mean-square deviation or standard deviation.

Definition 4: Finally, we define the *characteristic function:*

$$\chi(k) = \int dx \, e^{-ikx} w(x) \equiv \langle e^{-ikX} \rangle . \tag{1.2.5}$$

By taking its inverse Fourier transform, $w(x)$ can be expressed in terms of $\chi(k)$:

$$w(x) = \int \frac{dk}{2\pi} \, e^{ikx} \chi(k) . \tag{1.2.6}$$

Under the assumption that all the moments of the probability density $w(x)$ exist, it follows from Eq. (1.2.5) that the characteristic function is

$$\chi(k) = \sum_n \frac{(-ik)^n}{n!} \langle X^n \rangle . \tag{1.2.7}$$

If X has a discrete spectrum of values, i.e. the values $\xi_1, \xi_2, \ldots$ can occur with probabilities $p_1, p_2, \ldots$, the probability density has the form

$$w(x) = p_1 \delta(x - \xi_1) + p_2 \delta(x - \xi_2) + \ldots . \tag{1.2.8}$$

Often, the probability density will have discrete and continuous regions.

In the case of *multidimensional* systems (those with several components) $\mathbf{X} = (X_1, X_2, \ldots)$, let $\mathbf{x} = (x_1, x_2, \ldots)$ be the values taken on by $\mathbf{X}$. Then the probability density (also called the joint probability density) is $w(\mathbf{x})$ and it has the following significance: $w(\mathbf{x})d\mathbf{x} \equiv w(\mathbf{x})dx_1 dx_2 \ldots dx_N$ is the probability of finding $\mathbf{x}$ in the hypercubic element $\mathbf{x}, \mathbf{x} + d\mathbf{x}$. We will also use the term probability distribution or, for short, simply the distribution.
Definition 5: The mean value of a function $F(\mathbf{X})$ of the random variables $\mathbf{X}$ is defined by

$$\langle F(\mathbf{X}) \rangle = \int d\mathbf{x} \, w(\mathbf{x}) F(\mathbf{x}) . \tag{1.2.9}$$

Theorem: The probability density of $F(\mathbf{X})$
A function F of the random variables $\mathbf{X}$ is itself a random variable, which can take on the values f corresponding to a probability density $w_F(f)$. The

probability density $w_F(f)$ can be calculated from the probability density $w(\mathbf{x})$. We assert that:

$$w_F(f) = \langle \delta(F(\mathbf{X}) - f) \rangle \,. \tag{1.2.10}$$

Proof: We express the probability density $w_F(f)$ in terms of its characteristic function

$$w_F(f) = \int \frac{dk}{2\pi} e^{ikf} \sum_n \frac{(-ik)^n}{n!} \langle F^n \rangle \,.$$

If we insert $\langle F^n \rangle = \int d\mathbf{x}\, w(\mathbf{x}) F(\mathbf{x})^n$, we find

$$w_F(f) = \int \frac{dk}{2\pi} e^{ikf} \int d\mathbf{x}\, w(\mathbf{x}) e^{-ikF(\mathbf{x})}$$

and, after making use of the Fourier representation of the δ-function $\delta(y) = \int \frac{dk}{2\pi} e^{iky}$, we finally obtain

$$w_F(f) = \int d\mathbf{x}\, w(\mathbf{x}) \delta(f - F(\mathbf{x})) = \langle \delta(F(\mathbf{X}) - f) \rangle \,,$$

i.e. Eq. (1.2.10).

Definition 6: For multidimensional systems we define *correlations*

$$K_{ij} = \langle (X_i - \langle X_i \rangle)(X_j - \langle X_j \rangle) \rangle \tag{1.2.11}$$

of the random variables X_i and X_j. These indicate to what extent fluctuations (deviations from the mean value) of X_i and X_j are correlated.

If the probability density has the form

$$w(\mathbf{x}) = w_i(x_i) w'(\{x_k, k \neq i\}) \,,$$

where $w'(\{x_k, k \neq i\})$ does not depend on x_i, then $K_{ij} = 0$ for $j \neq i$, i.e. X_i and X_j are not correlated. In the special case

$$w(\mathbf{x}) = w_1(x_1) \cdots w_N(x_N) \,,$$

the stochastic variables $X_1, \ldots, X_N$ are completely uncorrelated.

Let $P_n(x_1, \ldots, x_{n-1}, x_n)$ be the probability density of the random variables $X_1, \ldots, X_{n-1}, X_n$. Then the probability density for a subset of these random variables is given by integration of P_n over the range of values of the remaining random variables; e.g. the probability density $P_{n-1}(x_1, \ldots, x_{n-1})$ for the random variables $X_1, \ldots, X_{n-1}$ is

$$P_{n-1}(x_1, \ldots, x_{n-1}) = \int dx_n\, P_n(x_1, \ldots, x_{n-1}, x_n) \,.$$

Finally, we introduce the concept of conditional probability and the conditional probability density.

Definition 7: Let $P_n(x_1, \ldots, x_n)$ be the probability (density). The conditional probability (density)

$$P_{k|n-k}(x_1, \ldots, x_k | x_{k+1}, \ldots, x_n)$$

is defined as the probability (density) of the random variables $x_1, \ldots, x_k$, if the remaining variables $x_{k+1}, \ldots, x_n$ have given values. We find

$$P_{k|n-k}(x_1, \ldots, x_k | x_{k+1}, \ldots, x_n) = \frac{P_n(x_1, \ldots, x_n)}{P_{n-k}(x_{k+1}, \ldots, x_n)} , \tag{1.2.12}$$

where $P_{n-k}(x_{k+1}, \ldots, x_n) = \int dx_1 \ldots dx_k P_n(x_1, \ldots, x_n)$.

Note concerning conditional probability: formula (1.2.12) is usually introduced as a definition in the mathematical literature, but it can be deduced in the following way, if one identifies the probabilities with statistical frequencies: $P_n(x_1, \ldots, x_k, x_{k+1}, \ldots, x_n)$ for fixed $x_{k+1}, \ldots, x_n$ determines the frequencies of the $x_1, \ldots, x_k$ with given values of $x_{k+1}, \ldots, x_n$. The probability density which corresponds to these frequencies is therefore proportional to $P_n(x_1, \ldots, x_k, x_{k+1}, \ldots, x_n)$. Since $\int dx_1 \ldots dx_k P_n(x_1, \ldots, x_k, x_{k+1}, \ldots, x_n) = P_{n-k}(x_{k+1}, \ldots, x_n)$, the conditional probability density normalized to one is then

$$P_{k|n-k}(x_1, \ldots, x_k | x_{k+1}, \ldots, x_n) = \frac{P_n(x_1, \ldots, x_n)}{P_{n-k}(x_{k+1}, \ldots, x_n)} .$$

1.2.2 The Central Limit Theorem

Let there be mutually independent random variables $X_1, X_2, \ldots, X_N$ which are characterized by common but independent probability distributions $w(x_1)$, $w(x_2), \ldots, w(x_N)$. Suppose that the mean value and the variance of $X_1, X_2, \ldots, X_N$ exist. We require the probability density for the sum

$$Y = X_1 + X_2 + \ldots + X_N \tag{1.2.13}$$

in the limit $N \to \infty$. As we shall see, the probability density for Y is given by a Gaussian distribution.

Examples of applications of this situation are

a) A system of *non-interacting particles*
    X_i = energy of the i-th particle, Y = total energy of the system
b) *The random walk*
    X_i = distance covered in the i-th step, Y = location after N steps.

In order to carry out the computation of the probability density of Y in a convenient way, it is expedient to introduce the random variable Z:

$$Z = \sum_i (X_i - \langle X \rangle)/\sqrt{N} = (Y - N\langle X \rangle)/\sqrt{N} , \tag{1.2.14}$$

where $\langle X \rangle \equiv \langle X_1 \rangle = \ldots = \langle X_N \rangle$ by definition.

From (1.2.10), the probability density $w_Z(z)$ of the random variables Z is given by

$$w_Z(z) = \int dx_1 \ldots dx_N\, w(x_1) \ldots w(x_N)\, \delta\left(z - \frac{x_1 + \ldots + x_N}{\sqrt{N}} + \sqrt{N}\langle X\rangle\right)$$

$$= \int \frac{dk}{2\pi} e^{ikz} \int dx_1 \ldots dx_N\, w(x_1) \ldots w(x_N) e^{-\frac{ik(x_1 + \ldots + x_N)}{\sqrt{N}} + ik\sqrt{N}\langle X\rangle}$$

$$= \int \frac{dk}{2\pi} e^{ikz + ik\sqrt{N}\langle X\rangle} \left(\chi\left(\frac{k}{\sqrt{N}}\right)\right)^N , \tag{1.2.15}$$

where $\chi(q)$ is the characteristic function of $w(x)$.

The representation (1.2.7) of the characteristic function in terms of the moments of the probability density can be reformulated by taking the logarithm of the expansion in moments,

$$\chi(q) = \exp\left[-iq\langle X\rangle - \frac{1}{2}q^2(\Delta x)^2 + \ldots q^3 + \ldots\right], \tag{1.2.16}$$

i.e. in general

$$\chi(q) = \exp\left[\sum_{n=1}^{\infty} \frac{(-iq)^n}{n!} C_n\right]. \tag{1.2.16'}$$

In contrast to (1.2.7), in (1.2.16') the logarithm of the characteristic function is expanded in a power series. The expansion coefficients C_n which occur in this series are called *cumulants of the nth order*. They can be expressed in terms of the moments (1.2.3); the three lowest take on the forms:

$$\begin{aligned} C_1 &= \langle X\rangle = \mu_1 \\ C_2 &= (\Delta x)^2 = \langle X^2\rangle - \langle X\rangle^2 = \mu_2 - \mu_1^2 \\ C_3 &= \langle X^3\rangle - 3\langle X^2\rangle\langle X\rangle + 2\langle X\rangle^3 = \mu_3 - 3\mu_1\mu_2 + 2\mu_1^3 . \end{aligned} \tag{1.2.17}$$

The relations (1.2.17) between the cumulants and the moments can be obtained by expanding the exponential function in (1.2.16) or in (1.2.16') and comparing the coefficients of the Taylor series with (1.2.7). Inserting (1.2.16) into (1.2.15) yields

$$w_Z(z) = \int \frac{dk}{2\pi} e^{ikz - \frac{1}{2}k^2(\Delta x)^2 + \ldots k^3 N^{-\frac{1}{2}} + \ldots} . \tag{1.2.18}$$

From this, neglecting the terms which vanish for large N as $1/\sqrt{N}$ or more rapidly, we obtain

$$w_Z(z) = \left(2\pi(\Delta x)^2\right)^{-1/2} e^{-\frac{z^2}{2(\Delta x)^2}} \tag{1.2.19}$$

and finally, using $W_Y(y)dy = W_Z(z)dz$ for the probability density of the random variables Y,

$$w_Y(y) = \left(2\pi N(\Delta x)^2\right)^{-1/2} e^{-\frac{(y - \langle X\rangle N)^2}{2(\Delta x)^2 N}} . \tag{1.2.20}$$

This is the *central limit theorem*: $w_Y(y)$ is a Gaussian distribution, although we did not in any way assume that $w(x)$ was such a distribution,

$$\text{mean value:} \qquad \langle Y \rangle = N \langle X \rangle \qquad\qquad (1.2.21a)$$

$$\text{standard deviation:} \qquad \Delta y = \Delta x \sqrt{N} \qquad\qquad (1.2.21b)$$

$$\text{relative deviation:} \quad \frac{\Delta y}{\langle Y \rangle} = \frac{\Delta x \sqrt{N}}{N \langle X \rangle} = \frac{\Delta x}{\langle X \rangle \sqrt{N}} \ . \qquad (1.2.21c)$$

The central limit theorem provides the mathematical basis for the fact that in the limiting case of large N, predictions about Y become sharp. From (1.2.21c), the relative deviation, i.e. the ratio of the standard deviation to the mean value, approaches zero in the limit of large N.

1.3 Ensembles in Classical Statistics

Although the correct theory in the atomic regime is based on quantum mechanics, and classical statistics can be derived from quantum statistics, it is more intuitive to develop classical statistics from the beginning, in parallel to quantum statistics. Later, we shall derive the classical distribution function within its range of validity from quantum statistics.

1.3.1 Phase Space and Distribution Functions

We consider N particles in three dimensions with coordinates $q_1, \ldots, q_{3N}$ and momenta $p_1, \ldots, p_{3N}$. Let us define *phase space*, also called Γ space, as the space which is spanned by the $6N$ coordinates and momenta. A microscopic state is represented by a point in the Γ space and the motion of the overall system by a curve in phase space (Fig. 1.2), which is also termed a phase-space orbit or phase-space trajectory.

As an example, we consider the *one-dimensional harmonic oscillator*

$$\begin{aligned} q &= q_0 \cos \omega t \\ p &= -m q_0 \omega \sin \omega t \ , \end{aligned} \qquad\qquad (1.3.1)$$

whose orbit in phase space is shown in Fig. 1.3.

For large N, the phase space is a space of many dimensions. As a rule, our knowledge of such a system is not sufficient to determine its position in phase space. As already mentioned in the introductory section 1.1, a macrostate characterized by macroscopic values such as that of its energy E, volume V, number of particles N etc., can be generated equally well by any one of a large number of microstates, i.e. by a large number of points in phase space. Instead of singling out just one of these microstates arbitrarily, we consider all of them, i.e. an ensemble of systems which all represent one and the same macrostate but which contains all of the corresponding possible microstates.

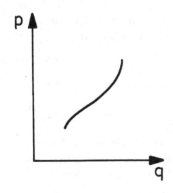

Fig. 1.2. A trajectory in phase space. Here, q and p represent the $6N$ coordinates and momenta $q_1, \ldots, q_{3N}$ and $p_1, \ldots, p_{3N}$.

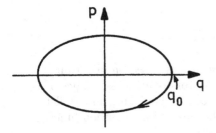

Fig. 1.3. The phase-space orbit of the one-dimensional harmonic oscillator.

The weight with which a point $(q,p) \equiv (q_1, \ldots, q_{3N}, p_1, \ldots, p_{3N})$ occurs at the time t is given by the probability density $\rho(q, p, t)$.

The introduction of this probability density is now not at all just an expression of our lack of knowledge of the detailed form of the microstates, but rather it has the following physical basis: every realistic macroscopic system, even with the best insulation from its surroundings, experiences an interaction with its environment. This interaction is to be sure so weak that it does not affect the macroscopic properties of the system, i.e. the macrostate remains unchanged, but it induces the system to change its microstate again and again and thus causes it for example to pass through a distribution of microstates during a measurement process. These states, which are occupied during a short time interval, are collected together in the distribution $\rho(q, p)$. This distribution thus describes not only the statistical properties of a fictitious ensemble of many copies of the system considered in its diverse microstates, but also each individual system. Instead of considering the sequential stochastic series of these microstates in terms of time-averaged values, we can observe the simultaneous time development of the whole ensemble. It will be a major task in the following chapter to determine the distribution functions which correspond to particular physical situations. To this end, knowledge of the equation of motion which we derive in the next section will prove to be very important. For large N, we know only the probability distribution $\rho(q, p, t)$. Here,

$$\rho(q,p,t)dqdp \equiv \rho(q_1,\ldots,q_{3N},p_1,\ldots,p_{3N},t)\prod_{i=1}^{3N}dq_idp_i \qquad (1.3.2)$$

is the probability of finding a system of the ensemble (or the individual systems in the course of the observation) at time t within the phase-space volume element $dqdp$ in the neighborhood of the point q,p in Γ space. $\rho(q,p,t)$ is called the distribution function. It must be positive, $\rho(q,p,t) \geq 0$, and normalizable. Here, q,p stand for the whole of the coordinates and momenta $q_1,\ldots,q_{3N},p_1,\ldots,p_{3N}$.

1.3.2 The Liouville Equation

We now wish to determine the time dependence of $\rho(q,p,t)$, beginning with the initial distribution $W(q_0,p_0)$ at time $t = 0$ on the basis of the classical Hamiltonian H. We shall assume that the system is closed. The following results are however also valid when H contains time-dependent external forces. We first consider a system whose coordinates in phase space at $t = 0$ are q_0 and p_0. The associated trajectory in phase space, which follows from the Hamiltonian equations of motion, is denoted by $q(t;q_0,p_0),p(t;q_0,p_0)$, with the intitial values of the trajectories given here explicitly. For a single trajectory, the probability density of the coordinates q and the momenta p has the form

$$\delta\big(q - q(t;q_0,p_0)\big)\delta\big(p - p(t;q_0,p_0)\big) . \qquad (1.3.3)$$

Here, $\delta(k) \equiv \delta(k_1)\ldots\delta(k_{3N})$. The initial values are however in general not precisely known; instead, there is a distribution of values, $W(q_0,p_0)$. In this case, the probability density in phase space at the time t is found by multiplication of (1.3.3) by $W(q_0,p_0)$ and integration over the initial values:

$$\rho(q,p,t) = \int dq_0 \int dp_0\, W(q_0,p_0)\delta\big(q-q(t;q_0,p_0)\big)\delta\big(p-p(t;q_0,p_0)\big) . \quad (1.3.3')$$

We wish to derive an equation of motion for $\rho(q,p,t)$. To this end, we use the Hamiltonian equations of motion

$$\dot{q}_i = \frac{\partial H}{\partial p_i}, \quad \dot{p}_i = -\frac{\partial H}{\partial q_i} . \qquad (1.3.4)$$

The velocity in phase space

$$\boldsymbol{v} = (\dot{q},\dot{p}) = \left(\frac{\partial H}{\partial p}, -\frac{\partial H}{\partial q}\right) \qquad (1.3.4')$$

fulfills the equation

$$\operatorname{div} \boldsymbol{v} \equiv \sum_i \left(\frac{\partial \dot{q}_i}{\partial q_i} + \frac{\partial \dot{p}_i}{\partial p_i}\right) = \sum_i \left(\frac{\partial^2 H}{\partial q_i \partial p_i} - \frac{\partial^2 H}{\partial p_i \partial q_i}\right) = 0 . \qquad (1.3.5)$$

That is, the motion in phase space can be treated intuitively as the "flow" of an incompressible "fluid".

Taking the time derivative of $(1.3.3')$, we find

$$\frac{\partial \rho(q, p, t)}{\partial t}$$

$$= -\sum_i \int dq_0 dp_0 W(q_0, p_0) \left(\dot{q}_i(t; q_0, p_0) \frac{\partial}{\partial q_i} + \dot{p}_i(t; q_0, p_0) \frac{\partial}{\partial p_i} \right)$$

$$\times \delta(q - q(t; q_0, p_0)) \delta(p - p(t; q_0, p_0)) . \quad (1.3.6)$$

Expressing the velocity in phase space in terms of (1.3.4), employing the δ-functions in (1.3.6), and finally using $(1.3.3')$ and (1.3.5), we obtain the following representations of the equation of motion for $\rho(q, p, t)$:

$$\frac{\partial \rho}{\partial t} = -\sum_i \left(\frac{\partial}{\partial q_i} \rho \dot{q}_i + \frac{\partial}{\partial p_i} \rho \dot{p}_i \right)$$

$$= -\sum_i \left(\frac{\partial \rho}{\partial q_i} \dot{q}_i + \frac{\partial \rho}{\partial p_i} \dot{p}_i \right) \qquad (1.3.7)$$

$$= \sum_i \left(-\frac{\partial \rho}{\partial q_i} \frac{\partial H}{\partial p_i} + \frac{\partial \rho}{\partial p_i} \frac{\partial H}{\partial q_i} \right) .$$

Making use of the Poisson bracket notation[4], the last line of Eq. (1.3.7) can also be written in the form

$$\frac{\partial \rho}{\partial t} = -\{H, \rho\} \qquad (1.3.8)$$

This is the *Liouville equation*, the fundamental equation of motion of the classical distribution function $\rho(q, p, t)$.

Additional remarks:

We discuss some equivalent representations of the Liouville equation and their consequences.

(i) The first line of the series of equations (1.3.7) can be written in abbreviated form as an equation of continuity

$$\frac{\partial \rho}{\partial t} = -\text{ div } \boldsymbol{v} \rho . \qquad (1.3.9)$$

One can imagine the motion of the ensemble in phase space to be like the flow of a fluid. Then (1.3.9) is the equation of continuity for the density and Eq. (1.3.5) shows that the fluid is incompressible.

---

[4] $\{u, v\} \equiv \sum_i \left[\frac{\partial u}{\partial p_i} \frac{\partial v}{\partial q_i} - \frac{\partial u}{\partial q_i} \frac{\partial v}{\partial p_i} \right]$

(ii) We once more take up the analogy of motion in phase space to fluid hydrodynamics: in our previous discussion, we considered the density at a fixed point q, p in Γ space. However, we could also consider the motion from the point of view of an observer moving with the "flow", i.e. we could ask for the time dependence of $\rho(q(t), p(t), t)$ (omitting the initial values of the coordinates, q_0 and p_0, for brevity). The second line of Eq. (1.3.7) can also be expressed in the form

$$\frac{d}{dt}\rho\big(q(t), p(t), t\big) = 0 . \qquad (1.3.10)$$

Hence, the distribution function is constant along a trajectory in phase space.
(iii) We now investigate the change of a volume element $d\Gamma$ in phase space. At $t = 0$, let a number dN of representatives of the ensemble be uniformly distributed within a volume element $d\Gamma_0$. Owing to the motion in phase space, they occupy a volume $d\Gamma$ at the time t. This means that the density ρ at $t = 0$ is given by $\frac{dN}{d\Gamma_0}$, while at time t, it is $\frac{dN}{d\Gamma}$. From (1.3.10), the equality of these two quantities follows, from which we find (Fig. 1.4) that their volumes are the same:

$$d\Gamma = d\Gamma_0 . \qquad (1.3.11)$$

Equation (1.3.8) is known in mechanics as the Liouville theorem.[5] There, it is calculated from the Jacobian with the aid of the theory of canonical transformations. Reversing this process, we can begin with Eq. (1.3.11) and derive Eq. (1.3.10) and die Liouville equation (1.3.8).

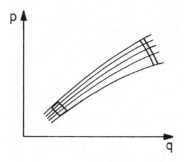

Fig. 1.4. The time dependence of an element in phase space; its volume remains constant.

---

[5] L. D. Landau and E. M. Lifshitz, *Course of Theoretical Physics I: Mechanics*, Eq. (46.5), Pergamon Press (Oxford, London, Paris 1960)

1.4 Quantum Statistics

1.4.1 The Density Matrix for Pure and Mixed Ensembles[6]

The density matrix is of special importance in the formulation of quantum statistics; it can also be denoted by the terms 'statistical operator' and 'density operator'.

Let a system be in the state $|\psi\rangle$. The observable A in this state has the mean value or expectation value

$$\langle A \rangle = \langle \psi| A |\psi\rangle \ . \tag{1.4.1}$$

The structure of the mean value makes it convenient to define the *density matrix* by

$$\rho = |\psi\rangle \langle \psi| \ . \tag{1.4.2}$$

We then have:

$$\langle A \rangle = \text{Tr}(\rho A) \tag{1.4.3a}$$

$$\text{Tr}\,\rho = 1 \ , \ \rho^2 = \rho \ , \ \rho^\dagger = \rho \ . \tag{1.4.3b,c,d}$$

Here, the definition of the trace (Tr) is

$$\text{Tr}\,X = \sum_n \langle n| X |n\rangle \ , \tag{1.4.4}$$

where $\{|n\rangle\}$ is an arbitrary complete orthonormal basis system. Owing to

$$\text{Tr}\,X = \sum_n \sum_m \langle n|m\rangle \langle m| X |n\rangle = \sum_m \sum_n \langle m| X |n\rangle \langle n|m\rangle$$
$$= \sum_m \langle m| X |m\rangle \ ,$$

the trace is independent of the basis used.

n.b. Proofs of (1.4.3a–c):

$$\text{Tr}\,\rho A = \sum_n \langle n|\psi\rangle \langle \psi| A |n\rangle = \sum_n \langle \psi| A |n\rangle \langle n|\psi\rangle = \langle \psi| A |\psi\rangle \ ,$$

$$\text{Tr}\,\rho = \text{Tr}\,\rho \mathbb{1} = \langle \psi| \mathbb{1} |\psi\rangle = 1 \ , \ \rho^2 = |\psi\rangle \langle \psi|\psi\rangle \langle \psi| = |\psi\rangle \langle \psi| = \rho \ .$$

If the systems or objects under investigation are all in one and the same state $|\psi\rangle$, we speak of a *pure ensemble*, or else we say that the systems are in a *pure state*.

---

[6] See e.g. F. Schwabl, *Quantum Mechanics*, 3rd edition, Springer, Heidelberg, Berlin, New York 2002 (corrected printing 2005), Chap. 20. In the following, this textbook will be abbreviated as 'QM I'.

Along with the statistical character which is inherent to quantum-mechanical systems, in addition a statistical distribution of states can be present in an ensemble. If an ensemble contains different states, we call it a *mixed ensemble*, a *mixture*, or we speak of a *mixed state*. We assume that the state $|\psi_1\rangle$ occurs with the probability p_1, the state $|\psi_i\rangle$ with the probability p_i, etc., with

$$\sum_i p_i = 1 \ .$$

The mean value or expectation value of A is then

$$\langle A \rangle = \sum_i p_i \langle \psi_i | A | \psi_i \rangle \ . \tag{1.4.5}$$

This mean value can also be represented in terms of the *density matrix* defined by

$$\rho = \sum_i p_i |\psi_i\rangle \langle \psi_i| \ . \tag{1.4.6}$$

We find:

$$\langle A \rangle = \text{Tr}\,\rho A \tag{1.4.7a}$$

$$\text{Tr}\,\rho = 1 \tag{1.4.7b}$$

$$\rho^2 \neq \rho \quad \text{and} \quad \text{Tr}\,\rho^2 < 1, \text{in the case that } p_i \neq 0 \text{ for more than one } i \tag{1.4.7c}$$

$$\rho^\dagger = \rho \ . \tag{1.4.7d}$$

The derivations of these relations and further remarks about the density matrices of mixed ensembles will be given in Sect. 1.5.2.

1.4.2 The Von Neumann Equation

From the *Schrödinger equation* and its adjoint

$$i\hbar \frac{\partial}{\partial t} |\psi, t\rangle = H |\psi, t\rangle , \quad -i\hbar \frac{\partial}{\partial t} \langle \psi, t| = \langle \psi, t| H \ ,$$

it follows that

$$i\hbar \frac{\partial}{\partial t} \rho = i\hbar \sum_i p_i \left(|\dot{\psi}_i\rangle \langle \psi_i| + |\psi_i\rangle \langle \dot{\psi}_i| \right)$$

$$= \sum_i p_i \left(H |\psi_i\rangle \langle \psi_i| - |\psi_i\rangle \langle \psi_i| H \right) \ .$$

From this, we find the *von Neumann equation*,

$$\frac{\partial}{\partial t}\rho = -\frac{i}{\hbar}[H,\rho] \; ; \tag{1.4.8}$$

it is the quantum-mechanical equivalent of the Liouville equation. It describes the time dependence of the density matrix in the Schrödinger representation. It holds also for a time-dependent H. It should not be confused with the equation of motion of Heisenberg operators, which has a positive sign on the right-hand side.

The expectation value of an observable A is given by

$$\langle A\rangle_t = \mathrm{Tr}\big(\rho(t)A\big) \; , \tag{1.4.9}$$

where $\rho(t)$ is found by solving the von Neumann equation (1.4.8). The time dependence of the expectation value is referred to by the index t.

We shall meet up with the von Neumann equation in the next chapter where we set up the equilibrium density matrices, and it is naturally of fundamental importance for all time-dependent processes.

We now treat the transformation to the *Heisenberg representation*. The formal solution of the Schrödinger equation has the form

$$|\psi(t)\rangle = U(t,t_0)\,|\psi(t_0)\rangle \; , \tag{1.4.10}$$

where $U(t,t_0)$ is a unitary operator and $|\psi(t_0)\rangle$ is the initial state at the time t_0. From this we find the time dependence of the density matrix:

$$\rho(t) = U(t,t_0)\rho(t_0)U(t,t_0)^\dagger \; . \tag{1.4.11}$$

(For a time-independent H, $U(t,t_0) = \mathrm{e}^{-\mathrm{i}H(t-t_0)/\hbar}$.)

The expectation value of an observable A can be computed both in the Schrödinger representation and in the Heisenberg representation

$$\langle A\rangle_t = \mathrm{Tr}\big(\rho(t)A\big) = \mathrm{Tr}\big(\rho(t_0)U(t,t_0)^\dagger AU(t,t_0)\big) = \mathrm{Tr}\big(\rho(t_0)A_H(t)\big) \; . \tag{1.4.12}$$

Here, $A_H(t) = U^\dagger(t,t_0)AU(t,t_0)$ is the operator in the Heisenberg representation. The density matrix $\rho(t_0)$ in the Heisenberg representation is time-independent.

*1.5 Additional Remarks

*1.5.1 The Binomial and the Poisson Distributions

We now discuss two probability distributions which occur frequently. Let us consider an interval of length L which is divided into two subintervals $[0,a]$ and $[a,L]$. We now distribute N distinguishable objects ('particles') in a completely random way over the two subintervals, so that the probability that a particle be found in the first or the second subinterval is given by $\frac{a}{L}$ or $\left(1 - \frac{a}{L}\right)$. The probability that n particles are in the interval $[0, a]$ is then given by the *binomial distribution*[7]

$$w_n = \left(\frac{a}{L}\right)^n \left(1 - \frac{a}{L}\right)^{N-n} \binom{N}{n}, \tag{1.5.1}$$

where the combinatorial factor $\binom{N}{n}$ gives the number of ways of choosing n objects from a set of N. The mean value of n is

$$\langle n \rangle = \sum_{n=0}^{N} n w_n = \frac{a}{L} N \tag{1.5.2a}$$

and its mean square deviation is

$$(\Delta n)^2 = \frac{a}{L} \left(1 - \frac{a}{L}\right) N . \tag{1.5.2b}$$

We now consider the limiting case $L \gg a$. Initially, w_n can be written using $\binom{N}{n} = \frac{N \cdot (N-1) \cdots (N-n+1)}{n!}$ in the form

$$w_n = \left(\frac{aN}{L}\right)^n \left(1 - \frac{a}{L}\right)^{N-n} \frac{1}{n!} 1 \cdot \left(1 - \frac{1}{N}\right) \cdots \left(1 - \frac{n-1}{N}\right)$$

$$= \overline{n}^n \frac{1}{n!} \left(1 - \frac{\overline{n}}{N}\right)^N \frac{1 \cdot (1 - \frac{1}{N}) \cdots (1 - \frac{n-1}{N})}{(1 - \frac{a}{L})^n}, \tag{1.5.3a}$$

where for the mean value (1.5.2a), we have introduced the abbreviation $\overline{n} = \frac{aN}{L}$. In the limit $\frac{a}{L} \to 0, N \to \infty$ for finite $\overline{n}$, the third factor in (1.5.3a) becomes $e^{-\overline{n}}$ and the last factor becomes equal to one, so that for the probability distribution, we find:

$$w_n = \frac{\overline{n}^n}{n!} e^{-\overline{n}} . \tag{1.5.3b}$$

This is the *Poisson distribution*, which is shown schematically in Fig. 1.5. The Poisson distribution has the following properties:

$$\sum_n w_n = 1 , \quad \langle n \rangle = \overline{n} , \quad (\Delta n)^2 = \overline{n} . \tag{1.5.4a,b,c}$$

The first two relations follow immediately from the derivation of the Poisson distribution starting from the binomial distribution. They are obtained in problem 1.5 together with 1.5.4c directly from 1.5.3b. The relative deviation

---

[7] A particular arrangement with n particles in the interval a and $N - n$ in $L - a$, e.g. the first particle in a, the second in $L - a$, the third in $L - a$, etc., has the probability $\left(\frac{a}{L}\right)^n \left(1 - \frac{b}{L}\right)^{N-n}$. From this we obtain w_n through multiplication by the number of combinations, i.e. the binomial coefficient $\binom{N}{n}$.

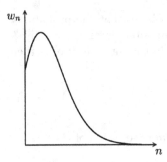

Fig. 1.5. The Poisson distribution

is therefore

$$\frac{\Delta n}{\overline{n}} = \frac{1}{\overline{n}^{1/2}} \, .$$

(1.5.5)

For numbers $\overline{n}$ which are not too large, e.g. $\overline{n} = 100$, $\Delta n = 10$ and $\frac{\Delta n}{\overline{n}} = \frac{1}{10}$. For macroscopic systems, e.g. $\overline{n} = 10^{20}$, we have $\Delta n = 10^{10}$ and $\frac{\Delta n}{\overline{n}} = 10^{-10}$. The relative deviation becomes extremely small. For large $\overline{n}$, the distribution w_n is highly concentrated around $\overline{n}$. The probability that no particles at all are within the subsystem, i.e. $w_0 = \mathrm{e}^{-10^{20}}$, is vanishingly small. The number of particles in the subsystem $[0, a]$ is not fixed, but however its relative deviation is very small for macroscopic subsystems.

In the figure below (Fig. 1.6a), the binomial distribution for $N = 5$ and $\frac{a}{L} = \frac{3}{10}$ (and thus $\overline{n} = 1.5$) is shown and compared to the Poisson distribution for $\overline{n} = 1.5$; in b) the same is shown for $N \equiv 10$, $\frac{a}{L} = \frac{3}{20}$ (i.e. again $\overline{n} = 1.5$). Even with these small values of N, the Poisson distribution already approximates the binomial distribution rather well. With $N = 100$, the curves representing the binomial and the Poisson distributions would overlap completely.

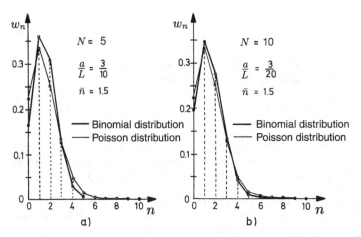

Fig. 1.6. Comparison of the Poisson distribution and the binomial distribution

*1.5.2 Mixed Ensembles and the Density Matrix of Subsystems

(i) Proofs of (1.4.7a–d)

$$\text{Tr } \rho A = \sum_n \sum_i p_i \langle \psi_i | A | n \rangle \langle n | \psi_i \rangle = \sum_i p_i \langle \psi_i | A | \psi_i \rangle = \langle A \rangle .$$

From this, (1.4.7b) also follows using $A = 1$.

$$\rho^2 = \sum_i \sum_j p_i p_j | \psi_i \rangle \langle \psi_i | \psi_j \rangle \langle \psi_j | \neq \rho .$$

For arbitrary $| \psi \rangle$, the expectation value of ρ

$$\langle \psi | \rho | \psi \rangle = \sum_i p_i | \langle \psi | \psi_i \rangle |^2 \geq 0$$

is positive definite. Since ρ is Hermitian, the eigenvalues P_m of ρ are positive and real:

$$\rho | m \rangle = P_m | m \rangle$$

$$\rho = \sum_{m=1}^{\infty} P_m | m \rangle \langle m | , \tag{1.5.6}$$

$$P_m \geq 0, \quad \sum_{m=1}^{\infty} P_m = 1, \quad \langle m | m' \rangle = \delta_{mm'} .$$

In this basis, $\rho^2 = \sum_m P_m^2 | m \rangle \langle m |$ and, clearly, $\text{Tr} \rho^2 = \sum_m P_m^2 < 1$, if more than only one state occurs. One can also derive (1.4.7c) directly from (1.4.6), with the condition that at least two different but not necessarily orthogonal states must occur in (1.4.6):

$$\text{Tr } \rho^2 = \sum_n \sum_{i,j} p_i p_j \langle \psi_i | \psi_j \rangle \langle \psi_j | n \rangle \langle n | \psi_i \rangle$$

$$= \sum_{i,j} p_i p_j | \langle \psi_i | \psi_j \rangle |^2 < \sum_i p_i \sum_j p_j = 1.$$

(ii) The criterion for a pure or a mixed state is – from Eq. (1.4.3c) and (1.4.7c) – given by $\text{Tr } \rho^2 = 1$ or $\text{Tr } \rho^2 < 1$.

(iii) We consider now a quantum-mechanical system which consists of two subsystems 1 and 2. Their combined state is taken to be

$$| \psi \rangle = \sum_n c_n | 1n \rangle | 2n \rangle , \tag{1.5.7}$$

where more than one c_n differs from zero. The associated density matrix is given by

$$\rho = |\psi\rangle \langle\psi| . \tag{1.5.8}$$

We now carry out measurements dealing with only subsystem 1, i.e. the operators corresponding to the observables A act only on the states $|1n\rangle$. We then find for the expectation value

$$\langle A \rangle = \mathrm{Tr}_1\mathrm{Tr}_2\rho A = \mathrm{Tr}_1[(\mathrm{Tr}_2\rho)A] . \tag{1.5.9}$$

Here, Tr_i refers to taking the trace over the subsystem i. According to Eq. (1.5.9), the density matrix which determines the outcome of these experiments is obtained by averaging ρ over subsystem 2:

$$\hat{\rho} = \mathrm{Tr}_2\rho = \sum_n |c_n|^2 |1n\rangle \langle 1n| . \tag{1.5.10}$$

This is the density matrix of a mixture, although the overall system is in a pure state.

The most general state of the subsystems 1 and 2 has the form[8]

$$|\psi\rangle = \sum_{n,m} c_{nm} |1n\rangle |2m\rangle . \tag{1.5.11}$$

Here, again, we find that

$$\hat{\rho} \equiv \mathrm{Tr}_2 |\psi\rangle \langle\psi| = \sum_n \sum_{n'} \sum_m c_{nm}c_{n'm}^* |1n\rangle \langle 1n'|$$

$$= \sum_m \left(\sum_n c_{nm} |1n\rangle\right)\left(\sum_{n'} c_{n'm}^* \langle 1n'|\right) \tag{1.5.12}$$

is in general a mixture. Since a macroscopic system will have spent some time in contact with some other systems, even when it is completely isolated, it will never be in a pure state, but always in a mixed state.

It may be instructive to consider the following special case: we write c_{nm} in the form $c_{nm} = |c_{nm}| e^{i\varphi_{nm}}$. In the case that the phases φ_{nm} are stochastic, from $\hat{\rho}$ we then obtain the density matrix

$$\hat{\hat{\rho}} = \prod_{\langle nm\rangle} \left(\int_0^{2\pi} d\frac{\varphi_{nm}}{2\pi}\right)\hat{\rho} = \sum_n \left(\sum_m |c_{nm}|^2\right) |1n\rangle \langle 1n| .$$

---

[8] As an aside, we note that it is possible to introduce a biorthogonal system (Schmidt basis) which brings the state (1.5.11) into the form (1.5.7); see QM I, problem 20.5.

Problems for Chapter 1

1.1 Prove the Stirling formula

$$x! \approx \sqrt{2\pi x}\, x^x\, e^{-x} \, ,$$

by starting from $N! = \int_0^\infty dx\, x^N\, e^{-x}$ and fitting the integrand $f(x) \equiv x^N e^{-x}$ up to second order to the function $g(x) = A\, e^{-(x-N)^2/a^2}$. $f(x)$ has a sharp maximum at $x_0 = N$.

1.2 Determine the probability $w(N, m)$ that in a system of N spins exactly m will be found to have the orientation "↑" and correspondingly $N - m$ have the orientation "↓". There is no external magnetic field and no interaction of the spins with one another, so that for each individual spin, the configurations ↑ and ↓ are equally probable.
(a) Verify

$$\sum_{m=0}^{N} w(N, m) = 1 \, .$$

(b) Calculate the mean value of m,

$$\langle m \rangle = \sum_{m=0}^{N} w(N, m)\, m \, ,$$

and its standard deviation, $\left(\langle m^2 \rangle - \langle m \rangle^2 \right)^{1/2}$.
The dimensionless magnetization is defined as $M = 2m - N$; give its mean value and its standard deviation.
(c) Calculate the distribution $w(N, M)$ for large N. Assume that $|M/N| \ll 1$.

1.3 Derive the central limit theorem for $w_i(x_i)$ instead of for $w(x_i)$.
Note: In the result, you need only replace $N\langle X \rangle$ by $\sum_i \langle X_i \rangle$ and $N(\Delta x)^2$ by $\sum_i (\Delta x_i)^2$.

1.4 The random walk: a particle moves at each step with equal probability by a unit distance either to the left or to the right.
(a) Calculate $\langle Y \rangle$ and $\langle Y^2 \rangle$ exactly after $N = N_+ + N_-$ steps, where $Y = N_+ - N_-$.
(b) What result would you obtain from the central limit theorem?

1.5 Verify the relations (1.5.4a-c) for the Poisson distribution, (1.5.3b):

$$w(n) = e^{-\overline{n}}\, \frac{\overline{n}^{\,n}}{n!} \, , \quad n \text{ an integer} \geq 0 \, .$$

1.6 The distribution function $\rho(E_1, \ldots, E_N)$ has the form

$$\rho = \prod_{i=1}^{N} f(E_i) \, .$$

Let the mean value and the standard deviation of the individual E_i-values be denoted by e and $\left\langle (\Delta E_i)^2 \right\rangle^{1/2} = \Delta$, respectively.
Compute the mean value and the standard deviation of $E = \sum_i E_i$.

1.7 Sketch the trajectory in phase space of a particle
(a) which moves with the energy E within a one-dimensional, infinitely high potential well (particle in a box):

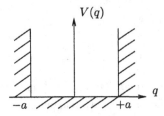

$V(q)$

$-a$ $+a$ q

(b) which falls from a height h under the influence of gravity, is inelastically reflected by the ground and rises again to a height $9/10h$, etc.

1.8 The gradient of the Hamiltonian is defined by $\nabla H(q,p) = \left(\frac{\partial H}{\partial q}, \frac{\partial H}{\partial p}\right)$. Compare with v of Eq. (1.3.4$'$) and show that $|v| = |\nabla H|$ and $v \perp \nabla H$.

1.9 An ion source emits ions of energy $E_1 = 5000 \pm 1.00$ eV from a surface of area 1mm^2 into a solid angle of $\Omega_1 = 1$ sterad. The ions are accelerated by electric fields to $E_2 = 10$ MeV and focused onto an area of 1cm^2. Calculate the opening angle of the ion beam on impact using the *Liouville theorem*.

Suggestion: Assume that the acceleration acts so rapidly that the different velocities within the beam do not lead to an additional broadening of the beam, i.e. $dx_2 = dx_1$; assume also that the width of the energy distribution remains unchanged, $dE_2 = dE_1$.

1.10 (a) Show that $\mathrm{Tr}(AB) = \mathrm{Tr}(BA)$.
(b) The operators ρ_ν are taken to be density matrices, so that they obey the conditions (1.4.7b-d), and $p_\nu \geq 0$, $\sum_\nu p_\nu = 1$. Show that $\sum_\nu p_\nu \rho_\nu$ also obeys these conditions.

1.11 Consider a beam of light which is propagating in the $+z$ direction. An arbitrary pure polarization state can be written as a linear combination

$$a\,|\!\uparrow\rangle + b\,|\!\downarrow\rangle\,,$$

where $|\!\uparrow\rangle$ represents the state which is polarized in the x direction and $|\!\downarrow\rangle$ the state polarized in the y direction.
(a) Calculate the density matrix: (i) for an arbitrary pure state, (ii) for the state polarized in the x direction, (iii) for the state polarized at 45°, and (iv) for the state polarized at 135°.
(b) What is the density matrix like for a mixed state, where e.g. 50% of the light is polarized along 45° and 50% along 135°, or 50% is polarized in the x direction and 50% in the y direction? The angles are those between the x axis and the direction of polarization.

1.12 A Galton board is a board with nails which is set upright; it has N horizontal rows of nails, all the same length, shifted so that the nails of successive rows are precisely in between those in the row above. In the center above the uppermost row is a funnel through which little balls can be released to fall down through the rows of nails. Below the bottom row are a series of compartments which catch the balls. What curve is represented by the height of the balls in the various compartments?

1.13 A container of volume V holds N particles. Consider the subvolume v and assume that the probability of finding a particular particle in this subvolume is given by v/V.
(a) Give the probability p_n of finding n particles within v.
(b) Calculate the mean value $\bar{n}$ and the mean square deviation $\overline{(n-\bar{n})^2}$.
(c) Show with the help of the Stirling formula that p_n corresponds approximately to a Gaussian distribution when N and n are large.
(d) Show in the limit $\frac{v}{V} \to 0$ and $V \to \infty$ with $\frac{N}{V} = $ const. that p_n approaches a Poisson distribution.

1.14 *The Gaussian distribution*: The Gaussian distribution is defined by the continuous probability density

$$w_G(x) = \frac{1}{\sqrt{2\pi\sigma^2}} e^{-(x-x_0)^2/2\sigma^2}.$$

For this distribution, compute $<X>$, Δx, $<X^4>$, and $<X-<X^3>>$.

1.15 *The log-normal distribution*: Let the statistical variables X have the property that $\log X$ obeys a Gaussian distribution with $<\log X>= \log x_0$.
(a) Show by transforming the Gaussian distribution that the probability density for X has the form

$$P(x) = \frac{1}{\sqrt{2\pi\sigma^2}} \frac{1}{x} e^{-\frac{(\log(x/x_0))^2}{2\sigma^2}}, \quad 0 < x < \infty.$$

(b) Show that

$$<X>= x_0 e^{\sigma^2/2}$$

and

$$<\log X >= \log x_0.$$

(c) Show that the log-normal distribution can be rewritten in the form

$$P(x) = \frac{1}{x_0\sqrt{2\pi\sigma^2}} (x/x_0)^{-1-\mu(x)}$$

with

$$\mu(x) = \frac{1}{2\sigma^2} \log \frac{x}{x_0};$$

it can thus be easily confused with a power law when analyzing data.

2. Equilibrium Ensembles

2.1 Introductory Remarks

As emphasized in the Introduction, a macroscopic system consists of $10^{19} - 10^{23}$ particles and correspondingly has an energy spectrum with spacings of $\Delta E \sim e^{-N}$. The attempt to find a detailed solution to the microscopic equations of motion of such a system is hopeless; furthermore, the required initial conditions or quantum numbers cannot even be specified. Fortunately, knowledge of the time development of such a microstate is also superfluous, since in each observation of the system (both of macroscopic quantities and of microscopic properties, e.g. the density correlation function, particle diffusion, etc.), one averages over a finite time interval. No system can be strictly isolated from its environment, and as a result it will undergo transitions into many different microstates during the measurement process. Figure 2.1 illustrates schematically how the system moves between various phase-space trajectories. Thus, a many-body system cannot be characterized by a single microstate, but rather by an ensemble of microstates. This statistical ensemble of microstates represents the macrostate which is specified by the macroscopic state variables $E, V, N, \ldots$[1] (see Fig. 2.1).

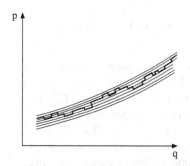

p

q

Fig. 2.1. A trajectory in phase space (schematic)

---

[1] A different justification of the statistical description is based on the ergodic theorem: nearly every microstate approaches arbitrarily closely to all the states of the corresponding ensemble in the course of time. This led Boltzmann to postulate that the time average for an isolated system is equal to the average over the states in the microcanonical ensemble (see Sect. 10.5.2).

Experience shows that every macroscopic system tends with the passage of time towards an equilibrium state, in which

$$\dot{\rho} = 0 = -\frac{i}{\hbar}[H, \rho] \tag{2.1.1}$$

must hold. Since, according to Eq. (2.1.1), in equilibrium the density matrix ρ commutes with the Hamiltonian H, it follows that in an equilibrium ensemble ρ can depend only on the *conserved quantities*. (The system changes its microscopic state continually even in equilibrium, but the distribution of microstates within the ensemble becomes time-independent.) Classically, the right-hand side of (2.1.1) is to be replaced by the Poisson bracket.

2.2 Microcanonical Ensembles

2.2.1 Microcanonical Distribution Functions and Density Matrices

We consider an *isolated system* with a fixed number of particles, a fixed volume V, and an energy lying within the interval $[E, E + \Delta]$ with a small Δ, whose Hamiltonian is $H(q, p)$ (Fig. 2.2). Its total momentum and total angular momentum may be taken to be zero.

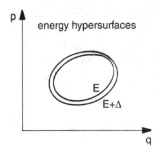

Fig. 2.2. Energy shell in phase space

We now wish to find the distribution function (density matrix) for this physical situation. It is clear from the outset that only those points in phase space which lie between the two hypersurfaces $H(q, p) = E$ and $H(q, p) = E + \Delta$ can have a finite statistical weight. The region of phase space between the hypersurfaces $H(q, p) = E$ and $H(q, p) = E + \Delta$ is called the *energy shell*. It is intuitively plausible that in equilibrium, no particular region of the energy shell should play a special role, i.e. that all points within the energy shell should have the same statistical weight. We can indeed derive this fact by making use of the conclusion following (2.1.1). If regions within the energy shell had different statistical weights, then the distribution function (density matrix) would depend on other quantities besides $H(q, p)$, and ρ would not commute with H (classically, the Poisson bracket would not vanish). Since for a given E, Δ, V, and N, the equilibrium distribution function depends only upon $H(q, p)$, it follows that every state within the energy shell, i.e. all of the points in Γ space with $E \leq H(q, p) \leq E + \Delta$, are equally probable. An ensemble with these properties is called a *microcanonical ensemble*. The associated *microcanonical distribution function* can be postulated to have the form

$$\rho_{MC} = \begin{cases} \frac{1}{\Omega(E)\Delta} & E \leq H(q, p) \leq E + \Delta \\ 0 & \text{otherwise}, \end{cases} \qquad (2.2.1)$$

where, as postulated, the normalization constant $\Omega(E)$ depends only on E, but not on q and p. $\Omega(E)\Delta$ is the volume of the energy shell.[2] In the limit $\Delta \to 0$, (2.2.1) becomes

$$\rho_{MC} = \frac{1}{\Omega(E)} \delta(E - H(q, p)). \qquad (2.2.1')$$

The normalization of the probability density determines $\Omega(E)$:

$$\int \frac{dq \, dp}{h^{3N} N!} \rho_{MC} = 1. \qquad (2.2.2)$$

The *mean value* of a quantity A is given by

$$\langle A \rangle = \int \frac{dq \, dp}{h^{3N} N!} \rho_{MC} \, A. \qquad (2.2.3)$$

The choice of the fundamental integration variables (whether q or q/const) is arbitrary at the present stage of our considerations and was made in (2.2.2) and (2.2.3) by reference to the limit which is found from quantum statistics. If the factor $(h^{3N} N!)^{-1}$ were not present in the normalization condition (2.2.2) and in the mean value (2.2.3), then ρ_{MC} would be replaced by $(h^{3N} N!)^{-1}\rho_{MC}$. All mean values would remain unchanged in this case; the difference however would appear in the entropy (Sect. 2.3). The factor $1/N!$ results from the indistinguishability of the particles. The necessity of including the factor $1/N!$ was discovered by Gibbs even before the development of quantum mechanics. Without this factor, an entropy of mixing of identical gases would erroneously appear (Gibbs' paradox). That is, the sum of the entropies of two identical ideal gases each consisting of N particles, $2S_N$, would be smaller than the entropy of one gas consisting of $2N$ particles. Mixing of ideal gases will be treated in Chap. 3, Sect. 3.6.3.4. We also refer to the calculation of the entropy of mixtures of ideal gases in Chap. 5 and the last paragraph of Appendix B.1.

---

[2] The surface area of the energy shell $\Omega(E)$ depends not only on the energy E but also on the spatial volume V and the number of particles N. For our present considerations, only its dependence on E is of interest; therefore, for clarity and brevity, we omit the other variables. We use a similar abbreviated notation for the partition functions which will be introduced in later sections, also. The complete dependences are collected in Table 2.1.

For the $6N$-dimensional volume element in phase space, we will also use the abbreviated notation

$$d\Gamma \equiv \frac{dq\,dp}{h^{3N}N!} \,.$$

From the normalization condition, (2.2.2), and the limiting form given in (2.2.1′), it follows that

$$\Omega\,(E) = \int \frac{dq\,dp}{h^{3N}N!}\,\delta\bigl(E - H(q,p)\bigr)\,. \tag{2.2.4}$$

After introducing coordinates on the energy shell and an integration variable along the normal $k_\perp$, (2.2.4) can also be given in terms of the surface integral:

$$\Omega\,(E) = \int_E \frac{dS}{h^{3N}N!}\,dk_\perp\,\delta\bigl(E - H(S_E) - |\boldsymbol{\nabla}H|k_\perp\bigr)$$

$$= \int \frac{dS}{h^{3N}N!}\,\frac{1}{|\boldsymbol{\nabla}H(q,p)|}\,. \tag{2.2.4′}$$

Here, dS is the differential element of surface area in the $(6N-1)$-dimensional hypersurface at energy E, and $\boldsymbol{\nabla}$ is the $6N$-dimensional gradient in phase space. In Eq. (2.2.4′) we have used $H(S_E) = E$ and performed the integration over $k_\perp$. According to Eq. (1.3.4′), it holds that $|\boldsymbol{\nabla}\,H(q,p)| = |\boldsymbol{v}|$ and the velocity in phase space is perpendicular to the gradient, i.e. $\boldsymbol{v} \perp \boldsymbol{\nabla}\,H(q,p)$. This implies that the velocity is always tangential to the surface of the energy shell; cf. problem 1.8.

Notes:

(i) Alternatively, the expression (2.2.4′) can be readily proven by starting with an energy shell of finite width Δ and dividing it into segments $dS\Delta k_\perp$. Here, dS is a surface element and $\Delta k_\perp$ is the perpendicular distance between the two hypersurfaces (Fig. 2.3). Since the gradient yields the variation perpendicular to an equipotential surface, we find $|\boldsymbol{\nabla}H(q,p)|\Delta k_\perp = \Delta$, where $\boldsymbol{\nabla}H(q,p)$ is to be computed on the hypersurface $H(q,p) = E$.

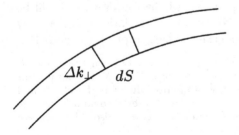

Fig. 2.3. Calculation of the volume of the energy shell

From this it follows that

$$\Omega\left(E\right)\Delta = \int \frac{dS}{h^{3N}\,N!}\,\Delta k_{\perp} = \int \frac{dS}{h^{3N}\,N!\,|\nabla H(q,p)|}\cdot\Delta\,,$$

i.e. we again obtain (2.2.4′).

(ii) Equation (2.2.4′) has an intuitively very clear significance. $\Omega\left(E\right)$ is given by the sum of the surface elements, each divided by the velocity in phase space. Regions with high velocity thus contribute less to $\Omega\left(E\right)$. In view of the ergodic hypothesis (see 10.5.2), this result is very plausible. See problem 1.8: $|v| = |\nabla H|$ and $v \perp \nabla H$.

As already mentioned, $\Omega\left(E\right)\Delta$ is the volume of the energy shell in classical statistical mechanics. We will occasionally also refer to $\Omega\left(E\right)$ as the "phase surface". We also define the volume inside the energy shell:

$$\bar{\Omega}(E) = \int \frac{dq\,dp}{h^{3N}\,N!}\,\Theta\big(E - H(q,p)\big)\,. \tag{2.2.5}$$

Clearly, the following relation holds:

$$\Omega\left(E\right) = \frac{d\bar{\Omega}(E)}{dE}\,. \tag{2.2.6}$$

Quantum mechanically, the definition of the microcanonical ensemble for an isolated system with the Hamiltonian H and associated energy eigenvalues E_n is:

$$\rho_{MC} = \sum_{n} p(E_n)\,|n\rangle\,\langle n|\,, \tag{2.2.7}$$

where, analogously to (2.2.1),

$$p(E_n) = \begin{cases} \frac{1}{\Omega\left(E\right)\Delta} & E \leq E_n \leq E + \Delta \\ 0 & \text{otherwise}\,. \end{cases} \tag{2.2.8}$$

In the *microcanonical density matrix* ρ_{MC}, all the energy eigenstates $|n\rangle$ whose energy E_n lies in the interval $[E, E + \Delta]$ contribute with equal weights. The normalization

$$\text{Tr}\,\rho_{MC} = 1 \tag{2.2.9a}$$

yields

$$\Omega\left(E\right) = \frac{1}{\Delta}\sum_{n}{}'1\,, \tag{2.2.9b}$$

where the summation is restricted to energy eigenstates within the energy shell. Thus $\Omega\left(E\right)\Delta$ is equal to the number of energy eigenstates within the energy shell $[E, E + \Delta]$. For the density matrix of the microcanonical ensemble, an abbreviated notation is also used:

$$\rho_{MC} = \Omega\left(E\right)^{-1}\delta(H - E) \tag{2.2.7′}$$

and

$$\Omega(E) = \mathrm{Tr}\ \delta(H - E)\ . \tag{2.2.9b'}$$

Equation (2.2.8) (and its classical analogue (2.2.1)) represent *the fundamental hypothesis of equilibrium statistical mechanics.* All the equilibrium properties of matter (whether isolated or in contact with its surroundings) can be deduced from them. The microcanonical density matrix describes an isolated system with given values of E, V, and N. The equilibrium density matrices corresponding to other typical physical situations, such as those of the canonical and the grand canonical ensembles, can be derived from it. As we shall see in the following examples, in fact essentially the whole volume within the hypersurface $H(q, p) = E$ lies at its surface. More precisely, comparison of $\bar{\Omega}(E)$ and $\Omega(E)\Delta$ shows that

$$\log\big(\Omega(E)\Delta\big) = \log \bar{\Omega}(E) + \mathcal{O}\!\left(\log \frac{E}{N\Delta}\right)\ .$$

Since $\log \Omega(E)\Delta$ and $\log \bar{\Omega}(E)$ are both proportional to N, the remaining terms can be neglected for large N; in this spirit, we can write

$$\Omega(E)\Delta = \bar{\Omega}(E)\ .$$

2.2.2 The Classical Ideal Gas

In this and in the next section, we present three simple examples for which $\Omega(E)$ can be calculated, and from which we can directly read off the characteristic dependences on the energy and the particle number. We shall now investigate the classical ideal gas, i.e. a classical system of N atoms between which there are no interactions at all; and we shall see from it how $\Omega(E)$ depends on the energy E and on the particle number N. Furthermore, we will make use of the results of this section later to derive the thermodynamics of the ideal gas. The Hamiltonian of the three-dimensional ideal gas is

$$H = \sum_{i=1}^{N} \frac{\mathbf{p}_i^2}{2m} + V_{\mathrm{wall}}\ . \tag{2.2.10}$$

Here, the $\mathbf{p}_i$ are the cartesian momenta of the particles and V_{wall} is the potential representing the wall of the container. The surface area of the energy shell is in this case

$$\Omega(E) = \frac{1}{h^{3N} N!} \int_V d^3x_1 \ldots \int_V d^3x_N \int d^3p_1 \ldots \int d^3p_N\ \delta\!\left(E - \sum_{i=1}^{N} \frac{\mathbf{p}_i^2}{2m}\right),$$

$$\tag{2.2.11}$$

where the integrations over x are restricted to the spatial volume V defined by the walls. It would be straightforward to calculate $\Omega(E)$ directly. We shall carry out this calculation here via $\bar{\Omega}(E)$, the volume inside the energy shell, which in this case is a hypersphere, in order to have both quantities available:

$$\bar{\Omega}(E) = \frac{1}{h^{3N}N!}$$
$$\times \int_V d^3x_1 \ldots \int_V d^3x_N \int d^3p_1 \ldots \int d^3p_N \, \Theta\!\left(E - \sum_i \mathbf{p}_i^2/2m\right). \quad (2.2.12)$$

Introducing the surface area of the d-dimensional unit sphere,[3]

$$(2\pi)^d K_d \equiv \int d\Omega_d = \frac{2\pi^{d/2}}{\Gamma(d/2)}, \quad (2.2.13)$$

we find, representing the momenta in spherical polar coordinates,

$$\bar{\Omega}(E) = \frac{V^N}{h^{3N}N!} \int d\Omega_{3N} \int_0^{\sqrt{2mE}} dp\, p^{3N-1}.$$

From this, we immediately obtain

$$\bar{\Omega}(E) = \frac{V^N(2\pi mE)^{\frac{3N}{2}}}{h^{3N}N!(\frac{3N}{2})!}, \quad (2.2.14)$$

where $\Gamma(\frac{3N}{2}) = (\frac{3N}{2} - 1)!$ was used, under the assumption – without loss of generality – of an even number of particles. For large N, Eq. (2.2.14) can be simplified by applying the Stirling formula (see problem 1.1).

$$N! \sim N^N e^{-N} (2\pi N)^{1/2}, \quad (2.2.15)$$

whereby it suffices to retain only the first two factors, which dominate the expression. Then

$$\bar{\Omega}(E) \approx \left(\frac{V}{N}\right)^N \left(\frac{4\pi mE}{3h^2 N}\right)^{\frac{3N}{2}} e^{\frac{5N}{2}}. \quad (2.2.16)$$

Making use of Eq. (2.2.6), we obtain from (2.2.14) and (2.2.16) the exact result for $\Omega(E)$:

$$\Omega(E) = \frac{V^N 2\pi m (2\pi mE)^{\frac{3N}{2}-1}}{h^{3N}N!(\frac{3N}{2}-1)!} \quad (2.2.17)$$

as well as an asymptotic expression which is valid in the limit of large N:

---

[3] The derivation of (2.2.13) will be given at the end of this section.

$$\Omega\left(E\right) \approx \left(\frac{V}{N}\right)^{N} \left(\frac{4\pi m E}{3h^{2}N}\right)^{\frac{3N}{2}} e^{\frac{5N}{2}} \frac{1}{E} \frac{3N}{2} \; . \tag{2.2.18}$$

In (2.2.16) and (2.2.18), the specific volume V/N and the specific energy E/N occur to the power N. We now compare $\bar{\Omega}(E)$, the volume inside the energy shell, with $\Omega\left(E\right)\Delta$, the volume of a spherical shell of thickness Δ, by considering the logarithms of these two quantities (due to the occurrence of the Nth powers):

$$\log\bigl(\Omega\left(E\right)\Delta\bigr) = \log\bar{\Omega}(E) + \mathcal{O}\Bigl(\log\frac{E}{N\Delta}\Bigr) \; . \tag{2.2.19}$$

Since $\log\Omega\left(E\right)\Delta$ and $\log\bar{\Omega}(E)$ are both proportional to N, the remaining terms can be neglected in the case that N is large. In this approximation, we find

$$\Omega\left(E\right)\Delta \approx \bar{\Omega}(E) \; , \tag{2.2.20}$$

i.e. nearly the whole volume of the hypersphere $H(q,p) \le E$ lies at its surface. This fact is due to the high dimensionality of the phase space, and it is to be expected that (2.2.20) remains valid even for systems with interactions.

We now prove the expression (2.2.13) for the surface area of the d-dimensional unit sphere. To this end, we compute the d-dimensional Gaussian integral

$$I = \int\limits_{-\infty}^{\infty} dp_{1} \ldots \int\limits_{-\infty}^{\infty} dp_{d} \, e^{-(p_{1}^{2}+\cdots+p_{d}^{2})} = (\sqrt{\pi})^{d} \; . \tag{2.2.21}$$

This integral can also be written in spherical polar coordinates:[4]

$$I = \int_{0}^{\infty} dp\, p^{d-1} \int d\Omega_{d}\, e^{-p^{2}} = \frac{1}{2}\int dt\, t^{\frac{d}{2}-1} e^{-t} \int d\Omega_{d} = \frac{1}{2}\Gamma\Bigl(\frac{d}{2}\Bigr)\int d\Omega_{d} \; , \tag{2.2.22}$$

where

$$\Gamma(z) = \int\limits_{0}^{\infty} dt\, t^{z-1} e^{-t} \tag{2.2.23}$$

is the *gamma function*. Comparison of the two expressions (2.2.21) and (2.2.22) yields

$$\int d\Omega_{d} = \frac{2\pi^{d/2}}{\Gamma(d/2)} \; . \tag{2.2.13'}$$

In order to gain further insights into how the volume of the energy shell depends upon the parameters of the microcanonical ensemble, we will calculate

---

[4] We denote an element of surface area on the d-dimensional unit sphere by $d\Omega_{d}$. For the calculation of the surface integral $\int d\Omega_{d}$, it is not necessary to use the detailed expression for $d\Omega_{d}$. The latter may be found in E. Madelung, *Die Mathematischen Hilfsmittel des Physikers*, Springer, Berlin, 7th edition (1964), p. 244.

$\Omega(E)$ for two other simple examples, this time quantum-mechanical systems; these are: (i) harmonic oscillators which are not coupled, and (ii) paramagnetic (not coupled) spins. Simple problems of this type can be solved for all ensembles with a variety of methods. Instead of the usual combinatorial method, we employ purely analytical techniques for the two examples which follow.

*2.2.3 Quantum-mechanical Harmonic Oscillators and Spin Systems

*2.2.3.1 Quantum-mechanical Harmonic Oscillators

We consider a system of N identical harmonic oscillators, which are either not coupled to each other at all, or else are so weakly coupled that their interactions may be neglected. Then the Hamiltonian for the system is given by:

$$H = \sum_{j=1}^{N} \hbar\omega \left(a_j^\dagger a_j + \frac{1}{2} \right) , \tag{2.2.24}$$

where $a_j^\dagger (a_j)$ are creation (annihilation) operators for the jth oscillator. Thus we have

$$\Omega(E) = \sum_{n_1=0}^{\infty} \cdots \sum_{n_N=0}^{\infty} \delta\left(E - \hbar\omega \sum_j \left(n_j + \frac{1}{2}\right) \right)$$

$$= \sum_{n_1=0}^{\infty} \cdots \sum_{n_N=0}^{\infty} \int \frac{dk}{2\pi} e^{ik\left(E - \sum_j \hbar\omega(n_j+\frac{1}{2})\right)} = \int \frac{dk}{2\pi} e^{ikE} \prod_{i=1}^{N} \frac{e^{-ik\hbar\omega/2}}{1 - e^{-ik\hbar\omega}} , \tag{2.2.25}$$

and finally

$$\Omega(E) = \int \frac{dk}{2\pi} e^{N\left(ik(E/N) - \log(2i\sin(k\hbar\omega/2))\right)} . \tag{2.2.26}$$

The computation of this integral can be carried out for large N using the saddle-point method.[5] The function

$$f(k) = ike - \log\left(2i\sin(k\hbar\omega/2)\right) \tag{2.2.27}$$

with $e = E/N$ has a maximum at the point

$$k_0 = \frac{1}{\hbar\omega i} \log \frac{e + \frac{\hbar\omega}{2}}{e - \frac{\hbar\omega}{2}} . \tag{2.2.28}$$

This maximum can be determined by setting the first derivative of (2.2.27) equal to zero

---

[5] N.G. de Bruijn, *Asymptotic Methods in Analysis*, (North Holland, 1970); P.M. Morse and H. Feshbach, *Methods of Theoretical Physics*, p. 434, (McGraw Hill, New York, 1953).

$$f'(k_0) = ie - \frac{\hbar\omega}{2}\cot\frac{k_0\hbar\omega}{2} = 0 \,.$$

Therefore, with

$$f(k_0) = ik_0 e - \log\left(2i/\sqrt{1-(2e/\hbar\omega)^2}\right)$$

$$= \frac{e}{\hbar\omega}\log\frac{e+\frac{\hbar\omega}{2}}{e-\frac{\hbar\omega}{2}} + \frac{1}{2}\log\left(\left(e+\frac{\hbar\omega}{2}\right)\left(e-\frac{\hbar\omega}{2}\right)\Big/(\hbar\omega)^2\right) \qquad (2.2.29)$$

and $f''(k_0) = \left(\frac{\hbar\omega}{2}\right)^2/\sin^2(k_0\hbar\omega/2)$, we find for $\Omega(E)$:

$$\Omega(E) = \frac{1}{2\pi}e^{Nf(k_0)}\int dk\, e^{N\frac{1}{2}f''(k_0)(k-k_0)^2} \,. \qquad (2.2.30)$$

The integral in this expression yields only a factor proportional to $\sqrt{N}$; thus, the number of states is given by

$$\Omega(E) = \exp\left\{N\left[\frac{e+\frac{1}{2}\hbar\omega}{\hbar\omega}\log\frac{e+\frac{1}{2}\hbar\omega}{\hbar\omega} - \frac{e-\frac{1}{2}\hbar\omega}{\hbar\omega}\log\frac{e-\frac{1}{2}\hbar\omega}{\hbar\omega}\right]\right\}. \qquad (2.2.31)$$

*2.2.3.2 Two-level Systems: the Spin-$\frac{1}{2}$ Paramagnet

As our third example, we consider a system of N particles which can occupy one of two states. The most important physical realization of such a system is a paramagnet in a magnetic field H ($h = -\mu_B H$), which has the Hamiltonian[6]

$$\mathcal{H} = -h\sum_{i=1}^{N}\sigma_i\,, \quad \text{with} \quad \sigma_i = \pm 1. \qquad (2.2.32)$$

The number of states of energy E is, from (2.2.1), given by

$$\Omega(E) = \sum_{\{\sigma_i=\pm 1\}}\delta\left(E+h\sum_{i=1}^{N}\sigma_i\right) = \int\frac{dk}{2\pi}\sum_{\{\sigma_i=\pm 1\}}e^{ik(E+h\sum_i\sigma_i)}$$

$$= \int\frac{dk}{2\pi}e^{ikE}(2\cos kh)^N = 2^N\int\frac{dk}{2\pi}e^{f(k)} \qquad (2.2.33)$$

with

$$f(k) = ikE + N\log\cos kh \,. \qquad (2.2.34)$$

The computation of the integral can again be accomplished by applying the saddle-point method. Using $f'(k) = iE - Nh\tan kh$ and $f''(k) = -Nh^2/\cos^2 kh$, we obtain

---

[6] In the literature of magnetism, it is usual to denote the magnetic field by **H** or H. To distinguish it from the Hamiltonian in the case of magnetic phenomena, we use the symbol $\mathcal{H}$ for the latter.

from the condition $f'(k_0) = 0$

$$k_0 h = \arctan \frac{iE}{Nh} = \frac{i}{2} \log \frac{1 + E/Nh}{1 - E/Nh} .$$

For the second derivative, we find

$$f''(k_0) = -\left(1 - (E/Nh)^2\right) Nh^2 \le 0 \quad \text{for} \quad -Nh \le E \le Nh .$$

Thus, using the abbreviation $e = E/Nh$, we have

$$\Omega(E) = 2^N \exp\left(-\frac{Ne}{2} \log \frac{1+e}{1-e} + N \log \frac{1}{\sqrt{1-e^2}}\right) \int \frac{dk}{2\pi} e^{-\frac{1}{2}\left(-f''(k_0)\right)(k-k_0)^2}$$

$$= \frac{2^N}{\sqrt{2\pi}} \exp\left(-\frac{Ne}{2} \log \frac{1+e}{1-e} + \frac{N}{2} \log \frac{1}{1-e^2} - \frac{1}{2} \log\left((1-e^2)Nh^2\right)\right)$$

$$= \frac{1}{\sqrt{2\pi}} \exp\left\{ -\frac{N}{2}(1+e) \log \frac{1+e}{2} - \frac{N}{2}(1-e) \log \frac{1-e}{2} - \right.$$

$$\left. -\frac{1}{2} \log(1-e^2) - \frac{1}{2} \log Nh^2 \right\} ,$$

$$\Omega(E) = \exp\left\{ -\frac{N}{2}\left[(1+e) \log \frac{1+e}{2} + (1-e) \log \frac{1-e}{2}\right] + \mathcal{O}(1, \log N) \right\} .$$

$$(2.2.35)$$

We have now calculated the number of states $\Omega(E)$ for three examples. The physical consequences of the characteristic energy dependences will be discussed after we have introduced additional concepts such as those of entropy and temperature.

2.3 Entropy

2.3.1 General Definition

Let an arbitrary density matrix ρ be given; then the *entropy* S is defined by

$$S = -k \operatorname{Tr}(\rho \log \rho) \equiv -k\langle \log \rho \rangle . \tag{2.3.1}$$

Here, we give the formulas only in their quantum-mechanical form, as we shall often do in this book. For classical statistics, the trace operation Tr is to be read as an integration over phase space. The physical meaning of S will become clear in the following sections. At this point, we can consider the entropy to be a measure of the size of the accessible part of phase space, and thus also of the uncertainty of the microscopic state of the system: the more states that occur in the density matrix, the greater the entropy S. For example, for M states which occur with equal probabilities $\frac{1}{M}$, the entropy is given by

$$S = -k \sum_1^M \frac{1}{M} \log \frac{1}{M} = k \log M .$$

For a pure state, $M = 1$ and the entropy is therefore $S = 0$. In the diagonal representation of ρ (Eq. 1.4.8), one can immediately see that the entropy is positive semidefinite:

$$S = -k \sum_n P_n \log P_n \geq 0 \tag{2.3.2}$$

since $x \log x \leq 0$ in the interval $0 < x \leq 1$ (see Fig. 2.4). The factor k in (2.3.1) is at this stage completely arbitrary. Only later, by identifying the temperature scale with the absolute temperature, do we find that it is then given by the Boltzmann constant $k = 1.38 \times 10^{-16}$ erg/K $= 1.38 \times 10^{-23}$ J/K. See Sect. 3.4. The value of the Boltzmann constant was determined by Planck in 1900.

The entropy is also a measure of the disorder and of the lack of information content in the density matrix. The more states contained in the density matrix, the smaller the weight of each individual state, and the less information about the system one has. Lower entropy means a higher information content. If for example a volume V is available, but the particles remain within a subvolume, then the entropy is smaller than if they occupied the whole of V. Correspondingly, the *information content* ($\propto \mathrm{Tr}\, \rho \log \rho$) of the density matrix is greater, since one knows that the particles are not anywhere within V, but rather only in the subvolume.

2.3.2 An Extremal Property of the Entropy

Let two density matrices, ρ and ρ_1, be given. The important inequality

$$\mathrm{Tr}\left(\rho(\log \rho_1 - \log \rho)\right) \leq 0 \ . \tag{2.3.3}$$

then holds. To prove (2.3.3), we use the diagonal representations of $\rho = \sum_n P_n \left|n\right\rangle \left\langle n\right|$ and $\rho_1 = \sum_\nu P_{1\nu} \left|\nu\right\rangle \left\langle \nu\right|$:

$$\mathrm{Tr}\left(\rho(\log \rho_1 - \log \rho)\right) = \sum_n P_n \left\langle n\right| (\log \rho_1 - \log P_n) \left|n\right\rangle =$$

$$= \sum_n P_n \left\langle n\right| \log \frac{\rho_1}{P_n} \left|n\right\rangle = \sum_n \sum_\nu P_n \left\langle n|\nu\right\rangle \left\langle \nu\right| \log \frac{P_{1\nu}}{P_n} \left|\nu\right\rangle \left\langle \nu|n\right\rangle =$$

$$\leq \sum_n \sum_\nu P_n \left\langle n|\nu\right\rangle \left\langle \nu\right| \left(\frac{P_{1\nu}}{P_n} - 1\right) \left|\nu\right\rangle \left\langle \nu|n\right\rangle = \sum_n P_n \left\langle n\right| \left(\frac{\rho_1}{P_n} - 1\right) \left|n\right\rangle =$$

$$= \mathrm{Tr}\,\rho_1 - \mathrm{Tr}\,\rho = 0 \ .$$

In an intermediate step, we used the basis $\left|\nu\right\rangle$ of ρ_1 as well as the inequality $\log x \leq x - 1$. This inequality is clear from Fig. 2.4. Formally, it follows from properties of the function $f(x) = \log x - x + 1$:

$$f(1) = 0, \quad f'(1) = 0, \quad f''(x) = -\frac{1}{x^2} < 0 \quad \text{(i.e. } f(x) \text{ is convex).}$$

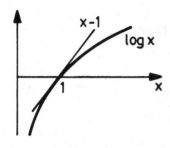

Fig. 2.4. Illustrating the inequality $\log x \leq x - 1$

2.3.3 Entropy of the Microcanonical Ensemble

For the entropy of the microcanonical ensemble, we obtain by referring to (2.3.1) and (2.2.7)

$$S_{MC} = -k\operatorname{Tr}\left(\rho_{MC}\log\rho_{MC}\right) = -k\operatorname{Tr}\left(\rho_{MC}\log\frac{1}{\Omega\left(E\right)\Delta}\right),$$

and, since the density matrix is normalized to 1, Eq. (2.2.9a), the final result:

$$S_{MC} = k\log\bigl(\Omega\left(E\right)\Delta\bigr) . \tag{2.3.4}$$

The entropy is thus proportional to the logarithm of the accessible phase space volume, or, quantum mechanically, to the logarithm of the number of accessible states.

We shall now demonstrate an interesting extremal property of the entropy. Of all the ensembles whose energy lies in the interval $[E, E + \Delta]$, the entropy of the microcanonical ensemble is greatest. To prove this statement, we set $\rho_1 = \rho_{MC}$ in (2.3.3) and use the fact that ρ, like ρ_{MC}, differs from zero only on the energy shell

$$S[\rho] \leq -k\operatorname{Tr}\left(\rho\log\rho_{MC}\right) = -k\operatorname{Tr}\left(\rho\log\frac{1}{\Omega\left(E\right)\Delta}\right) = S_{MC} . \tag{2.3.5}$$

Thus, we have demonstrated that the entropy is maximal for the microcanonical ensemble. We note also that for large N, the following representations of the entropy are all equivalent:

$$S_{MC} = k\log\Omega\left(E\right)\Delta = k\log\Omega\left(E\right)E = k\log\bar{\Omega}(E) . \tag{2.3.6}$$

This follows from the neglect of logarithmic terms in (2.2.19) and an analogous relation for $\Omega\left(E\right)E$.

We can now estimate the *density of states*. The spacing ΔE of the energy levels is given by

$$\Delta E = \frac{\Delta}{\Omega\left(E\right)\Delta} = \Delta\cdot e^{-S_{MC}/k} \sim \Delta\cdot e^{-N} . \tag{2.3.7}$$

The levels indeed lie enormously close together, i.e. at a high density, as already presumed in the Introduction. For this estimate, we used

$$S = k \log \Omega \left(E \right) \Delta \propto N ;$$

this can be seen from the classical results, (2.2.18) as well as (2.2.31) and (2.2.35).

2.4 Temperature and Pressure

The results for the microcanonical ensemble obtained thus far permit us to calculate the mean values of arbitrary operators. These mean values depend on the natural parameters of the microcanonical ensemble, $E, V,$ and N. The temperature and pressure have so far not made an appearance. In this section, we want to define these quantities in terms of the energy and volume derivatives of the entropy.

2.4.1 Systems in Contact: the Energy Distribution Function, Definition of the Temperature

We now consider the following physical situation: let a system be divided into two subsystems, which interact with each other, i.e. exchange of energy between the two subsystems is possible. The overall system is isolated. The division into two subsystems 1 and 2 is not necessarily spatial. Let the Hamiltonian of the system be $H = H_1 + H_2 + W$. Let further the interaction W be small in comparison to H_1 and H_2. For example, in the case of a spatial separation, the surface energy can be supposed to be small compared to the volume energy. The interaction is of fundamental importance, in that it allows the two subsystems to exchange energy. Let the overall system have the energy E, so that it is described by a microcanonical density matrix:

$$\rho_{MC} = \Omega_{1,2}(E)^{-1}\delta(H_1+H_2+W-E) \approx \Omega_{1,2}(E)^{-1}\delta(H_1+H_2-E) . \quad (2.4.1)$$

Here, W was neglected relative to H_1 and H_2, and $\Omega_{1,2}\left(E \right)$ is the phase-space surface of the overall system with a dividing wall (see remarks at the end of this section).

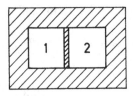

Fig. 2.5. An isolated system divided into subsystems 1 and 2 separated by a fixed diathermal wall (which permits the exchange of thermal energy)

$w\,(E_1)$ denotes the probability density for subsystem 1 to have the energy E_1. According to Eq. (1.2.10), $w\,(E_1)$ is given by

$$w\,(E_1) = \langle \delta(H_1 - E_1) \rangle$$
$$= \int d\Gamma_1 d\Gamma_2\, \Omega_{1,2}(E)^{-1} \delta(H_1 + H_2 - E)\delta(H_1 - E_1)$$
$$= \frac{\Omega_2(E - E_1)\Omega_1(E_1)}{\Omega_{1,2}(E)} . \qquad (2.4.2a)$$

Here, (2.4.1) was used and we have introduced the phase-space surfaces of subsystem 1, $\Omega_1(E_1) = \int d\Gamma_1\, \delta(H_1 - E_1)$, and subsystem 2, $\Omega_2(E - E_1) = \int d\Gamma_2\, \delta(H_2 - E + E_1)$. The most probable value of E_1, denoted as $\tilde{E}_1$, can be found from $\frac{dw\,(E_1)}{dE_1} = 0$:

$$\left(-\Omega_2'(E - E_1)\Omega_1(E_1) + \Omega_2(E - E_1)\Omega_1'(E_1) \right)\Big|_{\tilde{E}_1} = 0 .$$

Using formula (2.3.4) for the microcanonical entropy, we obtain

$$\frac{\partial}{\partial E_2} S_2(E_2)\Big|_{E - \tilde{E}_1} = \frac{\partial}{\partial E_1} S_1(E_1)\Big|_{\tilde{E}_1} . \qquad (2.4.3)$$

We now introduce the following *definition of the temperature*:

$$T^{-1} = \frac{\partial}{\partial E} S(E) . \qquad (2.4.4)$$

Then it follows from (2.4.3) that

$$T_1 = T_2 . \qquad (2.4.5)$$

In the most probable configuration, the temperatures of the two subsystems are equal. We are already using partial derivatives here, since later, several variables will occur. For the ideal gas, we can see immediately that the temperature increases proportionally to the energy per particle, $T \propto E/N$. This property, as well as (2.4.5), the equality of the temperatures of two systems which are in contact and in equilibrium, correspond to the usual concept of temperature.

Remarks:

The Hamiltonian has a lower bound and possesses a finite smallest eigenvalue E_0. In general, the Hamiltonian does not have an upper bound, and the density of the energy eigenvalues increases with increasing energy. As a result, the temperature cannot in general be negative, $(T \geq 0)$, and it increases with increasing energy. For spin systems there is also an upper limit to the energy. The density of states then again decreases as the upper limit is approached, so that in this energy range, $\Omega'/\Omega < 0$ holds. Thus in such systems there can be states with a negative absolute temperature (see Sect. 6.7.2). Due to the various possibilities for representing the entropy as given in (2.3.6), the temperature can also be written as $T = \left(k\frac{d}{dE} \log \bar{\Omega}(E) \right)^{-1}$.

Notes concerning $\Omega_{1,2}(E)$ in Eq. (2.4.1); may be skipped over in a first reading:

(i) In (2.4.1 and 2.4.2a), it must be taken into account that subsystems 1 and 2 are separated from each other. The normalization factor $\Omega_{1,2}(E)$ which occurs in (2.4.1) and (2.4.2a) is not given by

$$\int d\Gamma\, \delta(H - E) \equiv \int \frac{dq\, dp}{h^{3N} N!}\, \delta(H - E) \equiv \Omega(E) \,,$$

but instead by

$$
\begin{aligned}
\Omega_{1,2}(E) &= \int d\Gamma_1 d\Gamma_2\, \delta(H - E) \equiv \int \frac{dq_1\, dp_1}{N_1! h^{3N_1}} \frac{dq_2\, dp_2}{N_2! h^{3N_2}}\, \delta(H - E)\\
&= \int dE_1 \int d\Gamma_1 d\Gamma_2\, \delta(H - E)\delta(H_1 - E_1)\\
&= \int dE_1 \int d\Gamma_1 d\Gamma_2\, \delta(H_2 - E + E_1)\delta(H_1 - E_1)\\
&= \int dE_1\, \Omega_1(E_1)\Omega_2(E - E_1) \,.
\end{aligned}
\tag{2.4.2b}
$$

(ii) Quantum mechanically, one obtains the same result for (2.4.2a):

$$
\begin{aligned}
w\,(E_1) &= \langle \delta(H_1 - E_1)\rangle \equiv \mathrm{Tr}\,\left(\frac{1}{\Omega_{1,2}(E)}\delta(H_1 + H_2 - E)\delta(H_1 - E_1)\right)\\
&= \mathrm{Tr}_1 \mathrm{Tr}_2 \left(\frac{1}{\Omega_{1,2}(E)}\delta\big(H_2 - (E - E_1)\big)\delta(H_1 - E_1)\right)\\
&= \frac{\Omega_1(E_1)\Omega_2(E - E_1)}{\Omega_{1,2}(E)}
\end{aligned}
$$

and

$$
\begin{aligned}
\Omega_{1,2}(E) &= \mathrm{Tr}\,\delta(H_1 + H_2 - E) \equiv \int dE_1\, \mathrm{Tr}\,\big(\delta(H_1 + H_2 - E)\delta(H_1 - E_1)\big)\\
&= \int dE_1\, \mathrm{Tr}\,\big(\delta(H_2 - E + E_1)\delta(H_1 - E_1)\big) = \int dE_1\, \Omega_1(E_1)\Omega_2(E - E_1) \,.
\end{aligned}
$$

Here, we have used the fact that for the non-overlapping subsystems 1 and 2, the traces Tr_1 and Tr_2 taken over parts 1 and 2 are independent, and the states must be symmetrized (or antisymmetrized) only within the subsystems.

(iii) We recall that for quantum-mechanical particles which are in non-overlapping states (wavefunctions), the symmetrization (or antisymmetrization) has no effect on expectation values, and that therefore, in this situation, the symmetrization does not need to be carried out at all.[7] More precisely: if one considers the matrix elements of operators which act only on subsystem 1, their values are the same independently of whether one takes the existence of subsystem 2 into account, or bases the calculation on the (anti-)symmetrized state of the overall system.

---

[7] See e.g. G. Baym, *Lectures on Quantum Mechanics* (W.A. Benjamin, New York, Amsterdam 1969), p. 393

2.4.2 On the Widths of the Distribution Functions of Macroscopic Quantities

2.4.2.1 The Ideal Gas

For the *ideal gas*, from (2.2.18) one finds the following expression for the probability density of the energy E_1, Eq. (2.4.2a):

$$w\left(E_1\right) \propto \left(E_1/N_1\right)^{3N_1/2}\left(E_2/N_2\right)^{3N_2/2} . \tag{2.4.6}$$

In equilibrium, from the equality of the temperatures [Eq. (2.4.3)], i.e. from $\frac{\partial S(\tilde{E}_1)}{\partial E_1} = \frac{\partial S(\tilde{E}_2)}{\partial E_2}$, we obtain the condition $\frac{N_1}{\tilde{E}_1} = \frac{N_2}{E - \tilde{E}_1}$ and thus

$$\tilde{E}_1 = E\frac{N_1}{N_1 + N_2} . \tag{2.4.7}$$

If we expand the distribution function $w\left(E_1\right)$ around the most probable energy value $\tilde{E}_1$, using $\frac{dw\left(E_1\right)}{dE_1}\big|_{\tilde{E}_1} = 0$ and terminating the expansion after the quadratic term, we find

$$\log w\left(E_1\right) = \log w(\tilde{E}_1) + \frac{1}{2}\left(-\frac{3}{2}\frac{N_1}{\tilde{E}_1^2} - \frac{3}{2}\frac{N_2}{\tilde{E}_2^2}\right)\left(E_1 - \tilde{E}_1\right)^2 ,$$

and therefore

$$w\left(E_1\right) = w(\tilde{E}_1)\,\mathrm{e}^{-\frac{3}{4}\frac{N_1+N_2}{\tilde{E}_1\tilde{E}_2}\left(E_1-\tilde{E}_1\right)^2} = w(\tilde{E}_1)\,\mathrm{e}^{-\frac{3}{4}\frac{N}{N_1 N_2 \bar{e}^2}\left(E_1-\tilde{E}_1\right)^2} , \tag{2.4.8}$$

where $\frac{N_1}{\tilde{E}_1^2} + \frac{N_2}{\tilde{E}_2^2} = \frac{N_2}{\tilde{E}_1\tilde{E}_2} + \frac{N_1}{\tilde{E}_1\tilde{E}_2} = \frac{N}{\tilde{E}_1\tilde{E}_2}$ and $\bar{e} = E/N$ were used. Here, $\log w\left(E_1\right)$ rather than $w\left(E_1\right)$ was expanded, because of the occurrence of the powers of the particle numbers N_1 and N_2 in Eq. (2.4.6). This is also preferable since it permits the coefficients of the Taylor expansion to be expressed in terms of derivatives of the entropy. From (2.4.8), we obtain the relative mean square deviation:

$$\frac{\left\langle\left(E_1 - \tilde{E}_1\right)\right\rangle^2}{\tilde{E}_1^2} = \frac{1}{\tilde{E}_1^2}\frac{2}{3}\frac{\tilde{E}_1\tilde{E}_2}{\left(N_1 + N_2\right)} = \frac{2}{3}\frac{1}{N}\frac{N_2}{N_1} \approx 10^{-20} \tag{2.4.9}$$

and the relative width of the distribution, with $N_2 \approx N_1$,

$$\frac{\Delta E_1}{\tilde{E}_1} \sim \frac{1}{\sqrt{N}} . \tag{2.4.10}$$

For macroscopic systems, the distribution is very sharp. The most probable state occurs with a stupendously high probability. The sharpness of the distribution function becomes even more apparent if one expresses it in terms of the energy per particle, $e_1 = E_1/N_1$, including the normalization factor:

$$w_{e_1}\left(e_1\right) = \sqrt{\frac{3}{4\pi}\frac{NN_1}{N_2}}\,\bar{e}\,\mathrm{e}^{\frac{3NN_1}{4N_2\bar{e}^2}\left(e_1-\tilde{e}_1\right)^2} .$$

2.4.2.2 A General Interacting System

For *interacting systems* it holds quite generally that:
An arbitrary quantity A, which can be written as a volume integral over a density $A(\mathbf{x})$,

$$A = \int_V d^3x \, A(\mathbf{x}) \, . \tag{2.4.11}$$

Its average value depends on the volume as

$$\langle A \rangle = \int_V d^3x \langle A(\mathbf{x}) \rangle \sim V \, . \tag{2.4.12}$$

The mean square deviation is given by

$$(\Delta A)^2 = \left\langle \left(A - \langle A \rangle \right) \left(A - \langle A \rangle \right) \right\rangle$$
$$= \int_V d^3x \int_V d^3x' \left\langle \left(A(\mathbf{x}) - \langle A(\mathbf{x}) \rangle \right) \left(A(\mathbf{x}') - \langle A(\mathbf{x}') \rangle \right) \right\rangle \propto V l^3 \, .$$
$$\tag{2.4.13}$$

Both the integrals in (2.4.13) are to be taken over the volume V. The correlation function in the integral however vanishes for $|\mathbf{x} - \mathbf{x}'| > l$, where l is the range of the interactions (the correlation length). The latter is finite and thus the mean square deviation is likewise only of the order of V and not, as one might perhaps naively expect, quadratic in V. The relative deviation of A is therefore given by

$$\frac{\Delta A}{\langle A \rangle} \sim \frac{1}{V^{1/2}} \, . \tag{2.4.14}$$

2.4.3 External Parameters: Pressure

Let the Hamiltonian of a system depend upon an external parameter a: $H = H(a)$. This external parameter can for example be the volume V of the system. Using the volume in phase space, $\bar{\Omega}$, we can derive an expression for the total differential of the entropy dS. Starting from the phase-space volume

$$\bar{\Omega}(E, a) = \int d\Gamma \, \Theta(E - H(a)) \, , \tag{2.4.15}$$

we take its total differential

$$d\bar{\Omega}\,(E,a) = \int d\Gamma\,\delta\bigl(E - H(a)\bigr)\left(dE - \frac{\partial H}{\partial a}da\right)$$

$$= \Omega\,(E,a)\left(dE - \left\langle\frac{\partial H}{\partial a}\right\rangle da\right),\quad (2.4.16)$$

or

$$d\log\bar{\Omega} = \frac{\Omega}{\bar{\Omega}}\left(dE - \left\langle\frac{\partial H}{\partial a}\right\rangle da\right).\qquad (2.4.17)$$

We now insert $S(E,a) = k\log\bar{\Omega}\,(E,a)$ and (2.4.4), obtaining

$$dS = \frac{1}{T}\left(dE - \left\langle\frac{\partial H}{\partial a}\right\rangle da\right).\qquad (2.4.18)$$

From (2.4.18), we can read off the partial derivatives of the entropy in terms of E and a:[8]

$$\left(\frac{\partial S}{\partial E}\right)_a = \frac{1}{T}\quad;\quad \left(\frac{\partial S}{\partial a}\right)_E = -\frac{1}{T}\left\langle\frac{\partial H}{\partial a}\right\rangle.\qquad (2.4.19)$$

Introduction of the *pressure* (special case: $a = V$):

After the preceding considerations, we can turn to the derivation of pressure within the framework of statistical mechanics. We refer to Fig. 2.6 as a guide to this procedure. A movable piston at a distance L from the origin of the coordinate system permits variations in the volume $V = LA$, where A is the cross-sectional area of the piston. The influence of the walls of the container is represented by a wall potential. Let the spatial coordinate of the ith particle in the direction perpendicular to the piston be x_i. Then the total wall potential is given by

$$V_{\text{wall}} = \sum_{i=1}^{N} v(x_i - L).\qquad (2.4.20)$$

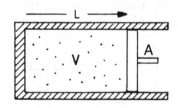

Fig. 2.6. The definition of pressure

Here, $v(x_i - L)$ is equal to zero for $x_i < L$ and is very large for $x_i \geq L$, so that penetration of the wall by the gas particles is prevented. We then obtain for the force on the molecules

---

[8] The symbol $\left(\frac{\partial S}{\partial E}\right)_a$ denotes the partial derivative of S with respect to the energy E, holding a constant, etc.

$$F = \sum_i F_i = \sum_i \left(-\frac{\partial v}{\partial x_i} \right) = \frac{\partial}{\partial L} \sum_i v(x_i - L) = \frac{\partial H}{\partial L} . \qquad (2.4.21)$$

The pressure is defined as the average force per unit area which the molecules exert upon the wall, from which we find using (2.4.21) that

$$P \equiv -\frac{\langle F \rangle}{A} = -\left\langle \frac{\partial H}{\partial V} \right\rangle \qquad (2.4.22)$$

In this case, the general relations (2.4.18) and (2.4.19) become

$$dS = \frac{1}{T}(dE + PdV) \qquad (2.4.23)$$

and

$$\frac{1}{T} = \left(\frac{\partial S}{\partial E} \right)_V , \quad \frac{P}{T} = \left(\frac{\partial S}{\partial V} \right)_E . \qquad (2.4.24)$$

Solving (2.4.23) for dE, we obtain

$$dE = TdS - PdV , \qquad (2.4.25)$$

a relation which we will later identify as the *First Law of Thermodynamics* [for a constant particle number; see Eqs. (3.1.3) and (3.1.3′)]. Comparison with phenomenological thermodynamics gives an additional justification for the identification of T with the temperature. As a result of

$$-PdV = \frac{\langle F \rangle}{A}dV = \langle FdL \rangle \equiv \delta W ,$$

the last term in (2.4.25) denotes the work δW which is performed on the system causing the change in volume.

We are now interested in the *pressure distribution in two subsystems*, which are separated from each other by a movable partition, keeping the particle numbers in each subsystem constant (Fig. 2.6′). The energies and volumes are additive

$$E = E_1 + E_2 , \quad V = V_1 + V_2 . \qquad (2.4.26)$$

The probability that subsystem 1 has the energy E_1 and the volume V_1 is given by

$$\omega\,(E_1, V_1) = \int d\Gamma_1 d\Gamma_2 \frac{\delta(H_1 + H_2 - E)}{\Omega_{1,2}(E,V)} \delta(H_1 - E_1)\Theta(q_1 \in V_1)\Theta(q_2 \in V_2)$$

$$= \frac{\Omega_1(E_1, V_1)\Omega_2(E_2, V_2)}{\Omega_{1,2}(E,V)} . \qquad (2.4.27a)$$

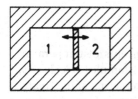

Fig. 2.6′. Two systems which are isolated from the external environment, separated by a movable wall which permits the exchange of energy.

In (2.4.27a), the function $\Theta(q_1 \in V_1)$ means that all the spatial coordinates of the sub-phase space 1 are limited to the volume V_1 and correspondingly, $\Theta(q_2 \in V_2)$. Here, both E_1 and V_1 are statistical variables, while in (2.4.2b), V_1 was a fixed parameter. Therefore, the normalization factor is given here by

$$\Omega_{1,2}(E, V) = \int dE_1 \int dV_1 \, \Omega_1(E_1, V_1)\Omega_2(E - E_1, V - V_1) \,. \qquad (2.4.27b)$$

In analogy to (2.4.3), the most probable state of the two systems is found by the condition of vanishing derivatives of (2.4.27a)

$$\frac{\partial \omega\,(E_1, V_1)}{\partial E_1} = 0 \quad \text{and} \quad \frac{\partial \omega\,(E_1, V_1)}{\partial V_1} = 0 \,.$$

From this, it follows that

$$\frac{\partial}{\partial E_1} \log \Omega_1(E_1, V_1) = \frac{\partial}{\partial E_2} \log \Omega_2(E_2, V_2) \Rightarrow T_1 = T_2$$

and $\qquad\qquad\qquad\qquad\qquad\qquad\qquad\qquad\qquad\qquad\qquad\qquad (2.4.28)$

$$\frac{\partial}{\partial V_1} \log \Omega_1(E_1, V_1) = \frac{\partial}{\partial V_2} \log \Omega_2(E_2, V_2) \Rightarrow P_1 = P_2 \,.$$

In systems which are separated by a movable wall and can exchange energy, the equilibrium temperatures and pressures are equal.

The microcanonical density matrix evidently depends on the energy E and on the volume V, as well as on the particle number N. If we regard these parameters likewise as variables, then the overall variation of S must be replaced by

$$dS = \frac{1}{T}dE + \frac{P}{T}dV - \frac{\mu}{T}dN \,. \qquad (2.4.29)$$

Here, we have defined the *chemical potential* μ by

$$\frac{\mu}{T} = k\frac{\partial}{\partial N} \log \Omega\,(E, V, N) \,. \qquad (2.4.30)$$

The chemical potential is related to the fractional change in the number of accessible states with respect to the change in the number of particles. Physically, its meaning is the change in energy per particle added to the system, as can be seen from (2.4.29) by solving that expression for dE.

2.5 Thermodynamic Properties of Some Non-interacting Systems

Now that we have introduced the thermodynamic concepts of temperature and pressure, we are in a position to discuss further the examples of a classical ideal gas, quantum-mechanical oscillators, and non-interacting spins treated in Sect. 2.2.2. In the following, we will derive the thermodynamic consequences of the phase-space surface or number of states $\Omega(E)$ which we calculated there for those examples.

2.5.1 The Ideal Gas

We first calculate the thermodynamic quantities introduced in the preceding sections for the case of an ideal gas. In (2.2.16), we found the phase-space volume in the limit of a large number of particles:

$$\bar{\Omega}(E) \equiv \int d\Gamma \,\Theta\big(E - H(q,p)\big) = \left(\frac{V}{N}\right)^N \left(\frac{4\pi mE}{3Nh^2}\right)^{\frac{3N}{2}} e^{\frac{5N}{2}} . \qquad (2.2.16)$$

If we insert (2.2.16) into (2.3.6), we obtain the entropy as a function of the energy and the volume:

$$S(E,V) = kN \log\left[\frac{V}{N}\left(\frac{4\pi mE}{3Nh^2}\right)^{\frac{3}{2}} e^{\frac{5}{2}}\right] . \qquad (2.5.1)$$

Eq. (2.5.1) is called the *Sackur–Tetrode equation*. It represents the starting point for the calculation of the temperature and the pressure. The *temperature* is, from (2.4.4), defined as the reciprocal of the partial energy derivative of the entropy, $T^{-1} = \left(\frac{\partial S}{\partial E}\right)_V = kN\frac{3}{2}E^{-1}$, from which the *caloric equation of state* of the ideal gas follows immediately:

$$E = \frac{3}{2}NkT . \qquad (2.5.2)$$

With (2.5.2), we can also find the entropy (2.5.1) as a function of T and V:

$$S(T,V) = kN \log\left[\frac{V}{N}\left(\frac{2\pi mkT}{h^2}\right)^{\frac{3}{2}} e^{\frac{5}{2}}\right] . \qquad (2.5.3)$$

The *pressure* is obtained from (2.4.24) by taking the volume derivative of (2.5.1)

$$P = T\left(\frac{\partial S}{\partial V}\right)_E = \frac{kTN}{V} . \qquad (2.5.4)$$

This is the *thermal equation of state* of the ideal gas, which is often written in the form

$$PV = NkT \ . \qquad\qquad (2.5.4')$$

The implications of the thermal equation of state are summarized in the diagrams of Fig. 2.7: Fig. 2.7a shows the PVT surface or surface of the equation of state, i.e. the pressure as a function of V and T. Figs. 2.7b,c,d are projections onto the PV-, the TV- and the PT-planes. In these diagrams, the isotherms ($T =$ const), the isobars ($P =$ const), and the isochores ($V =$ const) are illustrated. These curves are also drawn in on the PVT surface (Fig. 2.7a).

Remarks:

(i) It can be seen from (2.5.2) that the temperature increases with the energy content of the ideal gas, in accord with the usual concept of temperature.

(ii) The equation of state (2.5.4) also provides us with the possibility of measuring the temperature. The determination of the temperature of an ideal gas can be achieved by measuring its volume and its pressure.

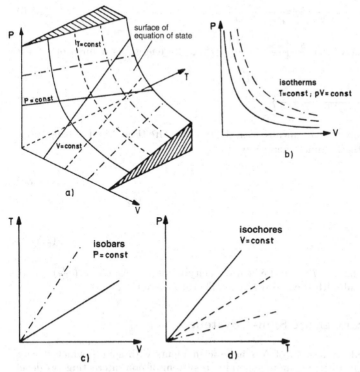

Fig. 2.7. The equation of state of the ideal gas: **(a)** surface of the equation of state, **(b)** P-V diagram, **(c)** T-V diagram, **(d)** P-T diagram

The temperature of any given body can be determined by bringing it into thermal contact with an ideal gas and making use of the fact that the two temperatures will equalize [Eq. (2.4.5)]. The relative sizes of the two systems (body and thermometer) must of course be chosen so that contact with the ideal gas changes the temperature of the body being investigated by only a negligible amount.

*2.5.2 Non-interacting Quantum Mechanical Harmonic Oscillators and Spins

2.5.2.1 Harmonic Oscillators

From (2.2.31) and (2.3.6), it follows for the entropy of non-coupled harmonic oscillators with $e = E/N$, that

$$S(E) = kN \left[\frac{e + \frac{1}{2}\hbar\omega}{\hbar\omega} \log \frac{e + \frac{1}{2}\hbar\omega}{\hbar\omega} - \frac{e - \frac{1}{2}\hbar\omega}{\hbar\omega} \log \frac{e - \frac{1}{2}\hbar\omega}{\hbar\omega} \right] , \qquad (2.5.5)$$

where a logarithmic term has been neglected. From Eq. (2.4.4), we obtain for the temperature

$$T = \left(\frac{\partial S}{\partial E} \right)^{-1} = \frac{\hbar\omega}{k} \left(\log \frac{e + \frac{1}{2}\hbar\omega}{e - \frac{1}{2}\hbar\omega} \right)^{-1} . \qquad (2.5.6)$$

From this, it follows via $\frac{E + \frac{1}{2}N\hbar\omega}{E - \frac{1}{2}N\hbar\omega} = e^{\frac{\hbar\omega}{kT}}$ that the energy as a function of the temperature is given by

$$E = N\hbar\omega \left\{ \frac{1}{e^{\hbar\omega/kT} - 1} + \frac{1}{2} \right\} . \qquad (2.5.7)$$

The energy increases monotonically with the temperature (Fig. 2.8). Limiting cases: For $E \to N\frac{\hbar\omega}{2}$ (the minimal energy), we find

$$T \to \frac{1}{\log \infty} = 0 , \qquad (2.5.8a)$$

and for $E \to \infty$

$$T \to \frac{1}{\log 1} = \infty . \qquad (2.5.8b)$$

We can also see that for $T \to 0$, the heat capacity tends to zero: $C_V = \left(\frac{\partial E}{\partial T} \right)_V \to 0$; this is in agreement with the Third Law of Thermodynamics.

2.5.2.2 A Paramagnetic Spin-$\frac{1}{2}$ System

Finally, we consider a system of N magnetic moments with spin $\frac{1}{2}$ which do not interact with each other; or, more generally, a system of non-interacting two-level systems. We refer here to Sect. 2.2.3.2. From (2.2.35), the entropy of such a system is given by

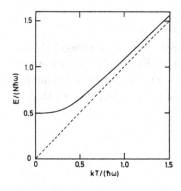

Fig. 2.8. Non-coupled harmonic oscillators: the energy as a function of the temperature.

$$S(E) = \frac{kN}{2}\left\{ -(1+e)\log\frac{1+e}{2} - (1-e)\log\frac{1-e}{2}\right\} \qquad (2.5.9)$$

with $e = E/Nh$. From this, we find for the temperature:

$$T = \left(\frac{\partial S}{\partial E}\right)^{-1} = \frac{2h}{k}\left(\log\frac{1-e}{1+e}\right)^{-1} . \qquad (2.5.10)$$

The entropy is shown as a function of the energy in Fig. 2.9, and the temperature as a function of the energy in Fig. 2.10. The ground-state energy is $E_0 = -Nh$. For $E \to -Nh$, we find from (2.5.10)

$$\lim_{E \to -Nh} T = 0 . \qquad (2.5.11)$$

The temperature increases with increasing energy beginning at $E_0 = -Nh$ monotonically until $E = 0$ is reached; this is the state in which the magnetic moments are completely disordered, i.e. there are just as many oriented parallel as antiparallel to the applied magnetic field h. The region $E > 0$, in which the temperature is negative (!), will be discussed later in Sect. 6.7.2.

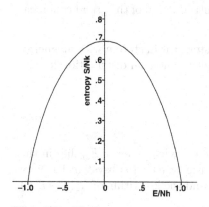

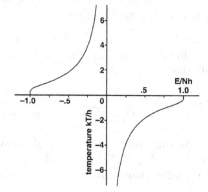

Fig. 2.9. The entropy as a function of the energy for a two-level system (spin$-\frac{1}{2}-$paramagnet)

Fig. 2.10. The temperature as a function of the energy for a two-level system (spin$-\frac{1}{2}-$paramagnet)

2.6 The Canonical Ensemble

In this section, the properties of a small subsystem 1 which is embedded in a large system 2, the heat bath,[9] will be investigated (Fig. 2.11). We first need to construct the density matrix, which we will derive from quantum mechanics in the following section. The overall system is taken to be isolated, so that it is described by a microcanonical ensemble.

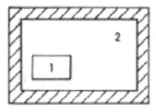

Fig. 2.11. A canonical ensemble. Subsystem 1 is in contact with the heat bath 2. The overall system is isolated.

2.6.1 The Density Matrix

The Hamiltonian of the total system

$$H = H_1 + H_2 + W \approx H_1 + H_2 \tag{2.6.1}$$

is the sum of the Hamiltonians H_1 and H_2 for systems 1 and 2 and the interaction term W. The latter is in fact necessary so that the two subsystems can come to equilibrium with each other; however, W is negligibly small compared to H_1 and H_2. Our goal is the derivation of the density matrix for subsystem 1 alone. We will give two derivations here, of which the second is shorter, but the first is more useful for the introduction of the grand canonical ensemble in the next section.

(i) Let $P_{E_{1n}}$ be the probability that subsystem 1 is in state n with an energy eigenvalue E_{1n}. Then for $P_{E_{1n}}$, using the microcanonical distribution for the total system, we find

$$P_{E_{1n}} = {\sum}' \frac{1}{\Omega_{1,2}(E)\Delta} = \frac{\Omega_2(E - E_{1n})}{\Omega_{1,2}(E)} . \tag{2.6.2}$$

The sum runs over all the states of subsystem 2 whose energy E_{2n} lies in the interval $E - E_{1n} \leq E_{2n} \leq E + \Delta - E_{1n}$. In the case that subsystem 1 is very much smaller than subsystem 2, we can expand the logarithm of $\Omega_2(E - E_{1n})$ in E_{1n}:

---

[9] A heat bath (or thermal reservoir) is a system which is so large that adding or subtracting a finite amount of energy to it does not change its temperature.

$$P_{E_{1n}} = \frac{\Omega_2(E - \tilde{E}_1 + \tilde{E}_1 - E_{1n})}{\Omega_{1,2}(E)}$$

$$\approx \frac{\Omega_2(E - \tilde{E}_1)}{\Omega_{1,2}(E)} e^{(\tilde{E}_1 - E_{1n})/kT} = Z^{-1}e^{-E_{1n}/kT} . \qquad (2.6.3)$$

This expression contains $T = \left(k \frac{\partial}{\partial E} \log \Omega_2(E - \tilde{E}_1) \right)^{-1}$, the temperature of the heat bath. The normalization factor Z, from (2.6.3), is given by

$$Z = \frac{\Omega_{1,2}(E)}{\Omega_2(\tilde{E}_2)} e^{-\tilde{E}_1/kT} . \qquad (2.6.4)$$

However, it is important that Z can be calculated directly from the properties of subsystem 1. The condition that the sum over all the $P_{E_{1n}}$ must be equal to 1 implies that

$$Z = \sum_n e^{-E_{1n}/kT} = \text{Tr}_1 e^{-H_1/kT} . \qquad (2.6.5)$$

Z is termed the partition function. The canonical density matrix is then given by the following equivalent representations

$$\rho_C = \sum_n P_{E_{1n}} |n\rangle \langle n| = Z^{-1} \sum_n e^{-E_{1n}/kT} |n\rangle \langle n| = Z^{-1}e^{-H_1/kT} . \quad (2.6.6)$$

(ii) The *second derivation* starts with the fact that the density matrix ρ for subsystem 1 can be obtained form the microcanonical density matrix by taking the trace over the degrees of freedom of system 2:

$$\rho_C = \text{Tr}_2 \, \rho_{MC} = \text{Tr}_2 \frac{\delta(H_1 + H_2 - E)}{\Omega_{1,2}(E)} = \frac{\Omega_2(E - H_1)}{\Omega_{1,2}(E)}$$

$$\equiv \frac{\Omega_2(E - \tilde{E}_1 + \tilde{E}_1 - H_1)}{\Omega_{1,2}(E)} \approx \frac{\Omega_2(E - \tilde{E}_1)}{\Omega_{1,2}(E)} e^{(\tilde{E}_1 - H_1)/kT} . \qquad (2.6.7)$$

This derivation is valid both in classical physics and in quantum mechanics, as is shown specifically in (2.6.9). Thus we have also demonstrated the validity of (2.6.6) with the definition (2.6.5) by this second route.

Expectation values of observables A which act only on the states of subsystem 1 are given by

$$\langle A \rangle = \text{Tr}_1 \text{Tr}_2 \, \rho_{MC} \, A = \text{Tr}_1 \rho_C \, A . \qquad (2.6.8)$$

Remarks:

(i) The classical distribution function:
The classical distribution function of subsystem 1 is obtained by integration of ρ_{MC} over Γ_2

$$\rho_C(q_1, p_1) = \int d\Gamma_2 \, \rho_{MC}$$

$$= \int d\Gamma_2 \, \frac{1}{\Omega_{1,2}(E)} \delta\big(E - H_1(q_1, p_1) - H_2(q_2, p_2)\big) \qquad (2.6.9)$$

$$= \frac{\Omega_2\big(E - H_1(q_1, p_1)\big)}{\Omega_{1,2}(E)} \ .$$

If we expand the logarithm of this expression with respect to H_1, we obtain

$$\rho_C(q_1, p_1) = Z^{-1} e^{-H_1(q_1, p_1)/kT} \qquad (2.6.10a)$$

$$Z = \int d\Gamma_1 \, e^{-H_1(q_1, p_1)/kT} \ . \qquad (2.6.10b)$$

Here, Z is called the *partition function*. Mean values of observables $A(q_1, p_1)$ which refer only to subsystem 1 are calculated in the classical case by means of

$$\langle A \rangle = \int d\Gamma_1 \, \rho_C(q_1, p_1) A(q_1, p_1) \ , \qquad (2.6.10c)$$

as one finds analogously to (2.6.8).

(ii) The energy distribution:
The energy distribution $\omega(E_1)$ introduced in Sect. 2.4.1 can also be calculated classically and quantum mechanically within the framework of the canonical ensemble (see problem 2.7):

$$\omega(E_1) = \frac{1}{\Delta_1} \int\limits_{E_1}^{E_1 + \Delta_1} dE_1' \sum_n \delta(E_1' - E_{1n}) P_{E_{1n}}$$

$$\approx \frac{\Omega_2(E - E_1)}{\Omega_{1,2}(E)} \frac{1}{\Delta_1} {\sum_n}' 1 = \frac{\Omega_2(E - E_1)\Omega_1(E_1)}{\Omega_{1,2}(E)} \ . \qquad (2.6.11)$$

This expression agrees with (2.4.2a).

(iii) The partition function (2.6.5) can also be written as follows:

$$Z = \int dE_1 \, \mathrm{Tr}_1 \, e^{-H_1/kT} \delta(H_1 - E_1) = \int dE_1 \, \mathrm{Tr}_1 \, e^{-E_1/kT} \delta(H_1 - E_1)$$

$$= \int dE_1 \, e^{-E_1/kT} \, \Omega_1(E_1) \ . \qquad (2.6.12)$$

(iv) In the derivation of the canonical density matrix, Eq. (2.6.7), we expanded the logarithm of $\Omega_2(E - H_1)$. We show that it was justified to terminate this expansion after the first term of the Taylor series:

$$\Omega_2(E - H_1) = \Omega_2(E - \tilde{E}_1 - (H_1 - \tilde{E}_1))$$

$$= \Omega_2(E - \tilde{E}_1) e^{-\frac{1}{kT}(H_1 - \tilde{E}_1) + \frac{1}{2}\left(\frac{\partial 1/T}{\partial \tilde{E}_2}\right)(H_1 - \tilde{E}_1)^2 + \dots}$$

$$= \Omega_2(E - \tilde{E}_1) e^{-\frac{1}{kT}(H_1 - \tilde{E}_1) - \frac{1}{2kT^2}\frac{\partial T}{\partial \tilde{E}_2}(H_1 - \tilde{E}_1)^2 + \dots}$$

$$= \Omega_2(E - \tilde{E}_1) e^{-\frac{1}{kT}(H_1 - \tilde{E}_1)(1 + \frac{1}{2TC}(H_1 - \tilde{E}_1) + \dots)} \ ,$$

where C is the heat capacity of the thermal bath. Since, owing to the large size of the thermal bath, $(H_1 - \tilde{E}_1) \ll TC$ holds (to be regarded as an inequality for the eigenvalues), it is in fact justified to ignore the higher-order corrections in the Taylor expansion.

(v) In later sections, we will be interested only in the (canonical) subsystem 1. The heat bath 2 enters merely through its temperature. We shall then leave off the index '1' from the relations derived in this section.

2.6.2 Examples: the Maxwell Distribution and the Barometric Pressure Formula

Suppose the subsystem to consist of one particle. The probability that its position and its momentum take on the values $\mathbf{x}$ and $\mathbf{p}$ is given by:

$$w(\mathbf{x}, \mathbf{p})\, d^3x\, d^3p = C\, e^{-\beta\left(\frac{\mathbf{p}^2}{2m} + V(\mathbf{x})\right)}\, d^3x\, d^3p \ . \tag{2.6.13}$$

Here, $\beta = \frac{1}{kT}$ and $V(\mathbf{x})$ refers to the potential energy, while $C = C'C''$ is a normalization factor[10]. Integration over spatial coordinates gives the *momentum distribution*

$$w(\mathbf{p})\, d^3p = C'\, e^{-\beta\frac{\mathbf{p}^2}{2m}}\, d^3p \ . \tag{2.6.14}$$

If we do not require the direction of the momentum, i.e. integrating over all angles, we obtain

$$w(p)\, dp = 4\pi C'\, e^{-\beta\frac{p^2}{2m}}\, p^2\, dp \ ; \tag{2.6.15}$$

this is the *Maxwell velocity distribution*. Integration of (2.6.13) over the momentum gives the *spatial distribution*:

$$w(\mathbf{x})\, d^3x = C''\, e^{-\beta V(\mathbf{x})}\, d^3x \ . \tag{2.6.16}$$

If we now set the potential $V(\mathbf{x})$ equal to the gravitational field $V(\mathbf{x}) = mgz$ and use the fact that the particle-number density is proportional to $w(\mathbf{x})$, we obtain [employing the equation of state for the ideal gas, (2.5.4'), which relates the pressure to the particle-number density] an expression for the altitude dependence of the pressure, the *barometric pressure formula*:

$$P(z) = P_0 e^{-mgz/kT} \tag{2.6.17}$$

(cf. also problem 2.15).

----

[10] $C' = \left(\frac{\beta}{2\pi m}\right)^{3/2}$ and $C'' = \left(\int d^3x\, e^{-\beta V(\mathbf{x})}\right)^{-1}$

2.6.3 The Entropy of the Canonical Ensemble and Its Extremal Values

From Eq. (2.6.6), we find for the entropy of the canonical ensemble

$$S_C = -k\langle \log \rho_C \rangle = \frac{1}{T}\bar{E} + k \log Z \qquad (2.6.18)$$

with

$$\bar{E} = \langle H \rangle . \qquad (2.6.18')$$

Now let ρ correspond to a different distribution with the same average energy $\langle H \rangle = \bar{E}$; then the inequality

$$S[\rho] = -k \operatorname{Tr} (\rho \log \rho) \leq -k \operatorname{Tr} \left(\rho \log \rho_C \right)$$
$$= -k \operatorname{Tr} \left(\rho \left(-\frac{H}{kT} - \log Z\right)\right) = \frac{1}{T}\langle H \rangle + k \log Z = S_C \qquad (2.6.19)$$

results. Here, the inequality in (2.3.3) was used along with $\rho_1 = \rho_C$. The canonical ensemble has the greatest entropy of all ensembles with the same average energy.

2.6.4 The Virial Theorem and the Equipartition Theorem

2.6.4.1 The Classical Virial Theorem and the Equipartition Theorem

Now, we consider a classical system and combine its momenta and spatial coordinates into $x_i = p_i, q_i$. For the average value of the quantity $x_i \frac{\partial H}{\partial x_j}$ we find the following relation:

$$\left\langle x_i \frac{\partial H}{\partial x_j} \right\rangle = Z^{-1} \int d\Gamma\, x_i \frac{\partial H}{\partial x_j} e^{-H/kT}$$
$$= Z^{-1} \int d\Gamma\, x_i \frac{\partial e^{-H/kT}}{\partial x_j}(-kT) = kT\, \delta_{ij}, \qquad (2.6.20)$$

where we have carried out an integration by parts. We have assumed that $\exp(-H(p,q)/kT)$ drops off rapidly enough for large p and q so that no boundary terms occur. This is the case for the kinetic energy and potentials such as those of harmonic oscillators. In the general case, one would have to take the wall potential into account. Eq. (2.6.20) contains the classical virial theorem as a special case, as well as the equipartition theorem.

Applying (2.6.20) to the spatial coordinates q_i, we obtain the classical *virial theorem*

$$\left\langle q_i \frac{\partial V}{\partial q_j} \right\rangle = kT\, \delta_{ij} . \qquad (2.6.21)$$

We now specialize to the case of harmonic oscillators, i.e.

$$V = \sum_i V_i \equiv \sum_i \frac{m\omega^2}{2} q_i^2 \ . \tag{2.6.22}$$

For this case, it follows from (2.6.21) that

$$\langle V_i \rangle = \frac{kT}{2} \ . \tag{2.6.23}$$

The potential energy of each degree of freedom has the average value $kT/2$.

Applying (2.6.20) to the momenta, we find the *equipartition theorem*. We take as the kinetic energy the generalized quadratic form

$$E_{kin} = \sum_{i,k} a_{ik} p_i p_k \ , \quad \text{with} \quad a_{ik} = a_{ki} \ . \tag{2.6.24}$$

For this form, we find $\frac{\partial E_{kin}}{\partial p_i} = \sum_k (a_{ik}p_k + a_{ki}p_k) = \sum_k 2a_{ik}p_k$ and therewith, after multiplication by p_i and summation over all i,

$$\sum_i p_i \frac{\partial E_{kin}}{\partial p_i} = \sum_k 2a_{ik} p_i p_k = 2E_{kin} \ . \tag{2.6.25}$$

Now we take the thermal average and find from (2.6.20)

$$\left\langle \sum_i p_i \frac{\partial H}{\partial p_i} \right\rangle = 2\langle E_{kin} \rangle = 3 NkT \ ; \tag{2.6.26}$$

i.e. the *equipartition theorem*. The average kinetic energy per degree of freedom is equal to $\frac{1}{2}kT$.

As previously mentioned, in the potential V, the interaction $\frac{1}{2} \sum_{m,n} v(|\mathbf{x}_{mn}|)$ (with $\mathbf{x}_{mn} = \mathbf{x}_m - \mathbf{x}_n$) of the particles with each other and in general their interaction with the wall, V_{wall}, must be taken into account. Then using (2.6.23) and (2.6.25), we find

$$PV = \frac{2}{3}\langle E_{kin} \rangle - \frac{1}{6} \sum_{m,n} \left\langle \mathbf{x}_{mn} \frac{\partial v(|\mathbf{x}_{mn}|)}{\partial \mathbf{x}_{mn}} \right\rangle \ . \tag{2.6.27}$$

The term PV results from the wall potential. The second term on the right-hand side is called the 'virial' and can be expanded in powers of $\frac{N}{V}$ (virial expansion, see Sect. 5.3).

*Proof of (2.6.27):
We begin with the Hamiltonian

$$H = \sum_n \frac{\mathbf{p}_n^2}{2m} + \frac{1}{2} \sum_{n,m} v(\mathbf{x}_n - \mathbf{x}_m) + V_{wall} \ , \tag{2.6.28}$$

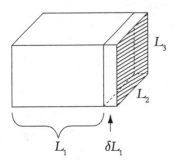

Fig. 2.12. Quantities related to the wall potential and the pressure: increasing the volume on displacing a wall by δL_1

and write for the pressure, using (2.4.22):

$$PV = -\left\langle \frac{\partial H}{\partial V} \right\rangle V = -\left\langle \frac{\partial H}{\partial L_1} \right\rangle \frac{V}{L_2 L_3} = -\frac{1}{3}\left\langle L_1 \frac{\partial H}{\partial L_1} + L_2 \frac{\partial H}{\partial L_2} + L_3 \frac{\partial H}{\partial L_3} \right\rangle . \quad (2.6.29)$$

Now, V_{wall} has the form (cf. Fig. 2.12)

$$V_{\text{wall}} = V_\infty \sum_i \{\Theta(x_{i1} - L_1) + \Theta(x_{i2} - L_2) + \Theta(x_{i3} - L_3)\} . \quad (2.6.30)$$

Here, V_∞ characterizes the barrier represented by the wall. The kinetic energy of the particles is much smaller than V_∞. Evidently, $\frac{\partial V_{\text{wall}}}{\partial L_1} = -V_\infty \sum_n \delta(x_{n1} - L_1)$ and therefore

$$\left\langle \sum_n x_{n1} \frac{\partial V_{\text{wall}}}{\partial x_{n1}} \right\rangle = \left\langle \sum_n x_{n1} V_\infty \delta(x_{n1} - L_1) \right\rangle = \left\langle \sum_n L_1 V_\infty \delta(x_{n1} - L_1) \right\rangle$$

$$= -\left\langle L_1 \frac{\partial V_{\text{wall}}}{\partial L_1} \right\rangle = -\left\langle L_1 \frac{\partial H}{\partial L_1} \right\rangle .$$

With this, (2.6.29) can be put into the form

$$PV = \frac{1}{3}\left\langle \sum_{n,\alpha} x_{n\alpha} \frac{\partial}{\partial x_{n\alpha}} V_{\text{wall}} \right\rangle = kTN - \frac{1}{3}\left\langle \sum_{n,\alpha} x_{n\alpha} \frac{\partial}{\partial x_{n\alpha}} v \right\rangle \quad (2.6.31)$$

$$= \frac{2}{3}\left\langle E_{kin} \right\rangle - \frac{1}{6}\left\langle \sum_\alpha \sum_{n \neq m} (x_{n\alpha} - x_{m\alpha}) \frac{\partial v}{\partial(x_{n\alpha} - x_{m\alpha})} \right\rangle . \quad (2.6.32)$$

In the first line, the virial theorem (2.6.21) was used, and we have abbreviated the sum of the pair potentials as v. In the second line, kT was substituted by (2.6.26) and the derivative of the pair potentials was written out explicitly, whereby for example

$$\left(x_1 \frac{\partial}{\partial x_1} + x_2 \frac{\partial}{\partial x_2} \right) v(\mathbf{x}_1 - \mathbf{x}_2) = (x_1 - x_2) \frac{\partial v(\mathbf{x}_1 - \mathbf{x}_2)}{\partial(x_1 - x_2)}$$

was used, and $x_1(x_2)$ refers to the x component of particle 1(2). With (2.6.32), we have proven (2.6.27).

*2.6.4.2 The Quantum-Statistical Virial Theorem

Starting from the Hamiltonian

$$H = \sum_n \frac{\mathbf{p}_n^2}{2m} + \sum_n V(\mathbf{x}_n - \mathbf{x}_{\text{wall}}) + \frac{1}{2} \sum_{n,m} v(\mathbf{x}_n - \mathbf{x}_m) \,, \qquad (2.6.33)$$

it follows that[11]

$$[H, \mathbf{x}_n \cdot \mathbf{p}_n] = -i\hbar \left(\frac{\mathbf{p}_n^2}{m} - \mathbf{x}_n \cdot \nabla_n V(\mathbf{x}_n - \mathbf{x}_{\text{wall}}) \right.$$
$$\left. - \sum_{n \neq m} \mathbf{x}_n \cdot \nabla_n v(\mathbf{x}_n - \mathbf{x}_m) \right). \quad (2.6.34)$$

Now, $\langle \psi | \, [H, \sum_n \mathbf{x}_n \cdot \mathbf{p}_n] \, | \psi \rangle = 0$ for energy eigenstates.

We assume the density matrix to be diagonal in the basis of the energy eigenstates; from this, it follows that

$$2\langle E_{kin} \rangle - \left\langle \sum_n \mathbf{x}_n \cdot \nabla_n V(\mathbf{x}_n - \mathbf{x}_{\text{wall}}) \right\rangle$$
$$- \left\langle \sum_n \sum_{m \neq n} \mathbf{x}_n \cdot \nabla_n v(\mathbf{x}_n - \mathbf{x}_m) \right\rangle = 0 \,. \quad (2.6.35)$$

With (2.6.31), we again obtain the virial theorem immediately

$$2\langle E_{kin} \rangle - 3PV - \frac{1}{2} \left\langle \sum_n \sum_m (\mathbf{x}_n - \mathbf{x}_m) \cdot \nabla v(\mathbf{x}_n - \mathbf{x}_m) \right\rangle = 0 \,. \quad (2.6.27)$$

Eq. (2.6.27) is called the *virial theorem of quantum statistics*. It holds both classically and quantum mechanically, while (2.6.21) and (2.6.26) are valid only classically.

From the virial theorem (2.6.27), we find for ideal gases:

$$PV = \frac{2}{3} \langle E_{kin} \rangle = \frac{2}{3} \sum_n \frac{m}{2} \langle \mathbf{v}_n^2 \rangle = \frac{1}{3} mN \langle \mathbf{v}^2 \rangle \,. \qquad (2.6.36)$$

For non-interacting classical particles, the mean squared velocity per particle, $\langle \mathbf{v}^2 \rangle$, can be computed using the Maxwell velocity distribution; then from (2.6.36), one again obtains the well-known equation of state of the classical ideal gas.

---

[11] See e.g. QM I, p. 218.

2.6.5 Thermodynamic Quantities in the Canonical Ensemble

2.6.5.1 A Macroscopic System: The Equivalence of the Canonical and the Microcanonical Ensemble

We assume that the smaller subsystem is also a *macroscopic* system. Then it follows from the preceding considerations on the width of the energy distribution function $\omega(E_1)$ that the average value of the energy $\bar{E}_1$ is equal to the most probable value $\tilde{E}_1$, i.e.

$$\bar{E}_1 = \tilde{E}_1 \ . \tag{2.6.37}$$

We now wish to investigate how statements about thermodynamic quantities in the microcanonical and the canonical ensembles are related. To this end, we rewrite the partition function (2.6.4) in the following manner:

$$Z = \frac{\Omega_{1,2}(E)}{\Omega_1(\tilde{E}_1)\Omega_2(E - \tilde{E}_1)} \Omega_1(\tilde{E}_1) e^{-\tilde{E}_1/kT} = \omega(\tilde{E}_1)^{-1}\Omega_1(\tilde{E}_1) e^{-\tilde{E}_1/kT} \ . \tag{2.6.38}$$

According to (2.4.8), the typical N_1-dependence of $\omega(E_1)$ is given by

$$\omega(E_1) \sim N_1^{-\frac{1}{2}} e^{-\frac{3}{4}(E_1 - \bar{E}_1)^2/N_1 \bar{e}^2} \ , \tag{2.6.39}$$

with the normalization factor determined by the condition $\int dE_1\, \omega(E_1) = 1$. From (2.4.14), the N_1-dependence takes the form of Eq. (2.6.39) even for interacting systems. We thus find from (2.6.38) that

$$Z = e^{-\tilde{E}_1/kT} \Omega_1(\tilde{E}_1) \sqrt{N_1} \ . \tag{2.6.40}$$

Inserting this result into Eq. (2.6.18), we obtain the following expression for the canonical entropy [using (2.6.37) and neglecting terms of the order of $\log N_1$]:

$$S_C = \frac{1}{T}\left(\bar{E}_1 - \tilde{E}_1 + kT \log \Omega_1(\tilde{E}_1)\right) = S_{MC}(\tilde{E}_1) \ . \tag{2.6.41}$$

From (2.6.41) we can see that the entropy of the canonical ensemble is equal to that of a microcanonical ensemble with the energy $\tilde{E}_1(= \bar{E}_1)$. *In both ensembles, one obtains identical results for the thermodynamic quantities.*

2.6.5.2 Thermodynamic Quantities

We summarize here how various thermodynamic quantities can be calculated for the canonical ensemble. Since the heat bath enters only through its temperature T, we leave off the index 1 which indicates subsystem 1. Then for the canonical density matrix, we have

$$\rho_C = e^{-\beta H}/Z \tag{2.6.42}$$

with the partition function

$$Z = \text{Tr } e^{-\beta H} , \tag{2.6.43}$$

where we have used the definition $\beta = \frac{1}{kT}$. We also define the *free energy*

$$F = -kT \log Z . \tag{2.6.44}$$

For the entropy, we obtain from (2.6.18)

$$S_C = \frac{1}{T}\left(\bar{E} + kT \log Z\right) . \tag{2.6.45}$$

The average energy is given by

$$\bar{E} = \langle H \rangle = -\frac{\partial}{\partial \beta} \log Z = kT^2 \frac{\partial}{\partial T} \log Z . \tag{2.6.46}$$

The pressure takes the form:

$$P = -\left\langle \frac{\partial H}{\partial V} \right\rangle = kT \frac{\partial \log Z}{\partial V} . \tag{2.6.47}$$

The derivation from Sect. 2.4.3, which gave $-\left\langle \frac{\partial H}{\partial V} \right\rangle$ for the pressure, is of course still valid for the canonical ensemble. From Eq. (2.6.45), it follows that

$$F = \bar{E} - TS_C . \tag{2.6.48}$$

Since the canonical density matrix contains T and V as parameters, F is likewise a function of these quantities. Taking the total differential of (2.6.44) by applying (2.6.43), we obtain

$$dF = -k\,dT \log \text{Tr } e^{-\beta H} - kT \frac{\text{Tr}\left(\left(\frac{dT}{kT^2}H - \frac{1}{kT}\frac{\partial H}{\partial V}dV\right)e^{-\beta H}\right)}{\text{Tr } e^{-\beta H}}$$

$$= -\frac{1}{T}\left(\bar{E} + kT \log Z\right)dT + \left\langle \frac{\partial H}{\partial V} \right\rangle dV$$

and, with (2.6.45)–(2.6.47),

$$dF(T,V) = -S_C dT - PdV . \tag{2.6.49}$$

From Eqs. (2.6.48) and (2.6.49) we find

$$d\bar{E} = TdS_C - PdV . \tag{2.6.50a}$$

This relation corresponds to (2.4.25) in the microcanonical ensemble. In the limiting case of macroscopic systems, $\bar{E} = \tilde{E} = E$ and $S_C = S_{MC}$.

The *First Law* of thermodynamics expresses the energy balance. The most general change in the energy of a system with a fixed number of particles is composed of the work $\delta W = -PdV$ performed on the system together with the quantity of heat δQ transferred to it:

$$dE = \delta Q + \delta W \ . \tag{2.6.50b}$$

Comparison with (2.6.50a) shows that the heat transferred is given by

$$\delta Q = TdS \tag{2.6.50c}$$

(this is the Second Law for transitions between equilibrium states).

The temperature and the volume occur in the canonical partition function and in the free energy as natural variables. The partition function is calculated for a Hamiltonian with a fixed number of particles.[12] As in the case of the microcanonical ensemble, however, one can here also treat the partition function or the free energy, in which the particle number is a parameter, as a function of N. Then the total change in F is given by

$$dF = -S_C dT - PdV + \left(\frac{\partial F}{\partial N}\right)_{T,V} dN \ , \tag{2.6.51}$$

and it follows from (2.6.48) that

$$d\bar{E} = TdS_C - PdV + \left(\frac{\partial F}{\partial N}\right)_{T,V} dN \ . \tag{2.6.52}$$

In the thermodynamic limit, (2.6.52) and (2.4.29) must agree, so that we find

$$\left(\frac{\partial F}{\partial N}\right)_{T,V} = \mu \ . \tag{2.6.53}$$

2.6.6 Additional Properties of the Entropy

2.6.6.1 Additivity of the Entropy

We now consider two subsystems in a common heat bath (Fig. 2.13). Assuming that each of these systems contains a large number of particles, the energy is additive. That is, the interaction energy, which acts only at the interfaces, is much smaller than the energy of each of the individual systems. We wish to show that the entropy is also additive. We begin this task with the two density matrices of the subsystems:

$$\rho_1 = \frac{e^{-\beta H_1}}{Z_1} \ , \qquad \rho_2 = \frac{e^{-\beta H_2}}{Z_2} \ . \tag{2.6.54a,b}$$

---

[12] Exceptions are photons and bosonic quasiparticles such as phonons and rotons in superfluid helium, for which the particle number is not fixed (Chap. 4).

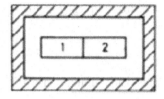

Fig. 2.13. Two subsystems 1 and 2 in one heat bath

The density matrix of the two subsystems together is

$$\rho = \rho_1 \rho_2 , \tag{2.6.54c}$$

where once again $W \ll H_1, H_2$ was employed. From

$$\langle \log \rho \rangle = \langle \log \rho_1 \rangle + \langle \log \rho_2 \rangle \tag{2.6.55}$$

it follows that the total entropy S is given by

$$S = S_1 + S_2 , \tag{2.6.56}$$

the sum of the entropies of the subsystems. Eq. (2.6.56) expresses the fact that the entropy is additive.

*2.6.6.2 The Statistical Meaning of Heat

Here, we want to add a few supplementary remarks that concern the statistical and physical meaning of heat transfer to a system. We begin with the average energy

$$\bar{E} = \langle H \rangle = \text{Tr } \rho H \tag{2.6.57a}$$

for an arbitrary density matrix and its total variation with a fixed number of particles

$$d\bar{E} = \text{Tr} \left(d\rho \, H + \rho \, dH \right) , \tag{2.6.57b}$$

where $d\rho$ is the variation of the density matrix and dH is the variation of the Hamiltonian (see the end of this section). The variation of the entropy

$$S = -k \, \text{Tr } \rho \log \rho \tag{2.6.58}$$

is given by

$$dS = -k \, \text{Tr} \left(d\rho \log \rho + \frac{\rho}{\rho} d\rho \right) . \tag{2.6.59}$$

Now we have

$$\text{Tr } d\rho = 0 , \tag{2.6.60}$$

since for all density matrices, $\text{Tr}\,\rho = \text{Tr}\,(\rho + d\rho) = 1$, from which it follows that

$$dS = -k\,\text{Tr}\left(\log\rho\,d\rho\right) . \tag{2.6.61}$$

Let the initial density matrix be the canonical one; then making use of (2.6.60), we have

$$dS = \frac{1}{T}\,\text{Tr}\left(H\,d\rho\right) . \tag{2.6.62}$$

If we insert this into (2.6.57b) and take the volume as the only parameter in H, i.e. $dH = \frac{\partial H}{\partial V}dV$, we again obtain (cf. (2.6.50a))

$$d\bar{E} = TdS + \left\langle\frac{\partial H}{\partial V}\right\rangle dV . \tag{2.6.63}$$

We shall now discuss the physical meaning of the general relation (2.6.57b):
1st term: this represents a change in the density matrix, i.e. a change in the occupation probabilities.
2nd term: the change of the Hamiltonian. This means a change in the energy as a result of influences which change the energy eigenvalues of the system.

Let ρ be diagonal in the energy eigenstates; then

$$\bar{E} = \sum_i p_i E_i , \tag{2.6.64}$$

and the variation of the average energy has the form

$$d\bar{E} = \sum_i dp_i E_i + \sum_i p_i dE_i . \tag{2.6.65}$$

Thus, the quantity of heat transferred is given by

$$\delta Q = \sum_i dp_i E_i . \tag{2.6.66}$$

A *transfer of heat* gives rise to a redistribution of the occupation probabilities of the states $|i\rangle$. Heating (heat input) increases the populations of the states at higher energies. Energy change by an *input of work* (work performed on the system) produces a change in the energy eigenvalues. In this process, the occupation numbers can change only in such a way as to keep the entropy constant.

When only the external parameters are varied, work is performed on the system, but no heat is put into it. In this case, although $d\rho$ may exhibit a change, there is no change in the entropy. This can be shown explicitly as follows: From Eq. (2.6.61), we have $dS = -k\text{Tr}\,(\log\rho d\rho)$. It then follows from the von Neumann Eq. (1.4.8), $\dot{\rho} = \frac{i}{\hbar}[\rho, H(V(t))]$, which is valid also for time-dependent Hamiltonians, e.g. one containing the volume $V(t)$:

$$\dot{S} = -k\,\mathrm{Tr}\left(\log\rho\,\dot{\rho}\right)$$
$$= -\frac{ik}{\hbar}\,\mathrm{Tr}\left(\log\rho\,[\rho, H]\right) = -\frac{ik}{\hbar}\,\mathrm{Tr}\left(H\,[\log\rho, \rho]\right) = 0\,. \tag{2.6.67}$$

The entropy does not change, and no heat is put into the system. An example which demonstrates this situation is the adiabatic reversible expansion of an ideal gas (Sect. 3.5.4.1). There, as a result of the work performed, the volume of the gas changes and with it the Hamilton function; furthermore, the temperature of the gas changes. These effects together lead to a change in the distribution function (density matrix), but however not of the entropy.

2.7 The Grand Canonical Ensemble

2.7.1 Systems with Particle Exchange

After considering systems in the preceding section which can exchange energy with a heat bath, we now wish to allow in addition the exchange of matter between subsystem 1 on the one hand and the heat bath 2 on the other; this will be a consistent generalization of the canonical ensemble (see Fig. 2.14). The overall system is isolated. The total energy, the total particle number and the overall volume are the sums of these quantities for the subsystems:

$$E = E_1 + E_2, \quad N = N_1 + N_2, \quad V = V_1 + V_2\,. \tag{2.7.1}$$

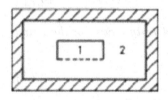

Fig. 2.14. Regarding the grand canonical ensemble: two subsystems 1 and 2, between which energy and particle exchange is permitted.

The probability distribution of the state variables E_1, N_1, and V_1 of subsystem 1 is found in complete analogy to Sect. 2.4.3,

$$\omega\left(E_1, N_1, V_1\right) = \frac{\Omega_1(E_1, N_1, V_1)\,\Omega_2(E - E_1, N - N_1, V - V_1)}{\Omega\left(E, N, V\right)}\,. \tag{2.7.2}$$

The attempt to find the maximum of this distribution leads again to equality of the logarithmic derivatives, in this case with respect to E, V and N. The first two relations were already seen in Eq. (2.4.28) and imply temperature and pressure equalization between the two systems. The third formula can be expressed in terms of the chemical potential which was defined in (2.4.29):

$$\mu = -kT\frac{\partial}{\partial N}\log \Omega\left(E, N, V\right) = -T\left(\frac{\partial S}{\partial N}\right)_{E,V}, \qquad (2.7.3)$$

and we obtain finally as a condition for the maximum probability the equalization of temperature, pressure, and chemical potential:

$$T_1 = T_2, \quad P_1 = P_2, \quad \mu_1 = \mu_2. \qquad (2.7.4)$$

2.7.2 The Grand Canonical Density Matrix

Next, we will derive the density matrix for the subsystem. The probability that in system 1 there are N_1 particles which are in the state $|n\rangle$ at the energy $E_{1n}(N_1)$ is given by:

$$p(N_1, E_{1n}(N_1), V_1) = \sum_{E-E_{1n}(N_1)\leq E_{2m}(N_2)\leq E-E_{1n}(N_1)+\Delta} \frac{1}{\Omega\left(E, N, V\right)\Delta}$$

$$= \frac{\Omega_2(E - E_{1n}, N - N_1, V_2)}{\Omega\left(E, N, V\right)}.$$

$$(2.7.5)$$

In order to eliminate system 2, we carry out an expansion in the variables E_{1n} and N_1 with the condition that subsystem 1 is much smaller than subsystem 2, analogously to the case of the canonical ensemble:

$$p(N_1, E_{1n}(N_1), V_1) = Z_G^{-1}e^{-(E_{1n}-\mu N_1)/kT}. \qquad (2.7.6)$$

We thus obtain the following expression for the *density matrix of the grand canonical ensemble*[13]:

$$\rho_G = Z_G^{-1}e^{-(H_1-\mu N_1)/kT}, \qquad (2.7.7)$$

where the *grand partition function* Z_G (or Gibbs distribution) is found from the normalization of the density matrix to be

$$Z_G = \text{Tr}\left(e^{-(H_1-\mu N_1)/kT}\right)$$

$$= \sum_{N_1}\text{Tr } e^{-H_1/kT+\mu N_1/kT} = \sum_{N_1}Z(N_1)\,e^{\mu N_1/kT}. \qquad (2.7.8)$$

---

[13] See also the derivation in second quantization, p. 69

The two *trace operations* Tr in Eq. (2.7.8) refer to *different spaces*. The trace after the second equals sign refers to a summation over all the diagonal matrix elements for a fixed particle number N_1, while the Tr after the first equals sign implies in addition the summation over all particle numbers $N_1 = 0, 1, 2, \ldots$. The average value of an operator A in the grand canonical ensemble is

$$\langle A \rangle = \mathrm{Tr}\,(\rho_G A) \, ,$$

where the trace is here to be understood in the latter sense.

In classical statistics, (2.7.7) remains unchanged for the distribution function, while $\mathrm{Tr} \longrightarrow \sum_{N_1} \int d\Gamma_{N_1}$ must be replaced by the $6N_1$-dimensional operator $d\Gamma_{N_1} = \frac{dq\,dp}{h^{3N_1} N_1!}$.

From (2.7.5), Z_G^{-1} can also be given in terms of

$$Z_G^{-1} = \frac{\Omega_2(E, N, V - V_1)}{\Omega\,(E, N, V)} = \mathrm{e}^{-PV_1/kT} \tag{2.7.9}$$

for $V_1 \ll V$; recall Eqns. (2.4.24) and (2.4.25).

From the density matrix, we find the entropy of the grand canonical ensemble,

$$S_G = -k\langle \log \rho_G \rangle = \frac{1}{T}(\bar{E} - \mu \bar{N}) + k \log Z_G \, . \tag{2.7.10}$$

Since the energy and particle reservoir, subsystem 2, enters only via its temperature and chemical potential, we dispense with the index 1 here and in the following sections.

The distribution function for the energy and the particle number is extremely narrow for macroscopic subsystems. The relative fluctuations are proportional to the square root of the average number of particles. Therefore, we have $\bar{E} = \tilde{E}$ and $\bar{N} = \tilde{N}$ for macroscopic subsystems. The grand canonical entropy, also, may be shown (cf. Sect. 2.6.5.1) in the limit of macroscopic subsystems to be identical with the microcanonical entropy, taken at the most probable values (with fixed volume V_1)

$$\tilde{E}_1 = \bar{E}_1, \quad \tilde{N}_1 = \bar{N}_1 \tag{2.7.11}$$

$$S_G = S_{MC}(\tilde{E}_1, \tilde{N}_1) \, . \tag{2.7.12}$$

2.7.3 Thermodynamic Quantities

In analogy to the free energy of the canonical ensemble, the *grand potential* is defined by

$$\Phi = -kT \log Z_G \, , \tag{2.7.13}$$

from which with (2.7.10) we obtain the expression

$$\Phi(T, \mu, V) = \bar{E} - TS_G - \mu\bar{N} .$$ (2.7.14)

The total differential of the grand potential is given by

$$d\Phi = \left(\frac{\partial \Phi}{\partial T}\right)_{V,\mu} dT + \left(\frac{\partial \Phi}{\partial V}\right)_{T,\mu} dV + \left(\frac{\partial \Phi}{\partial \mu}\right)_{V,T} d\mu .$$ (2.7.15)

The partial derivatives follow from (2.7.13) and (2.7.8):

$$\left(\frac{\partial \Phi}{\partial T}\right)_{V,\mu} = -k \log Z_G - kT\frac{1}{kT^2}\langle H - \mu N\rangle = \frac{1}{T}(\Phi - \bar{E} + \mu\bar{N}) = -S_G$$

$$\left(\frac{\partial \Phi}{\partial V}\right)_{T,\mu} = \left\langle\frac{\partial H}{\partial V}\right\rangle = -P \quad , \quad \left(\frac{\partial \Phi}{\partial \mu}\right)_{T,V} = -kT\frac{1}{kT}\langle N\rangle = -\bar{N} .$$

(2.7.16)

If we insert (2.7.16) into (2.7.15), we find

$$d\Phi = -S_G dT - PdV - \bar{N}d\mu .$$ (2.7.17)

From this, together with (2.7.14), it follows that

$$d\bar{E} = TdS_G - PdV + \mu d\bar{N} ;$$ (2.7.18)

this is again the *First Law*. As shown above, for macroscopic systems we can use simply E, N and S in (2.7.17) and (2.7.18) instead of the average values of the energy and the particle number and S_G; we shall do this in later chapters. For a constant particle number, (2.7.18) becomes identical with (2.4.25). The physical meaning of the First Law will be discussed in detail in Sect. 3.1. We have considered the fluctuations of physical quantities thus far only in Sect. 2.4.2. Of course, we could also calculate the autocorrelation function for energy and particle number in the grand canonical ensemble. This shows that these quantities are extensive and their relative fluctuations decrease inversely as the square root of the size of the system. We shall postpone these considerations to the chapter on thermodynamics, since there we can relate the correlations to thermodynamic derivatives.

We close this section with a tabular summary of the ensembles treated in this chapter.

Remark concerning Table 2.1: The thermodynamic functions which are found from the logarithm of the normalization factors are the entropy and the thermodynamic potentials F and Φ (see Chap. 3). The generalization to several different types of particles will be carried out in Chap. 5. To this end, one must merely replace N by $\{N_i\}$ and μ by $\{\mu_i\}$.

Table 2.1. The most important ensembles

Ensemble	microcanonical	canonical	grand canonical
Physical situation	isolated	energy exchange	energy and particle exchange
Density matrix	$\frac{1}{\Omega(E,V,N)} \times \delta(H-E)$	$\frac{1}{Z(T,V,N)} e^{-H/kT}$	$\frac{1}{Z_G(T,V,\mu)} \times e^{-(H-\mu N)/kT}$
Normalization	$\Omega(E,V,N) =$ Tr $\delta(H-E)$	$Z(T,V,N) =$ Tr $e^{-H/kT}$	$Z_G(T,V,\mu) =$ Tr $e^{-(H-\mu N)/kT}$
Independent variables	E,V,N	T,V,N	T,V,μ
Thermodynamic functions	S	F	Φ

2.7.4 The Grand Partition Function for the Classical Ideal Gas

As an example, we consider the special case of the classical ideal gas.

2.7.4.1 Partition Function

For the partition function for N particles, we obtain

$$
Z_N = \frac{1}{N!\,h^{3N}} \int_V dq_1 \dots dq_{3N} \int dp_1 \dots dp_{3N} \, e^{-\beta \sum p_i^2/2m}
$$

$$
= \frac{V^N}{N!} \left(\frac{2m\pi}{\beta h^2}\right)^{\frac{3N}{2}} = \frac{1}{N!} \left(\frac{V}{\lambda^3}\right)^N
\tag{2.7.19}
$$

with the *thermal wavelength*

$$
\lambda = h/\sqrt{2\pi mkT} .
\tag{2.7.20}
$$

Its name results from the fact that a particle of mass m and momentum h/λ will have a kinetic energy of the order of kT.

2.7.4.2 The Grand Partition Function

Inserting (2.7.19) into the grand partition function (2.7.8), we find

$$
Z_G = \sum_{N=0}^{\infty} e^{\beta\mu N} Z_N = \sum_{N=0}^{\infty} \frac{1}{N!} e^{\beta\mu N} \left(\frac{V}{\lambda^3}\right)^N = e^{zV/\lambda^3} ,
\tag{2.7.21}
$$

where the fugacity

$$z = e^{\beta\mu} \tag{2.7.22}$$

has been defined.

2.7.4.3 Thermodynamic Quantities

From (2.7.13) and (2.7.21), the grand potential takes on the simple form

$$\Phi \equiv -kT \log Z_G = -kTzV/\lambda^3 . \tag{2.7.23}$$

From the partial derivatives, we can compute the thermodynamic relations.[14]
Particle number

$$N = -\left(\frac{\partial\Phi}{\partial\mu}\right)_{T,V} = zV/\lambda^3 \tag{2.7.24}$$

Pressure

$$PV = -V\left(\frac{\partial\Phi}{\partial V}\right)_{T,\mu} = -\Phi = NkT \tag{2.7.25}$$

This is again the thermal equation of state of the ideal gas, as found in Sect. 2.5. For the *chemical potential*, we find from (2.7.22), (2.7.24), and (2.7.23)

$$\mu = -kT \log\left(\frac{V/N}{\lambda^3}\right) = -kT \log\frac{kT}{P\lambda^3} = kT \log P - kT \log\frac{kT}{\lambda^3} . \tag{2.7.26}$$

For the *entropy*, we find

$$\begin{aligned} S = -\left(\frac{\partial\Phi}{\partial T}\right)_{V,\mu} &= \frac{5}{2}kz\frac{V}{\lambda^3} + kT\left(-\frac{\mu}{kT^2}z\right)\frac{V}{\lambda^3} \\ &= kN\left(\frac{5}{2} + \log\frac{V/N}{\lambda^3}\right) , \end{aligned} \tag{2.7.27}$$

and for the *internal energy*, from (2.7.14), we obtain

$$E = \Phi + TS + \mu N = NkT\left(-1 + \frac{5}{2}\right) = \frac{3}{2}NkT . \tag{2.7.28}$$

---

[14] For the reasons mentioned at the end of the preceding section, we replace $\bar{E}$ and $\bar{N}$ in (2.7.16) and (2.7.17) by E and N.

*2.7.5 The Grand Canonical Density Matrix in Second Quantization

The derivation of ρ_G can be carried out most concisely in the formalism of the second quantization. In addition to the Hamiltonian H, expressed in terms of the field operators $\psi(\mathbf{x})$ (see Eq. (1.5.6d) in QM II[15]), we require the particle-number operator, Eq. (1.5.10)[15]

$$\hat{N} = \int_V d^3x\, \psi^\dagger(\mathbf{x})\psi(\mathbf{x}) \,. \tag{2.7.29}$$

The microcanonical density matrix for fixed volume V is

$$\rho_{MC} = \frac{1}{\Omega(E, N, V)}\delta(H - E)\delta(\hat{N} - N) \,. \tag{2.7.30}$$

Corresponding to the division of the overall volume into two subvolumes, $V = V_1 + V_2$, we have $H = H_1 + H_2$ and $\hat{N} = \hat{N}_1 + \hat{N}_2$ with $\hat{N}_i = \int_{V_i} d^3x\, \psi^\dagger(\mathbf{x})\psi(\mathbf{x})$, $i = 1, 2$. We find from (2.7.30) the probability that the energy and the particle number in subvolume 1 assume the values E_1 and N_1:

$$\begin{aligned}
\omega&(E_1, V_1, N_1) \\
&= \text{Tr}\frac{1}{\Omega(E, N, V)}\delta(H - E)\delta(\hat{N} - N)\delta(H_1 - E_1)\delta(\hat{N}_1 - N_1) \\
&= \text{Tr}\frac{1}{\Omega(E, N, V)}\delta(H_2 - (E - E_1))\delta(\hat{N}_2 - (N - N_1)) \\
&\quad \times \delta(H_1 - E_1)\delta(\hat{N}_1 - N_1) \\
&= \frac{\Omega_1(E_1, N_1, V_1)\Omega_2(E - E_1, N - N_1, V - V_1)}{\Omega(E, N, V)} \,.
\end{aligned} \tag{2.7.31}$$

The (grand canonical) density matrix for subsystem 1 is found by taking the trace of the density matrix of the overall system over subsystem 2, with respect to both the energy and the particle number:

$$\begin{aligned}
\rho_G &= \text{Tr}_2\frac{1}{\Omega(E, N, V)}\delta(H - E)\delta(\hat{N} - N) \\
&= \frac{\Omega_2(E - H_1, N - \hat{N}_1, V - V_1)}{\Omega(E, N, V)} \,.
\end{aligned} \tag{2.7.32}$$

Expansion of the logarithm of ρ_G in terms of H_1 and $\hat{N}_1$ leads to

$$\begin{aligned}
\rho_G &= Z_G^{-1}e^{-(H_1 - \mu\hat{N}_1)/kT} \\
Z_G &= \text{Tr}\, e^{-(H_1 - \mu\hat{N}_1)/kT} \,,
\end{aligned} \tag{2.7.33}$$

consistent with Equations (2.7.7) and (2.7.8), which were obtained by considering the probabilities.

---

[15] F. Schwabl, *Advanced Quantum Mechanics (QM II)*, 3rd ed., Springer Berlin, Heidelberg, New York 2005. This text will be cited in the rest of this book as QM II.

Problems for Chapter 2

2.1 Calculate $\Omega(E)$ for a spin system which is described by the Hamiltonian

$$\mathcal{H} = \mu_B H \sum_{i=1}^{N} S_i \ ,$$

where S_i can take on the values $S_i = \pm 1/2$

$$\Omega(E)\Delta = \sum_{E \leq E_n \leq E+\Delta} 1 \ .$$

Use a combinatorial method, rather than 2.2.3.2.

2.2 For a one-dimensional classical ideal gas, calculate $\langle p_1^2 \rangle$ and $\langle p_1^4 \rangle$.

$$\text{Formula:} \quad \int_0^{\pi} \sin^m x \cos^n x \, dx = \frac{\Gamma\left(\frac{m+1}{2}\right) \Gamma\left(\frac{n+1}{2}\right)}{\Gamma\left(\frac{n+m+2}{2}\right)} \ .$$

2.3 A particle is moving in one dimension; the distance between the walls of the container is changed by a piston at L. Compute the change in the phase-space volume $\bar{\Omega} = 2Lp$ ($p = $ momentum).
(a) For a slow, continuous motion of the piston.
(b) For a rapid motion of the piston between two reflections of the particle.

2.4 Assume that the entropy S depends on the volume $\bar{\Omega}(E)$ inside the energy shell: $S = f(\bar{\Omega})$. Show that from the additivity of S and the multiplicative character of $\bar{\Omega}$, it follows that $S = \text{const} \times \log \bar{\Omega}$.

2.5 (a) For a classical ideal gas which is enclosed within a volume V, calculate the free energy and the entropy, starting with the canonical ensemble.
(b) Compare them with the results of Sect. 2.2.

2.6 Using the assertion that the entropy $S = -k \operatorname{Tr}(\rho \log \rho)$ is maximal, show that with the conditions $\operatorname{Tr} \rho = 1$ and $\operatorname{Tr} \rho H = \bar{E}$ for ρ, the canonical density matrix results.
Hint: This is a variational problem with constraints, which can be solved using the method of Lagrange multipliers.

2.7 Show that for the energy distribution in the classical canonical ensemble

$$\omega(E_1) = \int d\Gamma_1 \, \rho_K \, \delta(H_1 - E_1)$$

$$= \Omega_1(E_1) \frac{\Omega_2(\tilde{E}_2)}{\Omega_{1,2}(E)} e^{\tilde{E}_1/kT} e^{-E_1/kT} \approx \frac{\Omega_2(E - E_1) \, \Omega_1(E_1)}{\Omega_{1,2}(E)} \ . \quad (2.7.34)$$

2.8 Consider a system of N classical, non-coupled one-dimensional harmonic oscillators and calculate for this system the entropy and the temperature, starting from the microcanonical ensemble.

2.9 Consider again the harmonic oscillators from problem 2.8 and calculate for this system the average value of the energy and the entropy, starting with the canonical ensemble.

2.10 In analogy to the preceding problems, consider N quantum-mechanical non-coupled one-dimensional harmonic oscillators and compute the average value of the energy $\bar{E}$ and the entropy, beginning with the canonical ensemble. Also investigate $\lim_{\hbar \to 0} \bar{E}$, $\lim_{\hbar \to 0} S$ and $\lim_{T \to 0} S$, and compare the limiting values you obtain with the results of problem 2.9.

2.11 For the Maxwell distribution, find
(a) the average value of the nth power of the velocity $\langle v^n \rangle$, (b) $\langle v \rangle$, (c) $\langle (v - \langle v \rangle)^2 \rangle$,
(d) $\left(\frac{m}{2} \right)^2 \langle (v^2 - \langle v^2 \rangle)^2 \rangle$, and (e) the most probable value of the velocity.

2.12 Determine the number of collisions of a molecule of an ideal gas with the wall of its container per unit area and unit time, when
(a) the angle between the normal to the wall and the direction of the velocity lies between Θ and $\Theta + d\Theta$;
(b) the magnitude of the velocity lies between v and $v + dv$.

2.13 Calculate the pressure of a Maxwellian gas with the velocity distribution

$$f(\mathbf{v}) = n \left(\frac{m\beta}{2\pi} \right)^{\frac{3}{2}} e^{-\frac{\beta m v^2}{2}} \ .$$

Suggestions: the pressure is produced by reflections of the particles from the walls of the container; it is therefore the average force on an area A of wall which acts over a time interval τ.

$$P = \frac{1}{\tau A} \int\limits_0^\tau dt \, F_x(t) \ .$$

If a particle is reflected from the wall with the velocity $\mathbf{v}$, its contribution is given from Newton's 2nd axiom in terms of $\int\limits_0^\tau dt \, F_x(t)$ by the momentum transferred per collision, $2mv_x$. Then $P = \frac{1}{\tau A} \sum 2mv_x$, whereby the sum extends over all particles which reach the area A within the time τ.
Result: $P = nkT$.

2.14 A simple model for thermalization: Calculate the average kinetic energy of a particle of mass m_1 with the velocity $\mathbf{v}_1$ due to contact with an ideal gas consisting of particles of mass m_2. As a simplification, assume that only elastic and linear collisions occur. The effect on the ideal gas can be neglected. It is helpful to use the abbreviations $M = m_1 + m_2$ and $m = m_1 - m_2$. How many collisions are required until, for $m_1 \neq m_2$, a temperature equal to the $(1 - e^{-1})$-fold temperature of the ideal gas is attained?

2.15 Using the canonical ensemble, calculate the average value of the particle-number density

$$n(\mathbf{x}) = \sum_{i=1}^{N} \delta(\mathbf{x} - \mathbf{x}_i)$$

for an ideal gas which is contained in an infinitely high cylinder of cross-sectional area A in the gravitational field of the Earth. The potential energy of a particle in the gravitational field is mgh. Also calculate
(a) the internal energy of this system,
(b) the pressure at the height (altitude) h, using the definition

$$P = \int_{h}^{\infty} \langle n(\mathbf{x}) \rangle mg \, dz \, ,$$

(c) the average distance $\langle z \rangle$ of an oxygen molecule and a helium atom from the surface of the Earth at a temperature of $0°C$, and
(d) the mean square deviation Δz for the particles in 2.15c.
At this point, we mention the three different derivations of the barometric pressure formula, each emphasizing different physical aspects, in R. Becker, *Theory of Heat*, 2nd ed., Sec. 27, Springer, Berlin 1967.

2.16 The potential energy of N non-interacting localized dipoles depends on their orientations relative to an applied magnetic field H:

$$\mathcal{H} = -\mu H_z \sum_{i=1}^{N} \cos \vartheta_i \, .$$

Calculate the partition function and show that the magnetization along the z-direction takes the form

$$M_z = \left\langle \sum_{i=1}^{N} \mu \cos \vartheta_i \right\rangle = N\mu \, L(\beta\mu H_z) \, ; \quad L(x) = \text{ Langevin function} \, .$$

Plot the Langevin function.
How large is the magnetization at high temperatures? Show that at high temperatures, the Curie law for the magnetic susceptibility holds:

$$\chi = \lim_{H_z \to 0} \left(\frac{\partial M_z}{\partial H_z} \right) \sim \text{const}/T \, .$$

2.17 Demonstrate the *equipartition theorem* and the *virial theorem* making use of the *microcanonical distribution*.

2.18 In the extreme relativistic case, the Hamilton function for N particles in three-dimensional space is $H = \sum_i |\mathbf{p}_i| c$. Compute the expectation value of H with the aid of the virial theorem.

2.19 Starting with the canonical ensemble of classical statistics, calculate the equation of state and the internal energy of a gas composed of N indistinguishable particles with the kinetic energy $\varepsilon(\mathbf{p}) = |\mathbf{p}| \cdot c$.

2.20 Show that for an ideal gas, the probability of finding a subsystem in the grand canonical ensemble with N particles is given by the Poisson distribution:

$$p_N = \frac{1}{N!} e^{-\bar{N}} \bar{N}^N ,$$

where $\bar{N}$ is the average value of N in the ideal gas.
Suggestions: Start from $p_N = e^{\beta(\Phi + N\mu)} Z_N$. Express Φ, μ, and Z_N in terms of $\bar{N}$.

2.21 (a) Calculate the grand partition function for a mixture of two ideal gases (2 chemical potentials!).
(b) Show that

$$PV = (N_1 + N_2) kT \qquad \text{and}$$

$$E = \frac{3}{2}(N_1 + N_2) kT$$

are valid, where N_1, N_2 and E are the average particle number and the average energy.

2.22 (a) Express $\bar{E}$ by taking an appropriate derivative of the grand partition function.
(b) Express $(\Delta E)^2$ in terms of a thermodynamic derivative of $\bar{E}$.

2.23 Calculate the density matrix in the x-representation for a free particle within a three-dimensional cube of edge length L:

$$\rho(x, x') = c \sum_n e^{-\beta E_n} \langle x|n \rangle \langle n|x' \rangle$$

where c is a normalization constant. Assume that L is so large that one can go to the limit of a continuous momentum spectrum

$$\sum_n \longrightarrow \int \frac{L^3 d^3 p}{(2\pi\hbar)^3} ; \quad \langle x|n \rangle \longrightarrow \langle x|p \rangle = \frac{1}{L^{3/2}} e^{i\mathbf{p}\mathbf{x}/\hbar} .$$

2.24 Calculate the canonical density matrix for a one-dimensional harmonic oscillator $H = -(\hbar^2/2m)(d^2/d^2x) + \frac{m\omega^2 x^2}{2}$ in the x-representation at low temperatures:

$$\rho(x, x') = c \sum_n e^{-\beta E_n} \langle x|n \rangle \langle n|x' \rangle ,$$

where c is the normalization constant.

$$\langle x|n \rangle = (\pi^{1/2} 2^n n! x_0)^{-1/2} e^{-(x/x_0)^2/2} H_n\left(\frac{x}{x_0}\right) ; \quad x_0 = \sqrt{\frac{\hbar}{\omega m}} .$$

The Hermite polynomials are defined in problem 2.27.
Suggestion: Consider which state makes the largest contribution.

2.25 Calculate the time average of q^2 for the example of problem 1.7, as well as its average value in the microcanonical ensemble.

2.26 Show that:

$$\int dq_1 \ldots dq_d \, f(q^2, \mathbf{q} \cdot \mathbf{k})$$

$$= (2\pi)^{-1} K_{d-1} \int_0^\infty dq \, q^{d-1} \int_0^\pi d\Theta (\sin \Theta)^{d-2} \, f(q^2, qk \cos \Theta) \,, \quad (2.7.35)$$

where $\mathbf{k} \in \mathcal{R}^d$ is a fixed vector and $q = |\mathbf{q}|$, $k = |\mathbf{k}|$, and $K_d = 2^{-d+1} \pi^{-d/2} \times (\Gamma(\frac{d}{2}))^{-1}$.

2.27 Compute the matrix elements of the canonical density matrix for a one-dimensional harmonic oscillator in the coordinate representation,

$$\rho_{x,x'} = \langle x | \rho | x' \rangle = \langle x | e^{-\beta H} | x' \rangle \,.$$

Hint: Use the completeness relation for the eigenfunctions of the harmonic oscillator and use the fact that the Hermite polynomials have the integral representation

$$H_n(\xi) = (-1)^n e^{\xi^2} \left(\frac{d}{d\xi} \right)^n e^{-\xi^2} = \frac{e^{\xi^2}}{\sqrt{\pi}} \int_{-\infty}^\infty (-2iu)^n e^{-u^2 + 2i\xi u} du \,.$$

Alternatively, the first representation for $H_n(x)$ and the identity from the next example can be used.
Result:

$$\rho_{x,x'} = \frac{1}{Z} \left[\frac{m\omega}{2\pi\hbar \sinh \beta\hbar\omega} \right]^{1/2}$$

$$\times \exp\left\{ -\frac{m\omega}{4\hbar} \left((x+x')^2 \tanh \frac{1}{2}\beta\hbar\omega + (x-x')^2 \operatorname{ctgh} \frac{1}{2}\beta\hbar\omega \right) \right\} \,. \quad (2.7.36)$$

2.28 Prove the following identity:

$$e^{\frac{\partial}{\partial x} \Pi \frac{\partial}{\partial x}} e^{-x\Delta x} = \frac{1}{\sqrt{\operatorname{Det}(1 + 4\Delta\Pi)}} e^{-x \frac{\Delta}{1+4\Delta\Pi} x} \,.$$

Here, Π and Δ ate two commuting symmetric matrices, e.g. $\frac{\partial}{\partial x} \Pi \frac{\partial}{\partial x} \equiv \frac{\partial}{\partial x_i} \Pi_{ik} \frac{\partial}{\partial x_k}$.

3. Thermodynamics

3.1 Thermodynamic Potentials and the Laws of Equilibrium Thermodynamics

3.1.1 Definitions

Thermodynamics treats the macroscopic properties of macroscopic systems. The fact that macroscopic systems can be completely characterized by a small number of variables, such as their energy E, volume V, and particle number N, and that all other quantities, e.g. the entropy, are therefore functions of only these variables, has far-reaching consequences.

In this section, we consider equilibrium states and transitions from one equilibrium state to another neighboring equilibrium state. In the preceding sections, we have already determined the change in the entropy due to changes in E, V and N, whereby the system goes from one equilibrium state E, V, N into a new equilibrium state $E + dE$, $V + dV$, $N + dN$. Building upon the differential entropy (2.4.29), we will investigate in the following the First Law and the significance of the quantities which occur in it. Beginning with the internal energy, we will then define the most important thermodynamic potentials and discuss their properties.

We assume the system we are considering to consist of one single type of particles of particle number N. We start with its entropy, which is a function of E, V, and N.

Entropy : $\qquad S = S(E, V, N)$

In (2.4.29), we found the differential entropy to be

$$dS = \frac{1}{T}\, dE + \frac{P}{T}\, dV - \frac{\mu}{T}\, dN .\qquad (3.1.1)$$

From this, we can read off the partial derivatives:

$$\left(\frac{\partial S}{\partial E}\right)_{V,N} = \frac{1}{T}, \qquad \left(\frac{\partial S}{\partial V}\right)_{E,N} = \frac{P}{T}, \qquad \left(\frac{\partial S}{\partial N}\right)_{E,V} = -\frac{\mu}{T}, \qquad (3.1.2)$$

which naturally agree with the definitions from equilibrium statistics. We can now imagine the equation $S = S(E, V, N)$ to have been solved for E and thereby obtain the energy E, which in thermodynamics is usually termed the internal energy, as a function of S, V, and N.

Internal Energy : $E = E(S, V, N)$

From (3.1.1), we obtain the differential relation

$$dE = TdS - PdV + \mu dN .\tag{3.1.3}$$

We are now in a position to interpret the individual terms in (3.1.3), keeping in mind all the various possibilities for putting energy into a system. This can be done by performing work, by adding matter (i.e. by increasing the number of particles), and through contact with other bodies, whereby heat is put into the system. The total change in the energy is thus composed of the following contributions:

$$dE \;=\; \underset{\substack{\downarrow \\ \text{heat input}}}{\delta Q} \;+\; \underset{\substack{\downarrow \\ \text{mechanical work}}}{\delta W} \;+\; \underset{\substack{\downarrow \\ \\ \text{energy increase through addition of matter} .}}{\delta E_N}\tag{3.1.3'}$$

The second term in (3.1.3) is the work performed on the system,

$$\delta W = -PdV ,\tag{3.1.4a}$$

while the third term gives the change in the energy on increasing the particle number

$$\delta E_N = \mu dN .\tag{3.1.4b}$$

The chemical potential μ has the physical meaning of the energy increase on adding one particle to the system (at constant entropy and volume). The first term must therefore be the energy change due to heat input δQ, i.e.

$$\delta Q = TdS .\tag{3.1.5}$$

Relation (3.1.3), the law of conservation of energy in thermodynamics, is called the *First Law of Thermodynamics*. It expresses the change in energy on going from one equilibrium state to another, nearby state an infinitesimal distance away. Equation (3.1.5) is the *Second Law* for such transitions. We will formulate the Second Law in a more general way later. In this connection, we will also clarify the question of under what conditions these relations of equilibrium thermodynamics can be applied to real thermodynamic processes which proceed at finite rates, such as for example the operation of steam engines or of internal combustion engines.

Remark:

It is important to keep the following in mind: δW and δQ do not represent changes of state variables. There are no state functions (functions of E, V and N) identifiable with W und Q. An object cannot be characterized by its 'heat or work content', but instead by its internal energy. Heat ($\sim$ energy transfer into an object through contact with other bodies) and work are ways of transferring energy from one body to another.

It is often expedient to consider other quantities – with the dimensions of energy – in addition to the internal energy itself. As the first of these, we define the free energy:

Free Energy (Helmholtz Free Energy) : $F = F(T, V, N)$
The free energy is defined by

$$F = E - TS \quad \left(= -kT \log Z(T, V, N)\right) ; \tag{3.1.6}$$

in parentheses, we have given its connection with the canonical partition function (Chap. 2). From (3.1.3), the differential free energy is found to be:

$$dF = -SdT - PdV + \mu dN \tag{3.1.7}$$

with the partial derivatives

$$\left(\frac{\partial F}{\partial T}\right)_{V,N} = -S , \quad \left(\frac{\partial F}{\partial V}\right)_{T,N} = -P , \quad \left(\frac{\partial F}{\partial N}\right)_{T,V} = \mu . \tag{3.1.8}$$

We can see from (3.1.8) that the internal energy can be written in terms of F in the form

$$E = F - T\left(\frac{\partial F}{\partial T}\right)_{V,N} = -T^2\left(\frac{\partial}{\partial T}\frac{F}{T}\right)_{V,N} . \tag{3.1.9}$$

From (3.1.7), it can be seen that the free energy is that portion of the energy which can be set free as work in an isothermal process; here we assume that the particle number N remains constant. In an isothermal volume change, the change of the free energy is given by $(dF)_{T,N} = -PdV = \delta W$, while $(dE)_{T,N} \neq \delta A$, since one would have to transfer heat into or out of the system in order to hold the temperature constant.

Enthalpy : $H = H(S, P, N)$
The enthalpy is defined as

$$H = E + PV . \tag{3.1.10}$$

From (3.1.3), it follows that

$$dH = TdS + VdP + \mu dN \tag{3.1.11}$$

and from this, its partial derivatives can be obtained:

$$\left(\frac{\partial H}{\partial S}\right)_{P,N} = T , \qquad \left(\frac{\partial H}{\partial P}\right)_{S,N} = V , \qquad \left(\frac{\partial H}{\partial N}\right)_{S,P} = \mu . \qquad (3.1.12)$$

For isobaric processes, $(dH)_{P,N} = TdS = \delta Q = dE + PdV$, thus the change in the enthalpy is equal to the change in the internal energy plus the energy change in the device supplying constant pressure (see Fig. 3.1). The weight F_G including the piston of area A holds the pressure constant at $P = F_G/A$. The change in the enthalpy is the sum of the change in the internal energy and the change in the potential energy of the weight. For a process at constant pressure, the heat δQ supplied to the system equals the increase in the system's enthalpy.

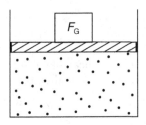

Fig. 3.1. The change in the enthalpy in isobaric processes; the weight F_G produces the constant pressure $P = F_G/A$, where A is the area of the piston.

Free Enthalpy (Gibbs' Free Energy) : $G = G(T, P, N)$
The Gibbs' free energy is defined as

$$G = E - TS + PV . \qquad (3.1.13)$$

Its differential follows from (3.1.3):

$$dG = -SdT + VdP + \mu dN . \qquad (3.1.14)$$

From Eq. (3.1.14), we can immediately read off

$$\left(\frac{\partial G}{\partial T}\right)_{P,N} = -S , \qquad \left(\frac{\partial G}{\partial P}\right)_{T,N} = V , \qquad \left(\frac{\partial G}{\partial N}\right)_{T,P} = \mu . \qquad (3.1.15)$$

The Grand Potential : $\Phi = \Phi(T, V, \mu)$
The grand potential is defined as

$$\Phi = E - TS - \mu N \qquad \left(= -kT \log Z_G(T, V, \mu)\right) ; \qquad (3.1.16)$$

in parentheses we give the connection to the grand partition function (Chap. 2). The differential expressions are

$$d\Phi = -SdT - PdV - Nd\mu , \qquad (3.1.17)$$

$$\left(\frac{\partial \Phi}{\partial T}\right)_{V,\mu} = -S , \qquad \left(\frac{\partial \Phi}{\partial V}\right)_{T,\mu} = -P , \qquad \left(\frac{\partial \Phi}{\partial \mu}\right)_{T,V} = -N . \qquad (3.1.18)$$

3.1.2 The Legendre Transformation

The transition from E to the thermodynamic *potentials* defined in (3.1.6), (3.1.10), (3.1.13), and (3.1.16) was carried out by means of so-called Legendre transformations, whose general structure will now be considered. We begin with a function Y which depends on the variables $x_1, x_2, \ldots$,

$$Y = Y(x_1, x_2, \ldots) . \tag{3.1.19}$$

The partial derivatives of Y in terms of the x_i are

$$a_i(x_1, x_2, \ldots) = \left(\frac{\partial Y}{\partial x_i}\right)_{\{x_j, j \neq i\}} . \tag{3.1.20a}$$

Our goal is now to replace the independent variable x_1 by the partial derivatives $\left(\frac{\partial Y}{\partial x_1}\right)$ as independent variables, i.e. for example to change from the independent variable S to T. This has a definite practical application, since the temperature is directly and readily measurable, while the entropy is not. The total differential of Y is given by

$$dY = a_1 dx_1 + a_2 dx_2 + \ldots \tag{3.1.20b}$$

From the rearrangement $dY = d(a_1 x_1) - x_1 da_1 + a_2 dx_2 + \ldots$, it follows that

$$d(Y - a_1 x_1) = -x_1 da_1 + a_2 dx_2 + \ldots \quad . \tag{3.1.21}$$

It is then expedient to introduce the function

$$Y_1 = Y - a_1 x_1 , \tag{3.1.22}$$

and to treat it as a function of the variables $a_1, x_2, \ldots$ (*natural* variables).[1] Thus, for example, the natural variables of the (Helmholtz) free energy are T, V, and N. The differential of $Y_1(a_1, x_2, \ldots)$ has the following form in terms of these independent variables:

$$dY_1 = -x_1 da_1 + a_2 dx_2 + \ldots \tag{3.1.21'a}$$

and its partial derivatives are

$$\left(\frac{\partial Y_1}{\partial a_1}\right)_{x_2, \ldots} = -x_1 , \qquad \left(\frac{\partial Y_1}{\partial x_2}\right)_{a_1, \ldots} = a_2 , \ldots \tag{3.1.21'b}$$

In this manner, one can obtain 8 thermodynamic potentials corresponding to the three pairs of variables. Table 3.1 collects the most important of these, i.e. the ones already introduced above.

---

[1] We make an additional remark here about the geometric significance of the Legendre transformation, referring to the case of a single variable: a curve can be represented either as a series of points $Y = Y(x_1)$, or through the family of its envelopes. In the latter representation, the intercepts of the tangential envelope lines on the ordinate as a function of their slopes a_1 are required. This geometric meaning of the Legendre transformation is the basis of the construction of $G(T, P)$ from $F(T, V)$ shown in Fig. 3.33. [If one simply eliminated x_1 in $Y = Y(x_1)$ in favor of a_1, then one would indeed obtain Y as a function of a_1, but it would no longer be possible to reconstruct $Y(x_1)$].

Table 3.1. Energy, entropy, and thermodynamic potentials

State function	Independent variables	Differentials
Energy E	$S, V, \{N_j\}$	$dE = TdS - PdV + \sum_j \mu_j dN_j$
Entropy S	$E, V, \{N_j\}$	$dS = \frac{1}{T}dE + \frac{P}{T}dV - \sum_j \frac{\mu_j}{T}dN_j$
Free Energy $F = E - TS$	$T, V, \{N_j\}$	$dF = -SdT - PdV + \sum_j \mu_j dN_j$
Enthalpy $H = E + PV$	$S, P, \{N_j\}$	$dH = TdS + VdP + \sum_j \mu_j dN_j$
Gibbs' Free Energy $G = E - TS + PV$	$T, P, \{N_j\}$	$dG = -SdT + VdP + \sum_j \mu_j dN_j$
Grand Potential $\Phi = E - TS - \sum_j \mu_j N_j$	$T, V, \{\mu_j\}$	$d\Phi = -SdT - PdV - \sum_j N_j d\mu_j$

This table contains the generalization to systems with several components (see Sect. 3.9). N_j and μ_j are the particle number and the chemical potential of the j-th component. The previous formulas are found as a special case when the index j and $\sum_j$ are omitted.

F, H, G and Φ are called thermodynamic potentials, since taking their derivatives with respect to the natural independent variables leads to the conjugate variables, analogously to the derivation of the components of force from the potential in mechanics. For the entropy, this notation is clearly less useful, since entropy does not have the dimensions of an energy. E, F, H, G and Φ are related to each other through Legendre transformations. The natural variables are also termed canonical variables. In a system consisting of only one chemical substance with a fixed number of particles, the state is completely characterized by specifying two quantities, e.g. T and V or V and P. All the other thermodynamic quantities can be calculated from the thermal and the caloric equations of state. If the state is characterized by T and V, then the pressure is given by the (thermal) *equation of state*

$$P = P(T, V) .$$

(The explicit form for a particular substance is found from statistical mechanics.) If we plot P against T and V in a three-dimensional graph, we obtain the *surface of the equation of state* (or *PVT* surface); see Fig. 2.7 and below in Sect. 3.8.

3.1.3 The Gibbs–Duhem Relation in Homogeneous Systems

In this section, we will concentrate on the important case of homogeneous thermodynamic systems.[2] Consider a system of this kind with the energy E, the volume V, and the particle number N. Now we imagine a second system which is completely similar in its properties but is simply larger by a factor α. Its energy, volume, and particle number are then αE, αV, and αN. Owing to the additivity of the entropy, it is given by

$$S(\alpha E, \alpha V, \alpha N) = \alpha S(E, V, N) .\qquad (3.1.23)$$

As a result, the entropy S is a homogeneous function of first order in E, V and N. Correspondingly, E is a homogeneous function of first order in S, V and N.

There are two types of state variables:

E, V, N, S, F, H, G, and Φ are called *extensive*, since they are proportional to α when the system is enlarged as described above. T, P, and μ are *intensive*, since they are independent of α; e.g. we find

$$T^{-1} = \frac{\partial S}{\partial E} = \frac{\partial \alpha S}{\partial \alpha E} \sim \alpha^0 ,$$

and this independence follows in a similar manner from the definitions of the other intensive variables, also. We wish to investigate the consequences of the homogeneity of S [Eq. (3.1.23)]. To this end, we differentiate (3.1.23) with respect to α and then set $\alpha = 1$:

$$\left(\frac{\partial S}{\partial \alpha E} E + \frac{\partial S}{\partial \alpha V} V + \frac{\partial S}{\partial \alpha N} N \right) \Big|_{\alpha=1} = S .$$

From this, we find using (3.1.2) that $-S + \frac{1}{T}E + \frac{P}{T}V - \frac{\mu}{T}N = 0$, that is

$$E = TS - PV + \mu N .\qquad (3.1.24)$$

This is the *Gibbs–Duhem* relation. Together with $dE = TdS - PdV + \mu dN$, we derive from Eq. (3.1.24)

$$SdT - VdP + Nd\mu = 0 ,\qquad (3.1.24')$$

the *differential* Gibbs–Duhem relation. It states that in a homogeneous system, T, P and μ cannot be varied independently, and it gives the relationship between the variations of these intensive quantities.[3] The following expressions can be derived from the Gibbs–Duhem relation:

---

[2] Homogeneous systems have the same specific properties in all spatial regions; they may also consist of several types of particles. Examples of inhomogeneous systems are those in a position-dependent potential and systems consisting of several phases which are in equilibrium, although in this case the individual phases can still be homogeneous.

[3] The generalization to systems with several components is given in Sect. 3.9, Eq. (3.9.7).

$$G(T, P, N) = \mu(T, P) \, N \tag{3.1.25}$$

and

$$\Phi(T, V, \mu) = -P(T, \mu) \, V \ . \tag{3.1.26}$$

Justification: from the definition (3.1.13), it follows immediately using (3.1.24) that $G = \mu N$, and from (3.1.15) we find $\mu = \left(\frac{\partial G}{\partial N}\right)_{T,P} = \mu + \left(\frac{\partial \mu}{\partial N}\right)_{T,P} N$; it follows that μ must be independent of N. We have thus demonstrated (3.1.25). Similarly, it follows from (3.1.16) that $\Phi = -PV$, and due to $-P = \left(\frac{\partial \Phi}{\partial V}\right)_{T,\mu}$, P must be independent of V.

Further conclusions following from homogeneity (in the canonical ensemble with independent variables T, V, and N) can be obtained starting with

$$P(T, V, N) = P(T, \alpha V, \alpha N) \quad \text{and} \quad \mu(T, V, N) = \mu(T, \alpha V, \alpha N) \tag{3.1.27a,b}$$

again by taking derivatives with respect to α around the point $\alpha = 1$:

$$\left(\frac{\partial P}{\partial V}\right)_{T,N} V + \left(\frac{\partial P}{\partial N}\right)_{T,V} N = 0 \quad \text{and} \quad \left(\frac{\partial \mu}{\partial V}\right)_{T,N} V + \left(\frac{\partial \mu}{\partial N}\right)_{T,V} N = 0 \ .$$
$$\tag{3.1.28a,b}$$

These two relations merely state that for intensive quantities, a volume increase is equivalent to a decrease in the number of particles.

3.2 Derivatives of Thermodynamic Quantities

3.2.1 Definitions

In this section, we will define the most important thermodynamic derivatives. In the following definitions, the particle number is always held constant.

The *heat capacity* is defined as

$$C = \frac{\delta Q}{dT} = T \frac{dS}{dT} \ . \tag{3.2.1}$$

It gives the quantity of heat which is required to raise the temperature of a body by $1\,\mathrm{K}$. We still have to specify which thermodynamic variables are held constant during this heat transfer. The most important cases are that the volume or the pressure is held constant. If the heat is transferred at constant volume, the heat capacity at constant volume is relevant:

$$C_V = T \left(\frac{\partial S}{\partial T}\right)_{V,N} = \left(\frac{\partial E}{\partial T}\right)_{V,N} \ . \tag{3.2.2a}$$

In rearranging $(\partial S/\partial T)_{V,N}$, we have used Eq. (3.1.1). If the heat transfer takes place under constant pressure, then the heat capacity at constant pressure from (3.2.1) must be used:

$$C_P = T\left(\frac{\partial S}{\partial T}\right)_{P,N} = \left(\frac{\partial H}{\partial T}\right)_{P,N} . \tag{3.2.2b}$$

For the rearrangement of the definition, we employed (3.1.11). If we divide the heat capacity by the mass of the substance or body, we obtain the *specific heat*, in general denoted as c, or c_V at constant volume or c_P at constant pressure. The specific heat is measured in units of J kg^{-1} K^{-1}. The specific heat may also be referred to 1 g and quoted in the (non-SI) units cal g^{-1} K^{-1}. The molar heat capacity (heat capacity per mole) gives the heat capacity of one mole of the substance. It is obtained from the specific heat referred to 1 g, multiplied by the molecular weight of the substance.

Remark: We will later show in general using Eq. (3.2.24) that the specific heat at constant pressure is larger than that at constant volume. The physical origin of this difference can be readily seen by writing the First Law for constant N in the form $\delta Q = dE + P dV$ and setting $dE = \left(\frac{\partial E}{\partial T}\right)_V dT + \left(\frac{\partial E}{\partial V}\right)_T dV = C_V dT + \left(\frac{\partial E}{\partial V}\right)_T dV$, that is

$$\delta Q = C_V dT + \left[P + \left(\frac{\partial E}{\partial V}\right)_T\right] dV .$$

In addition to the quantity of heat $C_V dT$ necessary for warming at constant volume, when V is increased, more heat is consumed by the work against the pressure, $P dV$, and by the change in the internal energy, $(\partial E/\partial V)_T\, dV$. For $C_P = \left(\frac{\delta Q}{dT}\right)_P$, it then follows from the last relation that

$$C_P = C_V + \left(P + \left(\frac{\partial E}{\partial V}\right)_T\right)\left(\frac{\partial V}{\partial T}\right)_P .$$

Further important thermodynamic derivatives are the compressibility, the coefficient of thermal expansion, and the thermal pressure coefficient. The *compressibility* is defined in general by

$$\kappa = -\frac{1}{V}\frac{dV}{dP} .$$

It is a measure of the relative volume decrease on increasing the pressure. For compression at a constant temperature, the *isothermal compressibility*, defined by

$$\kappa_T = -\frac{1}{V}\left(\frac{\partial V}{\partial P}\right)_{T,N} \tag{3.2.3a}$$

is the relevant quantity. For (reversible) processes in which no heat is transferred, i.e. when the entropy remains constant, the *adiabatic* (isentropic) *compressibility*

$$\kappa_S = -\frac{1}{V}\left(\frac{\partial V}{\partial P}\right)_{S,N} \qquad (3.2.3b)$$

must be introduced. The *coefficient of thermal expansion* is defined as

$$\alpha = \frac{1}{V}\left(\frac{\partial V}{\partial T}\right)_{P,N} . \qquad (3.2.4)$$

The definition of the *thermal pressure coefficient* is given by

$$\beta = \frac{1}{P}\left(\frac{\partial P}{\partial T}\right)_{V,N} . \qquad (3.2.5)$$

Quantities such as C, κ, and α are examples of so-called *susceptibilities*. They indicate how strongly an extensive quantity varies on changing (increasing) an intensive quantity.

3.2.2 Integrability and the Maxwell Relations

3.2.2.1 The Maxwell Relations

The Maxwell relations are expressions relating the thermodynamic derivatives; they follow from the integrability conditions. From the total differential of the function $Y = Y(x_1, x_2)$

$$dY = a_1 dx_1 + a_2 dx_2 , \qquad (3.2.6)$$

$$a_1 = \left(\frac{\partial Y}{\partial x_1}\right)_{x_2} , \qquad a_2 = \left(\frac{\partial Y}{\partial x_2}\right)_{x_1}$$

we find as a result of the commutatitivity of the order of the derivatives, $\left(\frac{\partial a_1}{\partial x_2}\right)_{x_1} = \frac{\partial^2 Y}{\partial x_2 \partial x_1} = \frac{\partial^2 Y}{\partial x_1 \partial x_2} = \left(\frac{\partial a_2}{\partial x_1}\right)_{x_2}$ the following *integrability condition*:

$$\left(\frac{\partial a_1}{\partial x_2}\right)_{x_1} = \left(\frac{\partial a_2}{\partial x_1}\right)_{x_2} . \qquad (3.2.7)$$

All together, there are 12 different Maxwell relations. The relations for fixed N are:

$$E: \quad \left(\frac{\partial T}{\partial V}\right)_S = -\left(\frac{\partial P}{\partial S}\right)_V , \qquad F: \quad \left(\frac{\partial S}{\partial V}\right)_T = \left(\frac{\partial P}{\partial T}\right)_V \qquad (3.2.8a,b)$$

$$H: \quad \left(\frac{\partial T}{\partial P}\right)_S = \left(\frac{\partial V}{\partial S}\right)_P \quad \text{or} \quad \left(\frac{\partial S}{\partial V}\right)_P = \left(\frac{\partial P}{\partial T}\right)_S \qquad (3.2.9)$$

$$G: \quad \left(\frac{\partial S}{\partial P}\right)_T = -\left(\frac{\partial V}{\partial T}\right)_P = -V\alpha . \qquad (3.2.10)$$

Here, we have labeled the Maxwell relations with the quantity from whose differential the relation is derived. There are also relations containing N and μ; of these, we shall require the following in this book:

$$F: \qquad \left(\frac{\partial \mu}{\partial V}\right)_{T,N} = -\left(\frac{\partial P}{\partial N}\right)_{T,V} . \qquad (3.2.11)$$

Applying this relation to homogeneous systems, we find from (3.1.28a) and (3.1.28b):

$$\left(\frac{\partial \mu}{\partial N}\right)_{T,V} = -\frac{V}{N}\left(\frac{\partial \mu}{\partial V}\right)_{T,N} = \frac{V}{N}\left(\frac{\partial P}{\partial N}\right)_{T,V}$$

$$= -\frac{V^2}{N^2}\left(\frac{\partial P}{\partial V}\right)_{T,N} = \frac{V}{N^2}\frac{1}{\kappa_T} . \qquad (3.2.12)$$

*3.2.2.2 Integrability Conditions, Exact and Inexact Differentials

It may be helpful at this point to show the connection between the integrability conditions and the results of vector analysis as they apply to classical mechanics. We consider a vector field $\mathbf{F}(\mathbf{x})$, which is defined within the simply-connected region G (this field could for example be a force field). Then the following statements are equivalent:

(I) $\mathbf{F}(\mathbf{x}) = -\boldsymbol{\nabla} V(\mathbf{x})$

with $V(\mathbf{x}) = -\int_{\mathbf{x}_0}^{\mathbf{x}} d\mathbf{x}' \mathbf{F}(\mathbf{x}')$, where $\mathbf{x}_0$ is an arbitrary fixed point of origin and the line integral is to be taken along an arbitrary path from $\mathbf{x}_0$ to $\mathbf{x}$. This means that $\mathbf{F}(\mathbf{x})$ can be derived from a potential.

(II) $\operatorname{curl} \mathbf{F} = 0$ at each point in G.

(III) $\oint d\mathbf{x}\, \mathbf{F}(\mathbf{x}) = 0$ along each closed path in G.

(IV) $\int_{\mathbf{x}_1}^{\mathbf{x}_2} d\mathbf{x}\, \mathbf{F}(\mathbf{x})$ is independent of the path.

Let us return to thermodynamics. We consider a system characterized by two independent thermodynamic variables x and y and a quantity whose differential variation is given by

$$dY = A(x,y)dx + B(x,y)dy . \qquad (3.2.13)$$

In the notation of mechanics, $\mathbf{F} = (A(x,y), B(x,y), 0)$. The existence of a state variable Y, i.e. a state function $Y(x,y)$ (Statement (I')) is equivalent to each of the three other statements (II',III', and IV').

(I')   A state function $Y(x,y)$ exists, with
$$Y(x,y) = Y(x_0, y_0) + \int_{(x_0,y_0)}^{(x,y)} (dx' A(x',y') + dy' B(x',y')) .$$

(II') $\left(\frac{\partial B}{\partial x}\right)_y = \left(\frac{\partial A}{\partial y}\right)_x .$

(III') $\oint (dx A(x,y) + dy B(x,y)) = 0$

(IV') $\int_{P_0}^{P_1} (dx A(x,y) + dy B(x,y))$ is independent of the path.

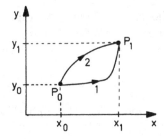

Fig. 3.2. Illustrating the path integrals III′ and IV′

The differential (3.2.13) is called an exact differential (or a perfect differential) when the coefficients A and B fulfill the integrability condition (II′).

3.2.2.3 The Non-integrability of δQ and δW

We can now prove that δQ and δW are not integrable. We first consider δW and imagine the independent thermodynamic variables to be V and T. Then the relation (3.1.4a) becomes

$$\delta W = -PdV + 0 \cdot dT \ . \tag{3.2.14}$$

The derivative of the pressure with respect to the temperature at constant volume is nonzero, $\left(\frac{\partial P}{\partial T}\right)_V \neq 0$, while of course the derivative of zero with respect to V gives zero. That is, the integrability condition is not fulfilled. Analogously, we write (3.1.5) in the form

$$\delta Q = TdS + 0 \cdot dV \ . \tag{3.2.15}$$

Again, we have $\left(\frac{\partial T}{\partial V}\right)_S = -\frac{\left(\frac{\partial S}{\partial V}\right)_T}{\left(\frac{\partial S}{\partial T}\right)_V} = -\frac{\left(\frac{\partial P}{\partial T}\right)_V}{\left(\frac{\partial S}{\partial T}\right)_V} \neq 0$, i.e. the integrability condition is not fulfilled. Therefore, there are no state functions $W(V, T, N)$ and $Q(V, T, N)$ whose differentials are equal to δW and δQ. This is the reason for the different notation used in the differential signs. The expressions relating the heat transferred to the system and the work performed on it to the state variables exist only in differential form. One can, of course, compute the integral $\int_1 \delta Q = \int_1 TdS$ along a given path (e.g. 1 in Fig. 3.2), and similarly for δW, but the values of these integrals depend not only on their starting and end points, but also on the details of the path which connects those points.

Remark:

In the case that a differential does not fulfill the integrability condition,

$$\delta Y = A(x, y)dx + B(x, y)dy \ ,$$

but can be converted into an exact differential through multiplication by a factor $g(x, y)$, then $g(x, y)$ is termed an *integrating factor*. Thus, $\frac{1}{T}$ is an integrating factor for δQ. In statistical mechanics, it is found quite naturally that the entropy is a state function, i.e. dS is an exact differential. In the historical development of thermodynamics, it was a decisive and nontrivial discovery that multiplication of δQ by $\frac{1}{T}$ yields an exact differential.

3.2.3 Jacobians

It is often necessary to transform from one pair of thermodynamic variables to a different pair. For the necessary recalculation of the thermodynamic derivatives, it is expedient to use Jacobians.

In the following, we consider functions of two variables: $f(u, v)$ and $g(u, v)$. We define the Jacobian determinant:

$$\frac{\partial(f, g)}{\partial(u, v)} = \begin{vmatrix} \left(\frac{\partial f}{\partial u}\right)_v & \left(\frac{\partial f}{\partial v}\right)_u \\ \left(\frac{\partial g}{\partial u}\right)_v & \left(\frac{\partial g}{\partial v}\right)_u \end{vmatrix} = \left(\frac{\partial f}{\partial u}\right)_v \left(\frac{\partial g}{\partial v}\right)_u - \left(\frac{\partial f}{\partial v}\right)_u \left(\frac{\partial g}{\partial u}\right)_v . \quad (3.2.16)$$

This Jacobian fulfills a series of important relations.

Let $u = u(x, y)$ and $v = v(x, y)$ be functions of x and y; then the following chain rule can be proved in an elementary fashion:

$$\frac{\partial(f, g)}{\partial(x, y)} = \frac{\partial(f, g)}{\partial(u, v)} \frac{\partial(u, v)}{\partial(x, y)} . \quad (3.2.17)$$

This relation is important for the changes of variables which are frequently needed in thermodynamics. Setting $g = v$, the definition (3.2.16) is simplified to

$$\frac{\partial(f, v)}{\partial(u, v)} = \left(\frac{\partial f}{\partial u}\right)_v . \quad (3.2.18)$$

Since a determinant changes its sign on interchanging two columns, we have

$$\frac{\partial(f, g)}{\partial(v, u)} = -\frac{\partial(f, g)}{\partial(u, v)} . \quad (3.2.19)$$

If we apply the chain rule (3.2.17) for $x = f$ and $y = g$, we find:

$$\frac{\partial(f, g)}{\partial(u, v)} \frac{\partial(u, v)}{\partial(f, g)} = 1 . \quad (3.2.20)$$

Setting $g = v$ in (3.2.20), we obtain with (3.2.18)

$$\left(\frac{\partial f}{\partial u}\right)_v = \frac{1}{\left(\frac{\partial u}{\partial f}\right)_v} . \quad (3.2.20')$$

Finally, from (3.2.18) we have

$$\left(\frac{\partial f}{\partial u}\right)_v = \frac{\partial(f,v)}{\partial(u,v)} = \frac{\partial(f,v)}{\partial(f,u)}\frac{\partial(f,u)}{\partial(u,v)} = -\frac{\left(\frac{\partial f}{\partial v}\right)_u}{\left(\frac{\partial u}{\partial v}\right)_f} . \tag{3.2.21}$$

Using this relation, one can thus transform a derivative at constant v into derivatives at constant u and f. The relations given here can also be applied to functions of more than two variables, provided the additional variables are held constant.

3.2.4 Examples

(i) We first derive some useful relations between the thermodynamic derivatives. Using Eqns. (3.2.21), (3.2.3a), and (3.2.4), we obtain

$$\left(\frac{\partial P}{\partial T}\right)_V = -\frac{\left(\frac{\partial V}{\partial T}\right)_P}{\left(\frac{\partial V}{\partial P}\right)_T} = \frac{\alpha}{\kappa_T} . \tag{3.2.22}$$

Thus, the thermal pressure coefficient $\beta = \frac{1}{P}\left(\frac{\partial P}{\partial T}\right)_V$ [Eq. (3.2.5)] is related to the coefficient of thermal expansion α and the isothermal compressibility κ_T. In problem 3.4, it is shown that

$$\frac{C_P}{C_V} = \frac{\kappa_T}{\kappa_S} \tag{3.2.23}$$

[cf. (3.2.3a,b)]. Furthermore, we see that

$$C_V = T\frac{\partial(S,V)}{\partial(T,V)} = T\frac{\partial(S,V)}{\partial(T,P)}\frac{\partial(T,P)}{\partial(T,V)} =$$
$$= T\left(\frac{\partial P}{\partial V}\right)_T\left[\left(\frac{\partial S}{\partial T}\right)_P\left(\frac{\partial V}{\partial P}\right)_T - \left(\frac{\partial S}{\partial P}\right)_T\left(\frac{\partial V}{\partial T}\right)_P\right] =$$
$$= C_P - T\frac{\left(\frac{\partial S}{\partial P}\right)_T\left(\frac{\partial V}{\partial T}\right)_P}{\left(\frac{\partial V}{\partial P}\right)_T} = C_P + T\frac{\left(\frac{\partial V}{\partial T}\right)_P^2}{\left(\frac{\partial V}{\partial P}\right)_T} .$$

Here, the Maxwell relation (3.2.10) was used. Thus we find for the heat capacities

$$C_P - C_V = \frac{TV\alpha^2}{\kappa_T} . \tag{3.2.24}$$

With $\kappa_T C_P - \kappa_T C_V = TV\alpha^2$ and $\kappa_T C_V = \kappa_S C_P$, it follows that the compressibilities obey the relation

$$\kappa_T - \kappa_S = \frac{TV\alpha^2}{C_P} . \tag{3.2.25}$$

It follows from (3.2.24) that the two heat capacities can become equal only when the coefficient of expansion α vanishes or κ_T becomes very large. The former occurs in the case of water at 4°C.

(ii) We now evaluate the thermodynamic derivatives for the *classical ideal gas*, based on Sect. 2.7 . For the enthalpy $H = E + PV$, it follows from Eqns. (2.7.25) and (2.7.28) that

$$H = \frac{5}{2}NkT \ . \tag{3.2.26}$$

Then, for the heat capacities, we find

$$C_V = \left(\frac{\partial E}{\partial T}\right)_V = \frac{3}{2}Nk \ , \qquad C_P = \left(\frac{\partial H}{\partial T}\right)_P = \frac{5}{2}Nk \ ; \tag{3.2.27}$$

and for the compressibilities,

$$\kappa_T = -\frac{1}{V}\left(\frac{\partial V}{\partial P}\right)_T = \frac{1}{P} \ , \qquad \kappa_S = \kappa_T\frac{C_V}{C_P} = \frac{3}{5P} \ , \tag{3.2.28}$$

finally, for the thermal expansion coefficient and the thermal pressure coefficient, we find

$$\alpha = \frac{1}{V}\left(\frac{\partial V}{\partial T}\right)_P = \frac{1}{T} \quad \text{and} \quad \beta = \frac{1}{P}\left(\frac{\partial P}{\partial T}\right)_V = \frac{1}{P}\frac{\alpha}{\kappa_T} = \frac{1}{T} \ . \tag{3.2.29a,b}$$

3.3 Fluctuations and Thermodynamic Inequalities

This section is concerned with fluctuations of the energy and the particle number, and belongs contextually to the preceding chapter. We are only now treating these phenomena because the final results are expressed in terms of thermodynamic derivatives, whose definitions and properties are only now at our disposal.

3.3.1 Fluctuations

1. We consider a canonical ensemble, characterized by the temperature T, the volume V, the fixed particle number N, and the density matrix

$$\rho = \frac{e^{-\beta H}}{Z} \ , \qquad Z = \text{Tr } e^{-\beta H} \ .$$

The average value of the energy [Eq. (2.6.37)] is given by

$$\bar{E} = \frac{1}{Z}\text{Tr } e^{-\beta H} H = \frac{1}{Z}\frac{\partial Z}{\partial(-\beta)} \ . \tag{3.3.1}$$

Taking the temperature derivative of (3.3.1),

$$\left(\frac{\partial \bar{E}}{\partial T}\right)_V = \frac{1}{kT^2}\frac{\partial \bar{E}}{\partial(-\beta)} = \frac{1}{kT^2}\left[\langle H^2\rangle - \langle H\rangle^2\right] = \frac{1}{kT^2}(\Delta E)^2 \ ,$$

we obtain after substitution of (3.2.2a) the following relation between the specific heat at constant volume and the mean square deviation of the internal energy:

$$C_V = \frac{1}{kT^2}(\Delta E)^2 \, . \tag{3.3.2}$$

2. Next, we start with the grand canonical ensemble, characterized by T, V, μ, and the density matrix

$$\rho_G = Z_G^{-1} e^{-\beta(H-\mu N)} \, , \qquad Z_G = \mathrm{Tr}\; e^{-\beta(H-\mu N)} \, .$$

The average particle number is given by

$$\bar{N} = \mathrm{Tr}\; \rho_G N = kT\, Z_G^{-1} \frac{\partial Z_G}{\partial \mu} \, . \tag{3.3.3}$$

Its derivative with respect to the chemical potential is

$$\left(\frac{\partial \bar{N}}{\partial \mu} \right)_{T,V} = \beta(\langle N^2 \rangle - \bar{N}^2) = \beta(\Delta N)^2 \, .$$

If we replace the left side by (3.2.12), we obtain the following relation between the isothermal compressibility and the mean square deviation of the particle number:

$$\kappa_T = -\frac{1}{V}\left(\frac{\partial V}{\partial P} \right)_{T,N} = \frac{V}{N^2}\left(\frac{\partial N}{\partial \mu} \right)_{T,V} = \frac{V}{N^2}\beta(\Delta N)^2 \, . \tag{3.3.4}$$

Eqns. (3.3.2) and (3.3.4) are fundamental examples of relations between susceptibilities (on the left-hand sides) and fluctuations, so called fluctuation-response theorems.

3.3.2 Inequalities

From the relations derived in 3.3.1, we derive (as a result of the positivity of the fluctuations) the following inequalities:

$$\kappa_T \geq 0 \, , \tag{3.3.5}$$

$$C_P \geq C_V \geq 0 \, . \tag{3.3.6}$$

In (3.3.6), we have used the fact that according to (3.2.24) and (3.3.5), C_P is larger than C_V. On decreasing the volume, the pressure increases. On increasing the energy, the temperature increases. The validity of these inequalities is a precondition for the stability of matter. If, for example, (3.3.5) were not valid, compression of the system would decrease its pressure; it would thus be further compressed and would finally collapse.

3.4 Absolute Temperature and Empirical Temperatures

The absolute temperature was defined in (2.4.4) as $T^{-1} = \left(\frac{\partial S(E,V,N)}{\partial E}\right)_{V,N}$.
Experimentally, one uses a temperature ϑ, which is for example given by the
length of a rod or a column of mercury, or the volume or the pressure of
a gas thermometer. We assume that the empirical temperature ϑ increases
monotonically with T, i.e. that ϑ also increases when we put heat into the
system. We now seek a method of determining the absolute temperature
from ϑ, that is, we seek the relation $T = T(\vartheta)$. To this end, we start with the
thermodynamic difference quotient $\left(\frac{\delta Q}{dP}\right)_T$:

$$\left(\frac{\delta Q}{dP}\right)_T = T\left(\frac{\partial S}{\partial P}\right)_T = -T\left(\frac{\partial V}{\partial T}\right)_P = -T\left(\frac{\partial V}{\partial \vartheta}\right)_P \frac{d\vartheta}{dT} . \qquad (3.4.1)$$

Here, we have substituted in turn $\delta Q = TdS$, the Maxwell relation (3.2.10),
and $T = T(\vartheta)$. It follows that

$$\frac{1}{T}\frac{dT}{d\vartheta} = -\frac{\left(\frac{\partial V}{\partial \vartheta}\right)_P}{\left(\frac{\delta Q}{dP}\right)_T} = -\left(\frac{\partial V}{\partial \vartheta}\right)_P\left(\frac{dP}{\delta Q}\right)_\vartheta . \qquad (3.4.2)$$

This expression is valid for any substance. The right-hand side can be measured experimentally and yields a function of ϑ. Therefore, (3.4.2) represents
an ordinary inhomogeneous differential equation for $T(\vartheta)$, whose integration
yields

$$T = \text{const} \cdot f(\vartheta) . \qquad (3.4.3)$$

We thus obtain a unique relation between the empirical temperature ϑ and
the absolute temperature. The constant can be chosen freely due to the arbitrary nature of the empirical temperature scale. The *absolute temperature
scale* is determined by defining the triple point of water to be $T_t = 273.16\,\text{K}$.

For magnetic thermometers, it follows from $\left(\frac{\delta Q}{dB}\right)_T = T\left(\frac{\partial S}{\partial B}\right)_T = T\left(\frac{\partial M}{\partial T}\right)_B$
(cf. Chap. 6), analogously,

$$\frac{1}{T}\frac{dT}{d\vartheta} = \left(\frac{\partial M}{\partial \vartheta}\right)_B\left(\frac{dB}{\delta Q}\right)_\vartheta . \qquad (3.4.4)$$

The absolute temperature

$$T = \left(\frac{\partial S}{\partial E}\right)_{V,N}^{-1} \qquad (3.4.5)$$

is positive, since the number of accessible states ($\propto \Omega(E)$) is a rapidly increasing function of the energy. The minimum value of the absolute temperature

is $T = 0$ (except for systems which have energetic upper bounds, such as an assembly of paramagnetic spins). This follows from the distribution of energy levels E in the neighborhood of the ground-state energy E_0. We can see from the models which we have already evaluated explicitly (quantum-mechanical harmonic oscillators, paramagnetic moments: Sects. 2.5.2.1 and 2.5.2.2) that $\lim_{E\to E_0} S'(E) = \infty$, and thus for these systems, which are generic with respect to their low-lying energy levels,

$$\lim_{E\to E_0} T = 0 .$$

We return once more to the determination of the temperature scale through Eq. (3.4.3) in terms of $T_t = 273.16\,\text{K}$. As mentioned in Sect. 2.3, the value of the Boltzmann constant is also fixed by this relation. In order to see this, we consider a system whose equation of state at T_t is known. Molecular hydrogen can be treated as an ideal gas at T_t and $P = 1$ atm. The density of H_2 under these conditions is

$$\rho = 8.989 \times 10^{-2}\text{g/liter} = 8.989 \times 10^{-5}\text{g/cm}^{-3} .$$

Its molar volume then has the value

$$V_M = \frac{2.016\ \text{g}}{8.989 \times 10^{-2}\ \text{g liters}^{-1}} = 22.414\ \text{liters} .$$

One mole is defined as: 1 mole corresponds to a mass equal to the atomic weight in g (e.g. a mole of H_2 has a mass of 2.016 g). From this fact, we can determine the *Boltzmann constant*:

$$k = \frac{PV}{NT} = \frac{1\ \text{atm}\ V_M}{N_A \times 273.16\ \text{K}} = 1.38066 \times 10^{-16}\ \text{erg/K}$$
$$= 1.38066 \times 10^{-23} J/\text{K} . \tag{3.4.6}$$

Here, Avogadro's number was used:

$$N_A \equiv \text{number of molecules per mole}$$
$$= \frac{2.016\ \text{g}}{\text{mass of}\ H_2} = \frac{2.016\ \text{g}}{2 \times 1.6734 \times 10^{-24}\text{g}} = 6.0221 \times 10^{23}\ \text{mol}^{-1} .$$

Further definitions of units and constants, e.g. the gas constant R, are given in Appendix I.

3.5 Thermodynamic Processes

In this section, we want to treat thermodynamic processes, i.e. processes which either during the whole course of their time development or at least in their initial or final stages can be sufficiently well described by thermodynamics.

3.5.1 Thermodynamic Concepts

We begin by introducing several concepts of thermodynamics which we will later use repeatedly (cf. Table 3.2).

Processes in which the pressure is held constant, i.e. $P = const$, are called *isobaric*; those in which the volume remains constant, $V = const$, are *isochoral*; those in which the entropy is constant, $S = const$, are *isentropic*; and those in which no heat is transferred, i.e. $\delta Q = 0$, are termed *adiabatic (thermally isolated)*.

Table 3.2. Some thermodynamic concepts

Concept	Definition
isobaric	$P = const.$
isochoral	$V = const.$
isothermal	$T = const.$
isentropic	$S = const.$
adiabatic	$\delta Q = 0$
extensive	proportional to the size of the system
intensive	independent of the size of the system

We mention here another definition of the terms *extensive* and *intensive*, which is equivalent to the one given in the section on the Gibbs–Duhem relation. We divide a system that is characterized by the thermodynamic variable Y into two parts, which are themselves characterized by Y_1 and Y_2. In the case that $Y_1 + Y_2 = Y$, Y is called extensive; when $Y_1 = Y_2 = Y$, it is termed intensive (see Fig. 3.3).

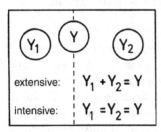

extensive: $Y_1 + Y_2 = Y$

intensive: $Y_1 = Y_2 = Y$

Fig. 3.3. The definition of extensive and intensive thermodynamic variables

Extensive variable include: V, N, E, S, the thermodynamic potentials, the electric polarization **P**, and the magnetization **M**.

Intensive variables include: P, μ, T, the electric field **E**, and the magnetic field **B**.

Quasistatic process: a quasistatic process takes place slowly with respect to the characteristic relaxation time of the system, i.e. the time within which the system passes from a nonequilibrium state to an equilibrium state, so that the system remains in equilibrium at each moment during such a process. Typical relaxation times are of the order of $\tau = 10^{-10} - 10^{-9}$ sec.

An *irreversible process* is one which cannot take place in the reverse direction, e.g. the transition from a nonequilibrium state to an equilibrium state (the initial state could also be derived from an equilibrium state with restrictions by lifting of those restrictions). Experience shows that a system which is not in equilibrium moves towards equilibrium; in this process, its entropy increases. The system then remains in equilibrium and does not return to the nonequilibrium state.

Reversible processes: reversible processes are those which can also occur in the reverse direction. An essential attribute of reversibility is that a process which takes place in a certain direction can be followed by the reverse process in such a manner that no changes in the surroundings remain.

The characterization of a thermodynamic state (with a fixed particle number N) can be accomplished by specifying two quantities, e.g. T and V, or P and V. The remaining quantities can be found from the thermal and the caloric equations of state. A system in which a quasistatic process is occurring, i.e. which is in thermal equilibrium at each moment in time, can be represented by a curve, for example in a P–V diagram (Fig. 2.7b).

A reversible process must in all cases be quasistatic. In non-quasistatic processes, turbulent flows and temperature fluctuations take place, leading to the irreversible production of heat. The intermediate states in a non-quasistatic process can furthermore not be sufficiently characterized by P and V. One requires for their characterization more degrees of freedom, or in other words, a space of higher dimensionality.

There are also quasistatic processes which are irreversibe (e.g. temperature equalization via a poor heat conductor, 3.6.3.1; or a Gay-Lussac experiment carried out slowly, 3.6.3.6). Even in such processes, equilibrium thermodynamics is valid for the individual components of the system.

Remark:

We note that thermodynamics rests on equilibrium statistical mechanics. In reversible processes, the course of events is so slow that the system is in equilibrium at each moment; in irreversible processes, this is true of at least the initial and final states, and thermodynamics can be applied to these states. In the following sections, we will clarify the concepts just introduced on the basis of some typical examples. In particular, we will investigate how the entropy changes during the course of a process.

3.5.2 The Irreversible Expansion of a Gas; the Gay-Lussac Experiment (1807)

The Gay-Lussac experiment[4] deals with the adiabatic expansion of a gas and is carried out as follows: a container of volume V which is insulated from its surroundings is divided by partition into two subvolumes, V_1 and V_2. Initially, the volume V_1 contains a gas at a temperature T, while V_2 is evacuated. The partition is then removed and the gas flows rapidly into V_2 (Fig. 3.4).

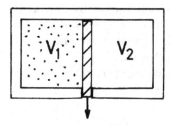

Fig. 3.4. The Gay-Lussac experiment

After the gas has reached equilibrium in the whole volume $V = V_1 + V_2$, its thermodynamic quantities are determined.

We first assume that this experiment is carried out using an ideal gas. The initial state is completely characterized by its volume V_1 and the temperature T. The entropy and the pressure before the expansion are, from (2.7.27) and (2.7.25), given by

$$ S = Nk \left(\frac{5}{2} + \log \frac{V_1/N}{\lambda^3} \right) \quad \text{and} \quad P = \frac{NkT}{V_1} , $$

with the thermal wavelength λ:

$$ \lambda = \frac{h}{\sqrt{2\pi mkT}} . $$

In the final state, the volume is now $V = V_1 + V_2$. The temperature is still equal to T, since the energy remains constant and the caloric equation of state of ideal gases, $E = \frac{3}{2}kTN$, contains no dependence on the volume. The entropy and the pressure after the expansion are:

$$ S' = Nk \left(\frac{5}{2} + \log \frac{V/N}{\lambda^3} \right) , \qquad P' = \frac{NkT}{V} . $$

We can see that in this process, there is an entropy production of

$$ \Delta S = S' - S = Nk \log \frac{V}{V_1} > 0 . \tag{3.5.1} $$

---

[4] Louis Joseph Gay-Lussac, 1778–1850. The goal of Gay-Lussac's experiments was to determine the volume dependence of the internal energy of gases.

It is intuitively clear that the process is irreversible. Since the entropy increases and no heat is transferred, ($\delta Q = 0$), the mathematical criterion for an irreversible process, Eq. (3.6.8) (which remains to be proved), is fulfilled. The initial and final states in the Gay-Lussac experiment are equilibrium states and can be treated with equilibrium thermodynamics. The intermediate states are in general not equilibrium states, and equilibrium thermodynamics can therefore make no statements about them. Only when the expansion is carried out as a quasistatic process can equilibrium thermodynamics be applied at each moment. This would be the case if the expansion were carried out by allowing a piston to move slowly (either by moving a frictionless piston in a series of small steps without performing work, or by slowing the expansion of the gas by means of the friction of the piston and transferring the resulting frictional heat back into the gas).

For an arbitrary isolated gas, the temperature change per unit volume at constant energy is given by

$$\left(\frac{\partial T}{\partial V}\right)_E = -\frac{\left(\frac{\partial E}{\partial V}\right)_T}{\left(\frac{\partial E}{\partial T}\right)_V} = -\frac{T\left(\frac{\partial S}{\partial V}\right)_T - P}{C_V} = \frac{1}{C_V}\left(P - T\left(\frac{\partial P}{\partial T}\right)_V\right) , \quad (3.5.2a)$$

where the Maxwell relation $\left(\frac{\partial S}{\partial V}\right)_T = \left(\frac{\partial P}{\partial T}\right)_V$ has been employed. This coefficient has the value 0 for an ideal gas, but for real gases it can have either a positive or a negative sign. The entropy production is, owing to $dE = TdS - PdV = 0$, given by

$$\left(\frac{\partial S}{\partial V}\right)_E = \frac{P}{T} > 0 , \quad (3.5.2b)$$

i.e. $dS > 0$. Furthermore, no heat is exchanged with the surroundings, that is, $\delta Q = 0$. Therefore, it follows that the inequality between the change in the entropy and the quantity of heat transferred

$$TdS > \delta Q \quad (3.5.3)$$

holds here.

The coefficients calculated from equilibrium thermodynamics (3.5.2a,b) can be applied to the whole course of the Gay-Lussac experiment if the process is carried out in a quasistatic manner. Yet it remains an irreversible process! By integration of (3.5.2a,b), one obtains the differences in temperature and entropy between the final and initial states. The result can by the way also be applied to the non-quasistatic irreversible process, since the two final states are identical. We shall return to the quasistatic, irreversible Gay-Lussac experiment in 3.6.3.6.

3.5.3 The Statistical Foundation of Irreversibility

How irreversible is the Gay-Lussac process? In order to understand why the Gay-Lussac experiment is irreversible, we consider the case that the volume increase δV fulfills the inequality $\delta V \ll V$, where V now means the initial volume (see Fig. 3.5).

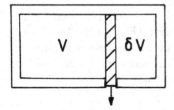

Fig. 3.5. Illustration of the Gay-Lussac experiment

In the expansion from V to $V + \delta V$, the phase-space surface changes from $\Omega(E, V)$ to $\Omega(E, V + \delta V)$, and therefore the entropy changes from $S(E, V)$ to $S(E, V + \delta V)$. After the gas has carried out this expansion, we ask what the probability would be of finding the system in only the subvolume V. Employing (1.3.2), (2.2.4), and (2.3.4), we find this probability to be given by

$$W(E, V) = \int_V \frac{dq\, dp}{N!\, h^{3N}} \frac{\delta(H - E)}{\Omega(E, V + \delta V)} = \frac{\Omega(E, V)}{\Omega(E, V + \delta V)} = \qquad (3.5.4)$$

$$= e^{-(S(E, V + \delta V) - S(E, V))/k} =$$

$$= e^{-\left(\frac{\partial S}{\partial V}\right)_E \delta V/k} = e^{-\frac{P}{T}\delta V/k} = e^{-\frac{\delta V}{V}N} \ll 1 \, ,$$

where in the last rearrangement, we have assumed an ideal gas. Due to the factor $N \approx 10^{23}$ in the exponent, the probability that the system will return spontaneously to the volume V is vanishingly small.

In general, it is found that for the probability, a *constraint* (a *restriction* C) occurs spontaneously:

$$W(E, C) = e^{-(S(E) - S(E, C))/k} \, . \qquad (3.5.5)$$

We find that $S(E, C) \ll S(E)$, since under the constraint, fewer states are accessible. The difference $S(E) - S(E, C)$ is macroscopic; in the case of the change in volume, it was proportional to $N\delta V/V$, and the probability $W(E, C) \sim e^{-N}$ is thus practically zero. The transition from a state with a constraint C to one without this restriction is irreversible, since the probability that the system will spontaneously search out a state with this constraint is vanishingly small.

3.5.4 Reversible Processes

In the first subsection, we consider the reversible isothermal and adiabatic expansion of ideal gases, which illustrate the concept of reversibility and are important in their own right as elements of thermodynamic processes.

3.5.4.1 Typical Examples: the Reversible Expansion of a Gas

In the reversible expansion of an ideal gas, work is performed on a spring by the expanding gas and energy is stored in the spring (Fig. 3.6). This energy can later be used to compress the gas again; the process is thus reversible. It can be seen as a reversible variation of the Gay-Lussac experiment. Such a process can be carried out isothermally or adiabatically.

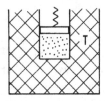

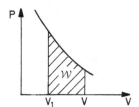

Fig. 3.6. The reversible isothermal expansion of a gas, where the work performed is stored by a spring. The work performed by the gas is equal to the area below the isotherm in the $P - V$ diagram.

a) Isothermal Expansion of a Gas, $T = $ const.
We first consider the isothermal expansion. Here, the gas container is in a heat bath at a temperature T. On expansion from the initial volume V_1 to the final volume V, the gas performs the work:[5]

$$\mathcal{W} = \int_{V_1}^{V} P dV = \int_{V_1}^{V} dV \frac{NkT}{V} = NkT \log \frac{V}{V_1} . \tag{3.5.6}$$

This work can be visualized as the area below the isotherm in the $P - V$ diagram (Fig. 3.6). Since the temperature remains constant, the energy of the ideal gas is also unchanged. Therefore, the heat bath must transfer a quantity of heat

---

[5] We distinguish the work performed by the system ($\mathcal{W}$) from work performed on the system (W), we use different symbols, implying opposite signs: $\mathcal{W} = -W$.

$$Q = W \tag{3.5.7}$$

to the system. The change in the entropy during this isothermal expansion is given according to (2.7.27) by:

$$\Delta S = Nk \log \frac{V}{V_1} . \tag{3.5.8}$$

Comparison of (3.5.6) with (3.5.8) shows us that the entropy increase and the quantity of heat taken up by the system here obey the following relation:

$$\Delta S = \frac{Q}{T} . \tag{3.5.9}$$

This process is reversible, since using the energy stored in the spring, one could compress the gas back to its original volume. In this compression, the gas would release the quantity of heat Q to the heat bath. The final state of the system and its surroundings would then again be identical to their original state. In order for the process to occur in a quasistatic way, the strength of the spring must be varied during the expansion or compression in such a way that it exactly compensates the gas pressure P (see the discussion in Sect. 3.5.4.2). One could imagine the storage and release of the energy from the work of compression or expansion in an idealized thought experiment to be carried out by the horizontal displacement of small weights, which would cost no energy.

We return again to the example of the irreversible expansion (Sect. 3.5.2). Clearly, by performing work in this case we could also compress the gas after its expansion back to its original volume, but then we would increase its energy in the process. The work required for this compression is finite and its magnitude is proportional to the change in volume; it cannot, in contrast to the case of reversible processes, in principle be made equal to zero.

b) Adiabatic Expansion of a Gas, $\Delta Q = 0$

We now turn to the adiabatic reversible expansion. In contrast to Fig. 3.6, the gas container is now insulated from its surroundings, and the curves in the P-V diagram are steeper. In every step of the process, $\delta Q = 0$, and since work is here also performed by the gas on its surroundings, it cools on expansion. It then follows from the First Law that

$$dE = -PdV .$$

If we insert the caloric and the thermal equations of state into this equation, we find:

$$\frac{dT}{T} = -\frac{2}{3} \frac{dV}{V} . \tag{3.5.10}$$

Integration of the last equation leads to the two forms of the adiabatic equation for an ideal gas:

$$T = T_1\left(V_1/V\right)^{2/3} \quad \text{and} \quad P = NkT_1\, V_1^{2/3}\, V^{-5/3} , \tag{3.5.11a,b}$$

where the equation of state was again used to obtain b.

We now once more determine the work $W(V)$ performed on expansion from V_1 to V. It is clearly less than in the case of the isothermal expansion, since no heat is transferred from the surroundings. Correspondingly, the area beneath the adiabats is smaller than that beneath the isotherms (cf. Fig. 3.7). Inserting Eq. (3.5.11b) yields for the work:

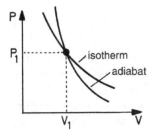

Fig. 3.7.   An isotherm and an adiabat passing through the initial point (P_1, V_1), with $P_1 = NkT_1/V_1$

$$W(V) = \int_{V_1}^{V} dV\ P = \frac{3}{2} NkT_1 \left(1 - \left(\frac{V}{V_1}\right)^{-2/3}\right) ; \tag{3.5.12}$$

geometrically, this is the area beneath the adiabats, Fig. 3.7. The change in the entropy is given by

$$\Delta S = Nk \log\left(\frac{V\,\lambda_1^3}{\lambda^3\,V_1}\right) = 0 , \tag{3.5.13}$$

and it is equal to zero. We are dealing here with a reversible process in an isolated systems, $(\Delta Q = 0)$, and find $\Delta S = 0$, i.e. the entropy remains unchanged. This is not surprising, since for each infinitesimal step in the process,

$$TdS = \delta Q = 0 \tag{3.5.14}$$

holds.

*3.5.4.2 General Considerations of Real, Reversible Processes

We wish to consider to what extent the situation of a reversible process can indeed be realized in practice. If the process can occur in both directions, what decides in which direction it in fact proceeds? To answer this question, in Fig. 3.8 we consider a process which takes place between the points 1 and 2.

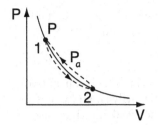
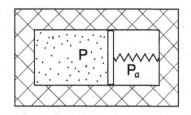

Fig. 3.8. A reversible process. P is the internal pressure of the system (solid line). P_a is the external pressure produced by the spring (dashed line).

The solid curve can be an isotherm or a polytrope (i.e. an equilibrium curve which lies between isotherms and adiabats). Along the path from 1 to 2, the working substance expands, and from 2 to 1, is is compressed again, back to its initial state 1 without leaving any change in the surroundings. At each moment, the pressure within the working substance is precisely compensated by the external pressure (produced here by a spring).

This quasistatic reversible process is, of course, an idealization. In order for the expansion to occur at all, the external pressure P_a^{Ex} must be somewhat lower than P during the expansion phase of the process. The external pressure is indicated in Fig. 3.8 by the dashed curve. This curve, which is supposed to characterize the real course of the process, is drawn in Fig. 3.8 as a dashed line, to indicate that a curve in the $P-V$ diagram cannot fully characterize the system. In the expansion phase with $P_a < P$, the gas near the piston is somewhat rarefied. This effectively reduces its pressure and the work performed by the gas is slightly less than would correspond to its actual pressure. Density gradients occur, i.e. there is a non-equilibrium state. The work obtained (which is stored as potential energy in the spring), $\int_1^2 dV\, P_a^{\text{Ex}}$, then obeys the inequality

$$\int\limits_1^2 dV\ P_a^{\text{Ex}} < \int_1^2 dV\ P < \int_1^2 dV\ P_a^{\text{Com}}\ . \tag{3.5.15}$$

For the compression, we must have $P_a^{\text{Com}} \gtrsim P$. On returning to point 1, the work $-\oint dV P_a = \oint dV P_a$ (which is equal to the area enclosed by the dashed curve) is performed. This work is given up to the heat bath in the form of a heat loss ΔQ_L. [Frictional losses; turbulent motions when the process is too rapid, which also produce heat.]

$$\Delta Q_L = \oint P_a\, dV > \left(\int\limits_1^2 P\, dV + \int\limits_2^1 P\, dV \right) = 0\ . \tag{3.5.16}$$

The inequality results from the fact that $P_a \gtrless P$, that is, the gas and the spring are not in equilibrium. On returning to point 1, the entropy is again

equal to the initial entropy, that is the change in the entropy $\Delta S = 0$. Therefore, the preceding inequality can also be written in the form

$$\Delta Q = -\Delta Q_L \leq T \Delta S , \qquad (3.5.17)$$

where ΔQ is the quantity of heat taken up by the system (which is negative).
These irreversible losses can in principle be made arbitrarily small by moving
the piston very slowly. The reversible process is the ideal limiting case of
extreme slowness.

Analogously, for processes with heat transfer, small temperature differences must be present. In order for the heat bath to give up heat to the
system, it must be slightly warmer; in order for it to take on heat from the system, it must be slightly cooler. After a whole cycle has been passed through,
heat will have been transferred from the warmer a to the cooler b (Fig. 3.9).

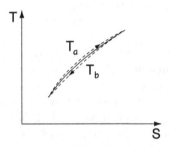

Fig. 3.9. Heat transfer

Strictly reversible processes are in fact not processes which proceed continuously in time, but rather a sequence of equilibrium states. All processes
which occur in practice as continuous variations with time are irreversible;
they contain equilibration processes between perturbed equilibrium states.
In spite of their unrealistic character in a strict sense, reversible processes
play a major role in thermodynamics. While in thermodynamics, statements
about irreversible processes can be made only in the form of inequalities
which determine the direction of the process, for reversible processes one can
make precise predictions, which can also be achieved in practice as limiting
cases. To be sure, thermodynamics can also deliver precise predictions for
irreversible processes, namely for the relation between their initial and final
states, as we have seen for the case of the irreversible adiabatic expansion.

3.5.5 The Adiabatic Equation

Here, we want to first discuss generally the adiabatic equation and then to
apply it to ideal gases. We start from Eq. (3.2.23),

$$\left(\frac{\partial P}{\partial V}\right)_S = \frac{C_P}{C_V}\left(\frac{\partial P}{\partial V}\right)_T , \qquad (3.5.18)$$

and define the ratio of the specific heats:

$$\kappa = \frac{C_P}{C_V} .$$

(3.5.19)

According to (3.3.6), $\kappa > 1$, and therefore for every substance, the slope of the adiabats, $P = P(V, S = \text{const.})$, is steeper than that of the isotherms, $P = P(V, T = \text{const.})$.

For a classical ideal gas, we find $\kappa = \text{const.}$[6] and $\left(\frac{\partial P}{\partial V}\right)_T = -\frac{NkT}{V^2} = -\frac{P}{V}$. It thus follows from (3.5.18)

$$\left(\frac{\partial P}{\partial V}\right)_S = -\kappa\frac{P}{V}$$

(3.5.20)

The solution of this differential equation is

$$PV^\kappa = \text{const} ,$$

and with the aid of the equation of state, we then find

$$TV^{\kappa-1} = \text{const} .$$

(3.5.21)

For a monatomic ideal gas, we have $\kappa = \frac{\frac{3}{2}+1}{\frac{3}{2}} = \frac{5}{3}$, where we have made use of (3.2.27).

3.6 The First and Second Laws of Thermodynamics

3.6.1 The First and the Second Law for Reversible and Irreversible Processes

3.6.1.1 Quasistatic and in Particular Reversible Processes

We recall the formulation of the First and Second Laws of Thermodynamics in Eqns. (3.1.3) and (3.1.5). In the case of reversible transitions between an equilibrium state and a neighboring, infinitesimally close equilibrium state, we have

$$dE = \delta Q - PdV + \mu dN$$

(3.6.1)

with

$$\delta Q = TdS .$$

(3.6.2)

---

[6] This is evident for a monatomic classical ideal gas from (3.2.27). For a molecular ideal gas as treated in Chap. 5, the specific heats are temperature independent only in those temperature regions where particular internal degrees of freedom are completely excited or not excited at all.

Equations (3.6.1) and (3.6.2) are the mathematical formulations of the First
and Second Laws. The Second Law in the form of Eq. (3.6.2) holds for reversible (and thus necessarily quasistatic) processes. It is also valid for quasistatic irreversible processes within those subsystems which are in equilibrium at every instant in time and in which only quasistatic transitions from an
equilibrium state to a neighboring equilibrium state take place. (An example
of this is the thermal equilibration of two bodies via a poor heat conductor
(see Sect. 3.6.3.1). The overall system is not in equilibrium, and the process
is irreversible. However, the equilibration takes place so slowly that the two
bodies within themselves are in equilibrium states at every moment in time).

3.6.1.2 Irreversible Processes

For arbitrary processes, the First Law holds in the form given in Eq. (3.1.3'):

$$dE = \delta Q + \delta W + \delta E_N ,\qquad(3.6.1')$$

where δQ, δW, and δE_N are the quantity of heat transferred, the work
performed on the system, and the increase in energy through addition of
matter.

In order to formulate the Second Law with complete generality, we recall the relation (2.3.4) for the entropy of the microcanonical ensemble and
consider the following situation: we start with two systems 1 and 2 which
are initially separated and are thus not in equilibrium with each other; their
entropies are S_1 and S_2. We now bring these two systems into contact. The
entropy of this nonequilibrium state is

$$S_{\text{initial}} = S_1 + S_2 .\qquad(3.6.3)$$

Suppose the two systems to be insulated from their environment and their
total energy, volume, and particle number to be given by E, V and N. Now the
overall system passes into the microcanonical equilibrium state corresponding
to these macroscopic values. Owing to the additivity of entropy, the total
entropy after equilibrium has been reached is given by

$$S_{1+2}(E,V,N) = S_1(\tilde{E}_1,\tilde{V}_1,\tilde{N}_1) + S_2(\tilde{E}_2,\tilde{V}_2,\tilde{N}_2) ,\qquad(3.6.4)$$

where $\tilde{E}_1,\tilde{V}_1,\tilde{N}_1$ ($\tilde{E}_2,\tilde{V}_2,\tilde{N}_2$) are the most probable values of these quantities in the subsystem 1 (2). Since the equilibrium entropy is a maximum
(Eq. 2.3.5), the following inequality holds:

$$S_1 + S_2 = S_{\text{initial}}\qquad(3.6.5)$$
$$\leq S_{1+2}(E,V,N) = S_1(\tilde{E}_1,\tilde{V}_1,\tilde{N}_1) + S_2(\tilde{E}_2,\tilde{V}_2,\tilde{N}_2) .$$

Whenever the initial density matrix of the combined systems 1+2 is not
already equal to the microcanonical density matrix, the inequality sign holds.

We now apply the inequality (3.6.5) to various physical situations.

(A) Let an isolated system be in a non-equilibrium state. We can decompose it into subsystems which are in equilibrium within themselves and apply the inequality (3.6.5). Then we find for the change ΔS in the total entropy

$$\Delta S > 0 . \tag{3.6.6}$$

This inequality expresses the fact that the entropy of an isolated systems can only increase and is also termed the Clausius principle.

(B) We consider two systems 1 and 2 which are in equilibrium within themselves but are not in equilibrium with each other. Let their entropy changes be denoted by ΔS_1 and ΔS_2. From the inequality (3.6.5), it follows that

$$\Delta S_1 + \Delta S_2 > 0 . \tag{3.6.7}$$

We now assume that system 2 is a heat bath, which is large compared to system 1 and which remains at the temperature T throughout the process. The quantity of heat transferred to system 1 is denoted by ΔQ_1. For system 2, the process occurs quasistatically, so that its entropy change ΔS_2 is related to the heat transferred, $-\Delta Q_1$, by

$$\Delta S_2 = -\frac{1}{T}\Delta Q_1 .$$

Inserting this into Eq. (3.6.7), we find

$$\Delta S_1 > \frac{1}{T}\Delta Q_1 . \tag{3.6.8}$$

In all the preceding relations, the quantities ΔS and ΔQ are by no means required to be small, but instead represent simply the change in the entropy and the quantity of heat transferred.

In the preceding discussion, we have considered the initial state and as final state a state of overall equilibrium. In fact, these inequalities hold also for portions of the relaxation process. Each intermediate step can be represented in terms of equilibrium states with constraints, whereby the limitations imposed by the constraints decrease in the course of time. At the same time, the entropy increases. Thus, for each infinitesimal step in time, the change in entropy of the isolated overall system is given by

$$dS \geq 0 . \tag{3.6.6'}$$

For the physical situation described under B, we have

$$dS_1 \geq \frac{1}{T}\delta Q_1 . \tag{3.6.8'}$$

We now *summarize* the content of the First and Second Laws.

The *First Law*:

$$dE = \delta Q + \delta W + \delta E_N \tag{3.6.9}$$

Change of energy = heat transferred + work performed + energy change due to transfer of matter; E is a state function.

The *Second Law*:

$$\delta Q \leq T dS \tag{3.6.10}$$

and S is a state function.
   a) For reversible changes: $\delta Q = T dS$.
   b) For irreversible changes: $\delta Q < T dS$.

Notes:

(i) The equals sign in Eq. (3.6.10) holds also for irreversible quasistatic processes in those subregions which are in equilibrium in each step of the process (see Sect. 3.6.3.1).
(ii) In (3.6.10), we have combined (3.6.6') and (3.6.8'). The situation of the isolated system (3.6.6) is included in (3.6.10), since in this case $\delta Q = 0$ (see the example 3.6.3.1).
(iii) In many processes, the particle number remains constant ($dN = 0$). Therefore, we often employ (3.6.9) considering only δQ and δW, without mentioning this expressly each time.

We now wish to apply the Second Law to a process which leads from a state A to a state B as indicated in Fig. 3.10. If we integrate (3.6.10), we obtain

$$\int_A^B dS \geq \int_A^B \frac{\delta Q}{T}$$

and from this,

$$S_B - S_A \geq \int_A^B \frac{\delta Q}{T} . \tag{3.6.11}$$

For reversible processes, the equals sign holds; for irreversible ones, the inequality. In a reversible process, the state of the system can be completely characterized at each moment in time by a point in the P–V-diagram. In an irreversible process leading from one equilibrium state (possibly with constraints) A to another equilibrium state B, this is not in general the case. This is indicated by the dashed line in Fig. 3.10.

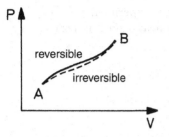

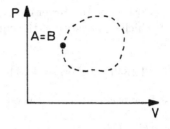

Fig. 3.10. The path of a process connecting two thermodynamic states A and B

Fig. 3.11. A cyclic process, represented by a closed curve in the $P - V$-diagram, which leads back to the starting point ($B = A$), whereby at least to some extent irreversible changes of state occur.

We consider the following special cases:

(i) An adiabatic process: For an adiabatic process ($\delta Q = 0$), it follows from (3.6.11) that

$$S_B \geq S_A \qquad \text{or} \qquad \Delta S \geq 0 . \tag{3.6.11'}$$

The entropy of a thermally isolated system cannot decrease. This statement is more general than Eq. (3.6.6), where completely isolated systems were assumed.

(ii) Cyclic processes: For a cyclic process, the final state is identical with the initial state, B = A (Fig. 3.11). Then we have $S_B = S_A$ and and it follows from Eq. (3.6.11) for a cyclic process that the inequality

$$0 \geq \oint \frac{\delta Q}{T} \tag{3.6.12}$$

holds, where the line integral $\oint$ is calculated along the closed curve of Fig. 3.11, corresponding to the actual direction of the process.

*3.6.2 Historical Formulations of the Laws of Thermodynamics and other Remarks

The First Law

There exists no perpetual motion machine of the first kind (A perpetual motion machine of the first kind refers to a machine which operates periodically and functions only as a source of energy). Energy is conserved and heat is only a particular form of energy, or more precisely, energy transfer. The recognition of the fact that heat is only a form of energy and not a unique material which can penetrate all material bodies was the accomplishment of Julius Robert Mayer (a physician, 1814–1878) in 1842.

James Prescott Joule (a brewer of beer) carried out experiments in the years 1843-1849 which demonstrated the equivalence of heat energy and the energy of work

$$1 \text{ cal} = 4.1840 \times 10^7 \text{ erg} = 4.1840 \text{ Joule} .$$

The First Law was mathematically formulated by Clausius:

$$\delta Q = dE + P dV .$$

The historical formulation quoted above follows from the First Law, which contains the conservation of energy and the statement that E is a state variable. Thus, if a machine has returned to its initial state, its energy must be the same as before and it can therefore not have given up any energy to its environment.

Second Law

Rudolf Clausius (1822–1888) in 1850 : *Heat can never pass on its own from a cooler reservoir to a warmer one.*
William Thomson (Lord Kelvin, 1824–1907) in 1851: *The impossibility of a perpetual motion machine of the second kind.* (A perpetual motion machine of the second kind refers to a periodically operating machine, which *only* extracts heat from a single reservoir and performs work.)
These formulations are equivalent to one another and to the mathematical formulation.

Equivalent formulations of the Second Law.

The existence of a perpetual motion machine of the second kind could be used to remove heat from a reservoir at the temperature T_1. The resulting work could then be used to heat a second reservoir at the higher temperature T_2. The correctness of Clausius' statement thus implies the correctness of Kelvin's statement.

If heat could flow from a colder bath to a warmer one, then one could use this heat in a Carnot cycle (see Sect. 3.7.2) to perform work, whereby part of the heat would once again be taken up by the cooler bath. In this overall process, only heat would be extracted from the cooler bath and work would be performed. One would thus have a perpetual motion machine of the second kind. The correctness of Kelvin's statement thus implies the correctness of Clausius' statement.

The two verbal formulations of the Second Law, that of Clausius and that of Kelvin, are thus equivalent. It remains to be demonstrated that Clausius' statement is equivalent to the differential form of the Second Law (Eq. 3.6.10). To this end, we note that it will be shown in Sect. 3.6.3.1 from (3.6.10) that heat passes from a warmer reservoir to a cooler one. Clausius' statement follows from (3.6.10). Now we must only demonstrate that the relation (3.6.10) follows from Clausius' statement. This can be seen as follows: if instead of (3.6.10), conversely $T dS < \delta Q$ would hold, then it would follow form the consideration of the quasistatic temperature equilibration that heat would be transported from a cooler to a warmer bath; i.e. that Clausius' statement is false. The correctness of Clausius' statement thus implies the correctness of the mathematical formulation of the Second Law (3.6.10).

All the formulations of the Second Law are equivalent. We have included these historical considerations here because precisely their verbal formulations show the connection to everyday consequences of the Second Law and because this type of reasoning is typical of thermodynamics.

The Zeroth Law

When two systems are in thermal equilibrium with a third system, then they are in equilibrium with one another.
Proof within statistical mechanics:
Systems 1, 2, and 3. Equilibrium of 1 with 3 implies that $T_1 = T_3$ and that of 2 with 3 that $T_2 = T_3$; it follows from this that $T_1 = T_2$, i.e. 1 and 2 are also in equilibrium with one another. The considerations for the pressure and the chemical potential are exactly analogous.
This fact is of course very important in practice, since it makes it possible to determine with the aid of thermometers and manometers whether two bodies are at the same temperature and pressure and will remain in equilibrium or not if they are brought into contact.

The Third Law

The Third Law (also called Nernst's theorem) makes statements about the temperature dependence of thermodynamic quantities in the limit $T \to 0$; it is discussed in the Appendix A.1. Its consequences are not as far-reaching as those of the First and Second Laws. The vanishing of specific heats as $T \to 0$ is a direct result of quantum mechanics. In this sense, its postulation in the era of classical physics can be regarded as visionary.

3.6.3 Examples and Supplements to the Second Law

We now give a series of examples which clarify the preceding concepts and general results, and which have also practical significance.

3.6.3.1 Quasistatic Temperature Equilibration

We consider two bodies at the temperatures T_1 and T_2 and with entropies S_1 and S_2. These two bodies are connected by a poor thermal conductor and are insulated from their environment (Fig. 3.12). The two temperatures are different: $T_1 \neq T_2$; thus, the two bodies are not in equilibrium with each other. Since the thermal conductor has a poor conductivity, all energy transfers occur slowly and each subsystem is in thermal equilibrium at each moment in time. Therefore, for a heat input δQ to body 1 and thus the equal but opposite heat transfer $-\delta Q$ from body 2, the Second Law applies to both subsystems in the form

$$dS_1 = \frac{\delta Q}{T_1} , \qquad dS_2 = -\frac{\delta Q}{T_2} . \qquad (3.6.13)$$

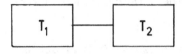

Fig. 3.12. Quasistatic temperature equilibration of two bodies connected by a poor conductor of heat

For the overall system, we have

$$dS_1 + dS_2 > 0 , \qquad (3.6.14)$$

since the total entropy increases during the transition to the equilibrium state. If we insert (3.6.13) into (3.6.14), we obtain

$$\delta Q \left(\frac{1}{T_1} - \frac{1}{T_2} \right) > 0 . \qquad (3.6.15)$$

We take $T_2 > T_1$; then it follows from (3.6.13) that $\delta Q > 0$, i.e. heat is transferred *from the warmer to the cooler container*. We consider here the differential substeps, since the temperatures change in the course of the process. The transfer of heat continues until the two temperatures have equalized; the total amount of heat transferred from 2 to 1, $\int \delta Q$, is positive.

Also in the case of a *non-quasistatic* temperature equilibration, heat is transferred from the warmer to the cooler body: if the two bodies mentioned above are brought into contact (again, of course, isolated from their environment, but without the barrier of a poor heat conductor), the final state is the same as in the case of the quasistatic process. Thus also in the non-quasistatic temperature equilibration, heat has passed from the warmer to the cooler body.

3.6.3.2 The Joule–Thomson Process

The Joule–Thomson process consists of the controlled expansion of a gas (cf. Fig. 3.13). Here, the stream of expanding gas is limited by a throttle valve. The gas volume is bounded to the left and the right of the throttle by the two sliding pistons S_1 and S_2, which produce the pressures P_1 and P_2 in the left and right chambers, with $P_1 > P_2$. The process is assumed to occur adiabatically, i.e. $\delta Q = 0$ during the entire process.

In the initial state (1), the gas in the left-hand chamber has the volume V_1 and the energy E_1. In the final state, the gas is entirely in the right-hand

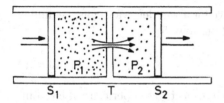

Fig. 3.13. A Joule–Thomson process, showing the sliding pistons S_1 and S_2 and the throttle valve T

chamber and has a volume V_2 and energy E_2. The left piston performs work on the gas, while the gas performs work on the right piston and thus on the environment. The difference of the internal energies is equal to the total work performed on the system:

$$E_2 - E_1 = \int_1^2 dE = \int_1^2 \delta W = \int_{V_1}^0 dV_1(-P_1) + \int_0^{V_2} dV_2(-P_2)$$

$$= P_1 V_1 - P_2 V_2 \ .$$

From this it follows that the enthalpy remains constant in the course of this process:

$$H_2 = H_1 \ , \tag{3.6.16}$$

where the definition $H_i = E_i + P_i V_i$ was used.

For cryogenic engineering it is important to know whether the gas is cooled by the controlled expansion. This is determined by the *Joule–Thomson coefficient*:

$$\left(\frac{\partial T}{\partial P} \right)_H = -\frac{\left(\frac{\partial H}{\partial P} \right)_T}{\left(\frac{\partial H}{\partial T} \right)_P} = -\frac{T \left(\frac{\partial S}{\partial P} \right)_T + V}{T \left(\frac{\partial S}{\partial T} \right)_P} = \frac{T \left(\frac{\partial V}{\partial T} \right)_P - V}{C_P} \ .$$

In the rearrangement, we have used (3.2.21), $dH = T dS + V dP$, and the Maxwell relation (3.2.10). Inserting the thermal expansion coefficient α, we find the following expression for the Joule–Thomson coefficient:

$$\left(\frac{\partial T}{\partial P} \right)_H = \frac{V}{C_P} (T\alpha - 1) \ . \tag{3.6.17}$$

For an ideal gas, $\alpha = \frac{1}{T}$; in this case, there is no change in the temperature on expansion. For a real gas, either cooling or warming can occur. When $\alpha > \frac{1}{T}$, the expansion leads to a cooling of the gas (positive Joule–Thomson effect). When $\alpha < \frac{1}{T}$, then the expansion gives rise to a warming (negative Joule–Thomson effect). The limit between these two effects is defined by the *inversion curve*, which is given by

$$\alpha = \frac{1}{T} \ . \tag{3.6.18}$$

We shall now calculate the inversion curve for a van der Waals gas, beginning with the *van der Waals equation of state* (Chap. 5)

$$P = \frac{kT}{v-b} - \frac{a}{v^2} \quad , \quad v = \frac{V}{N} \, . \tag{3.6.19}$$

We differentiate the equation of state with respect to temperature at constant pressure

$$0 = \frac{k}{v-b} - \frac{kT}{(v-b)^2} \left(\frac{\partial v}{\partial T} \right)_P + \frac{2a}{v^3} \left(\frac{\partial v}{\partial T} \right)_P \, .$$

In this expression, we insert the condition (3.6.18)

$$\alpha \equiv \frac{1}{v} \left(\frac{\partial v}{\partial T} \right)_P = \frac{1}{T}$$

for $\left(\frac{\partial v}{\partial T} \right)_P$ and thereby obtain $0 = \frac{k}{v} - \frac{k}{v-b} + \frac{2a}{v^3} \frac{1}{T} (v-b)$. Using the van-der-Waals equation again, we finally find for the inversion curve

$$0 = -\frac{b}{v} \left(P + \frac{a}{v^2} \right) + \frac{2a}{v^3} (v-b) \, ,$$

that is

$$P = \frac{2a}{bv} - \frac{3a}{v^2} \quad . \tag{3.6.20}$$

In the limit of low density, we can neglect the second term in (3.6.20) and the inversion curve is then given by

$$P = \frac{2a}{bv} = \frac{kT_{inv}}{v} \quad , \quad T_{inv} = \frac{2a}{bk} = 6.75 \, T_c \, . \tag{3.6.21}$$

Here, T_c is the critical temperature which follows from the van der Waals equation (5.4.13). For temperatures which are higher than the inversion temperature T_{inv}, the Joule–Thomson effect is always negative. The inversion temperature and other data for some gases are listed in Table I.4 in the Appendix.

The change in entropy in the Joule–Thomson process is determined by

$$\left(\frac{\partial S}{\partial P} \right)_H = -\frac{V}{T} \, , \tag{3.6.22}$$

as can be seen using $dH = TdS + VdP = 0$. Since the pressure decreases, we obtain for the entropy change $dS > 0$, although $\delta Q = 0$. The Joule–Thomson process is *irreversible*, since its initial state with differing pressures in the two chambers is clearly not an equilibrium state.

The complete inversion curve from the van der Waals theory is shown in Fig. 3.14a,b. Within the inversion curve, the expansion leads to cooling of the gas.

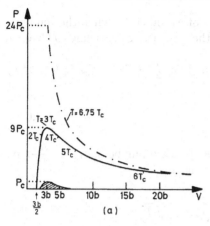

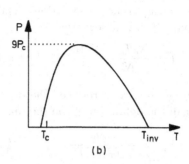

(a) The inversion curve for the Joule–Thomson effect (upper solid curve). The isotherm is for $T = 6.75\,T_c$ (dot-dashed curve). The shaded region is excluded, since in this region, the vapor and liquid phases are always both present.

(b) The inversion curve in the P-T diagram.

Fig. 3.14. The inversion curve for the Joule–Thomson effect

3.6.3.3 Temperature Equilibration of Ideal Gases

We will now investigate the thermal equilibration of two monatomic ideal gases (a and b). Suppose the two gases to be separated by a sliding piston and insulated from their environment (Fig. 3.15).

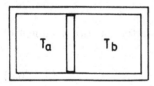

Fig. 3.15. The thermal equilibration of two ideal gases

The pressure of the two gases is taken to be equal, $P_a = P_b = P$, while their temperatures are different in the initial state, $T_a \neq T_b$. Their volumes and particle numbers are given by V_a, V_b and N_a, N_b, so that the total volume and total particle number are $V = V_a + V_b$ and $N = N_a + N_b$. The entropy of the initial state is given by

$$S = S_a + S_b = k \left\{ N_a \left(\frac{5}{2} + \log \frac{V_a}{N_a \lambda_a^3} \right) + N_b \left(\frac{5}{2} + \log \frac{V_b}{N_b \lambda_b^3} \right) \right\} . \quad (3.6.23)$$

The temperature after the establishment of equilibrium, when the temperatures of the two systems must approach the same value according to Chap. 2, will be denoted by T.

Owing to the conservation of energy, we have $\frac{3}{2}NkT = \frac{3}{2}N_akT_a + \frac{3}{2}N_bkT_b$, from which it follows that

$$T = \frac{N_aT_a + N_bT_b}{N_a + N_b} = c_aT_a + c_bT_b ,\tag{3.6.24}$$

where we have introduced the ratio of the particle numbers, $c_{a,b} = \frac{N_{a,b}}{N}$. We recall the definition of the thermal wavelengths

$$\lambda_{a,b} = \frac{h}{\sqrt{2\pi m_{a,b}kT_{a,b}}} , \qquad \lambda'_{a,b} = \frac{h}{\sqrt{2\pi m_{a,b}kT}} .$$

The entropy after the establishment of equilibrium is

$$S' = kN_a\left\{\frac{5}{2} + \log\frac{V_a'}{N_a\lambda_a'^3}\right\} + kN_b\left\{\frac{5}{2} + \log\frac{V_b'}{N_b\lambda_b'^3}\right\} ,$$

so that for the entropy increase, we find

$$S' - S = kN_a\log\frac{V_a'\lambda_a^3}{V_a\lambda_a'^3} + kN_b\log\frac{V_b'\lambda_b^3}{V_b\lambda_b'^3} .\tag{3.6.25}$$

We shall also show that the pressure remains unchanged. To this end, we add the two equations of state of the subsystems before the establishment of thermal equilibrium

$$V_aP = N_akT_a , \qquad V_bP = N_bkT_b \tag{3.6.26a}$$

and obtain using (3.6.24) the expression

$$(V_a + V_b)P = (N_a + N_b)kT .\tag{3.6.26b}$$

From the equations of state of the two subsystems after the establishment of equilibrium

$$V'_{a,b}P' = N_{a,b}kT \tag{3.6.26a'}$$

with $V_a' + V_b' = V$, it follows that

$$VP' = (N_a + N_b)kT ,\tag{3.6.26b'}$$

i.e. $P' = P$. Incidentally, in (3.6.24) and (3.6.26b'), the fact is used that the two monatomic gases have the same specific heat. Comparing (3.6.26b) and (3.6.26b'), we find the volume ratios

$$\frac{V'_{a,b}}{V_{a,b}} = \frac{T}{T_{a,b}} .$$

From this we obtain

$$S' - S = \frac{5}{2}k \log \frac{T^{N_a+N_b}}{T_a^{N_a}T_b^{N_b}} ,$$

which finally yields

$$S' - S = \frac{5}{2}kN \log \frac{T}{T_a^{c_a}T_b^{c_b}} = \frac{5}{2}kN \log \frac{c_aT_a + c_bT_b}{T_a^{c_a}T_b^{c_b}} . \tag{3.6.27}$$

Due to the convexity of the exponential function, we have

$$T_a^{c_a}T_b^{c_b} = \exp(c_a \log T_a + c_b \log T_b) \leq c_a \exp \log T_a + c_b \exp \log T_b$$
$$= c_aT_a + c_bT_b = T ,$$

and thus it follows from (3.6.27) that $S' - S \geq 0$, i.e. the entropy increases on thermal equilibration.

Note:

Following the equalization of temperatures, in which heat flows from the warmer to the cooler parts of the system, the volumes are given by:

$$V_a' = \frac{N_a}{N_a + N_b}V , \qquad V_b' = \frac{N_b}{N_a + N_b}V .$$

Together with Eq. (3.6.26b), this gives $V_a'/V_a = T/T_a$ and $V_b'/V_b = T/T_b$. The energy which is put into subsystem a is $\Delta E_a = \frac{3}{2}N_ak(T - T_a)$.
The enthalpy increase in subsystem a is given by $\Delta H_a = \frac{5}{2}N_ak(T - T_a)$. Since the process is isobaric, we have $\Delta Q_a = \Delta H_a$. The work performed on subsystem a is therefore equal to

$$\Delta W_a = \Delta E_a - \Delta Q_a = -N_ak(T - T_a) .$$

The warmer subsystem gives up heat. Since it would then be too rarefied for the pressure P, it will be compressed, i.e. it takes on energy through the work performed in this compression.

3.6.3.4 Entropy of Mixing

We now consider the process of mixing of two different ideal gases with the masses m_a and m_b.
The temperatures and pressures of the gases are taken to be the same,

$$T_a = T_b = T , \qquad P_a = P_b = P .$$

From the equations of state,

$$V_aP = N_akT , \qquad V_bP = N_bkT$$

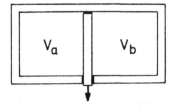

Fig. 3.16. The mixing of two gases

it follows that

$$\frac{N_a}{V_a} = \frac{N_b}{V_b} = \frac{N_a + N_b}{V_a + V_b} .$$

Using the thermal wavelength $\lambda_{a,b} = \dfrac{h}{\sqrt{2\pi m_{a,b} kT}}$, the entropy when the gases are separated by a partition is given by

$$S = S_a + S_b = k \left\{ N_a \left(\frac{5}{2} + \log \frac{V_a}{N_a \lambda_a^3} \right) + N_b \left(\frac{5}{2} + \log \frac{V_b}{N_b \lambda_b^3} \right) \right\} . \quad (3.6.28)$$

After removal of the partition and mixing of the gases, the value of the entropy is

$$S' = k \left\{ N_a \left(\frac{5}{2} + \log \frac{V_a + V_b}{N_a \lambda_a^3} \right) + N_b \left(\frac{5}{2} + \log \frac{V_a + V_b}{N_b \lambda_b^3} \right) \right\} . \quad (3.6.29)$$

From Eqns. (3.6.28) and (3.6.29), we obtain the difference in the entropies:

$$S' - S = k \log \frac{(N_a + N_b)^{N_a + N_b}}{N_a^{N_a} N_b^{N_b}} = k(N_a + N_b) \log \left(\frac{1}{c_a^{c_a} c_b^{c_b}} \right) > 0 ,$$

where we have used the relative particle numbers

$$c_{a,b} = \frac{N_{a,b}}{N_a + N_b} .$$

Since the argument of the logarithm is greater than 1, we find that the entropy of mixing is positive,

e.g. $N_a = N_b$, $S' - S = 2 k N_a \log 2 .$

The entropy of mixing always occurs when different gases interdiffuse, even when they consist of different isotopes of the same element. When, in contrast, the gases a and b are identical, the value of the entropy on removing the partition is

$$S'_{id} = k(N_a + N_b) \left\{ \frac{5}{2} + \log \frac{V_a + V_b}{(N_a + N_b)\lambda^3} \right\} \qquad (3.6.29')$$

and $\lambda = \lambda_a = \lambda_b$. We then have

$$S'_{id} - S = k \log \frac{(V_a + V_b)^{N_a+N_b} \, N_a^{N_a} N_b^{N_b}}{(N_a + N_b)^{N_a+N_b} \, V_a^{N_a} V_b^{N_b}} = 0$$

making use of the equation of state; therefore, no entropy of mixing occurs. This is due to the factor $1/N!$ in the basic phase-space volume element in Eqns. (2.2.2) and (2.2.3), which results from the indistinguishability of the particles. Without this factor, Gibbs' paradox would occur, i.e. we would find a positive entropy of mixing for identical gases, as mentioned following Eq. (2.2.3).

*3.6.3.5 Heating a Room

Finally, we consider an example, based on one given by Sommerfeld.[7] A room is to be heated from 0°C to 20°C. What quantity of heat is required? How does the energy content of the room change in the process?

If air can leave the room through leaks around the windows, for example, then the process is isobaric, but the number of air molecules in the room will decrease in the course of the heating process. The quantity of heat required depends on the increase in temperature through the relation

$$\delta Q = C_P dT , \qquad (3.6.30)$$

where C_P is the heat capacity at constant pressure. In the temperature range that we are considering, the rotational degrees of freedom of oxygen, O_2, and nitrogen, N_2, are excited (see Chap. 5), so that under the assumption that air is an ideal gas, we have

$$C_P = \frac{7}{2} Nk , \qquad (3.6.31)$$

where N is the overall number of particles.

The total amount of heat required is found by integrating (3.6.31) between the initial and final temperatures, T_1 and T_2:

$$Q = \int_{T_1}^{T_2} dT \, C_P . \qquad (3.6.32)$$

If we initially neglect the temperature dependence of the particle number, and thus the heat capacity (3.6.31), we find

$$Q = C_P(T_2 - T_1) = \frac{7}{2} N_1 k(T_2 - T_1) . \qquad (3.6.32')$$

---

[7] A. Sommerfeld, *Thermodynamics and Statistical Mechanics: Lectures on Theoretical Physics*, Vol. V, (Academic Press, New York, 1956)

Here, we have denoted the particle number at T_1 as N_1 and taken it to be constant. Equation (3.6.32') will be a good approximation, as long as $T_2 \approx T_1$.

If we wish to take into account the variation of the particle number within the room (volume V), we have to replace N in Eq. (3.6.31) by N from the equation of state, $N = PV/kT$, and it follows that

$$Q = \int_{T_1}^{T_2} dT \, \frac{7}{2} \frac{PV}{T} = \frac{7}{2} PV \log \frac{T_2}{T_1} = \frac{7}{2} N_1 kT_1 \log \frac{T_2}{T_1} \,. \tag{3.6.33}$$

With $\log \frac{T_2}{T_1} = \frac{T_2}{T_1} - 1 + \mathcal{O}\left(\left(\frac{T_2}{T_1} - 1\right)^2\right)$, we obtain from (3.6.33) for small temperature differences the approximate formula (3.6.32')

$$Q = \frac{7}{2} PV \frac{T_2 - T_1}{T_1} = 3.5 \left(10^6 \frac{\mathrm{dyn}}{\mathrm{cm}^2}\right) 10^6 (V \mathrm{m}^3) \frac{20}{273} = \frac{3.5 \times 2}{2.73} 10^{11} \mathrm{erg} \; (V \mathrm{m}^3)$$

$$= 6 \; \mathrm{kcal} \; (V \mathrm{m}^3).$$

It is instructive to compute the change in the energy content of the room on heating, taking into account the fact that the rotational degrees of freedom are fully excited, $T \gg \Theta_\mathrm{r}$ (see Chap. 5). Then the internal energy before and after the heating procedure is

$$E_i = \frac{5}{2} N_i kT_i - N_i k\Theta_\mathrm{r} \frac{1}{6} + N_i \varepsilon_\mathrm{el}$$

$$E_2 - E_1 = \frac{5}{2} k(N_2 T_2 - N_1 T_1) - \frac{1}{6} PV \Theta_\mathrm{r} \left(\frac{1}{T_2} - \frac{1}{T_1}\right) + PV \frac{\varepsilon_\mathrm{el}}{k} \left(\frac{1}{T_2} - \frac{1}{T_1}\right) \,.$$

$$\tag{3.6.34}$$

The first term is exactly zero, and the second one is positive; the third, dominant term is negative. The internal energy of the room actually decreases upon heating. The heat input is given up to the outside world, in order to increase the temperature in the room and thus the average kinetic energy of the remaining gas molecules.

Heating with a fixed particle number (a hermetically sealed room) requires a quantity of heat $Q = C_V(T_2 - T_1) \equiv \frac{5}{2} N_1 k(T_2 - T_1)$. For small temperature differences $T_2 - T_1$, it is then more favorable first to heat the room to the final temperature T_2 and then to allow the pressure to decrease. The point of intersection of the two curves (P, N) constant and P constant, with N variable (Fig. 3.17) at T_2^0 is determined by

$$\frac{T_2^0 - T_1}{T_1 \log \frac{T_2^0}{T_1}} = \frac{C_P}{C_V} \,.$$

A numerical estimate yields $T_2^0 = 1.9 \, T_1$ for the point of intersection in Fig. 3.17, i.e. at $T_1 = 273 \, \mathrm{K}$, $T_2^0 = 519 \, \mathrm{K}$.

For any process of space heating, isolated heating is more favorable. The difference in the quantities of heat required is

$$\Delta Q \approx (C_P - C_V)(T_2 - T_1) = \frac{1}{3.5} 6 \; \mathrm{kcal} \; (V \mathrm{m}^3) = 1.7 \; \mathrm{kcal} \; (V \mathrm{m}^3) \,.$$

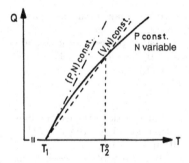

Fig. 3.17. The quantity of heat required for space heating: as an isobaric process (solid curve), isochore (dashed curve), or isobaric neglecting the decrease in particle number (dot-dashed curve).

All of the above considerations have neglected the heat capacity of the walls. They are applicable to a rapid heating of the air.

The change in pressure on heating a fixed amount of air by $20°C$ is, however,

$$\frac{\delta P}{P} = \frac{\delta T}{T} \sim \frac{20}{273} \sim 0.07 \; , \; \text{i.e. } \delta P \sim 0.07 \text{ bar} \sim 0.07 \text{ kg/cm}^2 \sim 700 \text{ kg/m}^2 \; !$$

*3.6.3.6 The Irreversible, Quasistatic Gay-Lussac Experiment

We recall the different versions of the Gay-Lussac experiment. In the irreversible form, we have $\Delta Q = 0$ and $\Delta S > 0$ (3.5.1). In the reversible case (isothermal or adiabatic), using (3.5.9) and (3.5.14), the corresponding relation for reversible processes is fulfilled.

It is instructive to carry out the Gay-Lussac experiment in a quasistatic, irreversible fashion. One can imagine that the expansion does not take place suddenly, but instead is slowed by friction of the piston to the point that the gas always remains in equilibrium. The frictional heat can then either be returned to the gas or given up to the environment. We begin by treating the first possibility. Since the frictional heat from the piston is returned to the gas, there is no change in the environment after each step in the process. The final result corresponds to the situation of the usual Gay-Lussac experiment. For the moment, we denote the gas by an index 1 and the piston, which initially takes up the frictional heat, by 2. Then the work which the gas performs on expansion by the volume change dV is given by

$$\delta W_{1 \to 2} = PdV \; .$$

This quantity of energy is passed by the piston to 1:

$$\delta Q_{2 \to 1} = \delta W_{1 \to 2} \; .$$

The energy change of the gas is $dE = \delta Q_{2 \to 1} - \delta W_{1 \to 2} = 0$. Since the gas is always in equilibrium at each instant, the relation $dE = TdS - PdV$ also holds and thus we have for the entropy increase of the gas:

$$TdS = \delta Q_{2 \to 1} > 0 \; .$$

The overall system of gas + piston transfers no heat to the environment and also performs no work on the environment, i.e. $\delta Q = 0$ and $\delta W = 0$. Since the entropy of the piston remains the same (for simplicity, we consider an ideal gas, whose temperature does not change), it follows that $TdS > \delta Q$.

Now we consider the situation that the frictional heat is passed to the outside world. This means that $\delta Q_{2\to1} = 0$ and thus $TdS = 0$, also $dS = 0$. The total amount of heat given off to the environment (heat loss δQ_L) is

$$\delta Q_L = \delta W_{1\to2} > 0\;.$$

Here, again, the inequality $-\delta Q_L < TdS$ is fulfilled, characteristic of the irreversible process. The final state of the gas corresponds to that found for the reversible adiabatic process. There, we found $\Delta S = 0$, $Q = 0$, and $W > 0$. Now, $\Delta S = 0$, while $Q_L > 0$ and is equal to the W of the adiabatic, reversible process, from Eq. (3.5.12).

3.6.4 Extremal Properties

In this section, we derive the extremal properties of the thermodynamic potentials. From these, we shall obtain the equilibrium conditions for multicomponent systems in various phases and then again the inequalities (3.3.5) and (3.3.6).

We assume in this section that no particle exchange with the environment occurs, i.e. $dN_i = 0$, apart from chemical reactions within the system. Consider the system in general not yet to be in equilibrium; then for example in an isolated system, the state is not characterized solely by E, V, and N_i , but instead we need additional quantities x_α, which give e.g. the concentrations of the independent components in the different phases or the concentrations of the components between which chemical reactions occur. Another situation not in equilibrium is that of spatial inhomogeneities.[8]

We now however assume that equilibrium with respect to the temperature and pressure is present, i.e. that the system is characterized by uniform (but variable) T and P values. This assumption may be relaxed somewhat. For the following derivation, it suffices that the system likewise be at the pressure P at the stage when work is being performed by the pressure P, and when it is exchanging heat with a reservoir at the temperature T, that it be at the temperature T. (This permits e.g. inhomogeneous temperature distributions during a chemical reaction in a subsystem.) Under these conditions, the First Law, Eq (3.6.9), is given by $dE = \delta Q - PdV$.

---

[8] As an example, one could imagine a piece of ice and a solution of salt in water at $P = 1$ atm and $-5°$C. Each component of this system is in equilibrium within itself. If one brings them into contact, then a certain amount of the ice will melt and some of the NaCl will diffuse into the ice until the concentrations are such that the ice and the solution are in equilibrium (see the section on eutectics). The initial state described here – a non-equilibrium state – is a typical example of an inhibited equilibrium. As long as barriers impede (inhibit) particle exchange, i.e. so long as only energy and volume changes are possible, this inhomogeneous state can be described in terms of equilibrium thermodynamics.

Our starting point is the Second Law, (3.6.10):

$$dS \geq \frac{\delta Q}{T} . \tag{3.6.35}$$

We insert the First Law into this equation and obtain

$$dS \geq \frac{1}{T}(dE + PdV) . \tag{3.6.36a}$$

We have used the principle of energy conservation from equilibrium thermodynamics here, which however also holds in non-equilibrium states. The change in the energy is equal to the heat transferred plus the work performed. The precondition is that during the process a particular, well-defined pressure is present.

If E, V are held constant, then according to Eq. (3.6.36a), we have

$$dS \geq 0 \qquad \text{for } E, V \text{ fixed} ; \tag{3.6.36b}$$

that is, an isolated system tends towards a maximum of the entropy. When a non-equilibrium state is characterized by a parameter x, its entropy has the form indicated in Fig. 3.18. It is maximal for the equilibrium value x_0. The parameter x could be e.g. the volume of the energy of a subsystem of the isolated system considered.

One refers to a process or variation as *virtual* – that is, possible in principle – if it is permitted by the conditions of a system. An inhomogeneous distribution of the energies of the subsystems with constant total energy would, to be sure, not occur spontaneously, but it is possible. In equilibrium, the entropy is maximal with respect to all virtual processes.

We now consider the free enthalpy or Gibbs' free energy,

$$G = E - TS + PV , \tag{3.6.37}$$

which we define with Eq. (3.6.37) for non-equilibrium states just as for equilibrium states. For the changes in such states, we find from (3.6.36a) that the inequality

$$dG \leq -SdT + VdP \tag{3.6.38a}$$

holds. For the case that T and P are held constant, it follows from (3.6.38a) that

$$dG \leq 0 \qquad \text{for } T \text{ and } P \text{ fixed,} \tag{3.6.38b}$$

i.e. the Gibbs' free energy G tends towards a minimum. In the neighborhood of the minimum (Fig. 3.19), we have for a virtual (in thought only) variation

$$\delta G = G(x_0 + \delta x) - G(x_0) = \frac{1}{2}G''(x_0)(\delta x)^2 . \tag{3.6.39}$$

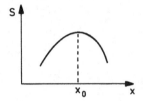

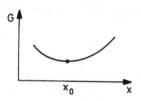

Fig. 3.18. The entropy as a function of a parameter x, with the equilibrium value x_0.

Fig. 3.19. The free enthalpy as a function of a parameter.

The first-order terms vanish, therefore in first order we find for δx:

$$\delta G = 0 \qquad \text{for } T \text{ and } P \text{ fixed.}^9 \tag{3.6.38c}$$

One terms this condition *stationarity*. Since G is minimal at x_0, we find

$$G''(x_0) > 0 . \tag{3.6.40}$$

Analogously, one can show for the free energy (Helmholtz free energy) $F = E - TS$ and for the enthalpy $H = E + PV$ that:

$$dF \leq -SdT - PdV \tag{3.6.41a}$$

and

$$dH \leq TdS + VdP . \tag{3.6.42a}$$

These potentials also tend towards minimum values at equilibrium under the condition that their natural variables are held constant:

$$dF \leq 0 \qquad \text{for } T \text{ and } V \text{ fixed} \tag{3.6.41b}$$

and

$$dH \leq 0 \qquad \text{for } S \text{ and } P \text{ fixed} . \tag{3.6.42b}$$

As conditions for equilibrium, it then follows that

$$\delta F = 0 \qquad \text{for } T \text{ and } V \text{ fixed} \tag{3.6.41c}$$

and

$$\delta H = 0 \qquad \text{for } S \text{ and } P \text{ fixed} . \tag{3.6.42c}$$

---

[9] This condition plays an important role in physical chemistry, since in chemical processes, the pressure and the temperature are usually fixed.

*3.6.5 Thermodynamic Inequalities Derived from Maximization of the Entropy

We consider a system whose energy is E and whose volume is V. We decompose this system into two equal parts and investigate a virtual change of the energy and the volume of subsystem 1 by δE_1 and δV_1. Correspondingly, the values for subsystem 2 change by $-\delta E_1$ and $-\delta V_1$. The overall entropy before the change is

$$S(E, V) = S_1 \left(\frac{E}{2}, \frac{V}{2} \right) + S_2 \left(\frac{E}{2}, \frac{V}{2} \right) . \tag{3.6.43}$$

Therefore, the change of the entropy is given by

$$
\begin{aligned}
\delta S = {} & S_1 \left(\frac{E}{2} + \delta E_1, \frac{V}{2} + \delta V_1 \right) + S_2 \left(\frac{E}{2} - \delta E_1, \frac{V}{2} - \delta V_1 \right) - S(E, V) \\
= {} & \left(\frac{\partial S_1}{\partial E_1} - \frac{\partial S_2}{\partial E_2} \right) \delta E_1 + \left(\frac{\partial S_1}{\partial V_1} - \frac{\partial S_2}{\partial V_2} \right) \delta V_1 \\
& + \frac{1}{2} \left(\frac{\partial^2 S_1}{\partial E_1^2} + \frac{\partial^2 S_2}{\partial E_2^2} \right) (\delta E_1)^2 + \frac{1}{2} \left(\frac{\partial^2 S_1}{\partial V_1^2} + \frac{\partial^2 S_2}{\partial V_2^2} \right) (\delta V_1)^2 \\
& + \left(\frac{\partial^2 S_1}{\partial E_1 \partial V_1} + \frac{\partial^2 S_2}{\partial E_2 \partial V_2} \right) \delta E_1 \delta V_1 + \dots
\end{aligned}
$$

$$\tag{3.6.44}$$

From the stationarity of the entropy, $\delta S = 0$, it follows that the terms which are linear in δE_1 and δV_1 must vanish. This means that in equilibrium the temperature T and the pressure P of the subsystems must be equal

$$T_1 = T_2 \ , \quad P_1 = P_2 \ ; \tag{3.6.45a}$$

this is a result that is already familiar to us from equilibrium statistics.

If we permit also virtual variations of the particle numbers, δN_1 and $-\delta N_1$, in the subsystems 1 and 2, then an additional term enters the second line of (3.6.44): $\left(\frac{\partial S_1}{\partial N_1} - \frac{\partial S_2}{\partial N_2} \right) \delta N_1$; and one obtains as an additional condition for equilibrium the equality of the chemical potentials:

$$\mu_1 = \mu_2 . \tag{3.6.45b}$$

Here, the two subsystems could also consist of different phases (e.g. solid and liquid).

We note that the second derivatives of S_1 and S_2 in (3.6.44) are both to be taken at the values $E/2$, $V/2$ and they are therefore equal. In the equilibrium state, the entropy is maximal, according to (3.6.36b). From this it follows that the coefficients of the quadratic form (3.6.44) obey the two conditions

$$\frac{\partial^2 S_1}{\partial E_1^2} = \frac{\partial^2 S_2}{\partial E_2^2} \leq 0 \tag{3.6.46a}$$

and

$$\frac{\partial^2 S_1}{\partial E_1^2}\frac{\partial^2 S_1}{\partial V_1^2} - \left(\frac{\partial^2 S_1}{\partial E_1 \partial V_1}\right)^2 \geq 0 . \tag{3.6.46b}$$

We now leave off the index 1 and rearrange the left side of the first condition:

$$\frac{\partial^2 S}{\partial E^2} = \left(\frac{\partial \frac{1}{T}}{\partial E}\right)_V = -\frac{1}{T^2 C_V} . \tag{3.6.47a}$$

The left side of the second condition, Eq. (3.6.46b), can be represented by a Jacobian, and after rearrangement,

$$\frac{\partial \left(\frac{\partial S}{\partial E}, \frac{\partial S}{\partial V}\right)}{\partial (E,V)} = \frac{\partial \left(\frac{1}{T}, \frac{P}{T}\right)}{\partial (E,V)} = \frac{\partial \left(\frac{1}{T}, \frac{P}{T}\right)}{\partial (T,V)}\frac{\partial (T,V)}{\partial (E,V)}$$

$$= -\frac{1}{T^3}\left(\frac{\partial P}{\partial V}\right)_T \frac{1}{C_V} = \frac{1}{T^3 V \kappa_T C_V} . \tag{3.6.47b}$$

If we insert the expressions (3.6.47a,b) into the inequalities (3.6.46a) and (3.6.46b), we obtain

$$C_V \geq 0 , \quad \kappa_T \geq 0 , \tag{3.6.48a,b}$$

which expresses the *stability* of the system. When heat is given up, the system becomes cooler. On compression, the pressure increases.

Stability conditions of the type of (3.6.48a,b) are expressions of *Le Chatelier's principle*: When a system is in a stable equilibrium state, every spontaneous change in its parameter leads to reactions which drive the system back towards equilibrium.

The inequalities (3.6.48a,b) were already derived in Sect. 3.3 on the basis of the positivity of the mean square deviations of the particle number and the energy. The preceding derivation relates them within thermodynamics to the stationarity of the entropy. The inequality $C_V \geq 0$ guarantees *thermal stability*. If heat is transferred to part of a system, then its temperature increases and it releases heat to its surroundings, thus again decreasing its temperature. If its specific heat were negative, then the temperature of the subsystem would decrease on input of heat, and more heat would flow in from its surroundings, leading to a further temperature decrease. The least input of heat would set off an instability. The inequality $\kappa_T \geq 0$ guarantees *mechanical stability*. A small expansion of the volume of a region results in a decrease in its pressure, so that the surroundings, at higher pressure, compress the region again. If however $\kappa_T < 0$, then the pressure would increase in the region and the volume element would continue to expand.

3.7 Cyclic Processes

The analysis of cyclic processes played an important role in the historical development of thermodynamics and in the discovery of the Second Law of thermodynamics. Even today, their understanding is interesting in principle and in addition, it has eminent practical significance. Thermodynamics makes statements concerning the efficiency of cyclic processes (periodically repeating processes) of the most general kind, which are of importance both for heat engines and thus for the energy economy, as well as for the energy balance of biological systems.

3.7.1 General Considerations

In cyclic processes, the working substance, i.e. the system, returns at intervals to its initial state (after each cycle). For practical reasons, in the steam engine, and in the internal combustion engine, the working substance is replenished after each cycle. We assume that the process takes place quasistatically; thus, we can characterize the state of the system by two thermodynamic variables, e.g. P and V or T and S. The process can be represented as a closed curve in the P-V or the T-S plane (Fig. 3.20).

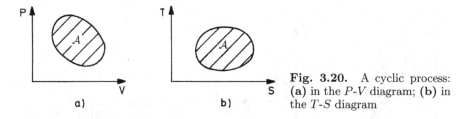

Fig. 3.20. A cyclic process: (a) in the P-V diagram; (b) in the T-S diagram

The work which is performed during one cycle is given by the line integral along the closed curve

$$W = -W = \oint PdV = \mathcal{A} , \qquad (3.7.1)$$

which is equal to the enclosed area $\mathcal{A}$ within the curve representing the cyclic process in the P-V diagram.

The heat taken up during one cycle is given by

$$Q = \oint TdS = \mathcal{A} . \qquad (3.7.2)$$

Since the system returns to its initial state after a cycle, thus in particular the internal energy of the working substance is unchanged, it follows from the principle of conservation of energy that

$$Q = W . \qquad (3.7.3)$$

The heat taken up is equal to the work performed on the surroundings. The direction of the cyclic path and the area in the P-V and T-S diagrams are thus the same. When the cyclic process runs in a clockwise direction (right-handed process), then

$$\circlearrowright \quad Q = W > 0 \qquad (3.7.4a)$$

and one refers to a work engine. In the case that the process runs counter-clockwise (left-handed process), we have

$$\circlearrowleft \quad Q = W < 0 \qquad (3.7.4b)$$

and the machine acts as a heat pump or a refrigerator.

3.7.2 The Carnot Cycle

The Carnot cycle is of fundamental importance; its P-V and T-S diagrams are shown in Fig. 3.21.

We initially discuss the process which runs clockwise, i.e. the work engine. The starting point is point A in the diagram. The cycle is divided into four operations: an isothermal expansion, an adiabatic expansion, an isothermal compression, and an adiabatic compression. The system is alternately connected to heat baths at temperatures T_2 and T_1, where $T_2 > T_1$, and in between it is insulated. The motion of the piston is shown in Fig. 3.22.

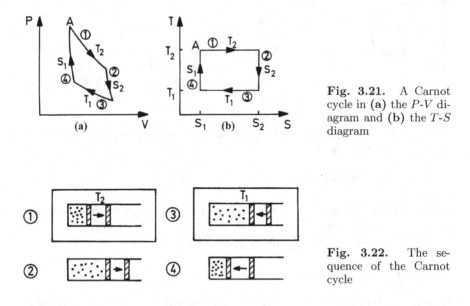

Fig. 3.21. A Carnot cycle in (a) the P-V diagram and (b) the T-S diagram

Fig. 3.22. The sequence of the Carnot cycle

1. Isothermal expansion: the system is brought into contact with the warmer heat bath at the temperature T_2. The quantity of heat

$$Q_2 = T_2(S_2 - S_1) \qquad (3.7.5a)$$

is taken up from the bath, while at the same time, work is performed on the surroundings.

2. Adiabatic expansion: the system is thermally insulated. Through an adiabatic expansion, work is performed on the outer world and the working substance cools from T_2 to the temperature T_1.

3. Isothermal compression: the working substance is brought into thermal contact with the heat bath at temperature T_1 and through work performed on it by the surroundings, it is compressed. The quantity of heat "taken up" by the working substance

$$Q_1 = T_1(S_1 - S_2) < 0 \qquad (3.7.5b)$$

is negative. That is, the quantity $|Q_1|$ of heat is given up to the heat bath.

4. Adiabatic compression: employing work performed by the outside world, the now once again thermally insulated working substance is compressed and its temperature is thereby increased to T_2.

After each cycle, the internal energy remains the same; therefore, the total work performed on the surroundings is equal to the quantity of heat taken up by the system, $Q = Q_1 + Q_2$; thus

$$\mathcal{W} = Q = (T_2 - T_1)(S_2 - S_1) . \qquad (3.7.5c)$$

The *thermal efficiency* (= work performed/heat taken up from the warmer heat bath) is defined as

$$\eta = \frac{\mathcal{W}}{Q_2} . \qquad (3.7.6a)$$

For the Carnot machine, we obtain

$$\eta_C = 1 - \frac{T_1}{T_2} , \qquad (3.7.6b)$$

where the index C stands for Carnot. We see that $\eta_C < 1$. The general validity of (3.7.6a) cannot be too strongly emphasized; it holds for any kind of working substance. Later, we shall show that there is no cyclic process whose efficiency is greater than that of the Carnot cycle.

The Inverse Carnot Cycle
Now, we consider the inverse Carnot cycle, in which the direction of the operations is counter-clockwise (Fig. 3.23). In this case, for the quantities of heat taken up from baths 2 and 1, we find

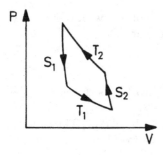

Fig. 3.23. The inverse Carnot cycle

$$Q_2 = T_2(S_1 - S_2) < 0$$
$$Q_1 = T_1(S_2 - S_1) > 0 \ . \tag{3.7.7a,b}$$

The overall quantity of heat taken up by the system, Q, and the work performed on the system, W, are then given by

$$Q = (T_1 - T_2)(S_2 - S_1) = -W < 0 \ . \tag{3.7.8}$$

Work is performed by the outside world on the system. The warmer reservoir is heated further, and the cooler one is cooled. Depending on whether the purpose of the machine is to heat the warmer reservoir or to cool the colder one, one defines the heating efficiency or the cooling efficiency.

The *heating efficiency* (= the heat transferred to bath 2/work performed) is

$$\eta_C^H = \frac{-Q_2}{W} = \frac{T_2}{T_2 - T_1} > 1 \ . \tag{3.7.9}$$

Since $\eta_C^H > 1$, this represents a more efficient method of heating than the direct conversion of electrical energy or other source of work into heat (this type of machine is called a heat pump). The formula however also shows that the use of heat pumps is reasonable only as long as $T_2 \approx T_1$; when the temperature of the heat bath (e.g. the Arctic Ocean) $T_1 \ll T_2$, it follows that $|Q_2| \approx |W|$, i.e. it would be just as effective to convert the work directly into heat.

The *cooling efficiency* (= the quantity of heat removed from the cooler reservoir/work performed) is

$$\eta_C^K = \frac{Q_1}{W} = \frac{T_1}{T_2 - T_1} \ . \tag{3.7.10}$$

For large-scale technical cooling applications, it is expedient to carry out the cooling process in several steps, i.e. as a cascade.

3.7.3 General Cyclic Processes

We now take up a general cyclic process (Fig. 3.24), in which heat exchange with the surroundings can take place at different temperatures, not necessarily only at the maximum and minimum temperature. We shall show, that

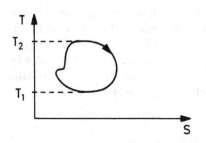

Fig. 3.24. The general cyclic process

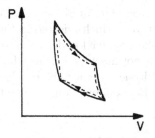

Fig. 3.25. The idealized (full curve) and real (dashed curve) sequence of the Carnot cycle

the efficiency η obeys the inequality

$$\eta \leq \eta_C , \qquad (3.7.11)$$

where η_C the efficiency of a Carnot cycle operating between the two extreme temperatures.

We decompose the process into sections with heat uptake ($\delta Q > 0$) and heat output ($\delta Q < 0$), and also allow irreversible processes to take place

$$W = Q = \oint \delta Q = \int_{\delta Q > 0} \delta Q + \int_{\delta Q < 0} \delta Q = \underset{>0}{Q_2} + \underset{<0}{Q_1} .$$

It follows from the Second Law that

$$0 \geq \oint \frac{\delta Q}{T} = \int_{\delta Q > 0} \frac{\delta Q}{T} + \int_{\delta Q < 0} \frac{\delta Q}{T} \geq \frac{Q_2}{T_2} + \frac{Q_1}{T_1} . \qquad (3.7.12)$$

Here, for the second inequality sign, we have used the inequality $T_1 \leq T \leq T_2$. We thus obtain

$$\frac{Q_1}{Q_2} \leq -\frac{T_1}{T_2} . \qquad (3.7.13)$$

From this, we find for the efficiency of this process the inequality

$$\eta = \frac{Q_1 + Q_2}{Q_2} = 1 + \frac{Q_1}{Q_2} \leq 1 - \frac{T_1}{T_2} = \eta_C , \qquad (3.7.14)$$

whereby (3.7.11) is proven. The efficiency η is only then equal to that of the Carnot cycle if the heat transfer occurs only at the minimum and the maximum temperatures and if the process is carried out reversibly (the second and first inequality signs in Eq. (3.7.12).

In the case of the real Carnot machine, also, there must be a small difference between the internal and the external pressure, in order to cause the process to take place at all (see Fig. 3.25). We recall the considerations at the end of Sect. 3.5, which referred to Fig. 3.9. This leads to the result that $\mathcal{W}$ is given by the area enclosed by the dashed curve. Therefore, the efficiency of the real Carnot machine is somewhat less than the maximum value given by (3.7.6b). Physics sets a universal limit here to the efficiency of industrially applicable heat engines, but also to that of biological systems.

3.8 Phases of Single-Component Systems

The different chemical substances within a system are called components. In the case of a single chemical substance, in contrast, one refers to a single-component system or a pure system. The components of a system can occur in different physical forms (structures), which are termed phases. In this section, we consider single-component systems.

3.8.1 Phase-Boundary Curves

Every substance can occur in several different phases: solid, liquid, gaseous. The solid and the liquid phases can further split into other phases with differing physical properties. Under which conditions can two phases occur in equilibrium with each other? The condition for equilibrium, (2.7.4), or also (3.6.45a,b) states that T, P and μ must be equal. Let $\mu_1(T, P)$ and $\mu_2(T, P,)$ be the chemical potentials of the first and the second phase; then we have

$$\mu_1(T, P) = \mu_2(T, P) . \tag{3.8.1}$$

From this, we obtain the phase boundary curve

$$P = P_0(T) . \tag{3.8.2}$$

The coexistence of two phases is possible along a curve in the P-T diagram. Examples of phase boundaries are (see Fig 3.26): solid–liquid: the melting curve; solid–gaseous: the sublimation curve; liquid–gaseous: the vapor pressure curve; also called the evaporation curve. Fig. 3.26 shows a phase diagram which is typical of most simple substances.

We first consider the process of evaporation on isobaric heating of the liquid, e.g. at the pressure P_0 in Fig. 3.27a. In the region 1, only the liquid is present; at a temperature $T(P_0)$ (point 2), the liquid evaporates, and in region 3, the substance is present in the gas phase.

For a complete characterization of the physical situation on the transition line, we represent the evaporation process at constant pressure P_0 in the T-V diagram (Fig. 3.27b). In region 1, only liquid is present, and an input of heat leads to an increase of the temperature and thermal expansion, until $T(P_0)$ is reached. Further input of heat then goes into the conversion of liquid into gas (region 2). Only when all of the liquid has evaporated does the temperature

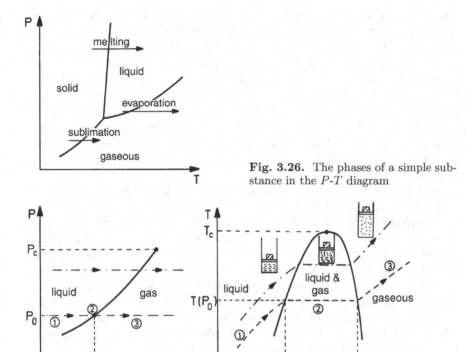

Fig. 3.26. The phases of a simple substance in the P-T diagram

Fig. 3.27. The evaporation process: **(a)** P-T diagram: the vapor-pressure curve; **(b)** T-V diagram: coexistence region, bounded by the coexistence curve (solid line), isobaric heating (dashed or dot-dashed)

once again increase (region 3). In the horizontal part of isobar 2, gas and liquid are present as the fractions c_G and c_L,

$$c_G + c_L = 1 \, . \tag{3.8.3}$$

The overall volume is

$$V = c_G V_G + c_L V_L = c_G V_G + (1 - c_G)V_L \, , \tag{3.8.4}$$

where V_G and V_L are the volumes of the pure gas and liquid phases at the evaporation temperature. It follows that

$$c_G = \frac{V - V_L}{V_G - V_L} \, . \tag{3.8.5}$$

If we consider the evaporation process at a different pressure, we find a similar behavior. The region of horizontal isobars is called the coexistence region, since here the liquid coexists with the gas. This region is bounded by the coexistence curve. When the pressure is increased, the difference between the liquid and the gas phase becomes less and the coexistence region narrows in the T-V diagram. The two branches of the coexistence curve join at the critical point, whose temperature and pressure T_c and P_c are called the critical temperature and critical pressure. For water, $T_c = 647.3$ K and $P_c = 221.36$ bar. The critical temperatures of some other substances are collected in Table I.4. At pressures above the critical pressure, there is no phase transition between a more dense liquid and a less dense gas phase. In this range, there is only a fluid phase which varies continuously with temperature. These facts are more clearly represented in a three-dimensional P-V-T diagram.

At temperatures below T_c, the liquid phase can be reached by isothermal compression. At temperatures above T_c, there is no phase transition from the gaseous to the liquid phase. This fact was first demonstrated with substances which are gaseous under normal conditions, O_2, N_2, ..., in gas compression experiments at extremely high pressures by Natterer[10] (cf. the values of T_c in Table I.4). The critical state was first investigated by Andrews[11] using CO_2.

In Fig. 3.28, the three-dimensional P-V-T diagram for a typical simple substance like CO_2 is drawn. The surface defined by $P = P(V, T)$ is called the surface of the equation of state or the PVT surface. The regions of coexistence of liquid-gas, solid-liquid, and solid-gas are clearly recognizable. This substance contracts upon solidifying. In Fig. 3.28, the projections on the P-T plane (i.e. the phase diagram) and on the P-V plane are also shown. The numerical values for CO_2 are given in the P-T diagram.

The situation already discussed for the liquid-gas transition is analogous in the cases of sublimation and melting; however, for these phase transitions there is no critical point. As can be seen from Figs. 3.26 and 3.28b, at the triple point, the solid, liquid, and gas phases coexist. In Fig. 3.28a,c, where the surface of the equation of state is plotted with respect to the extensive variable V, the triple point becomes a triple line (see Sect. 3.8.4).

In Fig. 3.29, the phase diagram is also shown for a case in which the substance expands upon solidifying, as is the case for water.

Notes:

a) It is usual to denote a gas in the neighborhood of the vapor-pressure curve as *vapor*. Vapors are simply gases which deviate noticeably from the state of an ideal gas, except at very low pressures. A vapor which is in equilibrium with its liquid is termed 'saturated'.

b) In the technical literature, the vapor-pressure curve is also called the *evaporation curve*, and the coexistence region is called the *saturation region*, while the coexistence curve is called the *saturation curve*. On evaporation of a liquid, there are also droplets of liquid floating in the vapor within the saturation region. These

---

[10] I. Natterer, Sitzungsberichte der kaiserlichen Akademie der Wissenschaften, mathem.-naturwiss. Classe, Vol. V, 351 (1850) and *ibid.*, Vol. VI, 557 (1851); and Sitzungsbericht der Wien. Akad. XII, 199 (1854)

[11] Th. Andrews, Philos. Trans. **159**, 11, 575 (1869)

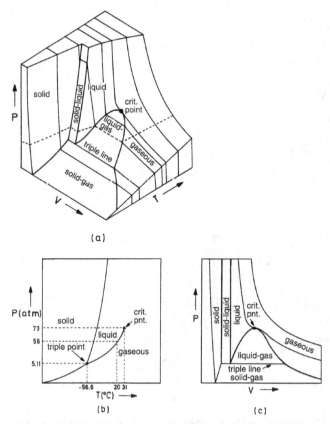

(a)

(b)

(c)

Fig. 3.28. CO_2: (a) The surface of the equation of state $P = P(V, T)$ of a substance which contracts on freezing. The isotherms are shown as solid curves and the isobars are dashed. (b) The P-T diagram (phase diagram). Here, the numerical values for CO_2 are given; the drawing is however not to scale. (c) The P-V diagram.

droplets are called wet vapor. This 'moist vapor' vanishes only when the water droplets evaporate, leaving a dry saturated vapor (V_G in Fig. 3.27b). The expression "saturated" is due to the fact that the least cooling of the vapor leads to the formation of water droplets, i.e. the vapor begins to condense. Vapor (gas) in the pure gas phase (region 3 in Fig. 3.27) is also termed 'superheated vapor'. It is incorrect to call clouds of floating solid or liquid particles 'vapor' or 'steam' (e.g. a "steaming" locomotive). Such clouds are correctly called fog or condensation clouds. Water vapor is invisible.

c) The right-hand branch of the coexistence curve (see Fig. 3.27b) is also called the *condensation boundary* and the left-hand branch the *boiling boundary*. Coming from the gas phase, the first liquid droplets form at the condensation boundary, and coming from the liquid phase, the first gas bubbles form at the boiling boundary.

d) To elucidate the concept of vapor pressure, we consider the following demonstration: take a cylindrical vessel containing a liquid, e.g. water, in its lower section. A movable, airtight piston is initially held directly above the water surface. If the piston is raised, keeping the temperature constant, then just enough water will

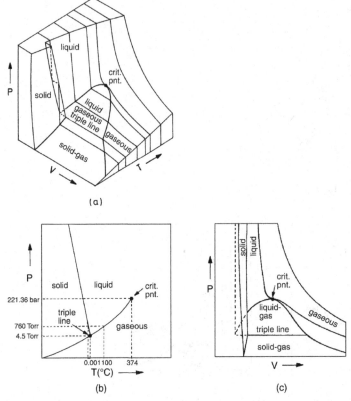

Fig. 3.29. H_2O: **(a)** The surface of the equation of state of a substance which expands on freezing. **(b)** The *P-T* diagram (phase diagram). Here, the numerical values for H_2O are shown. The diagram is however not drawn to scale. **(c)** The *P-V* diagram.

evaporate to produce a certain pressure in the free space which opens above the water surface, independent of the volume of this space. The vapor is saturated within this space. If the piston is again lowered, the vapor is not compressed, but rather just enough of it condenses into the liquid phase to keep the vapor phase saturated; cf. the isotherm in Fig. 3.29a.

e) The pressure (more precisely, the partial pressure; see p. 155) of the saturated vapor above its liquid is nearly independent of whether other, different gases are present above the liquid, e.g. air. Evaporation in this situation will be treated in more detail later in Sect. 3.9.4.1.

3.8.2 The Clausius–Clapeyron Equation

3.8.2.1 Derivation

According to the discussion of the preceding section, in general the volume and the entropy of the substance change upon passing through a phase boundary curve. The Clausius–Clapeyron equation gives a relation between this change and the slope of the phase-boundary curve. These quantities are related to each other because the equality of the chemical potentials (3.8.1) also implies the equality of the derivatives of the chemical potentials along the phase boundary curve, and the latter can be expressed in terms of the (specific) volumes and entropies.

In order to derive the Clausius–Clapeyron equation, we insert into the equilibrium condition (3.8.1) its solution, (3.8.2), i.e. the phase-boundary curve $P_0(T)$:

$$\mu_1(T, P_0(T)) = \mu_2(T, P_0(T)) \,,$$

and then take the derivative with respect to T,

$$\left(\frac{\partial \mu_1}{\partial T}\right)_P + \left(\frac{\partial \mu_1}{\partial P}\right)_T \frac{dP_0}{dT} = \left(\frac{\partial \mu_2}{\partial T}\right)_P + \left(\frac{\partial \mu_2}{\partial P}\right)_T \frac{dP_0}{dT} \,. \tag{3.8.6}$$

We recall the two thermodynamic relations $dG = -S dT + V dP + \mu dN$ and $G = \mu(T, P)N$, which are valid within each of the two homogeneous phases, from which it follows that

$$S = -\left(\frac{\partial \mu}{\partial T}\right)_P N \,, \qquad V = \left(\frac{\partial \mu}{\partial P}\right)_T N \,. \tag{3.8.7}$$

Applying this to the phases 1 and 2 with the chemical potentials μ_1 and μ_2, we obtain from (3.8.6)

$$\frac{dP_0}{dT} = \frac{\Delta S}{\Delta V} \,, \tag{3.8.8}$$

where the entropy and volume changes

$$\Delta S = S_2 - S_1 \quad \text{and} \quad \Delta V = V_2 - V_1 \tag{3.8.9a,b}$$

have been defined. Here, $S_{1,2}$ and $V_{1,2}$ are the entropies and volumes of the substance consisting of N molecules in the phases 1 and 2 along the boundary curve. ΔS and ΔV are the entropy and volume changes as a result of the phase transition of the whole substance. The *Clausius–Clapeyron equation* (3.8.8) expresses the slope of the phase-boundary curve in terms of the ratio of the entropy and volume changes in the phase transition. The latent heat Q_L is the quantity of heat which is required to convert the substance from phase 1 to phase 2:

$$Q_L = T\Delta S \,. \tag{3.8.10}$$

Inserting this definition into (3.8.8), we obtain the Clausius–Clapeyron equation in the following form:

$$\frac{dP_0}{dT} = \frac{Q_L}{T\Delta V} \,. \tag{3.8.11}$$

Remarks:

(i) Frequently, the right-hand side of the Clausius–Clapeyron equation, (3.8.8) or (3.8.11), is expressed in terms of the entropies (latent heats) and volumes of 1 g or 1 Mole of a substance.

(ii) In the transition from the low-temperature phase (1) to the high-temperature phase (2), ΔV can be either positive or negative; however, it *always* holds that $\Delta S > 0$. In this connection we recall the process of isobaric heating discussed in Sect. 3.8.1. In the coexistence region, the temperature T remains constant, since the heat put into the system is consumed by the phase transition. From (3.8.10), $Q_L = T \Delta S > 0$, it follows that $\Delta S > 0$. This can also be read off Fig. 3.34b, whose general form results from the concavity of G and $\left(\frac{\partial G}{\partial T} \right)_P = -S < 0$.

3.8.2.2 Example Applications of the Clausius–Clapeyron Equation:

We now wish to give some interesting examples of the application of the Clausius–Clapeyron equation.

(i) Liquid $\rightarrow$ gaseous: since, according to the previous considerations, $\Delta S > 0$ and the specific volume of the gas is larger than that of the liquid, $\Delta V > 0$, it follows that $\frac{dP_0}{dT} > 0$, i.e. the boiling temperature increases with increasing pressure (Table I.5 and Figs. 3.28(b) and 3.29(b)).
Table I.6 contains the heats of vaporization of some substances at their boiling points under standard pressure, i.e. 760 Torr. Note the high value for water.
(ii) Solid $\rightarrow$ liquid: in the transition to the high-temperature phase, we have always $\Delta S > 0$. Usually, $\Delta V > 0$; then it follows that $\frac{dT}{dP} > 0$. In the case of water, $\Delta V < 0$ and thus $\frac{dT}{dP} < 0$. The fact that ice floats on water implies via the Clausius–Clapeyron equation that its melting point decreases on increasing the pressure (Fig. 3.29).

Note: There are a few other substances which expand on melting, e.g. mercury and bismuth. The large volume increase of water on melting (9.1%) is related to the open structure of ice, containing voids (the bonding is due to the formation of hydrogen bonds between the oxygen atoms, cf. Fig. 3.30). Therefore, the liquid phase is more dense. Above 4°C above the melting point T_m, the density of water begins to decrease on cooling (water anomaly) since local ordering occurs already at temperatures above T_m.
While as a rule a solid material sinks within its own liquid phase (melt), ice floats on water, in such a way that about 9/10 of the ice is under the surface of the water. This fact together with the density anomaly of water plays a very important role in Nature and is fundamental for the existence of life on the Earth.
The volume change upon melting if ice is $V_L - V_S = (1.00 - 1.091)\,\text{cm}^3/\text{g} = -0.091\,\text{cm}^3\text{g}^{-1}$. The latent heat of melting per g is $Q = 80\,\text{cal/g} = 80 \times 42.7\,\text{atm cm}^3/\text{g}$. From this, it follows that the slope of the melting curve of ice near 0°C is

$$\frac{dP}{dT} = -\frac{80 \times 42.7}{273 \times 0.091} \frac{\text{atm}}{\text{K}} = -138\,\text{atm/K} . \qquad (3.8.12)$$

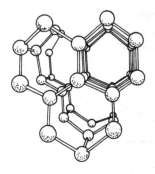

Fig. 3.30. The hexagonal structure of ice. The oxygen atoms are shown; they are connected to four neighbors via hydrogen bonds

The melting curve as a function of the temperature is very steep. It requires a pressure increase of 138 atm to lower the melting temperature by 1 K. This "freezing-point depression", small as it is, enters into a number of phenomena in daily life. If a piece of ice at somewhat below 0° C is placed under increased pressure, it at first begins to melt. The necessary heat of melting is taken from the ice itself, and it therefore cools to a somewhat lower temperature, so that the melting process is interrupted as long as no more heat enters the ice from its surroundings. This is the so-called *regelation* of ice (= the alternating melting and freezing of ice caused by changes in its temperature and pressure). Pressing together snow, which consists of ice crystals, to make a snowball causes the snow to melt to a small extent due to the increased pressure. When the pressure is released, it freezes again, and the snow crystals are glued together. The slickness of ice is essentially due to the fact that it melts at places where it is under pressure, so that between a sliding object and the surface of the ice there is a thin layer of liquid water, which acts like a lubricant, explaining e.g. the gliding motion of an ice skater. Part of the plasticity of glacial ice and its slow motion, like that of a viscous liquid, are also due to regelation of the ice. The lower portions of the glacier become movable as a result of the pressure from the weight of the ice above, but they freeze again when the pressure is released.

(iii) ^{3}He, liquid $\rightarrow$ solid: the phase diagram of ^{3}He is shown schematically in Fig. 3.31. At low temperatures, there is an interval where the melting curve falls. In this region, in the transition from liquid to solid (see the arrow in Fig. 3.31a), $\frac{dP}{dT} < 0$; furthermore, it is found experimentally that the volume of the solid phase is smaller than that of the liquid (as is the usual case), $\Delta V < 0$. We thus find from the Clausius–Clapeyron equation (3.8.8) $\Delta S > 0$, as expected from the general considerations in Remark (ii).

The **Pomeranchuk effect:** The fact that within the temperature interval mentioned above, the entropy increases on solidification is called the Pomeranchuk effect. It is employed for the purpose of reaching low temperatures (see Fig. 3.31b). Compression (dashed line) of liquid ^{3}He leads to its solidification and, because of $\Delta S > 0$, to the uptake of heat. This causes a decrease in the temperature of the substance. Compression therefore causes the phase transition to proceed along the melting curve (see arrow in Fig. 3.31b).

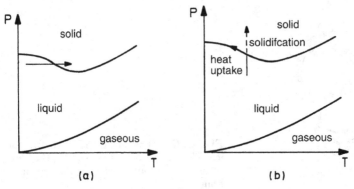

Fig. 3.31. The phase diagram of ^{3}He. **(a)** Isobaric solidification in the range where $\frac{dP}{dT} < 0$. **(b)** Pomeranchuk effect

This effect can be used to cool ^{3}He; with it, temperatures down to 2×10^{-3} K can be attained. The Pomeranchuk effect, however, has nearly no practical significance in low-temperature physics today. The currently most important methods for obtaining low temperatures are ^{3}He-^{4}He dilution ($2 \times 10^{-3} - 5 \times 10^{-3}$ K) and adiabatic demagnetization of copper nuclei ($1.5 \times 10^{-6} - 12 \times 10^{-6}$ K), where the temperatures obtained are shown in parentheses.

(iv) The sublimation curve: We consider a solid (1), which is in equilibrium with a classical, ideal gas (2). For the volumes of the two phases, we have $V_1 \ll V_2$; then it follows from the Clausius–Clapeyron equation (3.8.11) that

$$\frac{dP}{dT} = \frac{Q_L}{TV_2} ,$$

where Q_L represents the latent heat of sublimation. For V_2, we insert the ideal gas equation,

$$\frac{dP}{dT} = \frac{Q_L P}{kNT^2} . \tag{3.8.13}$$

This differential equation can be immediately integrated under the assumption that Q_L is independent of temperature:

$$P = P_0 \, e^{-q/kT} , \tag{3.8.14}$$

where $q = \frac{Q_L}{N}$ is the heat of sublimation per particle. Equation (3.8.14) yields the shape of the sublimation curve under the assumptions used.

The vapor pressure of most solid materials is rather small, and in fact in most cases, no observable decrease with time in the amount of these substances due to evaporation is detected. Only a very few solid materials exhibit a readily observable *sublimation* and have as a result a noticeable vapor pressure, which increases with increasing temperature; among them are some solid perfume substances. Numerical values for the vapor pressure over ice and iodine are given in Tables I.8 and I.9.

At temperatures well below $0°$ C and in dry air, one can observe a gradual disappearance of snow, which is converted directly into water vapor by sublimation. The reverse phenomenon is the direct formation of frost from water vapor in the air, or the condensation of snow crystals in the cool upper layers of the atmosphere. If iodine crystals are introduced into an evacuated glass vessel and a spot on the glass wall is cooled, then solid iodine condenses from the iodine vapor which forms in the vessel. Iodine crystals which are left standing in the open air, napthalene crystals ("moth balls"), and certain mercury salts, including "sublimate" ($HgCl_2$), among others, gradually vanish due to sublimation.

3.8.3 The Convexity of the Free Energy and the Concavity of the Free Enthalpy (Gibbs' Free Energy)

We now return again to the gas-liquid transition, in order to discuss some additional aspects of evaporation and the curvature of the thermodynamic potentials. The coexistence region and the coexistence curve are clearly visible in the T-V diagram. Instead, one often uses a P-V diagram. From the projection of the three-dimensional P-V-T diagram, we can see the shape drawn in Fig. 3.32. From the shape of the isotherms in the P-V diagram, the free energy can be determined analytically and graphically. Owing to $\left(\frac{\partial F}{\partial V}\right)_T = -P$, it follows for the free energy that

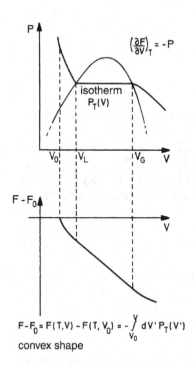

$F-F_0 = F(T,V) - F(T, V_0) = -\int_{V_0}^{V} dV' \, P_T(V')$

convex shape

Fig. 3.32. The isotherms $P_T(V)$ and the free energy as a function of the volume during evaporation; the thin line is the coexistence curve

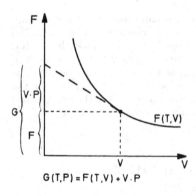

$$G(T,P) = F(T,V) + V \cdot P$$

Fig. 3.33. The determination of the free enthalpy from the free energy by construction

$$F(T,V) - F(T,V_0) = - \int_{V_0}^{V} dV' P_T(V') \,. \qquad (3.8.15)$$

One immediately sees that the isotherms in Fig. 3.32 lead qualitatively to the volume dependence of the free energy which is drawn below. The free energy is convex (curved upwards). The fundamental cause of this is the fact that the compressibility is positive:

$$\frac{\partial^2 F}{\partial V^2} = -\frac{\partial P}{\partial V} \propto \frac{1}{\kappa_T} > 0 \,,$$

while

$$\left(\frac{\partial^2 F}{\partial T^2}\right)_V = -\left(\frac{\partial S}{\partial T}\right)_V \propto -C_V < 0 \,. \qquad (3.8.16)$$

These inequalities are based upon the stability relations proved previously, (3.3.5, 3.3.6), and (3.6.48a,b).

The free enthalpy or Gibbs' free energy $G(T,P) = F + PV$ can be constructed from $F(T,V)$. Due to $P = -\left(\frac{\partial F}{\partial V}\right)_T$, $G(T,P)$ is obtained from $F(T,V)$ by constructing a tangent to $F(T,V)$ with the slope $-P$ (see Fig. 3.33). The intersection of this tangent with the ordinate has the co-ordinates

$$F(T,V) - V \left(\frac{\partial F}{\partial V}\right)_T = F + VP = G(T,P) \,. \qquad (3.8.17)$$

The result of this construction is drawn in Fig. 3.34.

The derivatives of the free enthalpy

$$\left(\frac{\partial G}{\partial P}\right)_T = V \quad \text{and} \quad \left(\frac{\partial G}{\partial T}\right)_P = -S$$

yield the volume and the entropy. They are discontinuous at a phase transition, which results in a kink in the curves. Here, $P_0(T)$ is the evaporation

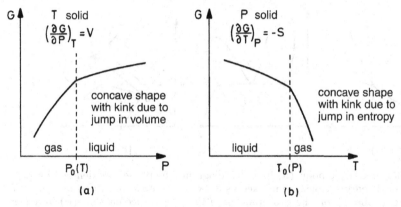

Fig. 3.34. The free enthalpy (Gibbs' free energy) as a function of **(a)** the pressure and **(b)** the temperature.

pressure at the temperature T, and $T_0(P)$ is the evaporation temperature at the pressure P. From this construction, one can also see that the free enthalpy is concave (Fig. 3.34). The curvatures are negative because $\kappa_T > 0$ and $C_P > 0$. The signs of the slopes result from $V > 0$ and $S > 0$. It is also readily seen from the figures that the entropy increases as a result of a transition to a higher-temperature phase, and the volume decreases as a result of a transition to a higher-pressure phase. These consequences of the stability conditions hold quite generally. In the diagrams (3.34a,b), the terms gas and liquid phases could be replaced by low-pressure and high-pressure or high-temperature and low-temperature phases.

On melting, the latent heat must be added to the system, on freezing (solidifying), it must be removed. When heat is put into or taken out of a system at constant pressure, it is employed to convert the solid phase to the liquid or *vice versa*. In the coexistence region, the temperature remains constant during these processes. This is the reason why in late Autumn and early Spring the temperature near the Earth remains close to zero degrees Celsius, the freezing point of water.

3.8.4 The Triple Point

At the triple point (Figs. 3.26 and 3.35), the solid, liquid and gas phases coexist in equilibrium. The condition for equilibrium of the gaseous, liquid and solid phases, or more generally for three phases 1, 2 and 3, is:

$$\mu_1(T, P) = \mu_2(T, P) = \mu_3(T, P) , \qquad (3.8.18)$$

and it determines the triple point pressure and the triple point temperature P_t, T_t.

In the P-T diagram, the triple point is in fact a single point. In the T-V diagram it is represented by the horizontal line drawn in Fig. 3.35b. Along this line, the three phases are in equilibrium. If the phase diagram is

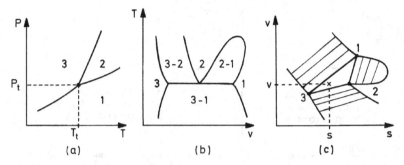

Fig. 3.35. The triple point **(a)** in a P-T diagram (the phases are denoted by 1, 2, 3. The coexistence regions are marked as 3-2 etc., i.e. denoting the coexistence of phase 3 and phase 2 on the two branches of the coexistence curve.); **(b)** in a T-v diagram; and **(c)** in a v-s diagram

represented in terms of two extensive variables, such as e.g. by V and S as in Fig. 3.35c, then the triple point becomes a triangular area as is visible in the figure. At each point on this triangle, the states of the three phases 1, 2, and 3 corresponding to the vertices of the triangle coexist with one another.

We now want to describe this more precisely. Let s_1, s_2 and s_3 be the entropies per particle in the phases 1, 2 and 3 just at the triple point, $s_i = -\left(\frac{\partial \mu_i}{\partial T}\right)_P \big|_{T_t, P_t}$, and correspondingly, v_1, v_2, v_3 are the specific volumes $v_i = \left(\frac{\partial \mu_i}{\partial P}\right)_T \big|_{T_t, P_t}$. The points (s_i, v_i) are shown in the s-v diagram as points 1, 2, 3. Clearly, every pair of phases can coexist with each other; the lines connecting the points 1 and 2 etc. yield the triangle with vertices 1, 2, and 3. The coexistence curves of two phases, e.g. 1 and 2, are found in the s-v diagram from $s_i(T) = -\left(\frac{\partial \mu_i}{\partial T}\right)_P \big|_{P_0(T)}$ and $v_i(T) = \left(\frac{\partial \mu_i}{\partial P}\right)_T \big|_{P_0(T)}$ with $i = 1$ and 2 along with the associated phase-boundary curve $P = P_0(T)$. Here, the temperature is a parameter; points on the two branches of the coexistence curves with the same value of T can coexist with each other. The diagram in 3.35c is only schematic. The (by no means parallel) lines within the two-phase coexistence areas show which of the pairs of single-component states can coexist with each other on the two branches of the coexistence line.

Now we turn to the interior of the triangular area in Fig. 3.35c. It is immediately clear that the three triple-point phases 1, 2, 3 can coexist with each other at the temperature T_t and pressure P_t in arbitrary quantities. This also means that a given amount of the substance can be distributed among these three phases in arbitrary fractions c_1, c_2, c_3 ($0 \le c_i \le 1$)

$$c_1 + c_2 + c_3 = 1 , \tag{3.8.19a}$$

and then will have the total specific entropy

$$c_1 s_1 + c_2 s_2 + c_3 s_3 = s \tag{3.8.19b}$$

and the total specific volume

$$c_1 v_1 + c_2 v_2 + c_3 v_3 = v \,. \tag{3.8.19c}$$

From (3.8.19a,b,c), it follows that s and v lie within the triangle in Fig. 3.35c. Conversely, every (heterogeneous) equilibrium state with the total specific entropy s and specific volume v can exist within the triangle, where c_1, c_2, c_3 follow from (3.8.19a–c). Eqns. (3.8.19a–c) can be interpreted by the following center-of-gravity rule: let a point (s, v) within the triangle in the v-s diagram (see Fig. 3.35c) be given. The fractions c_1, c_2, c_3 must be chosen in such a way that attributing masses c_1, c_2, c_3 to the vertices 1, 2, 3 of the triangle leads to a center of gravity at the position (s, v). This can be immediately understood if one writes (3.8.19b,c) in the two-component form:

$$c_1 \begin{pmatrix} v_1 \\ s_1 \end{pmatrix} + c_2 \begin{pmatrix} v_2 \\ s_2 \end{pmatrix} + c_3 \begin{pmatrix} v_3 \\ s_3 \end{pmatrix} = \begin{pmatrix} v \\ s \end{pmatrix} \,. \tag{3.8.20}$$

Remarks:

(i) Apart from the center-of-gravity rule, the linear equations can be solved algebraically:

$$c_1 = \frac{\begin{vmatrix} 1 & 1 & 1 \\ s & s_2 & s_3 \\ v & v_2 & v_3 \end{vmatrix}}{\begin{vmatrix} 1 & 1 & 1 \\ s_1 & s_2 & s_3 \\ v_1 & v_2 & v_3 \end{vmatrix}}, \quad c_2 = \frac{\begin{vmatrix} 1 & 1 & 1 \\ s_1 & s & s_3 \\ v_1 & v & v_3 \end{vmatrix}}{\begin{vmatrix} 1 & 1 & 1 \\ s_1 & s_2 & s_3 \\ v_1 & v_2 & v_3 \end{vmatrix}}, \quad c_3 = \frac{\begin{vmatrix} 1 & 1 & 1 \\ s_1 & s_2 & s \\ v_1 & v_2 & v \end{vmatrix}}{\begin{vmatrix} 1 & 1 & 1 \\ s_1 & s_2 & s_3 \\ v_1 & v_2 & v_3 \end{vmatrix}} \,.$$

(ii) Making use of the triple point gives a precise standard for a temperature and a pressure, since the coexistence of the three phases can be verified without a doubt. From Fig. 3.35c, it can also be seen that the triple point is not a point as a function of the experimentally controllable parameters, but rather the whole area of the triangle. The parameters which can be directly varied from outside the system are not P and T, but rather the volume V and the entropy S, which can be varied by performing work on the system or by transferring heat to it. If heat is put into the system at the point marked by a cross (Fig. 3.35c), then in the example of water, some ice would melt, but the state would still remain within the triangle. This explains why the triple point is insensitive to changes within wide limits and is therefore very suitable as a temperature fixed point.

(iii) For water, $T_t = 273.16\,\mathrm{K}$ and $P_t = 4.58\,\mathrm{Torr}$. As explained in Sect. 3.4, the absolute temperature scale is determined by the triple point of water. In order to reach the triple point, one simply needs to distill highly pure water

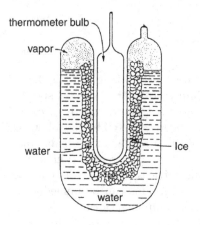

thermometer bulb

vapor

water

Ice

water

Fig. 3.36. A triple-point cell: ice, water, and water vapor are in equilibrium with each other. A freezing mixture in contact with the inner walls causes some water to freeze there. It is then replaced by the thermometer bulb, and a film of liquid water forms on the inner wall

into a container and to seal it off after removing all the air. One then has water and water vapor in coexistence (coexistence region 1-2 in Fig. 3.35c). Removing heat by means of a freezing mixture brings the system into the triple-point range. As long as all three phases are present, the temperature equals T_t (see Fig. 3.36).

3.9 Equilibrium in Multicomponent Systems

3.9.1 Generalization of the Thermodynamic Potentials

We consider a homogeneous mixture of n materials, or as one says in this connection, components, whose particle numbers are $N_1, N_2, \ldots, N_n$. We first need to generalize the thermodynamic relations to this situation. To this end, we refer to Chap. 2. Now, the phase-space volume and similarly the entropy are functions of the energy, the volume, and all of the particle numbers:

$$S = S(E, V, N_1, \ldots, N_n) . \qquad (3.9.1)$$

All the thermodynamic relations can be generalized to this case by replacing N and μ by N_i and μ_i and summing over i. We define the chemical potential of the ith material by

$$\mu_i = -T \left(\frac{\partial S}{\partial N_i} \right)_{E,V,\{N_{k \neq i}\}} \qquad (3.9.2a)$$

and, as before,

$$\frac{1}{T} = \left(\frac{\partial S}{\partial E} \right)_{V,\{N_k\}} \quad \text{and} \quad \frac{P}{T} = \left(\frac{\partial S}{\partial V} \right)_{E,\{N_k\}} . \qquad (3.9.2b,c)$$

Then for the differential of the entropy, we find

$$dS = \frac{1}{T}dE + \frac{P}{T}dV - \sum_{i=1}^{n}\frac{\mu_i}{T}dN_i \,, \tag{3.9.3}$$

and from it the *First Law*

$$dE = TdS - PdV + \sum_{i=1}^{n}\mu_i dN_i \tag{3.9.4}$$

for this mixture.

Die Gibbs–Duhem relation for *homogeneous mixtures* reads

$$E = TS - PV + \sum_{i=1}^{n}\mu_i N_i \,. \tag{3.9.5}$$

It is obtained analogously to Sect. 3.1.3, by differentiating

$$\alpha E = E(\alpha S, \alpha V, \alpha N_1, \ldots, \alpha N_n) \tag{3.9.6}$$

with respect to α. From (3.9.4) and (3.9.5), we find the differential form of the Gibbs–Duhem relation for mixtures

$$-SdT + VdP - \sum_{i=1}^{n}N_i d\mu_i = 0 \,. \tag{3.9.7}$$

It can be seen from this relation that of the $n+2$ variables $(T, P, \mu_1, \ldots, \mu_n)$, only $n+1$ are independent.

The *free enthalpy* (Gibbs' free energy) is defined by

$$G = E - TS + PV \,. \tag{3.9.8}$$

From the First Law, (3.9.4), we obtain its differential form:

$$dG = -SdT + VdP + \sum_{i=1}^{n}\mu_i dN_i \,. \tag{3.9.9}$$

From (3.9.9), we can read off

$$S = -\left(\frac{\partial G}{\partial T}\right)_{P,\{N_k\}} \,, \quad V = \left(\frac{\partial G}{\partial P}\right)_{T,\{N_k\}} \,, \quad \mu_i = \left(\frac{\partial G}{\partial N_i}\right)_{T,P,\{N_{k\neq i}\}} \,. \tag{3.9.10}$$

For homogeneous mixtures, using the Gibbs–Duhem relation (3.9.5) we find for the free enthalpy (3.9.8)

$$G = \sum_{i=1}^{n}\mu_i N_i \,. \tag{3.9.11}$$

Then we have

$$S = -\sum_{i=1}^{n} \left(\frac{\partial \mu_i}{\partial T} \right)_P N_i , \quad V = \sum_{i=1}^{n} \left(\frac{\partial \mu_i}{\partial P} \right)_T N_i .$$ (3.9.12)

The chemical potentials are intensive quantities and therefore depend only on T, P and the $n-1$ concentrations $c_1 = \frac{N_1}{N}, \ldots, c_{n-1} = \frac{N_{n-1}}{N}$ ($N = \sum_{i=1}^{n} N_i$, $c_n = 1 - c_1 - \ldots - c_{n-1}$).

The *grand canonical potential* is defined by

$$\Phi = E - TS - \sum_{i=1}^{n} \mu_i N_i .$$ (3.9.13)

For its differential, we find using the First Law (3.9.4)

$$d\Phi = -SdT - PdV - \sum_{i=1}^{n} N_i d\mu_i .$$ (3.9.14)

For homogeneous mixtures, we obtain using the Gibbs–Duhem relation (3.9.5)

$$\Phi = -PV .$$ (3.9.15)

The density matrix for mixtures depends on the total Hamiltonian and will be introduced in Chap. 5.

3.9.2 Gibbs' Phase Rule and Phase Equilibrium

We consider n chemically different materials (components), which can be in r phases (Fig. 3.37) and between which no chemical reactions are assumed to take place. The following *equilibrium conditions* hold:

Temperature T and *pressure* P must have uniform values in the whole system. Furthermore, for each component i, the *chemical potential* must be the same in each of the phases.

These equilibrium conditions can be derived directly by considering the microcanonical ensemble, or also from the stationarity of the entropy.

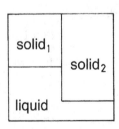

Fig. 3.37. Equilibrium between 3 phases

(i) As a first possibility, let us consider a microcanonical ensemble consisting of n chemical substances, and decompose it into r parts. Calculating the probability of a particular distribution of the energy, the volume and the particle numbers over these parts, one obtains for the most probable distribution the *equality of the temperature, pressure and the chemical potentials of each component*.

(ii) As a second possibility for deriving the equilibrium conditions, one can start from the maximization of the entropy in equilibrium, (3.6.36b)

$$dS \geq \frac{1}{T}\left(dE + PdV - \sum_{i=1}^{n} \mu_i dN_i\right), \tag{3.9.16}$$

and can then employ the resulting stationarity of the equilibrium state for fixed E, V, and $\{N_i\}$,

$$\delta S = 0 \tag{3.9.17}$$

with respect to virtual variations. One can then proceed as in Sect. 3.6.5, decomposing a system into two parts 1 and 2, and varying not only the energy and the volume, but also the particle numbers [see (3.6.44)]:

$$
\begin{aligned}
\delta S = &\left(\frac{\partial S_1}{\partial E_1} - \frac{\partial S_2}{\partial E_2}\right)\delta E_1 + \left(\frac{\partial S_1}{\partial V_1} - \frac{\partial S_2}{\partial V_2}\right)\delta V_1 \\
&+ \sum_i \left(\frac{\partial S_1}{\partial N_{i,1}} - \frac{\partial S_2}{\partial N_{i,2}}\right)\delta N_{i,1} + \dots
\end{aligned}
\tag{3.9.18}
$$

Here, $N_{i,1}$ ($N_{i,2}$) is the particle number of component i in the subsystem 1 (2).

From the condition of vanishing variation, the *equality of the temperatures and pressures* follow:

$$T_1 = T_2, \quad P_1 = P_2 \tag{3.9.19}$$

and furthermore $\frac{\partial S_1}{\partial N_{i,1}} = \frac{\partial S_2}{\partial N_{i,2}}$, i.e. the *equality of the chemical potentials*

$$\mu_{i,1} = \mu_{i,2} \quad \text{for } i = 1, \dots, n . \tag{3.9.20}$$

We have thus now derived the equilibrium conditions formulated at the beginning of this section, and we wish to apply them to n chemical substances in r phases (Fig. 3.37). In particular, we want to find out how many phases can coexist in equilibrium. Along with the equality of temperature and pressure in the whole system, from (3.9.20) the chemical potentials must also be equal,

$$\mu_1^{(1)} = \dots = \mu_1^{(r)},$$
$$\dots \tag{3.9.21}$$
$$\mu_n^{(1)} = \dots = \mu_n^{(r)}.$$

The upper index refers to the phases, and the lower one to the components. Equations (3.9.21) represent all together $n(r-1)$ conditions on the $2+(n-1)r$ variables $(T, P, c_1^{(1)}, \ldots, c_{n-1}^{(1)}, \ldots, c_1^{(r)}, \ldots, c_{n-1}^{(r)})$.

The number of quantities which can be varied (i.e. the *number of degrees of freedom* is therefore equal to $f = 2 + (n-1)r - n(r-1)$:

$$f = 2 + n - r \,. \tag{3.9.22}$$

This relation (3.9.22) is called *Gibbs' phase rule*.

In this derivation we have assumed that each substance is present in all r phases. We can easily relax this assumption. If for example substance 1 is not present in phase 1, then the condition on $\mu_1^{(1)}$ does not apply. The particle number of component 1 then also no longer occurs as a variable in phase 1. One thus has one condition and one variable less than before, and Gibbs' phase rule (3.9.22) still applies.[12]

Examples of Applications of Gibbs' Phase Rule:

(i) For single-component system, $n = 1$:

$$
\begin{array}{lllll}
r & = & 1, & f & = & 2 & T, \ P \text{ free} \\
r & = & 2, & f & = & 1 & P = P_0(T) \text{ Phase-boundary curve} \\
r & = & 3, & f & = & 0 & \text{Fixed point: triple point.}
\end{array}
$$

(ii) An example for a two-component system, $n = 2$, is a mixture of sal ammoniac and water, NH_4Cl+H_2O. The possible phases are: water vapor (it contains practically no NH_4Cl), the liquid mixture (solution), ice (containing some of the salt), the salt (containing some H_2O).

Possible coexisting phases are:

- liquid phase: $r = 1$, $f = 3$ (variables P, T, c)
- liquid phase + water vapor: $r = 2$, $f = 2$, variables P, T; the concentration is a function of P and T: $c = c(P,T)$.
- liquid phase + water vapor + one solid phase: $r = 3$, $f = 1$. Only one variable, e.g. the temperature, is freely variable.
- liquid phase + vapor + ice + salt: $r = 4$, $f = 0$. This is the eutectic point.

The phase diagram of the liquid and the solid phases is shown in Fig. 3.38. At the concentration 0, the melting point of pure ice can be seen, and at $c = 1$, that of the pure salt. Since the freezing point of a solution is lowered (see Chap. 5), we can understand the shape of the two branches of the freezing-point curve as a function of the concentration. The two branches meet at the eutectic point. In the regions ice-liq., ice and liquid, and in liq.-salt, liquid and salt coexist along the horizontal lines. The concentration of NH_4Cl in the ice

---

[12] The number of degrees of freedom is a statement about the intensive variables; there are however also variations of the extensive variables. For example, at a triple point, $f = 0$, the entropy and the volume can vary within a triangle (Sect. 3.8.4).

is considerably lower than in the liquid mixture which is in equilibrium with it. The solid phases often contain only the pure components; then the left-hand and the right-hand limiting lines are identical with the two vertical lines at $c = 0$ and $c = 1$. At the eutectic point, the liquid mixture is in equilibrium with the ice and with the salt. If the concentration of a liquid is less than that corresponding to the eutectic point, then ice forms on cooling the system. In this process, the concentration in the liquid increases until finally the eutectic concentration is reached, at which the liquid is converted to ice and salt. The resulting mixture of salt and ice crystals is called the eutectic. At the eutectic concentration, the liquid has its lowest freezing point.

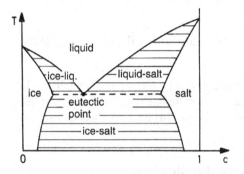

Fig. 3.38. The phase diagram of a mixture of sal ammoniac (ammonium chloride) and water. In the horizontally shaded regions, ice and liquid, liquid and solid salt, and finally ice and solid salt coexist with each other.

The phase diagram in Fig. 3.38 for the liquid and solid phases and the corresponding interpretation using Gibbs' phase rule can be applied to the following physical situations: (i) when the pressure is so low that also a gaseous phase (not shown) is present; (ii) without the gas phase at constant pressure P, in which case a degree of freedom is unavailable; or (iii) in the presence of air at the pressure P and vapor dissolved in it with the partial pressure cP.[13] The concentration of the vapor c in the air enters the chemical potential as $\log cP$ (see Chap. 5). It adjusts itself in such a way that the chemical potential of the water vapor is equal to the chemical potential in the liquid mixture. It should be pointed out that owing to the term $\log c$, the chemical potential of the vapor dissolved in the air is lower than that of the pure vapor. While at atmospheric pressure, boiling begins only at 100°C, and then the whole liquid phase is converted to vapor, here, even at very low temperatures a sufficient amount enters the vapor phase to permit the $\log c$ term to bring about the equalization of the chemical potentials.

The action of freezing mixtures becomes clear from the phase diagram 3.38. For example, if NaCl and ice at a temperature of 0°C are brought together, then they are not in equilibrium. Some of the ice will melt, and the salt will dissolve in the resulting liquid water. Its concentration is to be sure much too high to be in equilibrium with the ice, so that more ice melts. In the melting process,

---

[13] Gibbs' phase rule is clearly still obeyed: compared to (ii), there is one component (air) more and also one more phase (air-vapor mixture) present.

heat is taken up, the entropy increases, and thus the temperature is lowered. This process continues until the temperature of the eutectic point has been reached. Then the ice, hydrated salt, $NaCl \cdot 2H_2O$, and liquid with the eutectic concentration are in equilibrium with each other. For NaCl and H_2O, the eutectic temperature is $-21°C$. The resulting mixture is termed a freezing mixture. It can be used to hold the temperature constant at $-21°C$. Uptake of heat does not lead to an increase of the temperature of the freezing mixture, but rather to continued melting of the ice and dissolution of NaCl at a constant temperature.

Eutectic mixtures always occur when there is a miscibility gap between the two solid phases and the free energy of the liquid mixture is lower than that of the two solid phases (see problem 3.28). the melting point of the eutectic mixture is then considerably lower than the melting points of the two solid phases (see Table I.10).

3.9.3 Chemical Reactions, Thermodynamic Equilibrium and the Law of Mass Action

In this section we consider systems with several components, in which the particle numbers can change as a result of chemical reactions. We first determine the general condition for chemical equilibrium and then investigate mixtures of ideal gases.

3.9.3.1 The Condition for Chemical Equilibrium

Reaction equations, such as for example

$$2H_2 + O_2 \rightleftharpoons 2H_2O , \qquad (3.9.23)$$

can in general be written in the form

$$\sum_{j=1}^{n} \nu_j A_j = 0 , \qquad (3.9.24)$$

where the A_j are the chemical symbols and the stoichiometric coefficients ν_j are (small) integers, which indicate the participation of the components in the reaction. We will adopt the convention that left indicates positive and right negative.

The reaction equation (3.9.24) contains neither any information about the concentrations at which the A_j are present in thermodynamic and chemical equilibrium at a given temperature and pressure, nor about the direction in which the reaction will proceed. The change in the Gibbs free energy ($\equiv$ free enthalpy) with particle number at *fixed temperature* T and *fixed pressure* P for single-phase systems is[14]

---

[14] Chemical reactions in systems consisting of several phases are treated in M.W. Zemansky and R.H. Dittman, *Heat and Thermodynamics*, Mc Graw Hill, Auckland, Sixth Edition, 1987.

$$dG = \sum_{j=1}^{n} \mu_j dN_j \ . \tag{3.9.25}$$

In equilibrium, the N_j must be determined in such a way that G remains stationary,

$$\sum_{j=1}^{n} \mu_j dN_j = 0 \ . \tag{3.9.26}$$

If an amount dM participates in the reaction, then $dN_j = \nu_j dM$. The condition of stationarity then requires

$$\sum_{j=1}^{n} \mu_j \nu_j = 0 \ . \tag{3.9.27}$$

For every chemical reaction that is possible in the system, a relation of this type holds. It suffices for a fundamental understanding to determine the chemical equilibrium for a single reaction. The chemical potentials $\mu_j(T, P)$ depend not only on the pressure and the temperature, but also on the relative particle numbers (concentrations). The latter adjust themselves in such a way in chemical equilibrium that (3.9.27) is fulfilled.

In the case that substances which can react chemically are in thermal equilibrium, but not in chemical equilibrium, then from the change in Gibbs' free energy,

$$\delta G = \delta \left[\sum_j \mu_j(T, P) \nu_j M \right] \tag{3.9.25'}$$

we can determine the direction which the reaction will take. Since G is a minimum at equilibrium, we must have $\delta G \leq 0$; cf. Eq. (3.6.38b). The chemical composition is shifted towards the direction of smaller free enthalpy or lower chemical potentials.

Remarks:

(i) The condition for chemical equilibrium (3.9.27) can be interpreted to mean that the chemical potential of a compound is equal to the sum of the chemical potentials of its constituents.

(ii) The equilibrium condition (3.9.27) for the reaction (3.9.24) holds also when the system consists of several phases which are in contact with each other and between which the reactants can pass. This is shown by the equality of the chemical potential of each component in all of the phases which are in equilibrium with each other.

(iii) Eq. (3.9.27) can also be used to determine the equilibrium distribution of elementary particles which are transformed into one another by reactions.

For example, the distribution of electrons and positrons which are subject to pair annihilation, $e^- + e^+ \rightleftharpoons \gamma$, can be found (see problem 3.31). These applications of statistical mechanics are important in cosmology, in the description of the early stages of the Universe, and for the equilibria of elementary-particle reactions in stars.

3.9.3.2 Mixtures of Ideal Gases

To continue the evaluation of the equilibrium condition (3.9.27), we require information about the chemical potentials. In the following, we consider *reactions in* (classical) *ideal gases*. In Sect. 5.2, we show that the chemical potential of particles of type j in a mixture of ideal molecular gases can be written in the form

$$\mu_j = f_j(T) + kT \log c_j P , \tag{3.9.28a}$$

where $c_j = \frac{N_j}{N}$ holds and N is the total number of particles. The function $f_j(T)$ depends solely on temperature and contains the microscopic parameters of the gas of type j. From (3.9.27) and (3.9.28a), it follows that

$$\prod_j e^{\nu_j [f_j(T)/kT + \log(c_j P)]} = 1 . \tag{3.9.29}$$

According to Sect. 5.2, Eq. (5.2.4′) is valid:

$$f_j(T) = \varepsilon^0_{\text{el},j} - c_{P,j} T \log kT - kT \zeta_j . \tag{3.9.28b}$$

Inserting (3.9.28b) into (3.9.29) yields the product of the powers of the concentrations:

$$\prod_j c_j^{\nu_j} = K(T,P) \equiv e^{\sum_j \nu_j (\zeta_j - \frac{\varepsilon^0_{\text{el},j}}{kT})} (kT)^{\sum_j c_{P,j}\nu_j/k} P^{-\sum_j \nu_j} ; \tag{3.9.30}$$

where $\varepsilon^0_{\text{el},j}$ is the electronic energy, $c_{P,j}$ the specific heat of component j at constant pressure, and ζ_j is the chemical constant

$$\zeta_j = \log \frac{2m_j^{3/2}}{k\Theta_{\text{r},j}(2\pi\hbar^2)^{3/2}} . \tag{3.9.31}$$

Here, we have assumed that $\Theta_{\text{r}} \ll T \ll \Theta_{\text{v}}$, with Θ_{r} and Θ_{v} the characteristic temperatures for the rotational and vibrational degrees of freedom, Eqs. (5.1.11) and (5.1.17). Equation (3.9.30) is the *law of mass action* for the concentrations. The function $K(T,P)$ is also termed the mass action constant. The statement that $\prod_j c_j^{\nu_j}$ is a function of only T and P holds generally for *ideal mixtures* $\mu_j(T, P, \{c_i\}) = \mu_j(T, P, c_j = 1, c_i = 0(i \neq j)) + kT \log c_j$.

If, instead of the concentrations, we introduce the *partial pressures* (see remark (i) at the end of this section)

$$P_j = c_j P \,, \tag{3.9.32}$$

then we obtain

$$\prod_j P_j^{\nu_j} = K_P(T) \equiv e^{\sum_j \nu_j \left(\zeta_j - \frac{\varepsilon_{el,j}^0}{kT} \right)} (kT)^{\sum_j c_{P,j} \nu_j / k} \,, \tag{3.9.30'}$$

the *law of mass action* of Guldberg and Waage[15] for the *partial pressures*, with $K_P(T)$ independent of P.

We now find e.g. for the hydrogen-oxygen reaction of Eq. (3.9.23)

$$2H_2 + O_2 - 2H_2O = 0 \,,$$

with

$$\nu_{H_2} = 2 \,, \quad \nu_{O_2} = 1 \,, \quad \nu_{H_2O} = -2 \,, \tag{3.9.33}$$

the relation

$$K(T, P) = \frac{[H_2]^2 [O_2]}{[H_2O]^2} = const. \, e^{-q/kT} T^{\sum_j c_{P,j} \nu_j / k} P^{-1} \,. \tag{3.9.34}$$

Here, the concentrations $c_j = [A_j]$ are represented by the corresponding chemical symbols in square brackets, and we have used

$$q = 2\varepsilon_{H_2}^0 + \varepsilon_{O_2}^0 - 2\varepsilon_{H_2O}^0 > 0 \,,$$

the heat of reaction at absolute zero, which is positive for the oxidation of hydrogen. The degree of dissociation α is defined in terms of the concentrations:

$$[H_2O] = 1 - \alpha \,, \quad [O_2] = \frac{\alpha}{2} \,, \quad [H_2] = \alpha \,.$$

It then follows from (3.9.32) that

$$\frac{\alpha^3}{2(1-\alpha)^2} \sim e^{-q/kT} T^{\sum_j c_{P,j} \nu_j / k} P^{-1} \,, \tag{3.9.35}$$

from which we can calculate α; α decreases exponentially with falling temperature.

---

[15] The law of mass action was stated by Guldberg and Waage in 1867 on the basis of statistical considerations of reaction probabilities, and was later proved thermodynamically for ideal gases by Gibbs, who made it more specific through the calculation of $K(T, P)$.

The law of mass action makes important statements about the conditions under which the desired reactions can take place with optimum yields. It may be necessary to employ a catalyst in order to shorten the reaction time; however, what the equilibrium distribution of the reacting components will be is determined simply by the reaction equation and the chemical potentials of the constituents (components) – in the case of ideal gases, by Eq. (3.9.30).

The law of mass action has many applications in chemistry and technology. As just one example, we consider here the pressure dependence of the reaction equilibrium. From (3.9.30), it follows that the pressure derivative of $K(T, P)$ is given by

$$\frac{1}{K}\frac{\partial K}{\partial P} = \frac{\partial \log K}{\partial P} = -\frac{1}{P}\sum_i \nu_i \ , \tag{3.9.36a}$$

where $\nu = \sum_i \nu_i$ is the so called molar excess. From the equation of state of mixtures of ideal gases (Eq. (5.2.3)), $PV = kT\sum_i N_i$, we obtain for the changes ΔV and ΔN which accompany a reaction at constant T and P:

$$P\Delta V = kT\sum_i \Delta N_i \ . \tag{3.9.37a}$$

Let the number of individual reactions be $\Delta\mathcal{N}$, i.e. $\Delta N_i = \nu_i \Delta\mathcal{N}$, then it follows from (3.9.37a) that

$$-\frac{1}{P}\sum_i \nu_i = -\frac{\Delta V}{kT\Delta\mathcal{N}} \ . \tag{3.9.37b}$$

Taking $\Delta\mathcal{N} = L$ (the Loschmidt/Avagadro number), then ν_i moles of each component will react and it follows from (3.9.36a) and (3.9.37b) with the gas constant R that

$$\frac{1}{K}\frac{\partial K}{\partial P} = -\frac{\Delta V}{RT} \ . \tag{3.9.36b}$$

Furthermore, $\Delta V = \sum_i \nu_i V_{\text{mol}}$ is the volume change in the course of the reaction proceeding from right to left (for a reaction which is represented in the form (3.9.23)). (The value of the molar volume V_{mol} is the same for every ideal gas.) According to Eq. (3.9.36b) in connection with (3.9.30), a larger value of K leads to an increase in the concentrations c_j with positive ν_j, i.e. of those substances which are on the left-hand side of the reaction equation. Therefore, from (3.9.36b), a pressure increase leads to a shift of the equilibrium towards the side of the reaction equation corresponding to the smaller volume. When $\Delta V = 0$, the position of the equilibrium depends only upon the temperature, e.g. in the hydrogen chloride reaction $H_2 + Cl_2 \rightleftharpoons 2HCl$.

In a similar manner, one finds for the temperature dependence of $K(T, P)$ the result

$$\frac{\partial \log K}{\partial T} = \frac{\sum_i \nu_i h_i}{RT^2} = \frac{\Delta h}{RT^2} . \tag{3.9.38}$$

Here, h_i is the molar enthalpy of the substance i and Δh is the change of the overall molar enthalpy when the reaction runs its course one time from right to left in the reaction equation, c.f. problem 3.26.

An interesting and technically important application is Haber's synthesis of ammonia from nitrogen and hydrogen gas: the chemical reaction

$$N_2 + 3H_2 \rightleftharpoons 2NH_3 \tag{3.9.39}$$

is characterized by $1N_2 + 3H_2 - 2NH_3 \rightleftharpoons 0$ $(\nu = \sum_i \nu_i = 2)$:

$$\frac{c_{N_2} c_{H_2}^3}{c_{NH_3}^2} = K(T, P) = K_P(T) P^{-2} . \tag{3.9.40}$$

To obtain a high yield of NH_3, the pressure must be made as high as possible. Sommerfeld:[16] "The extraordinary success with which this synthesis is now carried out in industry is due to the complete understanding of the conditions for thermodynamic equilibrium (Haber), to the mastery of the engineering problems connected with high pressure (Bosch), and, finally, to the successful selection of catalyzers which promote high reaction rates (Mittasch)."

Remarks:

(i) The *partial pressures* introduced in Eq. (3.9.32), $P_j = c_j P$, with $c_j = N_j/N$, in accord with the equation of state of a mixture of ideal gases, Eq. (5.2.3), obey the equations

$$V P_j = N_j k T \qquad \text{and} \qquad P = \sum_i P_i . \tag{3.9.41}$$

(This fact is known as Dalton's Law: the non-interacting gases in the mixture produce partial pressures corresponding to their particle numbers, as if they each occupy the entire available volume.)

(ii) Frequently, the law of mass action is expressed in terms of the particle densities $\rho_i = N_i/V$:

$$\prod_i \rho_i^{\nu_i} = K_\rho(T) \equiv (kT)^{\sum_i \nu_i} K_P(T) . \tag{3.9.30'}$$

---

[16] A. Sommerfeld, *Thermodynamics and Statistical Mechanics: Lectures on Theoretical Physics*, Vol. V (Academic Press, New York, 1956), p. 86

(iii) Now we turn to the *direction* which a reaction will take. If a mixture is initially present with arbitrary densities, the direction in which the reaction will proceed can be read off the law of mass action. Let $\nu_1, \nu_2, \ldots, \nu_s$ be positive and $\nu_{s+1}, \nu_{s+2}, \ldots, \nu_n$ negative, so that the reaction equation (3.9.24) takes on the form

$$\nu_i A_i \rightleftharpoons \sum_{i=s+1}^{n} |\nu_i| A_i \,, \tag{3.9.24'}$$

Assume that the product of the particle densities obeys the inequality

$$\prod_i \rho_i^{\nu_i} \equiv \frac{\prod\limits_{i=1}^{s} \rho_i^{\nu_i}}{\prod\limits_{i=s+1}^{n} \rho_i^{|\nu_i|}} < K_\rho(T) \,, \tag{3.9.42}$$

i.e. the system is not in chemical equilibrium. If the chemical reaction proceeds from right to left, the densities on the left will increase, and the fraction in the inequality will become larger. Therefore, in the case (3.9.42), the reaction will proceed from right to left. If, in contrast, the inequality was initially reversed, with a $>$ sign, then the reaction would proceed from left to right.

(iv) All chemical reactions exhibit a heat of reaction, i.e. they are accompanied either by heat release (exothermic reactions) or by taking up of heat (endothermic reactions). We recall that for isobaric processes, $\Delta Q = \Delta H$, and the heat of reaction is equal to the change in the enthalpy; see the comment following Eq. (3.1.12). The temperature dependence of the reaction equilibrium follows from Eq. (3.9.38). A temperature increase at constant pressure shifts the equilibrium towards the side of the reaction equation where the enthalpy is higher; or, expressed differently, it leads to a reaction in the direction in which heat is taken up. As a rule, the electronic contribution $\mathcal{O}$ (eV) dominates. Thus, at low temperatures, the enthalpy-rich side is practically not present.

*3.9.4 Vapor-pressure Increase by Other Gases and by Surface Tension

3.9.4.1 The Evaporation of Water in Air

As discussed in detail in Sect. 3.8.1, a single-component system can evaporate only along its vapor-pressure curve $P_0(T)$, or, stated differently, only along the vapor-pressure curve are the gaseous and the liquid phases in equilibrium. If an additional gas is present, this means that there is one more degree of freedom in Gibbs' phase rule, so that a liquid can coexist with its vapor even outside of $P_0(T)$.

Here, we wish to investigate evaporation in the presence of additional gases and in particular that of water under an air atmosphere. To this end we assume that the other gas is dissolved in the liquid phase to only a negligible extent. If the chemical potential of the liquid were independent of the pressure, then the other gas would have no influence at all on the chemical potential of the liquid; the partial pressure of the vapor would then have to be identical with the vapor pressure of the pure substance – a statement which is frequently made. In fact, the total pressure acts on the liquid, which changes its chemical potential. The resulting increase of the vapor pressure will be calculated here.

To begin, we note that

$$\left(\frac{\partial \mu_L}{\partial P}\right)_T = \frac{V}{N} \tag{3.9.43}$$

is small, owing to the small specific volume $v_L = \frac{V}{N}$ of the liquid. When the pressure is changed by ΔP, the chemical potential of the liquid changes according to

$$\mu_L(T, P + \Delta P) = \mu_L(T, P) + v_L \Delta P + \mathcal{O}(\Delta P^2) \, . \tag{3.9.44}$$

From the Gibbs–Duhem relation, the chemical potential of the liquid is

$$\mu_L = e_L - T s_L + P v_L \, . \tag{3.9.45}$$

Here, e_L and s_L refer to the internal energy and the entropy per particle. When we can neglect the temperature and pressure dependence of e_L, s_L, and v_L, then (3.9.44) is valid with no further corrections.
The chemical potential of the vapor, assuming an ideal mixture[17], is

$$\mu_{vapor}(T, P) = \mu_0(T) + kT \log cP \, , \tag{3.9.46}$$

where c is the concentration of the vapor in the gas phase, $c = \frac{N_{vapor}}{N_{other} + N_{vapor}}$.
The vapor-pressure curve $P_0(T)$ without additional gases follows from

$$\mu_L(T, P_0) = \mu_0(T) + kT \log P_0 \, . \tag{3.9.47}$$

With an additional gas, the pressure is composed of the pressure of the other gas P_{other} and the partial pressure of the vapor, $P_{vapor} = cP$; all together, $P = P_{other} + P_{vapor}$. Then the equality of the chemical potentials in the liquid and the gaseous phases is expressed by

$$\mu_L(T, P_{other} + P_{vapor}) = \mu_0(T) + kT \log P_{vapor} \, .$$

Subtracting (3.9.47) from this, we find

---

[17] See Sect. 5.2

$$\mu_L(T, P_{\text{other}} + P_{\text{vapor}}) - \mu_L(T, P_0) = kT \log \left(\frac{P_{\text{vapor}} - P_0}{P_0} + 1 \right)$$

$$v_L(P_{\text{other}} + P_{\text{vapor}} - P_0) \approx kT \frac{P_{\text{vapor}} - P_0}{P_0}$$

$$v_L P_{\text{other}} = \left(\frac{kT}{P_0} - v_L \right)(P_{\text{vapor}} - P_0)$$

$$P_{\text{vapor}} - P_0 = \frac{v_L P_{\text{other}}}{v_G - v_L} = \frac{v_L}{v_G - v_L}(P - P_{\text{vapor}}) \,. \tag{3.9.48}$$

From the second term in Eq. 3.9.48, it follows that the increase in vapor pressure is given approximately by $P_{\text{vapor}} - P_0(T) \approx \frac{v_L}{v_G} P_{\text{other}}$, and the exact expression is found to be

$$P_{\text{vapor}} = P_0(T) + \frac{v_L}{v_G}(P - P_0(T)) \,. \tag{3.9.49}$$

The partial pressure of the vapor is increased relative to the vapor-pressure curve by $\frac{v_L}{v_G} \times (P - P_0(T))$. Due to the smallness of the factor $\frac{v_L}{v_G}$, the partial pressure is still to a good approximation the same as the vapor pressure at the temperature T. The most important result of these considerations is the following: while a liquid under the pressure P at the temperature T is in equilibrium with its vapor phase only for $P = P_0(T)$; that is, for $P > P_0(T)$ (or at temperatures below its boiling point) it exists only in liquid form, it is also in equilibrium in this region of (P, T) with its vapor when dissolved in another gas.

We now discuss the evaporation of water or the sublimation of ice under an atmosphere of air, see Fig. 3.39. The atmosphere predetermines a particular pressure P. At each temperature T below the evaporation temperature determined by this pressure $(P > P_0(T))$, just enough water evaporates to make its partial pressure equal that given by (3.9.49) (recall $P_{\text{vapor}} = cP$). The concentration of the water vapor is $c = (P_0(T) + \frac{v_L}{v_G}(P - P_0(T)))/P$.

In a free air atmosphere, the water vapor is transported away by diffusion or by convection (wind), and more and more water must evaporate (vaporize).[18] On

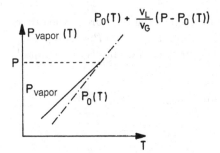

Fig. 3.39. The vapor pressure P_{vapor} lies above the vapor-pressure curve $P_0(T)$ (dot-dashed curve)

---

[18] As already mentioned, the above considerations are also applicable to sublimation. When one cools water at 1 atm below 0°C, it freezes to ice. This ice at

increasing the temperature, the partial pressure of the water increases, until finally it is equal to P. The vaporization which then results is called boiling.

For $P = P_0(T)$, the liquid is in equilibrium with its pure vapor. Evaporation then occurs not only at the liquid surface, but also within the liquid, in particular at the walls of its container. There, bubbles of vapor are formed, which then rise to the surface. Within these vapor bubbles, the vapor pressure is $P_0(T)$, corresponding to the temperature T. Since the vapor bubbles within the liquid are also subject to the hydrostatic pressure of the liquid, their temperature must in fact be somewhat higher than the boiling point under atmospheric pressure. If the liquid contains nucleation centers (such as the fat globules in milk), at which vapor bubbles can form more readily than in the pure liquid, then it will "boil over".

The increase in the vapor pressure by increased external pressure, or as one might say, by 'pressing on it', may seem surprising. The additional pressure causes an increase in the release of molecules from the liquid, i.e. an increase in the partial pressure.

3.9.4.2 Vapor-Pressure Increase by Surface Tension of Droplets

A further additional pressure is due to the surface tension and plays a role in the *evaporation of liquid droplets*. We consider a liquid droplet of radius r. When the radius is increased isothermally by an amount dr, the surface area increases by $8\pi r\, dr$, which leads to an energy increase of $\sigma 8\pi r\, dr$, where σ is the surface tension. Owing to the pressure difference p between the pressure within the droplet and the pressure of the surrounding atmosphere, there is a force $p\, 4\pi r^2$ which acts outwardly on the surface. The total change of the free energy is therefore

$$dF = \delta A = \sigma 8\pi r\, dr - p\, 4\pi r^2\, dr \ . \tag{3.9.50}$$

In equilibrium, the free energy of the droplet must be stationary, so that for the pressure difference we find the following dependence on the radius:

$$p = \frac{2\sigma}{r} \ . \tag{3.9.51}$$

Thus, small droplets have a higher vapor pressure than larger one. The vapor-pressure increase due to the surface tension is from Eq. (3.9.48) now seen to be

$$P_{\text{vapor}} - P_0(T) = \frac{2\sigma}{r} \frac{v_{\text{L}}}{v_{\text{G}} - v_{\text{L}}} \tag{3.9.52}$$

inversely proportional to the radius of the droplet. In a mixture of small and large droplets, the smaller ones are therefore consumed by the larger ones.

---

e.g. $-10°$C is to be sure as a single-component system not in equilibrium with the gas phase, but rather with the water vapor in the atmosphere at a partial pressure of about $P_0(-10°$C$)$, where $P_0(T)$ represents the sublimation curve. For this reason, frozen laundry dries, because ice sublimes in the atmosphere.

Remarks:

(i) Small droplets evaporate more readily than liquids with a flat surface, and conversely condensation occurs less easily on small droplets. This is the reason why extended solid cooled surfaces promote the condensation of water vapor more readily than small droplets do. The temperature at which the condensation of water from the atmosphere onto extended surfaces (dew formation) takes place is called the dew point. It depends on the partial pressure of water vapor in the air, i.e. its degree of saturation, and can be used to determine the amount of moisture in the air.

(ii) We consider the homogeneous condensation of a gas in free space without surfaces. The temperature of the gas is taken to be T and the vapor pressure at this temperature to be $P_0(T)$. We assume that the pressure P of the gas is greater than the vapor pressure; it is then referred to as supersaturated vapor. For each degree of supersaturation, then, a critical radius can be defined from (3.9.52):

$$r_{cr} = \frac{v_L}{v_G} \frac{2\sigma}{(P - P_0(T))} \ .$$

For droplets whose radius is smaller than r_{cr} the vapor is not supersaturated. Condensation can therefore not take place through the formation of very small droplets, since their vapor pressures would be higher than P. Some critical droplets must be formed through fluctuations in order that condensation can be initiated. Condensation is favored by additional attractive forces; for example, in the air, there are always electrically-charged dust particles and other impurities present, which as a result of their electrical forces promote condensation, i.e. they act as nucleation centers for condensation.

Problems for Chapter 3

3.1 Read off the partial derivatives of the internal energy E with respect to its natural variables from Eq. (3.1.3).

3.2 Show that

$$\delta g = \alpha dx + \beta \frac{x}{y} dy$$

is not an exact differential: a) using the integrability conditions and b) by integration from P_1 to P_2 along the paths C_1 and C_2. Show that $1/x$ is an integrating factor, $df = \delta g/x$.

3.3 Prove the chain rule (3.2.13) for Jacobians.

3.4 Derive the following relations:

$$\frac{C_P}{C_V} = \frac{\kappa_T}{\kappa_S} \ , \quad \left(\frac{\partial T}{\partial V}\right)_S = -\frac{T}{C_V}\left(\frac{\partial P}{\partial T}\right)_V \quad \text{and} \quad \left(\frac{\partial T}{\partial P}\right)_S = \frac{T}{C_P}\left(\frac{\partial V}{\partial T}\right)_P \ .$$

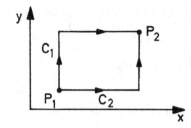

Fig. 3.40. Paths in the x-y diagram

3.5 Determine the work performed by an ideal gas, $W(V) = \int_{V_1}^{V} dV\, P$ during a reversible adiabatic expansion. From $\delta Q = 0$, it follows that $dE = -P dV$, and from this the adiabatic equations for an ideal gas can be obtained: $T = T_1\left(\frac{V_1}{V}\right)^{2/3}$ and $P = NkT_1 \frac{V_1^{2/3}}{V^{5/3}}$. They can be used to determine the work performed.

3.6 Show that the stability conditions (3.6.48a,b) follow from the maximalization of the entropy.

3.7 One liter of an ideal gas expands reversibly and isothermally at ($20°$C) from an initial pressure of 20 atm to 1 atm. How large is the work performed in Joules? What quantity of heat Q in calories must be transferred to the gas?

3.8 Show that the ratio of the entropy increase on heating of an ideal gas from T_1 to T_2 at constant pressure to that at constant volume is given by the ratio of the specific heats.

3.9 A thermally insulated system is supposed to consist of 2 subsystems (T_A, V_A, P) and (T_B, V_B, P), which are separated by a movable, diathermal piston (Fig. 3.41(a). The gases are ideal.
(a) Calculate the entropy change accompanying equalization of the temperatures (irreversible process).
(b) Calculate the work performed in a quasistatic temperature equalization; cf. Fig. 3.41(b).

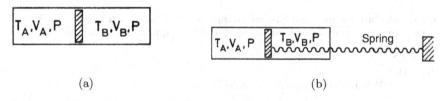

(a)   (b)

Fig. 3.41. For problem 3.9

3.10 Calculate the work obtained, $W = \oint P dV$, in a Carnot cycle using an ideal gas, by evaluating the ring integral.

3.11 Compare the cooling efficiency of a Carnot cycle between the temperatures T_1 and T_2 with that of two Carnot cycles operating between T_1 and T_3 and between T_3 and T_2 ($T_1 < T_3 < T_2$). Show that it is more favorable to decompose a cooling process into several smaller steps.

3.12 Discuss a Carnot cycle in which the working 'substance' is thermal radiation. For this case, the following relations hold: $E = \sigma V T^4$, $pV = \frac{1}{3}E$, $\sigma > 0$.
(a) Derive the adiabatic equation. **(b)** Compute C_V and C_P.

3.13 Calculate the efficiency of the Joule cycle (see Fig. 3.42):

   Result : $\eta = 1 - (P_2/P_1)^{(\kappa-1)/\kappa}$.

Compare this efficiency with that of the Carnot cycle (drawn in dashed lines), using an ideal gas as working substance.

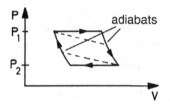

Fig. 3.42. The Joule cycle

3.14 Calculate the efficiency of the Diesel cycle (Fig. 3.43) Result:

$$\eta = 1 - \frac{1}{\kappa}\frac{(V_2/V_1)^\kappa - (V_3/V_1)^\kappa}{(V_2/V_1) - (V_3/V_1)} .$$

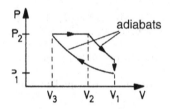

Fig. 3.43. The Diesel cycle

3.15 Calculate for an ideal gas the change in the internal energy, the work performed, and the quantity of heat transferred for the quasistatic processes along the following paths from 1 to 2 (see Fig. 3.44)
(a) 1-A-2
(b) 1-B-2
(c) 1-C-2. What is the shape of the $E(P,V)$ surface?

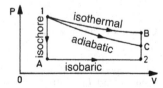

Fig. 3.44. For problem 3.15

3.16 Consider the socalled Stirling cycle, where a heat engine (with an ideal gas as working substance) performs work according to the following quasistatic cycle:
(a) isothermal expansion at the temperature T_1 from a volume V_1 to a volume V_2.
(b) cooling at constant volume V_2 from T_1 to T_2.
(c) isothermal compression at the temperature T_2 from V_2 to V_1.
(d) heating at constant volume from T_2 to T_1.
Determine the thermal efficiency η of this process!

3.17 The ratio of the specific volume of water to that of ice is 1.000:1.091 at 0°C and 1 atm. The heat of melting is 80 cal/g. Calculate the slope of the melting curve.

3.18 Integrate the Clausius–Clapeyron differential equation for the transition liquid-gas, by making the simplifying assumption that the heat of transition is constant, V_{liquid} can be neglected in comparison to V_{gas}, and that the equation of state for ideal gases is applicable to the gas phase.

3.19 Consider the neighborhood of the triple point in a region where the limiting curves can be approximated as straight lines. Show that $\alpha < \pi$ holds (see Fig. 3.45). *Hint:* Use $dP/dT = \Delta S/\Delta V$, and the fact that the slope of line 2 is greater than that of line 3.

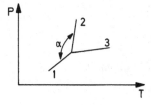

Fig. 3.45. The vicinity of a triple point

3.20 The latent heat of ice per unit mass is Q_L. A container holds a mixture of water and ice at the freezing point (absolute temperature T_0). An additional amount of the water in the container (of mass m) is to be frozen using a cooling apparatus. The heat output from the cooling apparatus is used to heat a body of heat capacity C and initial temperature T_0. What is the minimum quantity of heat energy transferred from the apparatus to the body? (Assume C to be temperature independent).

3.21 (a) Discuss the pressure dependence of the reaction $N_2 + 3H_2 \rightleftharpoons 2NH_3$ (ammonia synthesis). At what pressure is the yield of ammonia greatest?
(b) Discuss the thermal dissociation $2H_2O \rightleftharpoons 2H_2 + O_2$. Show that an increase in pressure works against the dissociation.

3.22 Give the details of the derivation of Eqs. (3.9.36a) and (3.9.36b).

3.23 Discuss the pressure and temperature dependence of the reaction

$$CO + 3H_2 \rightleftharpoons CH_4 + H_2O \ .$$

3.24 Apply the law of mass action to the reaction $H_2 + Cl_2 \rightleftharpoons 2HCl$.

3.25 Derive the law of mass action for the particle densities

$$\rho_j = N_j/V \qquad \text{(Eq. (3.9.30'))} .$$

3.26 Prove Eq. (3.9.38) for the temperature dependence of the mass-action constant.

Hint: Show that $H = G - T\frac{\partial G}{\partial T} = T^2 \frac{\partial}{\partial T}\left(\frac{G}{T}\right)$
and express the change in the free enthalpy

$$\Delta G = \sum_i \mu_i \nu_i$$

using Eq. (3.9.28), then insert the law of mass action (3.9.30) or (3.9.30').

3.27 The Pomeranchuk effect. The entropy diagram for solid and liquid He^3 has the shape shown below 3 K. Note that the specific volumes of both phases do not change within this temperature range. Draw $P(T)$ for the coexistence curves of the phases.

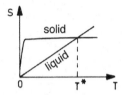

Fig. 3.46. The Pomeranchuk effect

3.28 The (specific) free energies f_α and f_β of two solid phases α and β with a miscibility gap and the (specific) free energy f_L of the liquid mixture are shown as functions of the concentration c in Fig. 3.47.

Discuss the meaning of the dashed and solid double tangents. On lowering the temperature, the free energy of the liquid phase is increased, i.e. f_L is shifted upwards relative to the two fixed branches of the free energy. Derive from this the shape of the eutectic phase diagram.

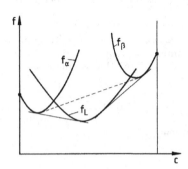

Fig. 3.47. Liquid mixture

3.29 A typical shape for the phase diagram of liquid and gaseous mixtures is shown in Fig. 3.48.

The components A and B are completely miscible in both the gas phase and the liquid phase. B has a higher boiling point than A. At a temperature in the interval $T_A < T < T_B$, the gas phase is therefore richer in A than the liquid phase. Discuss the boiling process for the initial concentration c_0

(a) in the case that the liquid remains in contact with the gas phase: show that vaporization takes place in the temperature interval T_0 to T_e.

(b) in the case that the vapor is pumped off: show that the vaporization takes place in the interval T_0 to T_B.

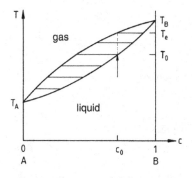

Fig. 3.48. Bubble point and dew point lines

Remark: The curve which is made by the boiling curve (evaporation limit) and the condensation curve together form the bubble point and dew point lines, a lens-shaped closed curve. Its shape is of decisive importance for the efficiency of distillation processes. This 'boiling lens' can also take on much more complex shapes than in Fig. 3.48, such as e.g. that shown in Fig. 3.49. A mixture with the concentration c_a is called azeotropic. For this concentration, the evaporation of the mixture occurs exactly at the temperature T_a and not in a temperature interval. The eutectic concentration is also special in this sense. Such a point occurs in an alcohol-water mixture at 96%, which limits the distillation of alcohol.[19]

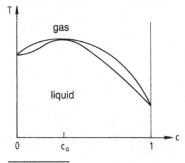

Fig. 3.49. Bubble point and dew point lines

---

[19] Detailed information about phase diagrams of mixtures can be found in M. Hansen, *Constitution of Binary Alloys*, McGraw Hill, 1983 und its supplements. Further detailed discussions of the shape of phase diagrams are to be found in L. D. Landau and E. M. Lifshitz, *Course of Theoretical Physics*, Vol. V, *Statistical Physics*, Pergamon Press 1980.

3.30 The free energy of the liquid phase, f_L, is drawn in Fig. (3.50) as a function of the concentration, as well as that of the gas phase, f_G. It is assumed that f_L is temperature independent and f_G shifts upwards with decreasing temperature (Fig. 3.50). Explain the occurrence of the 'boiling lens' in problem 3.29.

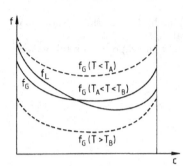

Fig. 3.50. Free energy

3.31 Consider the production of electron-positron pairs,

$$e^+ + e^- \rightleftharpoons \gamma \,.$$

Assume for simplicity that the chemical potential of the electrons and positrons is given in the nonrelativistic limit, taking the rest energy into account, by $\mu = mc^2 + kT \log \frac{\lambda^3 N}{V}$: Show that for the particle number densities $n_\pm$ of $e^\pm$ that

$$n_+ n_- = \lambda^{-6} e^{-\frac{2mc^2}{kT}}$$

holds and discuss the consequences.

3.32 Consider the boiling and condensation curves of a two-component liquid mixture. Take the concentrations in the gaseous and liquid phases to be c_G and c_L. Show that at the points where $c_G = c_L$ (the azeotropic mixture) i.e. where the boiling and condensation curves come together, for a fixed pressure P the following relation holds:

$$\frac{dT}{dc} = 0 \,,$$

and for fixed T

$$\frac{dP}{dc} = 0 \,,$$

thus the slopes are horizontal.
Method: Start from the differential Gibbs-Duhem relations for the gas and the liquid phases along the limiting curves.

3.33 Determine the temperature of the atmosphere as a function of altitude. How much does the temperature decrease per *km* of altitude? Compare your result for the pressure $P(z)$ with the barometric formula (see problem 2.15).

Method: Start with the force balance on a small volume of air. That gives

$$\frac{dP(z)}{dz} = -mg\,P(z)/k \cdot T(z) \,.$$

Assume that the temperature changes depend on the pressure changes of the air (ideal gas) adiabatically $\frac{dT(z)}{T(z)} = \frac{\gamma-1}{\gamma}\frac{dP(z)}{P}$. From this, one finds $\frac{dT(z)}{dz}$. Numerical values: $m = 29\,\text{g/mole}$, $\gamma = 1.41$.

3.34 In meteorology, the concept of a "homogeneous atmosphere" is used, where ρ is taken to be constant. Determine the pressure and the temperature in such an atmosphere as functions of the altitude. Calculate the entropy of the homogeneous atmosphere and compare it with that of an isothermal atmosphere with the same energy content. Could such a homogeneous atmosphere be stable?

4. Ideal Quantum Gases

In this chapter, we want to derive the thermodynamic properties of ideal quantum gases, i.e. non-interacting particles, on the basis of quantum statistics. This includes nonrelativistic fermions and bosons whose interactions may be neglected, quasiparticles in condensed matter, and relativistic quanta, in particular photons.

4.1 The Grand Potential

The calculation of the grand potential is found to be the most expedient way to proceed. In order to have a concrete system in mind, we start from the Hamiltonian for N non-interacting, nonrelativistic particles,

$$H = \sum_{i=1}^{N} \frac{1}{2m} \mathbf{p}_i^2 \ . \tag{4.1.1}$$

We assume the particles to be enclosed in a cube of edge length L and volume $V = L^3$, and apply periodic boundary conditions. The single-particle eigenfunctions of the Hamiltonian are then the momentum eigenstates $|\mathbf{p}\rangle$ and are given in real space by

$$\varphi_{\mathbf{p}}(\mathbf{x}) = \langle \mathbf{x} | \mathbf{p} \rangle = \frac{1}{\sqrt{V}} e^{i\mathbf{p}\cdot\mathbf{x}/\hbar} \ , \tag{4.1.2a}$$

where the momentum quantum numbers can take on the values

$$\mathbf{p} = \frac{2\pi\hbar}{L}(\nu_1, \nu_2, \nu_3) \ , \quad \nu_\alpha = 0, \pm 1, \ldots \ , \tag{4.1.2b}$$

and the single-particle kinetic energy is given by

$$\varepsilon_{\mathbf{p}} = \frac{\mathbf{p}^2}{2m} \ . \tag{4.1.2c}$$

For the complete characterization of the single-particle states, we must still take the spin s into account. It is integral for bosons and half-integral for fermions. The quantum number m_s for the z-component of the spins has

$2s+1$ possible values. We combine the two quantum numbers into one symbol, $p \equiv (\mathbf{p}, m_s)$ and find for the complete energy eigenstates

$$|p\rangle \equiv |\mathbf{p}\rangle \, |m_s\rangle \; . \tag{4.1.2d}$$

In the treatment which follows, we could start from arbitrary non-interacting Hamiltonians, which can also contain a potential and can depend on the spin, as is the case for electrons in a magnetic field. We then still denote the single-particle quantum numbers by p and the eigenvalue belonging to the energy eigenstate $|p\rangle$ by ε_p, but it need no longer be the same as (4.1.2c). These states form the basis of the N-particle states for bosons and fermions:

$$|p_1, p_2, \ldots, p_N\rangle = \mathcal{N} \sum_P (\pm 1)^P \, P \, |p_1\rangle \ldots |p_N\rangle \; . \tag{4.1.3}$$

Here, the sum runs over all the permutations P of the numbers 1 to N. The upper sign holds for bosons, $(+1)^P = 1$, the lower sign for fermions. $(-1)^P$ is equal to 1 for even permutations and -1 for odd permutations. The bosonic states are completely symmetric, the fermionic states are completely antisymmetric. As a result of the symmetrization operation, the state (4.1.3) is completely characterized by its *occupation numbers* n_p, which indicate how many of the N particles are in the state $|p\rangle$. For bosons, $n_p = 0, 1, 2, \ldots$ can assume all integer values from 0 to ∞. These particles are said to obey Bose–Einstein statistics. For fermions, each single-particle state can be occupied at most only once, $n_p = 0, 1$ (identical quantum numbers would yield zero due to the antisymmetrization on the right-hand side of (4.1.3)). Such particles are said to obey Fermi–Dirac statistics. The normalization factor in (4.1.3) is $\mathcal{N} = \frac{1}{\sqrt{N!}}$ for fermions and $\mathcal{N} = (N! \, n_{p_1}! \, n_{p_2}! \ldots)^{-1/2}$ for bosons.[1]

For an N-particle state, the sum of all the n_p obeys

$$N = \sum_p n_p \; , \tag{4.1.4}$$

and the energy eigenvalue of this N-particle state is

$$E(\{n_p\}) = \sum_p n_p \varepsilon_p \; . \tag{4.1.5}$$

We can now readily calculate the grand partition function (Sect. 2.7.2):

---

[1] Note: for bosons, the state (4.1.3) can also be written in the form $(N!/n_{p_1}! \, n_{p_2}! \ldots)^{-1/2} \sum_{P'} P' \, |p_1\rangle \ldots |p_N\rangle$, where the sum includes only those permutations P' which lead to different terms.

$$Z_G \equiv \sum_{N=0}^{\infty} \sum_{\substack{\{n_p\} \\ \sum_p n_p = N}} e^{-\beta(E(\{n_p\})-\mu N)} = \sum_{\{n_p\}} e^{-\beta \sum_p (\varepsilon_p - \mu) n_p}$$

$$= \prod_p \sum_{n_p} e^{-\beta(\varepsilon_p - \mu)n_p} = \begin{cases} \prod_p \dfrac{1}{1 - e^{-\beta(\varepsilon_p - \mu)}} & \text{for bosons} \\ \prod_p \left(1 + e^{-\beta(\varepsilon_p - \mu)}\right) & \text{for fermions .} \end{cases} \quad (4.1.6)$$

We give here some explanations relevant to (4.1.6). Here, $\sum_{\{n_p\}} \cdots \equiv \prod_p \sum_{n_p} \cdots$ refers to the multiple sum over all occupation numbers, whereby each occupation number n_p takes on the allowed values (0,1 for fermions and $0,1,2, \ldots$ for bosons). In this expression, $p \equiv (\mathbf{p}, m_s)$ runs over all values of $\mathbf{p}$ and m_s. The calculation of the grand partition function requires that one first sum over all the states allowed by a particular value of the particle number N, and then over all particle numbers, $N = 0, 1, 2, \ldots$. In the definition of Z_G, $\sum_{\{n_p\}}$ therefore enters with the constraint $\sum_p n_p = N$. Since however in the end we must sum over all N, the expression after the second equals sign is obtained; in it, the sum runs over all n_p independently of one another. Here, we see that it is most straightforward to calculate the grand partition function as compared to the other ensembles. For bosons, a product of geometric series is obtained in (4.1.6); the condition for their convergence requires that $\mu < \varepsilon_p$ for all p.

The grand potential follows from (4.1.6):

$$\Phi = -\beta^{-1} \log Z_G = \pm \beta^{-1} \sum_p \log \left(1 \mp e^{-\beta(\varepsilon_p - \mu)}\right) , \qquad (4.1.7)$$

from which we can derive all the thermodynamic quantities of interest. Here, and in what follows, the upper (lower) signs refer to bosons (fermions). For the average particle number, we therefore find

$$N \equiv -\left(\frac{\partial \Phi}{\partial \mu}\right)_\beta = \sum_p n(\varepsilon_p) , \qquad (4.1.8)$$

where we have introduced

$$n(\varepsilon_p) \equiv \frac{1}{e^{\beta(\varepsilon_p - \mu)} \mp 1} ; \qquad (4.1.9)$$

these are also referred to as the Bose or the Fermi distribution functions. We now wish to show that $n(\varepsilon_q)$ is the *average occupation number* of the state $|q\rangle$. To this end, we calculate the average value of n_q:

$$\langle n_q \rangle = \mathrm{Tr}(\rho_G n_q) = \frac{\sum_{\{n_p\}} e^{-\beta \sum_p n_p (\varepsilon_p - \mu)} n_q}{\sum_{\{n_p\}} e^{-\beta \sum_p n_p (\varepsilon_p - \mu)}} = \frac{\sum_{n_q} e^{-\beta n_q (\varepsilon_q - \mu)} n_q}{\sum_{n_q} e^{-\beta n_q (\varepsilon_q - \mu)}}$$

$$= -\frac{\partial}{\partial x} \log \sum_n e^{-xn} \bigg|_{x = \beta(\varepsilon_q - \mu)} = n(\varepsilon_q) ,$$

which demonstrates the correctness of our assertion. We now return to the calculation of the thermodynamic quantities. For the *internal energy*, we find from (4.1.7)

$$E = \left(\frac{\partial(\Phi\beta)}{\partial\beta}\right)_{\beta\mu} = \sum_p \varepsilon_p n(\varepsilon_p) , \qquad (4.1.10)$$

where in taking the derivative, the *product* $\beta\mu$ is held constant.

Remarks:

(i) In order to ensure that $n(\varepsilon_p) \geq 0$ for every value of p, for *bosons* we require that $\mu < 0$, and for an arbitrary energy spectrum, that $\mu < \min(\varepsilon_p)$.

(ii) For $e^{-\beta(\varepsilon_p - \mu)} \ll 1$ and $s = 0$, we obtain from (4.1.7)

$$\Phi = -\beta^{-1}\sum_p e^{-\beta(\varepsilon_p - \mu)} = -\frac{z}{\beta}\frac{V}{(2\pi\hbar)^3}\int d^3p\, e^{-\beta p^2/2m} = -\frac{zV}{\beta\lambda^3} , \qquad (4.1.11)$$

which is identical to the grand potential of a classical ideal gas, Eq. (2.7.23). Here, the dispersion relation $\varepsilon_p = \mathbf{p}^2/2m$ from Eq. (4.1.2c) was used for the right-hand side of (4.1.11). In

$$z = e^{\beta\mu} , \qquad (4.1.12)$$

we have introduced the fugacity, and $\lambda = \frac{h}{\sqrt{2\pi mkT}}$ (Eq. (2.7.20)) denotes the thermal wavelength. For $s \neq 0$, an additional factor of $(2s+1)$ would occur after the second and third equals signs in Eq. (4.1.11).

(iii) The calculation of the grand partition function becomes even simpler if we make use of the second-quantization formalism

$$Z_G = \text{Tr} \exp(-\beta(H - \mu\hat{N})) , \qquad (4.1.13a)$$

where the Hamiltonian and the particle number operator in second quantization[2] have the form

$$H = \sum_p \varepsilon_p a_p^\dagger a_p \qquad (4.1.13b)$$

and

$$\hat{N} = \sum_p a_p^\dagger a_p . \qquad (4.1.13c)$$

It then follows that

$$Z_G = \text{Tr} \prod_p e^{-\beta(\varepsilon_p - \mu)a_p^\dagger a_p} = \prod_p \sum_{n_p} e^{-\beta(\varepsilon_p - \mu)n_p} \qquad (4.1.13d)$$

and thus we once again obtain (4.1.6).

---

[2] See e.g. F. Schwabl, *Advanced Quantum Mechanics*, 3rd ed. (QM II), Springer, 2005, Chapter 1.

According to Eq. (4.1.2b) we may associate with each of the discrete $\mathbf{p}$ values a volume element of size $\Delta = 2\pi\hbar/L^3$. Hence, sums over $\mathbf{p}$ may be replaced by integrals in the limit of large V. For the Hamiltonian of free particles (4.1.1), this implies in (4.1.7) and (4.1.8)

$$\sum_p \ldots = g \sum_{\mathbf{P}} \ldots = g \frac{1}{\Delta} \sum_{\mathbf{P}} \Delta \ldots = g \frac{V}{(2\pi\hbar)^3} \int d^3p \ldots \qquad (4.1.14a)$$

with the degeneracy factor

$$g = 2s + 1 , \qquad (4.1.14b)$$

as a result of the spin-independence of the single-particle energy ε_p.

For the average particle number, we then find from (4.1.8)[3]

$$N = \frac{gV}{(2\pi\hbar)^3} \int d^3p\, n(\varepsilon_{\mathbf{p}}) = \frac{gV}{2\pi^2\hbar^3} \int_0^\infty dp\, p^2 n(\varepsilon_{\mathbf{p}})$$

$$= \frac{gVm^{3/2}}{2^{1/2}\pi^2\hbar^3} \int_0^\infty \frac{d\varepsilon\,\sqrt{\varepsilon}}{e^{\beta(\varepsilon-\mu)} \mp 1} , \qquad (4.1.15)$$

where we have introduced $\varepsilon = p^2/2m$ as integration variable. We also define the specific volume

$$v = V/N \qquad (4.1.16)$$

and substitute $x = \beta\varepsilon$, finally obtaining from (4.1.15)

$$\frac{1}{v} = \frac{1}{\lambda^3} \frac{2g}{\sqrt{\pi}} \int_0^\infty dx \frac{x^{1/2}}{e^x z^{-1} \mp 1} = \frac{g}{\lambda^3} \begin{cases} g_{3/2}(z) & \text{for bosons} \\ f_{3/2}(z) & \text{for fermions} . \end{cases} \qquad (4.1.17)$$

In this expression, we have introduced the generalized ζ-functions, which are defined by[4]

$$\left.\begin{matrix} g_\nu(z) \\ f_\nu(z) \end{matrix}\right\} \equiv \frac{1}{\Gamma(\nu)} \int_0^\infty dx \frac{x^{\nu-1}}{e^x z^{-1} \mp 1} . \qquad (4.1.18)$$

Similarly, from (4.1.7), we find

---

[3] For bosons, we shall see in Sect. 4.4 that in a temperature range where $\mu \to 0$, the term with $\mathbf{p} = 0$ must be treated separately in making the transition from the sum over momenta to the integral.

[4] The gamma function is defined as $\Gamma(\nu) = \int_0^\infty dt\, e^{-t} t^{\nu-1}$ [Re $\nu > 0$]. It obeys the relation $\Gamma(\nu + 1) = \nu\,\Gamma(\nu)$.

$$\Phi = \pm \frac{gV}{(2\pi\hbar)^3 \beta} \int d^3p \, \log\left(1 \mp e^{-\beta(\varepsilon_p - \mu)}\right)$$

$$= \pm \frac{gV m^{3/2}}{2^{1/2}\pi^2 \hbar^3 \beta} \int_0^\infty d\varepsilon \, \sqrt{\varepsilon} \log\left(1 \mp e^{-\beta(\varepsilon - \mu)}\right) , \qquad (4.1.19)$$

which, after integration by parts, leads to

$$\Phi = -PV = -\frac{2}{3}\frac{gV m^{3/2}}{2^{1/2}\pi^2 \hbar^3} \int_0^\infty \frac{d\varepsilon \, \varepsilon^{3/2}}{e^{\beta(\varepsilon - \mu)} \mp 1} = -\frac{gV kT}{\lambda^3} \begin{cases} g_{5/2}(z) \\ f_{5/2}(z) \end{cases} , \qquad (4.1.19')$$

where the upper lines holds for bosons and the lower line for fermions. The expression (3.1.26), $\Phi = -PV$, which is valid for homogeneous systems, was also used here. From (4.1.10) we obtain for the internal energy

$$E = \frac{gV}{(2\pi\hbar)^3} \int d^3p \, \varepsilon_{\mathbf{p}} n(\varepsilon_{\mathbf{p}}) = \frac{gV m^{3/2}}{2^{1/2}\pi^2 \hbar^3} \int_0^\infty \frac{d\varepsilon \, \varepsilon^{3/2}}{e^{\beta(\varepsilon - \mu)} \mp 1} . \qquad (4.1.20)$$

Comparison with (4.1.19') yields, remarkably, the same relation

$$PV = \frac{2}{3}E \qquad (4.1.21)$$

as for the classical ideal gas. Additional general relations follow from the homogeneity of Φ in T and μ. From (4.1.19'), (4.1.15), and (3.1.18), we obtain

$$P = -\frac{\Phi}{V} = -T^{5/2}\varphi\left(\frac{\mu}{T}\right) , \quad N = VT^{3/2}n\left(\frac{\mu}{T}\right) , \qquad (4.1.22\text{a,b})$$

$$S = -\left(\frac{\partial\Phi}{\partial T}\right)_{V,\mu} = VT^{3/2}s\left(\frac{\mu}{T}\right) , \quad \text{and} \quad \frac{S}{N} = \frac{s(\mu/T)}{n(\mu/T)} . \qquad (4.1.22\text{c,d})$$

Using these results, we can readily derive the adiabatic equation. The conditions $S = $ const. and $N = $ const., together with (4.1.22d), (4.1.22b) and (4.1.22a), yield $\mu/T = $ const., $VT^{3/2} = $ const., $PT^{-5/2} = $ const., and finally

$$PV^{5/3} = \text{const} . \qquad (4.1.23)$$

The adiabatic equation has the same form as that for the classical ideal gas, although most of the other thermodynamic quantities show different behavior, such as for example $c_P/c_V \neq 5/3$.

Following these preliminary general considerations, we wish to derive the equation of state from (4.1.22a). To this end, we need to eliminate μ/T from (4.1.22a) and replace it by the density N/V using (4.1.22b). The explicit computation is carried out in 4.2 for the classical limit, and in 4.3 and 4.4 for low temperatures where quantum effects predominate.

4.2 The Classical Limit $z = e^{\mu/kT} \ll 1$

We first formulate the equation of state in the nearly-classical limit. To do this, we expand the generalized ζ-functions g and f defined in (4.1.18) as power series in z:

$$\left.\begin{array}{c} g_\nu(z) \\ f_\nu(z) \end{array}\right\} = \frac{1}{\Gamma(\nu)} \int_0^\infty dx\, x^{\nu-1} e^{-x} z \sum_{k'=0}^\infty (\pm 1)^{k'} e^{-xk'} z^{k'} = \sum_{k=1}^\infty \frac{(\pm 1)^{k+1} z^k}{k^\nu} ,$$

$$(4.2.1)$$

where the upper lines (signs) hold for bosons and the lower for fermions. Then Eq. (4.1.17) takes on the form

$$\frac{\lambda^3}{v} = g \sum_{k=1}^\infty \frac{(\pm 1)^{k+1} z^k}{k^{3/2}} = g\left(z \pm \frac{z^2}{2^{3/2}} + \mathcal{O}(z^3)\right) . \tag{4.2.2}$$

This equation can be solved iteratively for z:

$$z = \frac{\lambda^3}{vg} \mp \frac{1}{2^{3/2}}\left(\frac{\lambda^3}{vg}\right)^2 + \mathcal{O}\left(\left(\frac{\lambda^3}{v}\right)^3\right) . \tag{4.2.3}$$

Inserting this in the series for Φ which follows from (4.1.19′) and (4.2.1),

$$\Phi = -\frac{gVkT}{\lambda^3}\left(z \pm \frac{z^2}{2^{5/2}} + \mathcal{O}(z^3)\right) , \tag{4.2.4}$$

we can eliminate μ in favor of N and obtain the equation of state

$$PV = -\Phi = NkT\left(1 \mp \frac{\lambda^3}{2^{5/2} gv} + \mathcal{O}\left(\left(\frac{\lambda^3}{v}\right)^2\right)\right) . \tag{4.2.5}$$

The symmetrization (antisymmetrization) of the wavefunctions causes a *reduction* (*increase*) in the pressure in comparison to the classical ideal gas. This acts like an attraction (repulsion) between the particles, which in fact are non-interacting (formation of clusters in the case of bosons, exclusion principle for fermions). For the chemical potential, we find from (4.1.12) and (4.2.3), and making use of $\frac{\lambda^3}{vg} \ll 1$, the following expansion:

$$\mu = kT \log z = kT\left[\log \frac{\lambda^3}{gv} \mp \frac{1}{2^{3/2}}\frac{\lambda^3}{gv} \cdots\right] , \tag{4.2.6}$$

i.e. $\mu < 0$. Furthermore, for the free energy $F = \Phi + \mu N$, we find from (4.2.5) and (4.2.6)

$$F = F_{\text{class}} \mp kT \frac{N\lambda^3}{2^{5/2} gv} , \tag{4.2.7a}$$

where

$$F_{\text{class}} = NkT \left(-1 + \log \frac{\lambda^3}{gv} \right)$$
(4.2.7b)

is the free energy of the classical ideal gas.

Remarks:

(i) The *quantum corrections* are proportional to $\hbar^3$, since λ is proportional to $\hbar$. These corrections are also called *exchange corrections*, as they depend only on the symmetry behavior of the wavefunctions (see also Appendix B).

(ii) The exchange corrections to the classical results at finite temperatures are of the order of λ^3/v. The classical equation of state holds for $z \ll 1$ or $\lambda \ll v^{1/3}$, i.e. in the extremely dilute limit. This limit is the more readily reached, the higher the temperature and the lower the density. The occupation number in the classical limit is given by (cf. Fig. 4.1)

$$n(\varepsilon_{\mathbf{p}}) \approx e^{-\beta \varepsilon_{\mathbf{p}}} e^{\beta \mu} = e^{-\beta \varepsilon_{\mathbf{p}}} \frac{\lambda^3}{gv} \ll 1 .$$
(4.2.8)

This classical limit (4.2.8) is equally valid for bosons and fermions. For comparison, the Fermi distribution at $T = 0$ is also shown. Its significance, as well as that of ε_F, will be discussed in Sect. 4.3 (Fig. 4.1).

(iii) Corresponding to the symmetry-dependent pressure change in (4.2.5), the exchange effects lead to a modification of the free energy (4.2.7a).

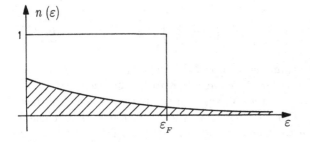

Fig. 4.1. The occupation number $n(\varepsilon)$ in the classical limit (shaded). For comparison, the occupation of a degenerate Fermi gas is also indicated

4.3 The Nearly-degenerate Ideal Fermi Gas

In this and the following section, we consider the opposite limit, in which quantum effects are predominant. Here, we must treat fermions and bosons separately in Sect. 4.4. We first recall the properties of the ground state of fermions, independently of their statistical mechanics.

4.3.1 Ground State, $T = 0$ (Degeneracy)

We first deal with the ground state of a system of N fermions. It is obtained at a temperature of zero Kelvin. In the ground state, the N lowest single-particle states $|p\rangle$ are each singly occupied. If the energy depends only on the momentum $\mathbf{p}$, every value of $\mathbf{p}$ occurs g-fold. For the dispersion relation (4.1.2c), all the momenta within a sphere (the Fermi sphere), whose radius is called the Fermi momentum p_F (Fig. 4.2), are thus occupied. The particle number is related to p_F as follows:

$$N = g \sum_{p \le p_F} 1 = g \frac{V}{(2\pi\hbar)^3} \int d^3p \, \Theta(p_F - p) = \frac{gVp_F^3}{6\pi^2\hbar^3} \,. \tag{4.3.1}$$

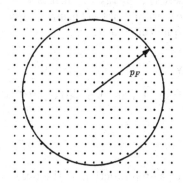

Fig. 4.2. The occupation of the momentum states within the Fermi sphere

From (4.3.1), we find the following relation between the particle density $n = \frac{N}{V}$ and the *Fermi momentum*:

$$p_F = \left(\frac{6\pi^2}{g} \right)^{1/3} \hbar \, n^{1/3} \,. \tag{4.3.2}$$

The single-particle energy corresponding to the Fermi momentum is called the *Fermi energy*:

$$\varepsilon_F = \frac{p_F^2}{2m} = \left(\frac{6\pi^2}{g} \right)^{2/3} \frac{\hbar^2}{2m} n^{2/3} \,. \tag{4.3.3}$$

For the ground-state energy, we find

$$E = \frac{gV}{(2\pi\hbar)^3} \int d^3p \, \frac{p^2}{2m} \Theta(p_F - p) = \frac{gVp_F^5}{20\pi^2\hbar^3 m} = \frac{3}{5} \varepsilon_F N \,. \tag{4.3.4}$$

From (4.1.21) and (4.3.4), the pressure of fermions at $T = 0$ is found to be

$$P = \frac{2}{5}\varepsilon_F n = \frac{1}{5}\left(\frac{6\pi^2}{g}\right)^{2/3} \frac{\hbar^2}{m} n^{5/3} \,. \qquad (4.3.5)$$

The degeneracy of the ground state is sufficiently small that the entropy and the product TS vanish at $T = 0$ (see also (4.3.19)). From this, and using (4.3.4) and (4.3.5), we obtain for the chemical potential using the Gibbs–Duhem relation $\mu = \frac{1}{N}(E + PV - TS)$:

$$\mu = \varepsilon_F \,. \qquad (4.3.6)$$

This result is also evident from the form of the ground state, which implies the occupation of all the levels up to the Fermi energy, from which it follows that the Fermi distribution of a system of N fermions at $T = 0$ becomes $n(\varepsilon) = \Theta(\varepsilon_F - \varepsilon)$. Clearly, one requires precisely the energy ε_F in order to put one additional fermion into the system. The existence of the Fermi energy is a result of the Pauli principle and is thus a quantum effect.

4.3.2 The Limit of Complete Degeneracy

We now calculate the thermodynamic properties in the limit of large μ/kT. In Fig. 4.3, the Fermi distribution function

$$n(\varepsilon) = \frac{1}{e^{(\varepsilon-\mu)/kT} + 1} \qquad (4.3.7)$$

is shown for low temperatures. In comparison to a step function at the position μ, it is broadened within a region kT. We shall see below that μ is equal to ε_F only at $T = 0$. For $T = 0$, the Fermi distribution function degenerates into a step function, so that one then speaks of a degenerate Fermi gas; at low T one refers to a nearly-degenerate Fermi gas.

It is expedient to replace the prefactors in (4.1.19′) and (4.1.15) with the Fermi energy (4.3.3)[5]; for the grand potential, one then obtains

$$\Phi = -N\varepsilon_F^{-3/2} \int\limits_0^\infty d\varepsilon\, \varepsilon^{3/2}\, n(\varepsilon) \,, \qquad (4.3.8)$$

and the formula for N becomes

$$1 = \frac{3}{2}\varepsilon_F^{-3/2} \int\limits_0^\infty d\varepsilon\, \varepsilon^{1/2}\, n(\varepsilon) \,. \qquad (4.3.9)$$

---

[5] In (4.3.8) and (4.3.14), Φ is expressed as usual in terms of its natural variables T, V and μ, since $N\varepsilon_F^{-3/2} \propto V$. In (4.3.14′), the dependence on μ has been substituted by T and N/V, using (4.3.13).

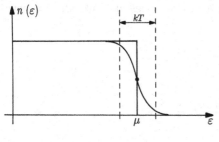

Fig. 4.3. The Fermi distribution function $n(\varepsilon)$ for low temperatures, compared with the step function $\Theta(\mu - \varepsilon)$.

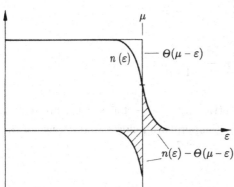

Fig. 4.4. The Fermi distribution function $n(\varepsilon)$, and $n(\varepsilon) - \Theta(\mu - \varepsilon)$.

There thus still remain integrals of the type

$$I = \int_0^\infty d\varepsilon \, f(\varepsilon) \, n(\varepsilon) \tag{4.3.10}$$

to be computed. The method of evaluation at low temperatures was given by Sommerfeld; I can be decomposed in the following manner:

$$I = \int_0^\mu d\varepsilon \, f(\varepsilon) + \int_0^\infty d\varepsilon \, f(\varepsilon) \big[n(\varepsilon) - \Theta(\mu - \varepsilon) \big]$$

$$\approx \int_0^\mu d\varepsilon \, f(\varepsilon) + \int_{-\infty}^\infty d\varepsilon \, f(\varepsilon) \big[n(\varepsilon) - \Theta(\mu - \varepsilon) \big] \tag{4.3.11}$$

and for $T \to 0$, the limit of integration in the second term can be extended to $-\infty$ to a good approximation, since for negative ε, $n(\varepsilon) = 1 + \mathcal{O}(e^{-(\mu-\varepsilon)/kT})$.[6] One can see immediately from Fig. 4.4 that $\big(n(\varepsilon) - \Theta(\mu-\varepsilon)\big)$ differs from zero only in the neighborhood of $\varepsilon = \mu$ and is antisymmetric around μ.[7] Therefore,

---

[6] If $f(\varepsilon)$ is in principle defined only for positive ε, one can e.g. define $f(-\varepsilon) = f(\varepsilon)$; the result depends on $f(\varepsilon)$ only for positive ε.

[7] $\frac{1}{e^x+1} - \Theta(-x) = 1 - \frac{1}{e^{-x}+1} - \Theta(-x) = -\left[\frac{1}{e^{-x}+1} - \Theta(x) \right]$.

we expand $f(\varepsilon)$ around the value μ in a Taylor series and introduce a new integration variable, $x = (\varepsilon - \mu)/kT$:

$$
I = \int_0^\mu d\varepsilon \, f(\varepsilon) + \int_{-\infty}^\infty dx \left[\frac{1}{e^x + 1} - \Theta(-x)\right] \times
$$

$$
\times \left(f'(\mu)\,(kT)^2\,x + \frac{f'''(\mu)}{3!}(kT)^4 x^3 + \dots\right)
$$

$$
= \int_0^\mu d\varepsilon \, f(\varepsilon) + 2(kT)^2 f'(\mu) \int_0^\infty dx \, \frac{x}{e^x + 1} +
$$

$$
+ \frac{2(kT)^4}{3!} f'''(\mu) \int_0^\infty dx \, \frac{x^3}{e^x + 1} + \dots
$$

(since $\left[\frac{1}{e^x+1} - \Theta(-x)\right]$ is antisymmetric and $= \frac{1}{e^x+1}$ for $x > 0$). From this, the general expansion in terms of the temperature follows, making use of the integrals computed in Appendix D., Eq. (D.7) [8]

$$
I = \int_0^\mu d\varepsilon \, f(\varepsilon) + \frac{\pi^2}{6}(kT)^2 f'(\mu) + \frac{7\pi^4}{360}(kT)^4 f'''(\mu) + \dots \ . \tag{4.3.12}
$$

Applying this expansion to Eq. (4.3.9), we find

$$
1 = \left(\frac{\mu}{\varepsilon_F}\right)^{3/2}\left\{1 + \frac{\pi^2}{8}\left(\frac{kT}{\mu}\right)^2 + \mathcal{O}(T^4)\right\}.
$$

This equation can be solved iteratively for μ, yielding the chemical potential as a function of T and N/V:

$$
\mu = \varepsilon_F\left\{1 - \frac{\pi^2}{12}\left(\frac{kT}{\varepsilon_F}\right)^2 + \mathcal{O}(T^4)\right\}, \tag{4.3.13}
$$

where ε_F is given by (4.3.3). The chemical potential decreases with increasing temperature, since then no longer all the states within the Fermi sphere are occupied. In a similar way, we find for (4.3.8)

$$
\Phi = -N\varepsilon_F^{-3/2}\left\{\frac{2}{5}\mu^{5/2} + \frac{\pi^2}{6}(kT)^2 \frac{3}{2}\mu^{1/2} + \dots\right\}, \tag{4.3.14}
$$

---

[8] This series is an asymptotic expansion in T. An asymptotic series for a function $I(\lambda)$, $I(\lambda) = \sum_{k=0}^m a_k \lambda^k + R_m(\lambda)$, is characterized by the following behavior of the remainder: $\lim_{\lambda\to 0} R_m(\lambda)/\lambda^m = 0$, $\lim_{m\to\infty} R_m(\lambda) = \infty$. For small values of λ, the function can be represented very accurately by a finite number of terms in the series. The fact that the integral in (4.3.10) for functions $f(\varepsilon) \sim \varepsilon^{1/2}$ etc. cannot be expanded in a Taylor series can be immediately recognized, since I diverges for $T < 0$.

from which, inserting (4.3.13),[9]

$$\Phi = -\frac{2}{5}N\varepsilon_F\left\{1 + \frac{5\pi^2}{12}\left(\frac{kT}{\varepsilon_F}\right)^2 + \mathcal{O}(T^4)\right\}$$
(4.3.14')

or using $P = -\Phi/V$, we obtain the equation of state. From (4.1.21), we find immediately the internal energy

$$E = \frac{3}{2}PV = \frac{3}{5}N\varepsilon_F\left\{1 + \frac{5\pi^2}{12}\left(\frac{kT}{\varepsilon_F}\right)^2 + \mathcal{O}(T^4)\right\}.$$
(4.3.15)

From this, we calculate the heat capacity at constant V and N:

$$C_V = Nk\frac{\pi^2}{2}\frac{T}{T_F},$$
(4.3.16)

where we have introduced the *Fermi temperature*

$$T_F = \varepsilon_F/k.$$
(4.3.17)

At low temperatures, $(T \ll T_F)$, the heat capacity is a linear function of the temperature (Fig. 4.5). This behavior can be qualitatively understood in a simple way: if one increases the temperature from zero to T, the energy of a portion of the particles increases by kT. The number of particles which are excited in this manner is limited to a shell of thickness kT around the Fermi sphere, i.e. it is given by NkT/ε_F. All together, the energy increase is

$$\delta E \sim kTN\frac{kT}{\varepsilon_F},$$
(4.3.16')

from which, as in (4.3.16), we obtain $C_V \sim kNT/T_F$. According to (4.3.14'), the pressure is given by

$$P = \frac{2}{5}\left(\frac{6\pi^2}{g}\right)^{2/3}\frac{\hbar^2}{2m}\left(\frac{N}{V}\right)^{5/3}\left[1 + \frac{5\pi^2}{12}\left(\frac{kT}{\varepsilon_F}\right)^2 + \dots\right].$$
(4.3.14'')

Due to the Pauli exclusion principle, there is a *pressure increase* at $T = 0$ relative to a classical ideal gas, as can be seen in Fig. 4.6. The isothermal compressibility is then

$$\kappa_T = -\frac{1}{V}\left(\frac{\partial V}{\partial P}\right)_T = \frac{3(V/N)}{2\varepsilon_F}\left[1 - \frac{\pi^2}{12}\left(\frac{kT}{\varepsilon_F}\right)^2 + \dots\right].$$
(4.3.18)

---

[9] If one requires the grand potential as a function of its natural variables, it is necessary to substitute $N\varepsilon_F^{-3/2} = Vg(2m)^{3/2}/6\pi^2\hbar^3$ in (4.3.14). For the calculation of C_V and the equation of state, it is however expedient to employ T, V, and N as variables.

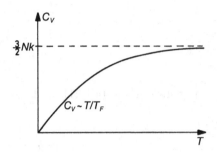

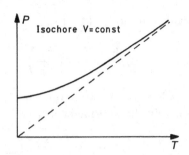

Fig. 4.5. The specific heat (heat capacity) of the ideal Fermi gas

Fig. 4.6. The pressure as a function of the temperature for the ideal Fermi gas (solid curve) and the ideal classical gas (dashed)

For the entropy, we find for $T \ll T_F$

$$S = kN\frac{\pi^2}{2}\frac{T}{T_F} \tag{4.3.19}$$

with $TS = E + PV - \mu N$ from (4.3.15), (4.3.14′) and (4.3.13) (cf. Appendix A.1, 'Third Law').

The chemical potential of an ideal Fermi gas with a fixed density can be found from Eq. (4.3.9) and is shown in Fig. 4.7 as a function of the temperature.

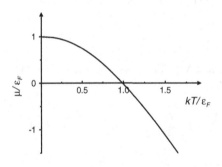

Fig. 4.7. The chemical potential of the ideal Fermi gas at fixed density as a function of the temperature.

Addenda:

(i) The Fermi temperature, also known as the degeneracy temperature,

$$T_F[\text{K}] = \frac{\varepsilon_F}{k} = 3.85 \times 10^{-38}\frac{1}{m[\text{g}]}\left(\frac{N}{V[\text{cm}^3]}\right)^{2/3} \tag{4.3.20}$$

characterizes the thermodynamic behavior of fermions (see Table 4.1). For $T \ll T_F$, the system is nearly degenerate, while for $T \gg T_F$, the classical limit applies. Fermi energies are usually quoted in electron volts (eV). Conversion to Kelvins is accomplished using $1\,\text{eV} \overset{\wedge}{=} 11605\,\text{K}$.

(ii) The *density of states* is defined as

$$\nu(\varepsilon) = \frac{Vg}{(2\pi\hbar)^3} \int d^3p \, \delta(\varepsilon - \varepsilon_{\mathbf{p}}) . \tag{4.3.21}$$

We note that $\nu(\varepsilon)$ is determined merely by the dispersion relation and not by statistics. The thermodynamic quantities do not depend on the details of the momentum dependence of the energy levels, but only on their distribution, i.e. on the density of states. Integrals over momentum space, whose integrands depend only on $\varepsilon_{\mathbf{p}}$, can be rearranged as follows:

$$\int d^3p \, f(\varepsilon_{\mathbf{p}}) = \int d\varepsilon \int d^3p \, f(\varepsilon)\delta(\varepsilon - \varepsilon_{\mathbf{p}}) = \frac{(2\pi\hbar)^3}{Vg} \int d\varepsilon \, \nu(\varepsilon)f(\varepsilon) .$$

For example, the particle number can be expressed in terms of the density of states in the form

$$N = \int_{-\infty}^{\infty} d\varepsilon \, \nu(\varepsilon)n(\varepsilon) . \tag{4.3.22}$$

For free electrons, we find from (4.3.21)

$$\nu(\varepsilon) = \frac{gV}{4\pi^2}\left(\frac{2m}{\hbar^2}\right)^{\frac{3}{2}} \varepsilon^{1/2} = \frac{3}{2}N\frac{\varepsilon^{1/2}}{\varepsilon_F^{3/2}} . \tag{4.3.23}$$

The dependence on $\varepsilon^{1/2}$ shown in Fig. 4.8 is characteristic of nonrelativistic, noninteracting material particles.

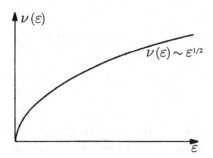

$$\nu(\varepsilon) \sim \varepsilon^{1/2}$$

Fig. 4.8. The density of states for free electrons in three dimensions

The derivations of the specific heat and the compressibility given above can be generalized to the case of arbitrary densities of states $\nu(\varepsilon)$ by evaluating (4.3.9) and (4.3.8) in terms of a general $\nu(\varepsilon)$. The results are

$$C_V = \frac{1}{3}\pi^2\nu(\varepsilon_F)k^2T + \mathcal{O}\big((T/T_F)^3\big) \tag{4.3.24a}$$

and

$$\kappa_T = \frac{V}{N^2}\nu(\varepsilon_F) + \mathcal{O}\big((T/T_F)^2\big) . \tag{4.3.24b}$$

The fact that only the value of the density of states at the Fermi energy is of importance for the low-temperature behavior of the system was to be expected after the discussion following equation (4.3.17). For (4.3.23), we find from (4.3.24a,b) once again the results (4.3.16) and (4.3.18).

(iii) Degenerate Fermi liquids: physical examples of degenerate Fermi liquids are listed in Table 4.1.

Table 4.1. Degenerate Fermi liquids: mass, density, Fermi temperature, Fermi energy

Particles	$m[g]$	$N/V[\text{cm}^{-3}]$	$T_F[\text{K}]$	$\varepsilon_F[\text{eV}]$
Metal electrons	0.91×10^{-27}	10^{24}	10^5	< 10
^{3}He, $P =$ 0–30 bar	5.01×10^{-24} $m^*/m = 2.8\text{–}5.5$	$(1.6\text{–}2.3) \times 10^{22}$	$1.7\text{–}1.1$	$(1.5\text{–}0.9) \times 10^{-4}$
Neutrons in the Nucleus	1.67×10^{-24}	0.11×10^{39} $\times \left(\frac{A-Z}{A}\right)$	5.3×10^{11} $\times \left(\frac{A-Z}{A}\right)^{\frac{2}{3}}$	$46\left(\frac{A-Z}{A}\right)^{\frac{2}{3}} \times 10^6$
Protons in the Nucleus	1.67×10^{-24}	$0.11 \times 10^{39} \frac{Z}{A}$	$5.3 \times 10^{11} \left(\frac{Z}{A}\right)^{\frac{2}{3}}$	$46\left(\frac{Z}{A}\right)^{\frac{2}{3}} \times 10^6$
Electrons in White Dwarf Stars	0.91×10^{-27}	10^{30}	3×10^9	3×10^5

(iv) Coulomb interaction: electrons in metals are not free, but rather they repel each other as a result of their Coulomb interactions

$$H = \sum_i \frac{p_i^2}{2m} + \frac{1}{2} \sum_{i \neq j} \frac{e^2}{r_{ij}} \, . \tag{4.3.25}$$

The following scaling of the Hamiltonian shows that the approximation of free electrons is particularly reasonable for *high densities*. To see this, we carry out the canonical transformation $r' = r/r_0$, $p' = p r_0$. The characteristic length r_0 is defined by $\frac{4\pi}{3} r_0^3 N = V$, i.e. $r_0 = \left(\frac{3V}{4\pi N}\right)^{1/3}$. In terms of these new variables, the Hamiltonian is

$$H = \frac{1}{r_0^2}\left(\sum_i \frac{p_i'^2}{2m} + r_0 \frac{1}{2} \sum_{i \neq j} \frac{e^2}{r_{ij}'}\right) \, . \tag{4.3.25'}$$

The Coulomb interaction becomes less and less important relative to the kinetic energy the smaller r_0, i.e. the more dense the gas becomes.

*4.3.3 Real Fermions

In this section, we will consider real fermionic many-body systems: the conduction electrons in metals, liquid ^{3}He, protons and neutrons in atomic nuclei, electrons in white dwarf stars, neutrons in neutron stars. All of these fermions interact; however, one can understand many of their properties without taking their interactions into account. In the following, we will deal with the parameters mass, Fermi energy, and temperature and discuss the modifications which must be made as a result of the interactions (see also Table 4.1).

a) The Electron Gas in Solids

The alkali metals Li, Na, K, Rb, and Cs are monovalent (with a body-centered cubic crystal structure); e.g. Na has a single $3s^1$ electron (Table 4.2). The noble metals (face-centered cubic crystal structure) are

Copper   Cu $4s^1 3d^{10}$
Silver   Ag $5s^1 4d^{10}$
Gold   Au $6s^1 5d^{10}$.

All of these elements have one valence electron per atom, which becomes a conduction electron in the metal. The number of these quasi-free electrons is equal to the number of atoms. The energy-momentum relation is to a good approximation parabolic, $\varepsilon_{\mathbf{p}} = \frac{\mathbf{p}^2}{2m}$.[10]

Table 4.2. Electrons in Metals; Element, Density, Fermi Energy, Fermi Temperature, $\gamma/\gamma_{\text{theor.}}$, Effective Mass

	N/V [cm^{-3}]	ε_F [eV]	T_F [K]	$\gamma/\gamma_{\text{theor.}}$	m^*/m
Li	4.6×10^{22}	4.7	5.5×10^4	2.17	2.3
Na	2.5	3.1	3.7	1.21	1.3
K	1.34	2.1	2.4	1.23	1.2
Rb	1.08	1.8	2.1	1.22	1.3
Cs	0.86	1.5	1.8	1.35	1.5
Cu	8.5	7	8.2	1.39	1.3
Ag	5.76	5.5	6.4	1.00	1.1
Au	5.9	5.5	6.4	1.13	1.1

---

[10] Remark concerning solid-state physics applications: for Na, we have $\frac{4\pi}{3}(\frac{p_F}{\hbar})^3 = \frac{4\pi^3 N}{V} = \frac{1}{2} V_{\text{Brill.}}$, where $V_{\text{Brill.}}$ is the volume of the first Brillouin zone. The Fermi sphere always lies within the Brillouin zone and thus never crosses the zone boundary, where there are energy gaps and deformations of the Fermi surface. The Fermi surface is therefore in practice spherical, $\Delta p_F/p_F \approx 10^{-3}$. Even in copper, where the $4s$ Fermi surface intersects the Brillouin zone of the fcc lattice, the Fermi surface remains in most regions spherical to a good approximation.

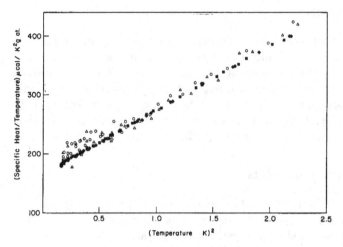

Fig. 4.9. The experimental determination of γ from the specific heat of gold (D. L. Martin, Phys. Rev. **141**, 576 (1966); *ibid.* **170**, 650 (1968))

Taking account of the electron-electron interactions requires many-body methods, which are not at our disposal here. The interaction of two electrons is weakened by screening from the other electrons; in this sense, it is understandable that the interactions can be neglected to a first approximation in treating many phenomena (e.g. Pauli paramagnetism; but not ferromagnetism).

The total specific heat of a metal is composed of a contribution from the electrons (Fig. 4.9) and from the phonons (lattice vibrations, see Sect 4.6):

$$\frac{C_V}{N} = \gamma T + D T^3 .$$

Plotting $\frac{C_V}{NT} = \gamma + D T^2$ vs. T^2, we can read γ off the ordinate. From (4.3.16), the theoretical value of γ is $\gamma_{\text{theor}} = \frac{\pi^2 k^2}{2 \varepsilon_F}$. The deviations between theory and experiment can be attributed to the fact that the electrons move in the potential of the ions in the crystal and are subject to the influence of the electron-electron interaction. The potential and the electron-electron interaction lead among other things to an effective mass m^* for the electrons, i.e. the dispersion relation is approximately given by $\varepsilon_{\mathbf{p}} = \frac{p^2}{2m^*}$. This *effective mass* can be larger or smaller than the mass of free electrons.

b) The Fermi Liquid ^{3}He

^{3}He has a nuclear spin of $I = \frac{1}{2}$, a mass $m = 5.01 \times 10^{-24}$g, a particle density of $n = 1.6 \times 10^{22}$ cm^{-3} at $P = 0$, and a mass density of 0.081 g cm^{-3}. It follows that $\varepsilon_F = 4.2 \times 10^{-4}$eV and $T_F = 4.9$ K. The interactions of the ^{3}He

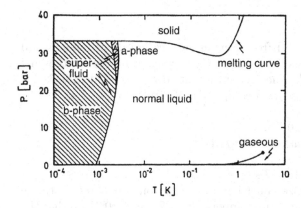

Fig. 4.10. The phase diagram of ^{3}He

atoms lead to an effective mass which at the pressures $P = 0$ and $P = 30$ bar is given by $m^* = 2.8\,m$ and $m^* = 5.5\,m$. Hence the Fermi temperature for $P = 30$, $T_F \approx 1\,K$, is reduced relative to a fictitious non-interacting ^{3}He gas. The particle densities at these pressures are $n = 1.6 \times 10^{23}\,cm^{-3}$ and $n = 2.3 \times 10^{22}\,cm^{-3}$. The interaction between the helium atoms is short-ranged, in contrast to the electron-electron interaction. The small mass of the helium atoms leads to large zero-point oscillations; for this reason, ^{3}He, like ^{4}He, remains a liquid at pressures below $\sim 30\,bar$, even at $T \to 0$. ^{3}He and ^{4}He are termed quantum liquids. At $10^{-3}\,K$, a phase transition into the superfluid state takes place ($l = 1$, $s = 1$) with formation of BCS pairs.[11] In the superconductivity of metals, the Cooper pairs formed by the electrons have $l = 0$ and $s = 0$. The relatively complex phase diagram of ^{3}He is shown in Fig. 4.10.[11]

c) Nuclear Matter

A further example of many-body systems containing fermions are the neutrons and protons in the nucleus, which both have masses of about $m = 1.67 \times 10^{-24}$g. The nuclear radius depends on the nucleon number A via $R = 1.3 \times 10^{-13}A^{1/3}$cm. The nuclear volume is $V = \frac{4\pi}{3}R^3 = \frac{4\pi}{3}(1.3)^3 \times 10^{-39}A\,cm^3 = 9.2 \times 10^{-39}A\,cm^3$. A is the overall number of nucleons and Z the number of protons in the nucleus. Nuclear matter[12] occurs not only within large atomic nuclei, but also in neutron stars, where however also the gravitational interactions must be taken into account.

---

[11] D. Vollhardt and P. Wölfle, *The Superfluid Phases of Helium 3*, Taylor & Francis, London, 1990
[12] A. L. Fetter and J. D. Walecka, *Quantum Theory of Many-Particle Systems*, McGraw-Hill, New York 1971

d) White Dwarfs

The properties of the (nearly) free electron gas are indeed of fundamental importance for the stability of the white dwarfs which can occur at the final stages of stellar evolution.[13] The first such white dwarf to be identified, Sirius B, was predicted by Bessel as a companion of Sirius.

Mass $\approx M_\odot = 1.99 \times 10^{33}$g

Radius $0.01 R_\odot$, $R_\odot = 7 \times 10^{10}$cm

Density $\approx 10^7 \rho_\odot = 10^7$g/cm^3, $\rho_\odot = 1$g/cm^3

$\rho_{\text{Sirius B}} \approx 0.69 \times 10^5$g/cm^3

Central temperature $\approx 10^7$K $\approx T_\odot$

White dwarfs consist of ionized nuclei and free electrons. Helium can still be burned in white dwarfs. The Fermi temperature is $T_F \approx 3 \cdot 10^9$K, so that the electron gas is highly degenerate. The high zero-point pressure of the electron gas opposes the gravitational attraction of the nuclei which compresses the star. The electrons can in fact be regarded as free; their Coulomb repulsion is negligible at these high pressures.

*e) The Landau Theory of Fermi Liquids

The characteristic temperature dependences found for ideal Fermi gases at low temperatures remain in effect in the presence of interactions. This is the result of Landau's Fermi liquid theory, which is based on physical arguments that can also be justified in terms of microscopic quantum-mechanical many-body theory. We give only a sketch of this theory, including its essential results, and refer the reader to more detailed literature[14]. One first considers

---

[13] An often-used classification of the stars in astronomy is based on their positions in the Hertzsprung–Russell diagram, in which their magnitudes are plotted against their colors (equivalent to their surface temperatures). Most stars lie on the so called main sequence. These stars have masses ranging from about one tenth of the Sun's mass up to a sixty-fold solar mass in the evolutionary stages in which hydrogen is converted to helium by nuclear fusion ('burning'). During about 90% of their evolution, the stars stay on the main sequence – as long as nuclear fusion and gravitational attraction are in balance. When the fusion processes come to an end as their 'fuel' is exhausted, gravitational forces become predominant. In their further evolution, the stars become red giants and finally contract to one of the following end stages: in stars with less than 1.4 solar masses, the compression process is brought to a halt by the increase of the Fermi energy of the electrons, and a white dwarf is formed, consisting mainly of helium and electrons. Stars with two- or threefold solar masses end their contraction after passing through intermediate phases as neutron stars. Above three or four solar masses, the Fermi energy of the neutrons is no longer able to stop the compression process, and a black hole results.

[14] A detailed description of Landau's Fermi liquid theory can be found in D. Pines and P. Nozières, *The Theory of Quantum Liquids*, W. A. Benjamin, New York 1966, as well as in J. Wilks, *The Properties of Liquid and Solid Helium*, Clarendon Press, Oxford, 1967. See also J. Wilks and D. S. Betts, *An Introduction to Liquid Helium*, Oxford University Press, 2nd ed., Oxford, (1987).

the ground state of the ideal Fermi gas, and the ground state with an additional particle (of momentum $\mathbf{p}$); then the interaction is 'switched on'. The ideal ground state becomes a modified ground state and the state with the additional particle becomes the modified ground state plus an excited quantum (a quasiparticle of momentum $\mathbf{p}$). The energy of the quantum, $\varepsilon(\mathbf{p})$, is shifted relative to $\varepsilon_0(\mathbf{p}) \equiv \mathbf{p}^2/2m$. Since every non-interacting single-particle state is only singly occupied, there are also no multiply-occupied quasiparticle states; i.e. the quasiparticles also obey Fermi–Dirac statistics.

When several quasiparticles are excited, their energy also depends upon the number $\delta n(\mathbf{p})$ of the other excitations

$$\varepsilon(\mathbf{p}) = \varepsilon_0(\mathbf{p}) + \sum_{\mathbf{p}'} \mathcal{F}(\mathbf{p}, \mathbf{p}') \delta n(\mathbf{p}') . \tag{4.3.26}$$

The average occupation number takes a similar form to that of ideal fermions, owing to the fermionic character of the quasiparticles:

$$n_{\mathbf{p}} = \frac{1}{e^{(\varepsilon(\mathbf{p})-\mu)/kT} + 1} , \tag{4.3.27}$$

where, according to (4.3.26), $\varepsilon(\mathbf{p})$ itself depends on the occupation number. This relation is usually derived in the present context by maximizing the entropy expression found in problem 4.2, which can be obtained from purely combinatorial considerations. At low temperatures, the quasiparticles are excited only near the Fermi energy, and due to the occupied states and energy conservation, the phase space for scattering processes is severely limited. Although the interactions are by no means necessarily weak, the scattering rate vanishes with temperature as $\frac{1}{\tau} \sim T^2$, i.e. the quasiparticles are practically stable particles.

The interaction between the quasiparticles can be written in the form

$$\mathcal{F}(\mathbf{p}, \boldsymbol{\sigma}; \mathbf{p}', \boldsymbol{\sigma}') = f^s(\mathbf{p}, \mathbf{p}') + \boldsymbol{\sigma} \cdot \boldsymbol{\sigma}' f^a(\mathbf{p}, \mathbf{p}') \tag{4.3.28a}$$

with the Pauli spin matrices $\boldsymbol{\sigma}$. Since only momenta in the neighborhood of the Fermi momentum contribute, we introduce

$$f^{s,a}(\mathbf{p}, \mathbf{p}') = f^{s,a}(\chi) \tag{4.3.28b}$$

and

$$F^{s,a}(\chi) = \nu(\varepsilon_F) f^{s,a}(\chi) = \frac{V m^* p_F}{\pi^2 \hbar^3} f^{s,a}(\chi) , \tag{4.3.28c}$$

where χ is the angle between $\mathbf{p}$ and $\mathbf{p}'$ and $\nu(\varepsilon_F)$ is the density of states. A series expansion in terms of Legendre polynomials leads to

$$F^{s,a}(\chi) = \sum_l F_l^{s,a} P_l(\cos\chi) = 1 + F_1^{s,a} \cos\chi + \ldots . \tag{4.3.28d}$$

The F_l^s and F_l^a are the spin-symmetric and spin-antisymmetric Landau parameters; the F_l^a result from the exchange interaction.

Due to the Fermi character of the quasiparticles, which at low temperatures can be excited only near the Fermi energy, it is clear from the qualitative estimate (4.3.16′) that the specific heat of the Fermi liquid will also have a linear temperature dependence. In detail, one obtains for the specific heat, the compressibility, and the magnetic susceptibility:

$$C_V = \frac{1}{3}\pi^2 \nu(\varepsilon_F)\, k^2 T \,, \tag{4.3.29a}$$

$$\kappa_T = \frac{V}{N^2} \frac{\nu(\varepsilon_F)}{1 + F_0^s} \,, \tag{4.3.29b}$$

$$\chi = \mu_B^2 \frac{\nu(\varepsilon_F) N}{1 + F_0^a} \,, \tag{4.3.29c}$$

with the density of states $\nu(\varepsilon_F) = \frac{V m^* p_F}{\pi^2 \hbar^3}$ and the effective mass ratio

$$\frac{m^*}{m} = 1 + \frac{1}{3} F_1^s \,. \tag{4.3.29d}$$

The structure of the results is the same as for ideal fermions.

4.4 The Bose–Einstein Condensation

In this section, we investigate the low-temperature behavior of a nonrelativistic ideal Bose gas of spin $s = 0$, i.e. $g = 1$ and

$$\varepsilon_{\mathbf{p}} = \frac{\mathbf{p}^2}{2m} \,. \tag{4.4.1}$$

In their ground state, non-interacting bosons all occupy the energetically lowest single-particle state; their low-temperature behavior is therefore quite different from that of fermions. Between the high-temperature phase, where the bosons are distributed over the whole spectrum of momentum values, corresponding to the Bose distribution function, and the phase in which the $(\mathbf{p} = 0)$ state is macroscopically occupied (at $T = 0$, all the particles are in this state), a phase transition takes place. This so called Bose–Einstein condensation of an ideal Bose gas was predicted by Einstein[15] on the basis of the statistical considerations of Bose, nearly seventy years before it was observed experimentally.

We first refer to the results of Sect 4.1, where we found for the particle density, i.e. for the reciprocal of the specific volume, in Eq. (4.1.17):

$$\frac{\lambda^3}{v} = g_{3/2}(z) \tag{4.4.2a}$$

---

[15] A. Einstein, Sitzber. Kgl. Preuss. Akad. Wiss. **1924**, 261, (1924), ibid. **1925**, 3 (1925); S. Bose, Z. Phys. **26**, 178 (1924)

with $\lambda = \hbar\sqrt{2\pi/mkT}$ and, using (4.2.1),

$$g_{3/2}(z) = \frac{2}{\sqrt{\pi}} \int_0^\infty dx \, \frac{x^{1/2}}{e^x z^{-1} - 1} = \sum_{k=1}^\infty \frac{z^k}{k^{3/2}} \, . \qquad (4.4.2b)$$

According to Remark (i) in Sect. 4.1, the fugacity of bosons $z = e^{\mu/kT}$ is limited to $z \leq 1$. The maximum value of the function $g_{3/2}(z)$, which is shown in Fig. 4.11, is then given by $g_{3/2}(1) = \zeta(3/2) = 2.612$.

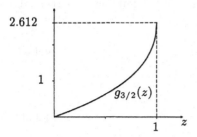

Fig. 4.11. The function $g_{3/2}(z)$.

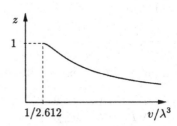

Fig. 4.12. The fugacity z as a function of v/λ^3

In the following, we take the particle number and the volume, and thus the specific volume v, to be fixed at given values. Then from Eq. (4.4.2a), we can calculate z as a function of T, or, more expediently, of $v\lambda^{-3}$. On lowering the temperature, $\frac{v}{\lambda^3}$ decreases and z therefore increases, until finally at $\frac{v}{\lambda^3} = \frac{1}{2.612}$ it reaches its maximum value $z = 1$ (Fig. 4.12). This defines a characteristic temperature

$$kT_c(v) = \frac{2\pi\hbar^2/m}{(2.612\,v)^{2/3}} \, . \qquad (4.4.3)$$

When z approaches 1, we must be more careful in taking the limit of $\sum_{\mathbf{p}} \to \int d^3p$ used in (4.1.14a) and (4.1.15). This is also indicated by the fact that (4.4.2a) would imply for $z = 1$ that at temperatures below $T_c(v)$, the density $\frac{1}{v}$ must decrease with decreasing temperature. From (4.4.2a), there would appear to no longer be enough space for all the particles. Clearly, we have to treat the ($\mathbf{p} = 0$) term in the sum in (4.1.8), which diverges for $z \to 1$, separately:

$$N = \frac{1}{z^{-1} - 1} + \sum_{\mathbf{p} \neq 0} n(\varepsilon_{\mathbf{p}}) = \frac{1}{z^{-1} - 1} + \frac{V}{(2\pi\hbar)^3} \int d^3p \, n(\varepsilon_{\mathbf{p}}) \, .$$

The $\mathbf{p} = 0$ state for fermions did not require any special treatment, since the average occupation numbers can have at most the value 1. Even for bosons, this modification is important only at $T < T_c(v)$ and leads at $T = 0$ to the complete occupation of the $\mathbf{p} = 0$ state, in agreement with the ground state which we described above.

We thus obtain for bosons, instead of (4.4.2a):

$$N = \frac{1}{z^{-1} - 1} + N \frac{v}{\lambda^3} g_{3/2}(z) \,, \tag{4.4.4}$$

or, using Eq. (4.4.3),

$$N = \frac{1}{z^{-1} - 1} + N \left(\frac{T}{T_c(v)} \right)^{3/2} \frac{g_{3/2}(z)}{g_{3/2}(1)} \,. \tag{4.4.4'}$$

The overall particle number N is thus the sum of the number of particles in the ground state

$$N_0 = \frac{1}{z^{-1} - 1} \tag{4.4.5a}$$

and the numbers in the excited states

$$N' = N \left(\frac{T}{T_c(v)} \right)^{3/2} \frac{g_{3/2}(z)}{g_{3/2}(1)} \,. \tag{4.4.5b}$$

For $T > T_c(v)$, Eq. (4.4.4') yields a value for z of $z < 1$. The first term on the right-hand side of (4.4.4') is therefore finite and can be neglected relative to N. Our initial considerations thus hold here; in particular, z follows from

$$g_{3/2}(z) = 2.612 \left(\frac{T_c(v)}{T} \right)^{3/2} \qquad \text{for } T > T_c(v) \,. \tag{4.4.5c}$$

For $T < T_c(v)$, from Eq. (4.4.4'), $z = 1 - \mathcal{O}(1/N)$, so that all of the particles which are no longer in excited states can find sufficient 'space' to enter the ground state. When z is so close to 1, we can set $z = 1$ in the second term and obtain

$$N_0 = N \left(1 - \left(\frac{T}{T_c(v)} \right)^{3/2} \right) \,.$$

Defining the condensate fraction in the thermodynamic limit by

$$\nu_0 = \lim_{\substack{N \to \infty \\ v \text{ fixed}}} \frac{N_0}{N} \,, \tag{4.4.6}$$

we find in summary

$$\nu_0 = \begin{cases} 0 & T > T_c(v) \\ 1 - \left(\frac{T}{T_c(v)} \right)^{3/2} & T < T_c(v) \,. \end{cases} \tag{4.4.7}$$

This phenomenon is called the *Bose–Einstein condensation*. Below $T_c(v)$, the ground state $\mathbf{p} = 0$ is *macroscopically* occupied. The temperature dependence of ν_0 and $\sqrt{\nu_0}$ is shown in Fig. 4.13. The quantities ν_0 and $\sqrt{\nu_0}$ are

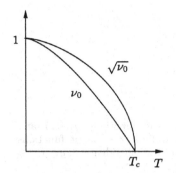
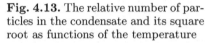
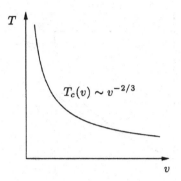

Fig. 4.13. The relative number of particles in the condensate and its square root as functions of the temperature

Fig. 4.14. The transition temperature as a function of the specific volume

characteristic of the condensation or the ordering of the system. For reasons which will become clear later, one refers to $\sqrt{\nu_0}$ as the order parameter. In the neighborhood of T_c, $\sqrt{\nu_0}$ goes to zero as

$$\sqrt{\nu_0} \propto \sqrt{T_c - T} \ . \tag{4.4.7'}$$

In Fig. 4.14, we show the transition temperature as a function of the specific volume. The higher the density (i.e. the smaller the specific volume), the higher the transition temperature $T_c(v)$ at which the Bose–Einstein condensation takes place.

Remark: One might ask whether the next higher terms in the sum $\sum_{\mathbf{p}} n(\varepsilon_{\mathbf{p}})$ could not also be macroscopically occupied. The following estimate however shows that $n(\varepsilon_{\mathbf{p}}) \ll n(0)$ for $p \neq 0$. Consider e.g. the momentum $\mathbf{p} = \left(\frac{2\pi\hbar}{L}, 0, 0\right)$, for which

$$\frac{1}{V} \frac{1}{e^{\beta p_1^2/2m} z^{-1} - 1} < \frac{1}{V} \frac{1}{e^{\beta p_1^2/2m} - 1} < \frac{2m}{V \beta p_1^2} \sim \mathcal{O}(V^{-1/3})$$

holds, while $\frac{1}{V} \frac{1}{z^{-1}-1} \sim \mathcal{O}(1)$.

There is no change in the grand potential compared to the integral representation (4.1.19'), since for the term with $\mathbf{p} = 0$ in the thermodynamic limit, it follows that

$$\lim_{V \to \infty} \frac{1}{V} \log(1 - z(V)) = \lim_{V \to \infty} \frac{1}{V} \log \frac{1}{V} = 0 \ .$$

Therefore, the pressure is given by (4.1.19') as before, where z for $T > T_c(v)$ follows from (4.4.5c), and for $T < T_c(v)$ it is given by $z = 1$. Thus finally the pressure of the ideal Bose gas is

$$P = \begin{cases} \dfrac{kT}{\lambda^3} g_{5/2}(z) & T > T_c \\[3mm] \dfrac{kT}{\lambda^3} 1.342 & T < T_c \end{cases} , \tag{4.4.8}$$

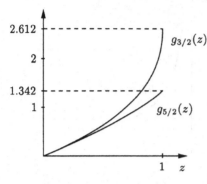

Fig. 4.15. The functions $g_{3/2}(z)$ and $g_{5/2}(z)$. In the limit $z \to 0$, the functions become asymptotically identical, $g_{3/2}(z) \approx g_{5/2}(z) \approx z$.

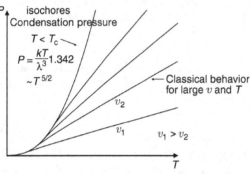

Fig. 4.16. The equation of state of the ideal Bose gas. The isochores are shown for decreasing values of v. For $T < T_c(v)$, the pressure is $P = \frac{kT}{\lambda^3} 1.342$.

with $g_{5/2}(1) = \zeta\left(\frac{5}{2}\right) = 1.342$. If we insert z from (4.4.4) here, we obtain the equation of state. For $T > T_c$, using (4.4.5c), we can write (4.4.8) in the form

$$P = \frac{kT}{v} \frac{g_{5/2}(z)}{g_{3/2}(z)} . \qquad (4.4.9)$$

The functions $g_{5/2}(z)$ and $g_{3/2}(z)$ are drawn in Fig. 4.15. The shape of the equation of state can be qualitatively seen from them. For small values of z, $g_{5/2}(z) \approx g_{3/2}(z)$, so that for large v and high T, we obtain again from (4.4.9) the classical equation of state (see Fig. 4.16). On approaching $T_c(v)$, it becomes increasingly noticeable that $g_{5/2}(z) < g_{3/2}(z)$. At $T_c(v)$, the isochores converge into the curve $P = \frac{kT}{\lambda^3} 1.342$, which represents the pressure for $T < T_c(v)$. All together, this leads to the *equation of state* corresponding to the *isochores* in Fig. 4.16.

For the entropy, we find[16]

$$S = \left(\frac{\partial PV}{\partial T}\right)_{V,\mu} = \begin{cases} Nk \left(\frac{5}{2}\frac{v}{\lambda^3} g_{5/2}(z) - \log z\right) & T > T_c \\ \\ Nk \frac{5}{2}\frac{g_{5/2}(1)}{g_{3/2}(1)} \left(\frac{T}{T_c}\right)^{3/2} & T < T_c \end{cases} , \qquad (4.4.10)$$

---

[16] Note that $\frac{d}{dz} g_\nu(z) = \frac{1}{z} g_{\nu-1}(z)$.

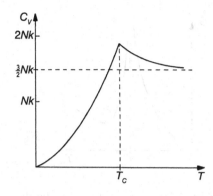

Fig. 4.17. The heat capacity $= N\times$ the specific heat of an ideal Bose gas

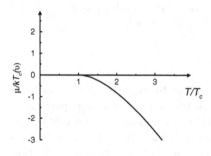

Fig. 4.18. The chemical potential of the ideal Bose gas at a fixed density as a function of the temperature

and, after some calculation, we obtain for the heat capacity at constant volume

$$C_V = T\left(\frac{\partial S}{\partial T}\right)_{N,V} = Nk \begin{cases} \dfrac{15}{4}\dfrac{v}{\lambda^3}g_{5/2}(z) - \dfrac{9}{4}\dfrac{g_{3/2}(z)}{g_{1/2}(z)} & T > T_c \\[3mm] \dfrac{15}{4}\dfrac{g_{5/2}(1)}{g_{3/2}(1)}\left(\dfrac{T}{T_c}\right)^{3/2} & T < T_c. \end{cases} \qquad (4.4.11)$$

The entropy and the specific heat vary as $T^{3/2}$ at low T. Only the excited states contribute to the entropy and the internal energy; the entropy of the condensate is zero. At $T = T_c$, the specific heat of the ideal Bose gas has a cusp (Fig. 4.17).

From Eq. (4.4.4) or from Fig. 4.12, one can obtain the chemical potential, shown in Fig. 4.18 as a function of the temperature.

At $T_\lambda = 2.18\,\mathrm{K}$, the so called lambda point, ^{4}He exhibits a phase transition into the superfluid state (see Fig. 4.19). If we could neglect the interactions of the helium atoms, the temperature of a Bose–Einstein condensation would be $T_c(v) = 3.14\,\mathrm{K}$, using the specific volume of helium in (4.4.3). The interactions are however very important, and it would be incorrect to identify the phase transition into the superfluid state with the Bose–Einstein

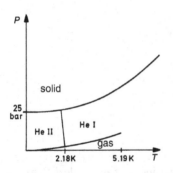
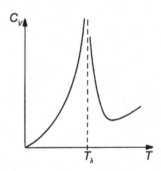

Fig. 4.19. The phase diagram of ^{4}He (schematic). Below 2.18 K, a phase transition from the normal liquid He I phase into the superfluid He II phase takes place

Fig. 4.20. The experimental specific heat of ^{4}He, showing the characteristic lambda anomaly

condensation treated above. The superfluid state in three-dimensional helium is indeed also created by a condensation (macroscopic occupation) of the $\mathbf{p} = 0$ state, but at $T = 0$, the fraction of condensate is only 8%. The specific heat (Fig. 4.20) exhibits a λ anomaly (which gives the transition its name), i.e. an approximately logarithmic singularity. The typical excitation spectrum and the hydrodynamic behavior as described by the two-fluid model are compatible only with an interacting Bose system (Sect. 4.7.1).

Another Bose gas, which is more ideal than helium and in which one can likewise expect a Bose–Einstein condensation – which has been extensively searched for experimentally – is atomic hydrogen in a strong magnetic field (the spin polarization of the hydrogen electrons prevents recombination to molecular H_2). Because of the difficulty of suppressing recombination of H to H_2, over a period of many years it however proved impossible to prepare atomic hydrogen at a sufficient density. The development of atom traps has recently permitted remarkable progress in this area.

The Bose–Einstein condensation was first observed, 70 years after its original prediction, in a gas consisting of around 2000 spin-polarized ^{87}Rb atoms, which were enclosed in a quadrupole trap.[17,18] The transition temperature is at 170×10^{-9}K. One might at first raise the objection that at low temperatures the alkali atoms should form a solid; however, a metastable gaseous state can be maintained within the trap even at temperatures in the nanokelvin range. In the initial experiments, the condensed state could be kept for about ten seconds. Similar results were obtained with a gas consisting of 2×10^5

---

[17] M. H. Anderson, J. R. Ensher, M. R. Matthews, C. E. Wieman, and E. A. Cornell, Science **269**, 198 (1995)

[18] See also G. P. Collins, Physics Today, August 1995, 17.

spin-polarized ^{7}Li atoms.[19] In this case, the condensation temperature is $T_c \approx 400 \times 10^{-9}$ K. In ^{87}Rb, the s-wave scattering length is positive, while in ^{7}Li, it is negative. However, even in ^{7}Li, the gas phase does not collapse into a condensed phase, in any case not within the spatially inhomogeneous atom trap.[19] Finally, it also proved possible to produce and maintain a condensate containing more than 10^8 atoms of atomic hydrogen, with a transition temperature of about 50 μK, for up to 5 seconds.[20]

4.5 The Photon Gas

4.5.1 Properties of Photons

We next want to determine the thermal properties of the radiation field. To start with, we list some of the characteristic properties of photons.

(i) Photons obey the dispersion relation $\varepsilon_{\mathbf{p}} = c|\mathbf{p}| = \hbar ck$ and are bosons with a spin $s = 1$. Since they are completely relativistic particles ($m = 0$, $v = c$), their spins have only two possible orientations, i.e. parallel or antiparallel to $\mathbf{p}$, corresponding to right-hand or left-hand circularly polarized light (0 and π are the only angles which are Lorentz invariant). The degeneracy factor for photons is therefore $g = 2$.

(ii) The mutual interactions of photons are practically zero, as one can see from the following argument: to lowest order, the interaction consists of the scattering of two photons γ_1 and γ_2 into the final states γ_3 and γ_4; see Fig. 4.21a. In this process, for example photon γ_1 decays into a virtual electron-positron pair, photon γ_2 is absorbed by the positron, the electron emits photon γ_3 and recombines with the positron to give photon γ_4. The scattering cross-section for this process is extremely small, of order $\sigma \approx 10^{-50}$ cm^2. The mean collision time can be calculated from the scattering cross-section as follows: in the time Δt, a photon traverses the distance $c\Delta t$. We thus consider the cylinder shown in Fig. 4.21b, whose basal area is equal to the scattering cross-section and whose length is the velocity of light $\times \Delta t$. A photon interacts within the time Δt with all other photons which are in the volume $c\sigma\,\Delta t$, roughly speaking. Let N be the total number of photons within the volume V (which depends on the temperature and which we still have to determine; see the end of Sect. 4.5.4). Then a photon interacts with $c\sigma\,N/V$ particles per unit time. Thus the mean collision time (time between two collisions on average) τ is determined by

$$\tau = \frac{(V/N)}{c\sigma} = 10^{40} \frac{\text{sec}}{\text{cm}^3} \frac{V}{N} \;.$$

[19] C. C. Bradley, C. A. Sackett, J. J. Tollett, and R. G. Hulet, Phys. Rev. Lett. **75**, 1687 (1995)

[20] D. Kleppner, Th. Greytak *et al.*, Phys. Rev. Lett. **81**, 3811 (1998)

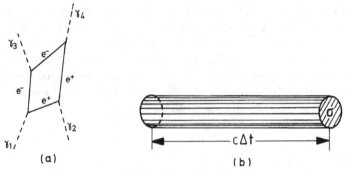

Fig. 4.21. (a) Photon-photon scattering (dashed lines: photons; solid lines: electron and positron). **(b)** Scattering cross-section and mean collision time

The value of the mean collision time is approximately $\tau \approx 10^{31}$ sec at room temperature and $\tau \approx 10^{18}$ sec at the temperature of the Sun's interior (10^7K). Even at the temperature in the center of the Sun, the interaction of the photons it negligible. In comparison, the age of the Universe is $\sim 10^{17}$ sec. Photons do indeed constitute an ideal quantum gas.

The interaction with the surrounding matter is crucial in order to establish equilibrium within the radiation field. The establishment of equilibrium in the photon gas is brought about by absorption and emission of photons by matter. In the following, we will investigate the radiation field within a *cavity* of volume V and temperature T, and without loss of generality of our considerations, we take the quantization volume to be cubical in shape (the shape is irrelevant for short wavelengths, and the long waves have a low statistical weight).

(iii) The number of photons is not conserved. Photons are emitted and absorbed by the material of the cavity walls. From the quantum-field description of photons it follows that each wavenumber and polarization direction corresponds to a harmonic oscillator. The *Hamiltonian* thus has the form

$$ H = \sum_{\mathbf{p},\lambda} \varepsilon_{\mathbf{p}} \hat{n}_{\mathbf{p},\lambda} \equiv \sum_{\mathbf{p},\lambda} \varepsilon_{\mathbf{p}} a^{\dagger}_{\mathbf{p},\lambda} a_{\mathbf{p},\lambda} , \qquad \mathbf{p} \neq 0 , \tag{4.5.1} $$

where $\hat{n}_{\mathbf{p},\lambda} = a^{\dagger}_{\mathbf{p},\lambda} a_{\mathbf{p},\lambda}$ is the occupation number operator for the momentum $\mathbf{p}$ and the direction of polarization λ; also, $a^{\dagger}_{\mathbf{p},\lambda}$, $a_{\mathbf{p},\lambda}$ are the creation and annihilation operators for a photon in the state $\mathbf{p},\lambda$. We note that in the Hamiltonian of the radiation field, there is no zero-point energy, which is automatically accomplished in quantum field theory by defining the Hamiltonian in terms of normal-ordered products.[21]

---

[21] C. Itzykson, J.-B. Zuber, *Quantum Field Theory*, McGraw-Hill; see also QM II.

4.5.2 The Canonical Partition Function

The canonical partition function is given by ($n_{\mathbf{p},\lambda} = 0, 1, 2, \ldots$):

$$Z = \operatorname{Tr} e^{-\beta H} = \sum_{\{n_{\mathbf{p},\lambda}\}} e^{-\beta \sum_{\mathbf{p}} \varepsilon_{\mathbf{p}} n_{\mathbf{p},\lambda}} = \left[\prod_{\mathbf{p} \neq 0} \frac{1}{1 - e^{-\beta \varepsilon_{\mathbf{p}}}} \right]^2 . \tag{4.5.2}$$

Here, there is no condition on the number of photons, since it is not fixed. In (4.5.2), the power 2 enters due to the two possible polarizations λ. With this expression, we find for the free energy

$$F(T, V) = -kT \log Z = 2kT \sum_{\mathbf{p} \neq 0} \log \left(1 - e^{-\varepsilon_{\mathbf{p}}/kT} \right)$$

$$= \frac{2V}{\beta} \int \frac{d^3 p}{(2\pi\hbar)^3} \log(1 - e^{-\beta \varepsilon_{\mathbf{p}}}) = \frac{V(kT)^4}{\pi^2 (\hbar c)^3} \int_0^\infty dx\, x^2 \log(1 - e^{-x}) . \tag{4.5.3}$$

The sum has been converted to an integral according to (4.1.14a). For the integral in (4.5.3), we find after integration by parts

$$\int_0^\infty dx\, x^2 \log(1 - e^{-x}) = -\frac{1}{3} \int_0^\infty \frac{dx\, x^3}{e^x - 1} = -2 \sum_{n=1}^\infty \frac{1}{n^4} \equiv -2\zeta(4) = -\frac{\pi^4}{45} ,$$

where $\zeta(n)$ is Riemann's ζ-function (Eqs. (D.2) and (D.3)), so that for F, we have finally

$$F(T, V) = -\frac{V(kT)^4}{(\hbar c)^3} \frac{\pi^2}{45} = -\frac{4\sigma}{3c} V T^4 \tag{4.5.4}$$

with the *Stefan–Boltzmann constant*

$$\sigma \equiv \frac{\pi^2 k^4}{60\hbar^3 c^2} = 5.67 \times 10^{-8} \text{ J sec}^{-1} \text{ m}^{-2} \text{ K}^{-4} . \tag{4.5.5}$$

From (4.5.4), we obtain the entropy:

$$S = -\left(\frac{\partial F}{\partial T} \right)_V = \frac{16\sigma}{3c} V T^3 , \tag{4.5.6a}$$

the internal energy (caloric equation of state)

$$E = F + TS = \frac{4\sigma}{c} V T^4 , \tag{4.5.6b}$$

and the pressure (thermal equation of state)

$$P = -\left(\frac{\partial F}{\partial V}\right)_T = \frac{4\sigma}{3c}T^4 , \tag{4.5.6c}$$

and finally the heat capacity

$$C_V = T\left(\frac{\partial S}{\partial T}\right)_V = \frac{16\sigma}{c}VT^3 . \tag{4.5.7}$$

Because of the relativistic dispersion, for photons

$$E = 3PV$$

holds instead of $\frac{3}{2}PV$. Eq. (4.5.6b) is called the *Stefan–Boltzmann* law: the internal energy of the radiation field increases as the fourth power of the temperature. The radiation pressure (4.5.6c) is very low, except at extremely high temperatures. At 10^5 K, the temperature produced by the a nuclear explosion, it is $P = 0.25$ bar, and at 10^7 K, the Sun's central temperature, it is $P = 25 \times 10^6$ bar.

4.5.3 Planck's Radiation Law

We now wish to discuss some of the characteristics of the radiation field. The *average occupation number* of the state $(\mathbf{p}, \lambda)$ is given by

$$\langle n_{\mathbf{p},\lambda}\rangle = \frac{1}{e^{\varepsilon_{\mathbf{p}}/kT} - 1} \tag{4.5.8a}$$

with $\varepsilon_{\mathbf{p}} = \hbar\omega_{\mathbf{p}} = cp$, since

$$\langle n_{\mathbf{p},\lambda}\rangle \equiv \frac{\operatorname{Tr} e^{-\beta H}\hat{n}_{\mathbf{p},\lambda}}{\operatorname{Tr} e^{-\beta H}} = \frac{\displaystyle\sum_{n_{\mathbf{p},\lambda}=0}^{\infty} n_{\mathbf{p},\lambda}e^{-n_{\mathbf{p},\lambda}\varepsilon_{\mathbf{p}}/kT}}{\displaystyle\sum_{n_{\mathbf{p},\lambda}=0}^{\infty} e^{-n_{\mathbf{p},\lambda}\varepsilon_{\mathbf{p}}/kT}}$$

can be evaluated analogously to Eq. (4.1.9). The average occupation number (4.5.8a) corresponds to that of atomic or molecular free bosons, Eq. (4.1.9), with $\mu = 0$.

The number of occupied states in a differential element d^3p within a fixed volume is therefore (see (4.1.14a)):

$$\langle n_{\mathbf{p},\lambda}\rangle \frac{2V}{(2\pi\hbar)^3} d^3p , \tag{4.5.8b}$$

and in the interval $[p, p + dp]$, it is

$$\langle n_{\mathbf{p},\lambda}\rangle \frac{V}{\pi^2\hbar^3} p^2 dp . \tag{4.5.8c}$$

It follows from this that the number of occupied states in the interval $[\omega, \omega + d\omega]$ is equal to

$$\frac{V}{\pi^2 c^3} \frac{\omega^2 d\omega}{e^{\hbar\omega/kT} - 1} .$$

(4.5.8d)

The *spectral energy density* $u(\omega)$ is defined as the energy per unit volume and frequency, i.e. as the product of (4.5.8d) with $\hbar\omega/V$:

$$u(\omega) = \frac{\hbar}{\pi^2 c^3} \frac{\omega^3}{e^{\hbar\omega/kT} - 1} .$$

(4.5.9)

This is the famous *Planck radiation law* (1900), which initiated the development of quantum mechanics.

We now want to discuss these results in detail. The occupation number (4.5.8a) for photons diverges for $p \to 0$ as $1/p$ (see Fig. 4.22), since the energy of the photons goes to zero when $\mathbf{p} \to 0$. Because the density of states in three dimensions is proportional to ω^2, this divergence is irrelevant to the energy content of the radiation field. The spectral energy density is shown in Fig 4.22.

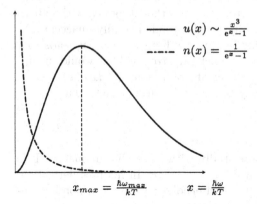

$$\text{---} \quad u(x) \sim \frac{x^3}{e^x - 1}$$

$$\text{--·--} \quad n(x) = \frac{1}{e^x - 1}$$

$$x_{max} = \frac{\hbar\omega_{max}}{kT} \qquad x = \frac{\hbar\omega}{kT}$$

Fig. 4.22. The photon number as a function of $\hbar\omega/kT$ (dot-dashed curve). The spectral energy density as a function of $\hbar\omega/kT$ (solid curve).

As a function of $\hbar\omega$, it shows a maximum at

$$\hbar\omega_{\max} = 2.82 \, kT ,$$

(4.5.10)

i.e. around three times the thermal energy. The maximum shifts proportionally to the temperature. Equation (4.5.10), *Wien's displacement law* (1893), played an important role in the historical development of the theory of the radiation field, leading to the discovery of Planck's quantum of action. In Fig. 4.23, we show $u(\omega, T)$ for different temperatures T.

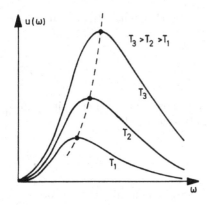

Fig. 4.23. Planck's law for three temperatures, $T_1 < T_2 < T_3$

We now consider the *limiting cases* of Planck's radiation law:

(i) $\hbar\omega \ll kT$: for low frequencies, we find using (4.5.9) that

$$u(\omega) = \frac{kT\omega^2}{\pi^2 c^3} \; ; \tag{4.5.11}$$

the *Rayleigh–Jeans* radiation law. This is the classical low-energy limit. This result of classical physics represented one of the principal problems in the theory of the radiation field. Aside from the fact that it agreed with experiment only for very low frequencies, it was also fundamentally unacceptable: for according to (4.5.11), in the high-frequency limit $\omega \to \infty$, it leads to a divergence in $u(\omega)$, the so called ultraviolet catastrophe. This would in turn imply an infinite energy content of the cavity radiation, $\int_0^\infty d\omega \, u(\omega) = \infty$.

(ii) $\hbar\omega \gg kT$: In the high-frequency limit, we find from (4.5.9) that

$$u(\omega) = \frac{\hbar\omega^3}{\pi^2 c^3} e^{-\hbar\omega/kT} \; . \tag{4.5.12}$$

The energy density decreases exponentially with increasing frequency. This empirically derived relation is known as *Wien's law*. In his first derivation, Planck farsightedly obtained (4.5.9) by interpolating the corresponding entropies between equations (4.5.11) and (4.5.12).

Often, the energy density is expressed in terms of the wavelength λ: starting from $\omega = ck = \frac{2\pi c}{\lambda}$, we obtain $d\omega = -\frac{2\pi c}{\lambda^2} d\lambda$. Therefore, the energy per unit volume in the interval $[\lambda, \lambda + d\lambda]$ is given by

$$\frac{dE_\lambda}{V} = u\left(\omega = \frac{2\pi c}{\lambda}\right)\left|\frac{d\omega}{d\lambda}\right| d\lambda = \frac{16\pi^2 \hbar c \, d\lambda}{\lambda^5 \left(e^{\frac{2\pi\hbar c}{kT\lambda}} - 1\right)} \; , \tag{4.5.13}$$

where we have inserted (4.5.9). The energy density as a function of the wavelength $\frac{dE_\lambda}{d\lambda}$ has its maximum at the value $\lambda_{\max}$, determined by

$$\frac{2\pi\hbar c}{kT\lambda_{\max}} = 4.965 \; . \tag{4.5.14}$$

We will now calculate the radiation which emerges from an opening in the *cavity* at the temperature T. To do this, we first note that the radiation within the cavity is completely isotropic. The emitted thermal radiation at a frequency ω into a solid angle $d\Omega$ is therefore $u(\omega)\frac{d\Omega}{4\pi}$. The radiation energy which emerges per unit time onto a unit surface is

$$I(\omega, T) = \frac{1}{4\pi} \int d\Omega\, c\, u(\omega) \cos\vartheta = \frac{1}{4\pi} \int_0^{2\pi} d\varphi \int_0^1 d\eta\, \eta\, c\, u(\omega) = \frac{c}{4} u(\omega) \;. \quad (4.5.15)$$

The integration over the solid angle $d\Omega$ extends over only one hemisphere (see Fig. 4.24). The total radiated power per unit surface (the energy flux) is then

$$I_E(T) = \int d\omega\, I(\omega, T) = \sigma T^4 \;, \quad (4.5.16)$$

where again the Stefan–Boltzmann constant σ from Eq. (4.5.5) enters the expression.

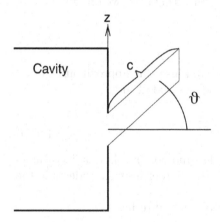

Fig. 4.24. The radiation emission per unit surface area from a cavity radiator (black body)

A body which completely absorbs all the radiation falling upon it is called a *black body*. A small opening in the wall of a cavity whose walls are good absorbers is the ideal realization of a black body. The emission from such an opening calculated above is thus the radiation emitted by a black body. As an approximation, Eqns. (4.5.15,16) are also used to describe the radiation from celestial bodies.

Remark: The Universe is pervaded by the so called *cosmic background radiation* discovered by Penzias and Wilson, which corresponds according to Planck's law to a temperature of 2.73 K. It is a remainder from the earliest times of the Universe, around 300,000 years after the Big Bang, when the temperature of the cosmos had already cooled to about 3000 K. Previous to this time, the radiation was in thermal equilibrium with the matter. At temperatures of 3000 K and below, the electrons bond to atomic nuclei to form atoms, so that the cosmos became transparent to this radiation and it was practically decoupled from the matter in the Universe. The expansion of the Universe by a factor of about one thousand then led to a corresponding increase of all wavelengths due to the red shift, and thus to a Planck distribution at an effective temperature of 2.73 K.

*4.5.4 Supplemental Remarks

Let us now interpret the properties of the photon gas in a physical sense and compare it with other gases.

The *mean photon number* is given by

$$N = 2\sum_{\mathbf{p}}' \frac{1}{e^{cp/kT} - 1} = \frac{V}{\pi^2 c^3} \int_0^\infty \frac{d\omega\, \omega^2}{e^{\hbar\omega/kT} - 1}$$

$$= \frac{V(kT)^3}{\pi^2 c^3 \hbar^3} \int_0^\infty \frac{dx\, x^2}{e^x - 1} = \frac{2\zeta(3)}{\pi^2} V \left(\frac{kT}{\hbar c}\right)^3 ,$$

where the value $\mathbf{p} = 0$ is excluded in $\sum_{\mathbf{p}}'$. Inserting $\zeta(3)$, we obtain

$$N = 0.244\, V \left(\frac{kT}{\hbar c}\right)^3 . \tag{4.5.17}$$

Combining this with (4.5.6c) and (4.5.6a) and inserting approximate numerical values shows a formal similarity to the classical ideal gas:

$$PV = 0.9\, NkT \tag{4.5.18}$$

$$S = 3.6\, Nk , \tag{4.5.19}$$

where N is however always given by (4.5.17) and does not have a fixed value. The pressure per particle is of about the same order of magnitude as in the classical ideal gas.

The thermal wavelength of the photon gas is found to be

$$\lambda_T = \frac{2\pi}{k_{max}} = \frac{2\pi\hbar c}{2.82\, kT} = \frac{0.510}{T[\mathrm{K}]} [\mathrm{cm}] . \tag{4.5.20}$$

With the numerical factor 0.510, λ_T is obtained in units of cm. Inserting into (4.5.17), we find

$$N = 0.244 \left(\frac{2\pi}{2.82}\right)^3 \frac{V}{\lambda_T^3} = 2.70 \frac{V}{\lambda_T^3} . \tag{4.5.21}$$

For the classical ideal gas, $\frac{V}{N\lambda_T^3} \gg 1$; in contrast, the average spacing of the photons $(V/N)^{1/3}$ is, from (4.5.21), of the order of magnitude of λ_T, and therefore, they must be treated quantum mechanically.

At room temperature, i.e. $T = 300\,\mathrm{K}$, $\lambda_T = 1.7 \times 10^{-3}\,\mathrm{cm}$ and the density is $\frac{N}{V} = 5.5 \times 10^8\,\mathrm{cm}^{-3}$. At the temperature of the interior of the Sun, i.e. $T \approx 10^7\,\mathrm{K}$, $\lambda_T = 5.1 \times 10^{-8}\,\mathrm{cm}$ and the density is $\frac{N}{V} = 2.0 \times 10^{22}\,\mathrm{cm}^{-3}$. In comparison, the wavelength of visible light is in the range $\lambda = 10^{-4}\,\mathrm{cm}$.

Note: If the photon had a finite rest mass m, then we would have $g = 3$. In that case, a factor of $\frac{3}{2}$ would enter the Stefan–Boltzmann law. The experimentally demonstrated validity of the Stefan–Boltzmann law implies that either $m = 0$, or that the longitudinal photons do not couple to matter.

The chemical potential: The chemical potential of the photon gas can be computed from the Gibbs–Duhem relation $E = TS - PV + \mu N$, since we are dealing with a homogeneous system:

$$\mu = \frac{1}{N}(E - TS + PV) = \frac{1}{N}\left(4 - \frac{16}{3} + \frac{4}{3}\right)\frac{\sigma V T^3}{3c} \equiv 0 . \tag{4.5.22}$$

The chemical potential of the photon gas is identical to 0 for all temperatures, because the number of photons is not fixed, but rather adjusts itself to the temperature and the volume. Photons are absorbed and emitted by the surrounding matter, the walls of the cavity. In general, the chemical potential of particles and quasiparticles such as phonons, whose particle numbers are not subject to a conservation law, is zero. For example we consider the free energy of a fictitious constant number of photons (phonons etc.), $F(T, V, N_{\mathrm{Ph}})$. since the number of photons (phonons) is not fixed, it will adjust itself in such a way that the free energy is minimized, $\left(\frac{\partial F}{\partial N_{\mathrm{Ph}}}\right)_{T,V} = 0$. This is however just the expression for the chemical potential, which therefore vanishes: $\mu = 0$. We could have just as well started from the maximization of the entropy, $\left(\frac{\partial S}{\partial N_{\mathrm{Ph}}}\right)_{E,V} = -\frac{\mu}{T} = 0$.

*4.5.5 Fluctuations in the Particle Number of Fermions and Bosons

Now that we have become acquainted with the statistical properties of various quantum gases, that is of fermions and bosons (including photons, whose particle-number distribution is characterized by $\mu = 0$), we now want to investigate the fluctuations of their particle numbers. For this purpose, we begin with the grand potential

$$\Phi = -\beta^{-1}\log\sum_{\{n_p\}} e^{-\beta\sum_p n_p(\varepsilon_p - \mu)} . \tag{4.5.23}$$

Taking the derivative of Φ with respect to ε_q yields the mean value of n_q:

$$\frac{\partial \Phi}{\partial \varepsilon_q} = \frac{\displaystyle\sum_{\{n_p\}} n_q e^{-\beta\sum_p n_p(\varepsilon_p - \mu)}}{\displaystyle\sum_{\{n_p\}} e^{-\beta\sum_p n_p(\varepsilon_p - \mu)}} = \langle n_q \rangle . \tag{4.5.24}$$

The second derivative of Φ yields the mean square deviation

$$\frac{\partial^2 \Phi}{\partial \varepsilon_q^2} = -\beta \left\{ \langle n_q^2 \rangle - \langle n_q \rangle^2 \right\} \equiv -\beta (\Delta n_q)^2 \ . \tag{4.5.25}$$

Thus, using $\frac{e^x}{e^x \mp 1} = 1 \pm \frac{1}{e^x \mp 1}$, we obtain

$$(\Delta n_q)^2 = -\beta^{-1} \frac{\partial \langle n_q \rangle}{\partial \varepsilon_q} = \frac{e^{\beta(\varepsilon_q - \mu)}}{\left(e^{\beta(\varepsilon_q - \mu)} \mp 1 \right)^2} = \langle n_q \rangle \left(1 \pm \langle n_q \rangle \right) \ . \tag{4.5.26}$$

For fermions, the mean square deviation is always small. In the range of occupied states, where $\langle n_q \rangle = 1$, Δn_q is zero; and in the region of small $\langle n_q \rangle$, $\Delta n_q \approx \langle n_q \rangle^{1/2}$.

Remark: For bosons, the fluctuations can become very large. In the case of large occupation numbers, we have $\Delta n_q \sim \langle n(q) \rangle$ and the relative deviation approaches one. This is a consequence of the tendency of bosons to cluster in the same state. These strong fluctuations are also found in a spatial sense. If N bosons are enclosed in a volume of L^3, then the mean number of bosons in a subvolume a^3 is given by $\bar{n} = N a^3 / L^3$. In the case that $a \ll \lambda$, where λ is the extent of the wavefunctions of the bosons, one finds the mean square deviation of the particle number $(\Delta N_{a^3})^2$ within the subvolume to be[22]

$$(\Delta N_{a^3})^2 = \bar{n}(\bar{n} + 1) \ .$$

For comparison, we recall the quite different behavior of *classical particles*, which obey a Poisson distribution (see Sect. 1.5.1). The probability of finding n particles in the subvolume a^3 for $a/L \ll 1$ and $N \to \infty$ is then

$$P_n = e^{-\bar{n}} \frac{\bar{n}^n}{n!}$$

with $\bar{n} = N a^3 / L^3$, from which it follows that

$$(\Delta n)^2 = \overline{n^2} - \bar{n}^2 = \sum_n P_n n^2 - \bar{n}^2 = \bar{n} \ .$$

The deviations of the counting rates of bosons from the Poisson law have been experimentally verified using intense photon beams.[23]

4.6 Phonons in Solids

4.6.1 The Harmonic Hamiltonian

We recall the mechanics of a linear chain consisting of N particles of mass m which are coupled to their nearest neighbors by springs of force constant f. In the harmonic approximation, its Hamilton function takes on the form

---

[22] A detailed discussion of the tendency of bosons to cluster in regions where their wavefunctions overlap may be found in E. M. Henley and W. Thirring, *Elementary Quantum Field Theory*, McGraw Hill, New York 1962, p. 52ff.
[23] R. Hanbury Brown and R. Q. Twiss, Nature **177**, 27 (1956).

$$H = W_0 + \sum_n \left[\frac{m}{2} \dot{u}_n^2 + \frac{f}{2} (u_n - u_{n-1})^2 \right] . \tag{4.6.1}$$

One obtains expression (4.6.1) by starting from the Hamilton function of N particles whose positions are denoted by x_n. Their equilibrium positions are x_n^0, where for an infinite chain or a finite chain with periodic boundary conditions, the equilibrium positions have exact translational invariance and the distance between neighboring equilibrium positions is given by the lattice constant $a = x_{n+1}^0 - x_n^0$. One then introduces the displacements from the equilibrium positions, $u_n = x_n - x_n^0$, and expands in terms of the u_n. The quantity W_0 is given by the value of the overall potential energy $W(\{x_n\})$ of the chain in the equilibrium positions. Applying the canonical transformation

$$u_n = \frac{1}{\sqrt{Nm}} \sum_k e^{ikan} Q_k , \quad m \dot{u}_n = \sqrt{\frac{m}{N}} \sum_k e^{-ikan} P_k , \tag{4.6.2}$$

we can transform H into a sum of uncoupled harmonic oscillators

$$H = W_0 + \sum_k \frac{1}{2} (P_k P_{-k} + \omega_k^2 Q_k Q_{-k}) , \tag{4.6.1'}$$

where the frequencies are related to the wavenumber via

$$\omega_k = 2\sqrt{\frac{f}{m}} \sin \frac{ka}{2} . \tag{4.6.3}$$

The Q_k are called normal coordinates and the P_k normal momenta. The Q_k and P_k are conjugate variables, which we will take to be quantum-mechanical operators in what follows. In the quantum representation, commutation rules hold:

$$[u_n, m\dot{u}_{n'}] = i\hbar\delta_{nn'} , \quad [u_n, u_{n'}] = [m\dot{u}_n, m\dot{u}_{n'}] = 0$$

which in turn imply that

$$[Q_k, P_{k'}] = i\hbar\delta_{kk'} , \quad [Q_k, Q_{k'}] = [P_k, P_{k'}] = 0 ;$$

furthermore, we have $Q_k^\dagger = Q_{-k}$ and $P_k^\dagger = P_{-k}$. Finally, by introducing the creation and annihilation operators

$$Q_k = \sqrt{\frac{\hbar}{2\omega_k}} (a_k + a_{-k}^\dagger) , \quad P_k = -i\sqrt{\frac{\hbar\omega_k}{2}} (a_{-k} - a_k^\dagger) , \tag{4.6.4}$$

we obtain

$$H = W_0 + \sum_k \hbar\omega_k \left(\hat{n}_k + \frac{1}{2} \right) \tag{4.6.1''}$$

with the occupation (number) operator

$$\hat{n}_k = a_k^\dagger a_k \qquad (4.6.5)$$

and $[a_k, a_{k'}^\dagger] = \delta_{kk'}$, $[a_k, a_{k'}] = [a_k^\dagger, a_{k'}^\dagger] = 0$.

In this form, we can readily generalize the Hamiltonian to three dimensions. In a three-dimensional crystal with one atom per unit cell, there are three lattice vibrations for each wavenumber, one longitudinal (l) and two transverse (t_1, t_2) (see Fig. 4.25). If the unit cell contains s atoms, there are $3s$ lattice vibrational modes. These are composed of the three acoustic modes, whose frequencies vanish at $k = 0$, and the $3(s - 1)$ optical phonon modes, whose frequencies are finite at $k = 0$.[24]

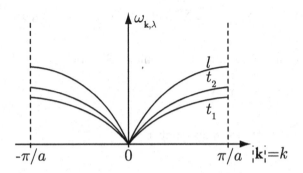

Fig. 4.25. The phonon frequencies in a crystal with one atom per unit cell

We shall limit ourselves to the simple case of a single atom per unit cell, i.e. to Bravais-lattice crystals. Then, according to our above considerations, the Hamiltonian is given by:

$$H = W_0(V) + \sum_{\mathbf{k},\lambda} \hbar\omega_{\mathbf{k},\lambda} \left(\hat{n}_{\mathbf{k},\lambda} + \frac{1}{2} \right) . \qquad (4.6.6)$$

Here, we have characterized the lattice vibrations in terms of their wavevector $\mathbf{k}$ and their polarization λ. The associated frequency is $\omega_{\mathbf{k},\lambda}$ and the operator for the occupation number is $\hat{n}_{\mathbf{k},\lambda}$. The potential energy $W_0(V)$ in the equilibrium lattice locations of the crystal depends on its lattice constant, or, equivalently when the number of particles is fixed, on the volume. For brevity, we combine the wavevector and the polarization into the form $k \equiv (\mathbf{k}, \lambda)$. In a lattice with a total of N atoms, there are $3N$ vibrational degrees of freedom.

---

[24] See e.g. J. M. Ziman, *Principles of the Theory of Solids*, 2nd edition, Cambridge University Press, 1972.

4.6.2 Thermodynamic Properties

In analogy to the calculation for photons, we find for the free energy

$$F = -kT \log Z = W_0(V) + \sum_k \left[\frac{\hbar \omega_k}{2} + kT \log \left(1 - e^{-\hbar \omega_k / kT} \right) \right] . \quad (4.6.7)$$

The internal energy is found from

$$E = -T^2 \left(\frac{\partial}{\partial T} \frac{F}{T} \right)_V , \quad (4.6.8)$$

thus

$$E = W_0(V) + \sum_k \frac{\hbar \omega_k}{2} + \sum_k \hbar \omega_k \frac{1}{e^{\hbar \omega_k / kT} - 1} . \quad (4.6.8')$$

It is again expedient for the case of phonons to introduce the normalized density of states

$$g(\omega) = \frac{1}{3N} \sum_k \delta(\omega - \omega_k) , \quad (4.6.9)$$

where the prefactor has been chosen so that

$$\int_0^\infty d\omega \, g(\omega) = 1 . \quad (4.6.10)$$

Using the density of states, the internal energy can be written in the form:

$$E = W_0(V) + E_0 + 3N \int_0^\infty d\omega \, g(\omega) \frac{\hbar \omega}{e^{\hbar \omega / kT} - 1} , \quad (4.6.11)$$

where we have used $E_0 = \sum_k \hbar \omega_k / 2$ to denote the zero-point energy of the phonons. For the thermodynamic quantities, the precise dependence of the phonon frequencies on wavenumber is not important, but instead only their distribution, i.e. the density of states.

Now, in order to determine the thermodynamic quantities such as the internal energy, we first have to calculate the density of states, $g(\omega)$. For small k, the frequency of the longitudinal phonons is $\omega_{k,l} = c_l k$, and that of the transverse phonons is $\omega_{k,t} = c_t k$, the latter doubly degenerate; here, c_l and c_t are the longitudinal and transverse velocities of sound. Inserting these expressions into (4.6.9), we find

$$g(\omega) = \frac{V}{3N} \frac{1}{2\pi^2} \int dk \, k^2 [\delta(\omega - c_l k) + 2\delta(\omega - c_t k)] = \frac{V}{N} \frac{\omega^2}{6\pi^2} \left(\frac{1}{c_l^3} + \frac{2}{c_t^3} \right) . \quad (4.6.12)$$

Equation (4.6.12) applies only to low frequencies, i.e. in the range where the phonon dispersion relation is in fact linear. In this frequency range, the density of states is proportional to ω^2, as was also the case for photons. Using (4.6.12), we can now compute the thermodynamic quantities for low temperatures, since in this temperature range. only low-frequency phonons are thermally excited. In the high-temperature limit, as we shall see, the detailed shape of the phonon spectrum is unimportant; instead, only the total number of vibrational modes is relevant. We can therefore treat this case immediately, also (Eq. 4.6.14). At *low* temperatures only low frequencies contribute, since frequencies $\omega \gg kT/\hbar$ are suppressed by the exponential function in the integral (4.6.11). Thus the low-frequency result (4.6.12) for $g(\omega)$ can be used. Corresponding to the calculation for photons, we find

$$E = W_0(V) + E_0 + \frac{V\pi^2 k^4}{30\hbar^3}\left(\frac{1}{c_l^3} + \frac{2}{c_t^3}\right)T^4 . \qquad (4.6.13)$$

At high temperatures, i.e. temperatures which are much higher than $\hbar\omega_{max}/k$, where ω_{max} is the maximum frequency of the phonons, we find for all frequencies at which $g(\omega)$ is nonvanishing that $\left(e^{\hbar\omega/kT} - 1\right)^{-1} \approx \frac{kT}{\hbar\omega}$, and therefore, it follows from (4.6.11) and (4.6.10) that

$$E = W_0(V) + E_0 + 3NkT . \qquad (4.6.14)$$

Taking the derivative with respect to temperature, we obtain from (4.6.13) and (4.6.14) in the low-temperature limit

$$C_V \sim T^3 ; \qquad (4.6.15)$$

this is *Debye's law*. In the high-temperature limit, we have

$$C_V \approx 3Nk , \qquad (4.6.16)$$

the law of *Dulong–Petit*. At low temperatures, the specific heat is proportional to T^3, while at high temperatures, it is equal to the number of degrees of freedom times the Boltzmann constant.

In order to determine the specific heat over the whole range of temperatures, we require the normalized density of states $g(\omega)$ for the whole frequency range. The typical shape of $g(\omega)$ for a Bravais crystal[24] is shown in Fig. 4.26. At small values of ω, the ω^2 behavior is clearly visible. Above the maximum frequency, $g(\omega)$ becomes zero. In intermediate regions, the density of states exhibits characteristic structures, so called van Hove singularities[24] which result from the maxima, minima, and saddle points of the phonon dispersion relation; their typical form is shown in Fig. 4.27.

An interpolation formula which is adequate for many purposes can be obtained by approximating the density of states using the *Debye approximation*:

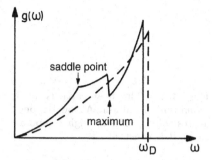

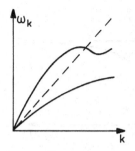

Fig. 4.26. The phonon density of states $g(\omega)$. Solid curve: a realistic density of states; dashed curve: the Debye approximation

Fig. 4.27. A phonon dispersion relation with maxima, minima, and saddle points, which express themselves in the density of states as van Hove singularities

$$g_D(\omega) = \frac{3\omega^2}{\omega_D^3}\,\Theta(\omega_D - \omega)\,,\tag{4.6.17a}$$

with

$$\frac{1}{\omega_D^3} = \frac{1}{18\pi^2}\frac{V}{N}\left(\frac{1}{c_l^3} + \frac{2}{c_t^3}\right)\,.\tag{4.6.17b}$$

With the aid of (4.6.17a), the low-frequency expression (4.6.12) is extended to cover the whole range of frequencies and is cut off at the so called Debye frequency ω_D, which is chosen in such a way that (4.6.10) is obeyed. The Debye approximation is also shown in Fig. 4.26.

Inserting (4.6.17a) into (4.6.11), we obtain

$$E = W_0(V) + E_0 + 3NkT\,D\left(\frac{\hbar\omega_D}{kT}\right)\tag{4.6.18}$$

with

$$D(x) = \frac{3}{x^3}\int_0^x \frac{dy\,y^3}{e^y - 1}\,.\tag{4.6.19}$$

Taking the temperature derivative of (4.6.18), we obtain an expression for the specific heat, which interpolates between the two limiting cases of the Debye and the Dulong-Petit values (see Fig. 4.28).

*4.6.3 Anharmonic Effects and the Mie–Grüneisen Equation of State

So far, we have treated only the harmonic approximation. In fact, the Hamiltonian for phonons in a crystal also contains anharmonic terms, e.g.

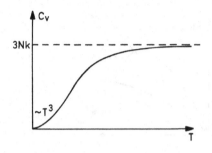

Fig. 4.28. The heat capacity of a monatomic insulator. At low temperatures, $C_V \sim T^3$; at high temperatures, it is constant

$$H_{\text{int}} = \sum_{k_1, k_2} c(k_1, k_2) Q_{k_1} Q_{k_2} Q_{-k_1-k_2}$$

with coefficients $c(k_1, k_2)$. Terms of this type and higher powers arise from the expansion of the interaction potential in terms of the displacements of the lattice components. These nonlinear terms are responsible for (i) the thermal expansion of crystals, (ii) the occurrence of a linear term in the specific heat at high T, (iii) phonon damping, and (iv) a finite thermal conductivity. These terms are also decisive for structural phase transitions. A systematic treatment of these phenomena requires perturbation-theory methods. The anharmonic terms have the effect that the frequencies ω_k depend on the lattice constants, i.e. on the volume V of the crystal. This effect of the anharmonicity can be taken into account approximately by introducing a minor extension to the harmonic theory of the preceding subsection for the derivation of the equation of state.

We take the volume derivative of the free energy F. In addition to the potential energy W_0 of the equilibrium configuration, also ω_k, owing to the anharmonicities, depends on the volume V; therefore, we find for the pressure

$$P = -\left(\frac{\partial F}{\partial V}\right)_T = -\frac{\partial W_0}{\partial V} - \sum_k \hbar \omega_k \left(\frac{1}{2} + \frac{1}{e^{\hbar \omega_k / kT} - 1}\right) \frac{\partial \log \omega_k}{\partial V} . \quad (4.6.20)$$

For simplicity, we assume that the logarithmic derivative of ω_k with respect to the volume is the same for all wavenumbers (the Grüneisen assumption):

$$\frac{\partial \log \omega_k}{\partial V} = \frac{1}{V} \frac{\partial \log \omega_k}{\partial \log V} = -\gamma \frac{1}{V} . \quad (4.6.21)$$

The material constant γ which occurs here is called the Grüneisen constant. The negative sign indicates that the frequencies become smaller on expansion of the lattice. We now insert (4.6.21) into (4.6.20) and compare with (4.6.8′), thus obtaining, with $E_{\text{Ph}} = E - W_0$, the *Mie–Grüneisen* equation of state:

$$P = -\frac{\partial W_0}{\partial V} + \gamma \frac{E_{\text{Ph}}}{V} . \quad (4.6.22)$$

This formula applies to insulating crystals in which there are no electronic excitations and the thermal behavior is determined solely by the phonons.

From the Mie–Grüneisen equation of state, the various thermodynamic derivatives can be obtained, such as the thermal pressure coefficient (3.2.5)

$$\beta = \left(\frac{\partial P}{\partial T}\right)_V = \gamma C_V(T)/V \tag{4.6.23}$$

and the linear expansion coefficient (cf. Appendix I., Table I.3)

$$\alpha_l = \frac{1}{3V}\left(\frac{\partial V}{\partial T}\right)_P , \tag{4.6.24}$$

which, owing to $\left(\frac{\partial P}{\partial T}\right)_V = -\left(\frac{\partial V}{\partial T}\right)_P / \left(\frac{\partial V}{\partial P}\right)_T \equiv \frac{\left(\frac{\partial V}{\partial T}\right)_P}{\kappa_T V}$, can also be given in the form

$$\alpha_l = \frac{1}{3}\beta\kappa_T . \tag{4.6.25}$$

In this last relation, at low temperatures the compressibility can be replaced by

$$\kappa_T(0) = -\frac{1}{V}\left(\frac{\partial V}{\partial P}\right)_{T=0} = \left(V\frac{\partial^2 W_0}{\partial V^2}\right)^{-1} . \tag{4.6.26}$$

At low temperatures, from Eqns. (4.6.23) and (4.6.25), the coefficient of thermal expansion and the thermal pressure coefficient of an insulator, as well as the specific heat, are proportional to the third power of the temperature:

$$\alpha \propto \beta \propto T^3 .$$

As a result of the thermodynamic relationship of the specific heats (3.2.24), we find $C_P - C_V \propto T^7$. Therefore, at temperatures below the Debye temperature, the isobaric and the isochoral specific heats are practically equal.

In analogy to the phonons, one can determine the thermodynamic properties of other quasiparticles. *Magnons* in antiferromagnetic materials likewise have a linear dispersion relation at small values of k and therefore, their contribution to the specific heat is also proportional to T^3. Magnons in ferromagnets have a quadratic dispersion relation $\sim k^2$, leading to a specific heat $\sim T^{3/2}$.

4.7 Phonons und Rotons in He II

4.7.1 The Excitations (Quasiparticles) of He II

At the conclusion of our treatment of the Bose–Einstein condensation in 4.4, we discussed the phase diagram of ^{4}He. In the He II phase below $T_\lambda = 2.18\,\mathrm{K}$, ^{4}He undergoes a condensation. States with the wavenumber 0 are occupied macroscopically. In the language of second quantization, this means that the expectation value of the field operator $\psi(\mathbf{x})$ is finite. The order parameter here is $\langle\psi(\mathbf{x})\rangle$.[25] The excitation spectrum is then quite different from that of a system of free bosons. We shall not enter into the quantum-mechanical theory here, but instead use the experimental results as starting point. At low temperatures, only the lowest excitations are relevant. In Fig. 4.29, we show the excitations as determined by neutron scattering.

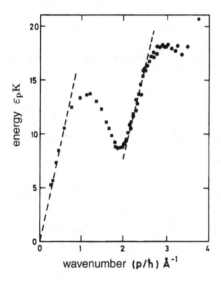

Fig. 4.29. The quasiparticle excitations in superfluid ^{4}He: phonons and rotons after Henshaw and Woods.[26]

The excitation spectrum exhibits the following characteristics: for small values of p, the excitation energy depends linearly on the momentum

$$\varepsilon_\mathbf{p} = cp \, . \tag{4.7.1a}$$

In this region, the excitations are called phonons, whose velocity of sound is $c = 238\,\mathrm{m/sec}$. A second characteristic of the excitation spectrum is a minimum at $p_0 = 1.91\,\text{Å}^{-1}\hbar$. In this range, the excitations are called rotons, and they can be represented by

$$\varepsilon_\mathbf{p} = \Delta + \frac{(|\mathbf{p}| - p_0)^2}{2\mu} \, , \tag{4.7.1b}$$

---

[25] We have

$$a_0\,|\phi_0(N)\rangle = \sqrt{N}\,|\phi_0(N-1)\rangle \approx \sqrt{N}\,|\phi_0(N)\rangle$$
$$a_0^\dagger\,|\phi_0(N)\rangle = \sqrt{N+1}\,|\phi_0(N+1)\rangle \approx \sqrt{N}\,|\phi_0(N)\rangle \, ,$$

since due to the macroscopic occupation of the ground state, $N \gg 1$. See for example QM II, Sect. 3.2.2.

[26] D. G. Henshaw and A. D. Woods, Phys. Rev. **121**, 1266 (1961)

with an effective mass $\mu = 0.16\,m_{\text{He}}$ and an energy gap $\Delta/k = 8.6\,\text{K}$. These properties of the dispersion relations will make themselves apparent in the thermodynamic properties.

4.7.2 Thermal Properties

At low temperatures, the number of excitations is small, and their interactions can be neglected. Since the ^{4}He atoms are bosons, the quasiparticles in this system are also bosons.[27] We emphasize that the quasiparticles in Eqns. (4.7.1a) and (4.7.1b) are collective density excitations, which have nothing to do with the motions of individual helium atoms.

As a result of the Bose character and due to the fact that the number of quasiparticles is not conserved, i.e. the chemical potential is zero, we find for the mean occupation number

$$n(\varepsilon_{\mathbf{p}}) = \left(e^{\beta\varepsilon_{\mathbf{p}}} - 1\right)^{-1} . \tag{4.7.2}$$

From this, the free energy follows:

$$F(T, V) = \frac{kTV}{(2\pi\hbar)^3} \int d^3p \, \log\left(1 - e^{-\beta\varepsilon_{\mathbf{p}}}\right) , \tag{4.7.3a}$$

and for the average number of quasiparticles

$$N_{\text{QP}}(T, V) = \frac{V}{(2\pi\hbar)^3} \int d^3p \, n(\varepsilon_{\mathbf{p}}) \tag{4.7.3b}$$

and the internal energy

$$E(T, V) = \frac{V}{(2\pi\hbar)^3} \int d^3p \, \varepsilon_{\mathbf{p}} n(\varepsilon_{\mathbf{p}}) . \tag{4.7.3c}$$

At low temperatures, only the phonons and rotons contribute in (4.7.3a) through (4.7.3c), since only they are thermally excited. The contribution of the phonons in this limit is given by

$$F_{\text{ph}} = -\frac{\pi^2 V (kT)^4}{90(\hbar c)^3} , \quad \text{or} \quad E_{\text{ph}} = \frac{\pi^2 V (kT)^4}{30(\hbar c)^3} . \tag{4.7.4a,b}$$

From this, we find for the heat capacity at constant volume:

$$C_V = \frac{2\pi^2 V k^4 T^3}{15(\hbar c)^3} . \tag{4.7.4c}$$

---

[27] In contrast, in interacting fermion systems there can be both Fermi and Bose quasiparticles. The particle number of bosonic quasiparticles is in general not fixed. Additional quasiparticles can be created; since the changes in the angular momentum of every quantum-mechanical system must be integral, these excitations must have integral spins.

Due to the gap in the roton energy (4.7.1b), the roton occupation number at low temperatures $T \leq 2\,\mathrm{K}$ can be approximated by $n(\varepsilon_{\mathbf{p}}) \approx \mathrm{e}^{-\beta\varepsilon_{\mathbf{p}}}$, and we find for the average number of rotons

$$
\begin{aligned}
N_{\mathrm{r}} &\approx \frac{V}{(2\pi\hbar)^3} \int d^3p\, \mathrm{e}^{-\beta\varepsilon_{\mathbf{p}}} = \frac{V}{2\pi^2\hbar^3} \int_0^\infty dp\, p^2\, \mathrm{e}^{-\beta\varepsilon_{\mathbf{p}}} \\
&= \frac{V}{2\pi^2\hbar^3} \mathrm{e}^{-\beta\Delta} \int_0^\infty dp\, p^2\, \mathrm{e}^{-\beta(p-p_0)^2/2\mu} \\
&\approx \frac{V}{2\pi^2\hbar^3} \mathrm{e}^{-\beta\Delta} p_0^2 \int_{-\infty}^\infty dp\, \mathrm{e}^{-\beta(p-p_0)^2/2\mu} = \frac{V p_0^2}{2\pi^2\hbar^3} (2\pi\mu kT)^{1/2}\, \mathrm{e}^{-\beta\Delta} .
\end{aligned}
$$

$$(4.7.5a)$$

The contribution of the rotons to the internal energy is

$$
E_{\mathrm{r}} \approx \frac{V}{(2\pi\hbar)^3} \int d^3p\, \varepsilon_p\, \mathrm{e}^{-\beta\varepsilon_p} = -\frac{\partial}{\partial\beta} N_{\mathrm{r}} = \left(\Delta + \frac{kT}{2}\right) N_{\mathrm{r}} ,
\qquad (4.7.5b)
$$

from which we obtain the specific heat

$$
C_{\mathrm{r}} = k \left(\frac{3}{4} + \frac{\Delta}{kT} + \left(\frac{\Delta}{kT}\right)^2 \right) N_{\mathrm{r}} ,
\qquad (4.7.5c)
$$

where from (4.7.5a), N_{r} goes exponentially to zero for $T \to 0$. In Fig. 4.30, the specific heat is drawn in a log-log plot as a function of the temperature. The straight line follows the T^3 law from Eq. (4.7.4c). Above 0.6 K, the roton contribution (4.7.5c) becomes apparent.

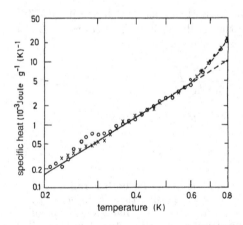

Fig. 4.30. The specific heat of helium II under the saturated vapor pressure (Wiebes, Niels-Hakkenberg and Kramers).

*4.7.3 Superfluidity and the Two-Fluid Model

The condensation of helium and the resulting quasiparticle dispersion relation (Eq. 4.7.1a,b, Fig. 4.29) have important consequences for the dynamic behavior of ^{4}He in its He II phase. Superfluidity and its description in terms of the two-fluid model are among them. To see this, we consider the flow of helium through a tube in two different inertial frames. In frame K, the tube is at rest and the liquid is flowing at the velocity $-\mathbf{v}$. In frame K_0, we suppose the helium to be at rest, while the tube moves with the velocity $\mathbf{v}$ (see Fig. 4.31).

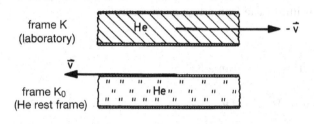

frame K
(laboratory)

frame K_0
(He rest frame)

Fig. 4.31. Superfluid helium in the rest frame of the tube, K, and in the rest frame of the liquid, K_0

The total energies (E, E_0) and the total momenta $(\mathbf{P}, \mathbf{P}_0)$ of the liquid in the two frames (K, K_0) are related by a *Galilei transformation*.

$$\mathbf{P} = \mathbf{P}_0 - M\mathbf{v} \tag{4.7.6a}$$

$$E = E_0 - \mathbf{P}_0 \cdot \mathbf{v} + \frac{M\mathbf{v}^2}{2} . \tag{4.7.6b}$$

Here, we have used the notation

$$\sum_i \mathbf{p}_i = \mathbf{P} , \quad \sum_i \mathbf{p}_{i0} = \mathbf{P}_0 , \quad \sum_i m_i = M . \tag{4.7.6c}$$

One can derive (4.7.6a,b) by applying the Galilei transformation for the individual particles

$$\mathbf{x}_i = \mathbf{x}_{i0} - \mathbf{v}t \quad , \quad \mathbf{p}_i = \mathbf{p}_{i0} - m\mathbf{v} .$$

This gives for the total momentum

$$\mathbf{P} = \sum \mathbf{p}_i = \sum (\mathbf{p}_{i0} - m\mathbf{v}) = \mathbf{P}_0 - M\mathbf{v}$$

and for the total energy

$$E = \sum_i \frac{1}{2m}\mathbf{p}_i^2 + \sum_{\langle i,j \rangle} V(\mathbf{x}_i - \mathbf{x}_j) = \sum_i \frac{m}{2}\left(\frac{\mathbf{p}_{i0}}{m} - \mathbf{v}\right)^2 + \sum_{\langle i,j \rangle} V(\mathbf{x}_{i0} - \mathbf{x}_{j0})$$

$$= \sum_i \frac{\mathbf{p}_{i0}^2}{2m} - \mathbf{P}_0 \cdot \mathbf{v} + \frac{M}{2}\mathbf{v}^2 + \sum_{\langle i,j \rangle} V(\mathbf{x}_{i0} - \mathbf{x}_{j0}) = E_0 - \mathbf{P}_0 \cdot \mathbf{v} + \frac{M}{2}\mathbf{v}^2 .$$

In an ordinary liquid, any flow which might initially be present is damped by friction. Seen from the frame K_0, this means that in the liquid, excitations are created which move along with the walls of the tube, so that in the course of time more and more of the liquid is pulled along with the moving tube. Seen from the tube frame K, this process implies that the flowing liquid is slowed down. The energy of the liquid must simultaneously decrease in order for such excitations to occur at all. We now need to investigate whether for the particular excitation spectrum of He II, Fig. 4.29, the flowing liquid can reduce its energy by the creation of excitations.

Is it energetically favorable to excite quasiparticles? We first consider helium at the temperature $T = 0$, i.e. in its ground state. In the ground state, energy and momentum in the frame K_0 are given by

$$E_0^g \quad \text{and} \quad \mathbf{P}_0 = 0 . \tag{4.7.7a}$$

It follows for these quantities in the frame K:

$$E^g = E_0^g + \frac{M\mathbf{v}^2}{2} \quad \text{and} \quad \mathbf{P} = -M\mathbf{v} . \tag{4.7.7b}$$

If a quasiparticle with momentum $\mathbf{p}$ and energy $\varepsilon_{\mathbf{p}}$ is created, the energy and the momentum in the frame K_0 have the values

$$E_0 = E_0^g + \varepsilon_{\mathbf{p}} \quad \text{and} \quad \mathbf{P}_0 = \mathbf{p} , \tag{4.7.7c}$$

and from (4.7.6a,b) we find for the energy in the frame K:

$$E = E_0^g + \varepsilon_{\mathbf{p}} - \mathbf{p} \cdot \mathbf{v} + \frac{M\mathbf{v}^2}{2} \quad \text{and} \quad \mathbf{P} = \mathbf{p} - M\mathbf{v} . \tag{4.7.7d}$$

The excitation energy in K (the tube frame) is thus

$$\Delta E = \varepsilon_{\mathbf{p}} - \mathbf{p} \cdot \mathbf{v} . \tag{4.7.8}$$

ΔE is the energy change of the liquid due to the appearance of an excitation in frame K. Only when $\Delta E < 0$ does the flowing liquid reduce its energy. Since $\varepsilon - \mathbf{pv}$ is a minimum when $\mathbf{p}$ is parallel to $\mathbf{v}$, the inequality

$$v > \frac{\varepsilon}{p} \tag{4.7.9a}$$

must be obeyed for an excitation to occur. From (4.7.9a) we find the critical velocity (Fig. 4.32)

$$v_c = \left(\frac{\varepsilon}{p}\right)_{\min} \approx 60 \, \text{m/sec} . \tag{4.7.9b}$$

If the flow velocity is smaller than v_c, no quasiparticles will be excited and the liquid flows unimpeded and loss-free through the tube. This phenomenon

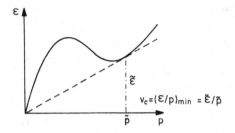

Fig. 4.32. Quasiparticles and the critical velocity

is called superfluidity. The occurrence of a finite critical velocity is closely connected to the shape of the excitation spectrum, which has a finite group velocity at $\mathbf{p} = 0$ and is everywhere greater than zero (Fig. 4.32).

The value (4.7.9b) of the critical velocity is observed for the motion of ions in He II. The critical velocity for flow in capillaries is much smaller than v_c, since vortices occur already at lower velocities; we have not considered these excitations here.

A corresponding argument holds also for the formation of additional excitations at nonzero temperatures. At finite temperatures, thermal excitations of quasiparticles are present. What effect do they have? The quasiparticles will be in equilibrium with the moving tube and will have the average velocity of the frame K_0, $\mathbf{v}$. The condensate, i.e. the superfluid component, is at rest in K_0. The quasiparticles have momentum $\mathbf{p}$ and an excitation energy of $\varepsilon_{\mathbf{p}}$ in K_0. The mean number of these quasiparticles is $n(\varepsilon_{\mathbf{p}} - \mathbf{p} \cdot \mathbf{v})$. (One has to apply the equilibrium distribution functions in the frame in which the quasiparticle gas is at rest! – and there, the excitation energy is $\varepsilon_{\mathbf{p}} - \mathbf{p} \cdot \mathbf{v}$). The momentum of the quasiparticle gas in K_0 is given by

$$\mathbf{P}_0 = \frac{V}{(2\pi\hbar)^3} \int d^3p\, \mathbf{p}\, n(\varepsilon_{\mathbf{p}} - \mathbf{p} \cdot \mathbf{v}) \ . \tag{4.7.10}$$

For low velocities, we can expand (4.7.10) in terms of $\mathbf{v}$. Using $\int d^3p\, \mathbf{p}\, n(\varepsilon_{\mathbf{p}}) = 0$ and terminating the expansion at first order in $\mathbf{v}$, we find

$$\mathbf{P}_0 \approx \frac{-V}{(2\pi\hbar)^3} \int d^3p\, \mathbf{p}(\mathbf{p} \cdot \mathbf{v})\frac{\partial n}{\partial \varepsilon_{\mathbf{p}}} = \frac{-V}{(2\pi\hbar)^3}\, \mathbf{v}\, \frac{1}{3} \int d^3p\, p^2 \frac{\partial n}{\partial \varepsilon_{\mathbf{p}}} \ ,$$

where $\int d^3p\, p_i p_j\, f(|\mathbf{p}|) = \frac{1}{3}\delta_{ij} \int d^3p\, \mathbf{p}^2 f(|\mathbf{p}|)$ was used. At low T, it suffices to take the phonon contribution in this equation into account, i.e.

$$\mathbf{P}_{0,\text{ph}} = -\frac{4\pi V}{(2\pi\hbar)^3}\, \mathbf{v}\, \frac{1}{3c^5} \int\limits_0^\infty d\varepsilon\, \varepsilon^4 \frac{\partial n}{\partial \varepsilon} \ . \tag{4.7.11}$$

After integration by parts and replacement of $4\pi \int d\varepsilon\, \varepsilon^2/c^3$ by $\int d^3p$, one obtains

$$\mathbf{P}_{0,\text{ph}} = \frac{V}{(2\pi\hbar)^3} \mathbf{v} \frac{4}{3c^2} \int d^3 p\, \varepsilon_{\mathbf{p}} n(\varepsilon_{\mathbf{p}}) \ .$$

We write this result in the form

$$\mathbf{P}_{0,\text{ph}} = V\rho_{n,\text{ph}} \mathbf{v} \ , \tag{4.7.12}$$

where we have defined the *normal fluid density* by

$$\rho_{n,\text{ph}} = \frac{4}{3}\frac{E_{\text{ph}}}{Vc^2} = \frac{2\pi^2}{45}\frac{(kT)^4}{\hbar^3 c^5} \ ; \tag{4.7.13}$$

compare (4.7.4b). In (4.7.13), the phonon contribution to ρ_n is evaluated. The contribution of the rotons is given by

$$\rho_{n,\text{r}} = \frac{p_0^2}{3kT}\frac{N_{\text{r}}}{V} \ . \tag{4.7.14}$$

Eq. (4.7.14) follows from (4.7.10) using similar approximations as in the determination of N_{r} in Eq. (4.7.5a). One calls $\rho_n = \rho_{n,\text{ph}} + \rho_{n,\text{r}}$ the mass density of the normal component. Only this portion of the density reaches equilibrium with the walls.

Using (4.7.10) and (4.7.12), the total momentum per unit volume, $\mathbf{P}_0/V$, is found to be given by

$$\mathbf{P}_0/V = \rho_n \mathbf{v} \ . \tag{4.7.15}$$

We now carry out a Galilei transformation from the frame K_0, in which the condensate is at rest, to a frame in which the condensate is moving at the velocity $\mathbf{v}_s$. The quasiparticle gas, i.e. the normal component, has the velocity $\mathbf{v}_n = \mathbf{v} + \mathbf{v}_s$ in this reference frame. The momentum is found from (4.7.15) by adding $\rho\mathbf{v}_s$ due to the Galilei transformation:

$$\mathbf{P}/V = \rho\mathbf{v}_s + \rho_n\mathbf{v} \ .$$

If we substitute $\mathbf{v} = \mathbf{v}_n - \mathbf{v}_s$, we can write the momentum in the form

$$\mathbf{P}/V = \rho_s\mathbf{v}_s + \rho_n\mathbf{v}_n \ , \tag{4.7.16}$$

where the *superfluid density* is defined by

$$\rho_s = \rho - \rho_n \ . \tag{4.7.17}$$

Similarly, the free energy in the frame K_0 can be calculated, and from it, by means of a Galilei transformation, the free energy per unit volume of the flowing liquid in the frame in which the superfluid component is moving at $\mathbf{v}_s$ (problem 4.23):

$$F(T, V, \mathbf{v}_s, \mathbf{v}_n)/V = F(T, V)/V + \frac{1}{2}\rho_s\mathbf{v}_s^2 + \frac{1}{2}\rho_n\mathbf{v}_n^2 \ , \tag{4.7.18}$$

where the free energy of the liquid at rest, $F(T, V)$ is given by (4.7.3a) and the relations which follow it.

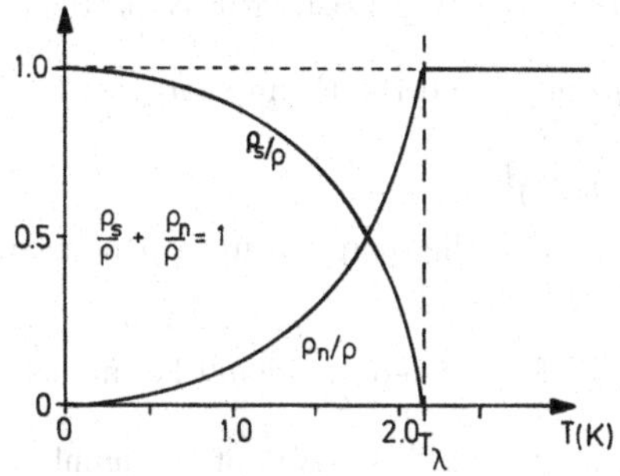

Fig. 4.33. The superfluid and the normal density ρ_s and ρ_n in He II as functions of the temperature, measured using the motion of a torsional oscillator by Andronikaschvili.

The hydrodynamic behavior of the helium in the He II phase is as would be expected if the helium consisted of two fluids, a normal fluid with the density ρ_n, which reaches equilibrium with obstacles such as the inner wall of a tube in which it is flowing, and a superfluid with the density ρ_s, which flows without resistance. When $T \to 0$, $\rho_s \to \rho$ and $\rho_n \to 0$; for $T \to T_\lambda$ $\rho_s \to 0$ and $\rho_n \to \rho$. This theoretical picture, the two–fluid model of Tisza and Landau, was experimentally confirmed by Andronikaschvili, among others (Fig. 4.33). It provides the theoretical basis for the fascinating macroscopic properties of superfluid helium.

Problems for Chapter 4

4.1 Demonstrate the validity of equations (4.3.24a) and (4.3.24b).

4.2 Show that the entropy of an ideal Bose (Fermi) gas can be formulated as follows:

$$S = k \sum_{\mathbf{p}} \Big(-\langle n_{\mathbf{p}} \rangle \log \langle n_{\mathbf{p}} \rangle \pm \big(1 \pm \langle n_{\mathbf{p}} \rangle\big) \log \big(1 \pm \langle n_{\mathbf{p}} \rangle\big) \Big) .$$

Consider this expression in the classical limit, also, as well as in the limit $T \to 0$.

4.3 Calculate C_V, C_P, κ_T, and α for ideal Bose and Fermi gases in the limit of extreme dilution up to the order $\hbar^3$.

4.4 Estimate the Fermi energies (in eV) and the Fermi temperatures (in K) for the following systems (in the free–particle approximation: $\varepsilon_F = \frac{\hbar^2}{2m}\left(\frac{N}{V}\right)^{2/3}\left(\frac{6\pi^2}{g}\right)^{2/3}$):

(**a**) Electrons in metal
(**b**) Neutrons in a heavy nucleus
(**c**) ^{3}He in liquid ^{3}He ($V/N = 46.2\,\text{Å}^3$).

4.5 Consider a one–dimensional electron gas $(S = 1/2)$, consisting of N particles confined to the interval $(0, L)$.
(a) What are the values of the Fermi momentum p_F and the Fermi energy ε_F?
(b) Calculate, in analogy to Sect. 4.3, $\mu = \mu(T, N/L)$.
Result: $p_F = \frac{\pi \hbar N}{L}$, $\mu = \varepsilon_F \left[1 + \frac{\pi^2}{12} \left(\frac{kT}{\varepsilon_F} \right)^2 + \mathcal{O}(T^4) \right]$.
Give a qualitative explanation of the different sign of the temperature dependence when compared to the three–dimensional case.

4.6 Calculate the chemical potential $\mu(T, N/V)$ for a two–dimensional Fermi gas.

4.7 Determine the mean square deviation $(\Delta N)^2 = \langle N^2 \rangle - \langle N \rangle^2$ of the number of electrons for an electron gas in the limit of zero temperature.

4.8 Calculate the isothermal compressibility (Eq. (4.3.18)) of the electron gas at low temperatures, starting from the formula (4.3.14') for the pressure, $P = \frac{2}{5} \frac{\varepsilon_F N}{V} + \frac{\pi^2}{6} \frac{(kT)^2}{\varepsilon_F} \frac{N}{V}$. Compare with the mean square deviation of the particle number found in problem 4.7.

4.9 Compute the free energy of the nearly degenerate Fermi gas, as well as α and C_P.

4.10 Calculate for a completely relativistic Fermi gas $(\varepsilon_p = pc)$
(a) the grand potential Φ
(b) the thermal equation of state
(c) the specific heat C_V.
Consider also the limiting case of very low temperatures.

4.11 (a) Calculate the ground state energy of a relativistic electron gas, $E_p = \sqrt{(m_e c^2)^2 + (pc)^2}$, in a white dwarf star, which contains N electrons and $N/2$ helium nuclei (at rest), and give the zero–point pressure for the two limiting cases

$$x_F \ll 1 : P_0 = \frac{m_e c^2}{v^5} x_F^2$$

$$x_F \gg 1 : P_0 = \frac{m_e c^2}{v^4} x_F \left(1 - \frac{1}{x_F^2} \right) ;$$

$x_F = \frac{p_F}{m_e c}$. How does the pressure depend on the radius R of the star?
(b) Derive the relation between the mass M of the star and its radius R for the two cases $x_F \ll 1$ and $x_F \gg 1$, and show that a white dwarf can have no greater mass than

$$M_0 = \frac{9 m_p}{64} \sqrt{\frac{3\pi}{\alpha^3}} \left(\frac{\hbar c}{\gamma m_p^2} \right)^{3/2} .$$

$\alpha \sim 1$,

$G = 6.7 \times 10^{-8} \mathrm{dyn\,cm^2 g^{-2}}$ Gravitational constant

$m_p = 1.7 \times 10^{-24} \mathrm{g}$ Proton mass

(c) If a star of a given mass $M = 2 m_p N$ is compressed to a (finite) radius R, then its energy is reduced by the self–energy E_g of gravitation, which for a homogeneous mass distribution has the form $E_g = -\alpha GM^2/R$, where α is a number of the order of 1. From

$$\frac{dE_0}{dV} + \frac{dR}{dV}\frac{dE_g}{dR} = 0$$

you can determine the equilibrium radius, with $dE_0 = -P_0(R)\,4\pi R^2 dR$ as the differential of the ground–state energy.

4.12 Show that in a two–dimensional ideal Bose gas, there can be no Bose–Einstein condensation.

4.13 Prove the formulas (4.4.10) and (4.4.11) for the entropy and the specific heat of an ideal Bose gas.

4.14 Compute the internal energy of the ideal Bose gas for $T < T_c(v)$. From the result, determine the specific heat (heat capacity) and compare it with Eq. (4.4.11).

4.15 Show for bosons with $\varepsilon_p = ap^s$ and $\mu = 0$ that the specific heat at low temperatures varies as $T^{3/s}$ in three dimensions. In the special case of $s = 2$, this yields the specific heat of a ferromagnet where these bosons are spin waves.

4.16 Show that the maximum in Planck's formula for the energy distribution $u(\omega)$ is at $\omega_{\max} = 2.82\,\frac{kT}{\hbar}$; see (4.5.10).

4.17 Confirm that the energy flux $I_E(T)$ which is emitted by a black body of temperature T into one hemisphere is given by (Eq. (4.5.16)), $I_E(T) \equiv \frac{\text{energy emitted}}{\text{cm}^2\,\text{sec}} = \frac{cE}{4V} = \sigma T^4$, starting from the energy current density

$$\mathbf{j}_E = \frac{1}{V}\sum_{\mathbf{p},\lambda} c\frac{\mathbf{p}}{p}\varepsilon_\mathbf{p}\langle n_{\mathbf{p},\lambda}\rangle .$$

The energy flux I_E per unit area through a surface element of $d\mathbf{f}$ is $\mathbf{j}_E \frac{d\mathbf{f}}{|d\mathbf{f}|}$.

4.18 The energy flux which reaches the Earth from the Sun is equal to $b = 0,136$ Joule $\text{sec}^{-1}\,\text{cm}^{-2}$ (without absorption losses, for perpendicular incidence). b is called the solar constant.
(a) Show that the total emission from the Sun is equal to 4×10^{26} Joule sec^{-1}.
(b) Calculate the surface temperature of the Sun under the assumption that it radiates as a black body ($T \sim 6000\,\text{K}$).
$R_S = 7 \times 10^{10}\text{cm}$, $R_{SE} = 1\,\text{AU} = 1.5 \times 10^{13}\text{cm}$

4.19 Phonons in a solid: calculate the contribution of the so called optical phonons to the specific heat of a solid, taking the dispersion relation of the vibrations to be $\varepsilon(k) = \omega_E$ (Einstein model).

4.20 Calculate the frequency distribution corresponding to Equation (4.6.17a) for a one- or a two–dimensional lattice. How does the specific heat behave at low temperatures in these cases? (examples of low–dimensional systems are selenium (one–dimensional chains) and graphite (layered structure)).

4.21 The pressure of a solid is given by $P = -\frac{\partial W_0}{\partial V} + \gamma\frac{E_{ph}}{V}$ (see (4.6.22)). Show, under the assumption that $W_0(V) = (V-V_0)^2/2\chi_0 V_0$ for $V \sim V_0$ and $\chi_0 C_V T \ll V_0$, that the thermal expansion (at constant $P \sim 0$) can be expressed as

$$\alpha \equiv \frac{1}{V}\left(\frac{\partial V}{\partial T}\right) = \frac{\gamma\chi_0 C_V}{V_0} \quad \text{and} \quad C_P - C_V = \frac{\gamma^2\chi_0 C_V^2 T}{V_0}.$$

4.22 Specific heat of metals: compare the contributions of phonons and electrons. Show that the linear contribution to the specific heat becomes predominant only at $T < T^* = 0.140\theta_D\sqrt{\theta_D/T_F}$. Estimate T^* for typical values of θ_D and T_F.

4.23 Superfluid helium: show that in a coordinate frame in which the superfluid component is at rest, the free energy $F = E - TS$ is given by

$$\Phi_v + \rho_n v^2, \quad \text{where} \quad \Phi_v = \frac{1}{\beta}\sum_p \log\left[1 - e^{-\beta(\varepsilon_p - \mathbf{p}\cdot\mathbf{v})}\right].$$

Expand Φ_v and show also that in the system in which the superfluid component is moving at a velocity $\mathbf{v}_s$,

$$F = \Phi_0 + \frac{\rho_n v_n^2}{2} + \frac{\rho_s v_s^2}{2}; \quad \mathbf{v}_n = \mathbf{v} + \mathbf{v}_s.$$

Hint: In determining the free energy F, note that the distribution function n for the quasiparticles with energy ε_p is equal to $n(\varepsilon_p - \mathbf{p}\cdot\mathbf{v})$.

4.24 Ideal Bose and Fermi gases in the canonical ensemble:
(a) Calculate the canonical partition function for ideal Bose and Fermi gases.
(b) Calculate the average occupation number in the canonical ensemble.
Suggestion: instead of Z_N, compute the quantity

$$Z(x) = \sum_{N=0}^{\infty} x^N Z_N$$

and determine Z_N using $Z_N = \frac{1}{2\pi i}\oint\frac{Z(x)}{x^{N+1}}dx$, where the path in the complex x plane encircles the origin, but includes no singularities of $Z(x)$. Use the saddle–point method for evaluating the integral.

4.25 Calculate the chemical potential μ for the atomic limit of the Hubbard model,

$$H = U\sum_{i=1}^{N} n_{i\uparrow}n_{i\downarrow},$$

where $n_{i\uparrow} = c_{i\uparrow}^\dagger c_{i\uparrow}$ is the number operator for electrons in the state i (at lattice site i) and $\sigma = +\frac{1}{2}$. (In the general case, which is *not* under consideration here, the Hubbard model is given by:

$$H = \sum_{ij\sigma} t_{ij}c_{i\sigma}^\dagger c_{j\sigma} + U\sum_i n_{i\uparrow}n_{i\downarrow} \ .)$$

5. Real Gases, Liquids, and Solutions

In this chapter, we consider real gases, that is we take the interactions of the atoms or molecules and their structures into account. In the first section, the extension from the classical ideal gas will involve only including the internal degrees of freedom. In the second section, we consider mixtures of such ideal gases. The following sections take the interactions of the molecules into account, leading to the virial expansion and the van der Waals theory of the liquid and the gaseous phases. We will pay special attention to the transition between these two phases. In the final section, we investigate mixtures. This chapter also contains references to every-day physics. It touches on bordering areas with applications in physical chemistry, biology, and technology.

5.1 The Ideal Molecular Gas

5.1.1 The Hamiltonian and the Partition Function

We consider a gas consisting of N molecules, enumerated by the index n. In addition to their translational degrees of freedom, which we take to be classical as before, we now must consider the internal degrees of freedom (rotation, vibration, electronic excitation). The mutual interactions of the molecules will be neglected. The overall Hamiltonian contains the translational energy (kinetic energy of the molecular motion) and the Hamiltonian for the internal degrees of freedom $H_{i,n}$, summed over all the molecules:

$$H = \sum_{n=1}^{N} \left(\frac{\mathbf{p}_n^2}{2m} + H_{i,n} \right) . \tag{5.1.1}$$

The eigenvalues of $H_{i,n}$ are the internal energy levels $\varepsilon_{i,n}$. The partition function is given by

$$Z(T,V,N) = \frac{V^N}{(2\pi\hbar)^{3N} N!} \int d^3 p_1 \ldots d^3 p_N \, e^{-\sum_n \mathbf{p}_n^2/2mkT} \prod_n \sum_{\varepsilon_{i,n}} e^{-\varepsilon_{i,n}/kT} .$$

The classical treatment of the translational degrees of freedom, represented by the partition integral over momenta, is justified when the specific volume is much larger than the cube of the thermal wavelength $\lambda = 2\pi\hbar/\sqrt{2\pi mkT}$ (Chap. 4). Since the internal energy levels $\varepsilon_{i,n} \equiv \varepsilon_i$ are identical for all of the molecules, it follows that

$$Z(T,V,N) = \frac{1}{N!}[Z_{tr}(1)\,Z_i]^N = \frac{1}{N!}\left[\frac{V}{\lambda^3}Z_i\right]^N , \qquad (5.1.2)$$

where $Z_i = \sum_{\varepsilon_i} e^{-\varepsilon_i/kT}$ is the partition function over the internal degrees of freedom and $Z_{tr}(1)$ is the translational partition integral for a single molecule. From (5.1.2), we find the free energy, using the Stirling approximation for large N:

$$F = -kT\log Z \approx -NkT\left[1 + \log\frac{V}{N\lambda^3} + \log Z_i\right] . \qquad (5.1.3)$$

From (5.1.3), we obtain the equation of state

$$P = -\left(\frac{\partial F}{\partial V}\right)_{T,N} = \frac{NkT}{V} , \qquad (5.1.4)$$

which is the same as that of a monatomic gas, since the internal degrees of freedom do not depend on V. For the entropy, we have

$$S = -\left(\frac{\partial F}{\partial T}\right)_{V,N} = Nk\left[\frac{5}{2} + \log\frac{V}{N\lambda^3} + \log Z_i + T\frac{\partial\log Z_i}{\partial T}\right] , \qquad (5.1.5a)$$

and from it, we obtain the internal energy,

$$E = F + TS = NkT\left[\frac{3}{2} + T\frac{\partial\log Z_i}{\partial T}\right] . \qquad (5.1.5b)$$

The caloric equation of state (5.1.5b) is altered by the internal degrees of freedom compared to that of a monatomic gas. Likewise, the internal degrees of freedom express themselves in the heat capacity at constant volume,

$$C_V = \left(\frac{\partial E}{\partial T}\right)_{V,N} = Nk\left[\frac{3}{2} + \frac{\partial}{\partial T}T^2\frac{\partial\log Z_i}{\partial T}\right] . \qquad (5.1.6)$$

Finally, we give also the chemical potential for later applications:

$$\mu = \left(\frac{\partial F}{\partial N}\right)_{T,V} = -kT\log\left(\frac{V}{N\lambda^3}Z_i\right) ; \qquad (5.1.5c)$$

it agrees with $\mu = \frac{1}{N}(F + PV)$, since we are dealing with a homogeneous system.

To continue the evaluation, we need to investigate the contributions due to the internal degrees of freedom. The energy levels of the internal degrees of freedom are composed of three contributions:

$$\varepsilon_i = \varepsilon_{el} + \varepsilon_{rot} + \varepsilon_{vib} . \tag{5.1.7}$$

Here, ε_{el} refers to the electronic energy including the Coulomb repulsion of the nuclei relative to the energy of widely separated atoms. ε_{rot} is the rotational energy and ε_{vib} is the vibrational energy of the molecules.

We consider diatomic molecules containing two different atoms (e.g. HCl; for identical atoms, cf. Sect. 5.1.4). Then the rotational energy has the form[1]

$$\varepsilon_{rot} = \frac{\hbar^2 l(l+1)}{2I} , \tag{5.1.8a}$$

where l is the angular momentum quantum number and $I = m_{red} R_0^2$ the moment of inertia, depending on the reduced mass m_{red} and the distance between the atomic nuclei, R_0.[2] The vibrational energy ε_{vib} takes the form[1]

$$\varepsilon_{vib} = \hbar\omega \left(n + \frac{1}{2} \right) , \tag{5.1.8b}$$

where ω is the frequency of the molecular vibration and $n = 0, 1, 2, \ldots$. The electronic energy levels ε_{el} can be compared to the dissociation energy ε_{Diss}. Since we want to consider non-dissociated molecules, i.e. we require that $kT \ll \varepsilon_{Diss}$, and on the other hand the excitation energies of the lowest electronic levels are of the same order of magnitude as ε_{Diss}, it follows from the condition $kT \ll \varepsilon_{Diss}$ that the electrons must be in their ground state, whose energy we denote by ε_{el}^0. Then we have

$$Z_i = \exp \left(-\frac{\varepsilon_{el}^0}{kT} \right) Z_{rot} Z_{vib} . \tag{5.1.9}$$

We now consider in that order the rotational part Z_{rot} and the vibrational part Z_{vib} of the partition function.

5.1.2 The Rotational Contribution

Since the rotational energy ε_{rot} (5.1.8a) does not depend on the quantum number m (the z component of the angular momentum), the sum over m just yields a factor $(2l+1)$, and only the sum over l remains, which runs over all the natural numbers

$$Z_{rot} = \sum_{l=0}^{\infty} (2l+1) \exp \left(-\frac{l(l+1)\Theta_r}{2T} \right) . \tag{5.1.10}$$

---

[1] In general, the moment of inertia I and the vibration frequency ω depend[2] on l. The latter dependence leads to a coupling of the rotational and the vibrational degrees of freedom. For the following evaluation we have assumed that these dependences are weak and can be neglected.

[2] See e.g. QM I

Here, we have introduced the characteristic temperature

$$\Theta_{\rm r} = \frac{\hbar^2}{Ik} \, .\tag{5.1.11}$$

We next consider two limiting cases:

$T \ll \Theta_{\rm r}$: At low temperatures, only the smallest values of l contribute in (5.1.10)

$$Z_{\rm rot} = 1 + 3\,{\rm e}^{-\Theta_{\rm r}/T} + 5\,{\rm e}^{-3\Theta_{\rm r}/T} + \mathcal{O}\left({\rm e}^{-6\Theta_{\rm r}/T}\right) \, .\tag{5.1.12}$$

$T \gg \Theta_{\rm r}$: At high temperatures, the sum must be carried out over all l values, leading to

$$Z_{\rm rot} = 2\frac{T}{\Theta_{\rm r}} + \frac{1}{3} + \frac{1}{30}\frac{\Theta_{\rm r}}{T} + \mathcal{O}\left(\left(\frac{\Theta_{\rm r}}{T}\right)^2\right) \, .\tag{5.1.13}$$

To prove (5.1.13), one uses the Euler–MacLaurin summation formula[3]

$$\sum_{l=0}^{\infty} f(l) = \int_0^{\infty} dl\, f(l) + \frac{1}{2}f(0) + \sum_{k=1}^{n-1} \frac{(-1)^k B_k}{(2k)!} f^{(2k-1)}(0) + {\rm Rest}_n \, ,\tag{5.1.14}$$

for the special case that $f(\infty) = f'(\infty) = \ldots = 0$. The first Bernoulli numbers B_n are given by $B_1 = \frac{1}{6}, B_2 = \frac{1}{30}$. The first term in (5.1.14) yields just the classical result

$$\int_0^{\infty} dl\, f(l) = \int_0^{\infty} dl\, (2l+1)\exp\left(-\frac{l(l+1)}{2}\frac{\Theta_{\rm r}}{T}\right) = 2\int_0^{\infty} dx\,{\rm e}^{-x\frac{\Theta_{\rm r}}{T}} = 2\frac{T}{\Theta_{\rm r}} \, ,\tag{5.1.15}$$

which one would also obtain by treating the rotational energy classically instead of quantum-mechanically.[4] The further terms are found via

$$f(0) = 1 \, , \quad f'(0) = 2 - \frac{\Theta_{\rm r}}{2T} \, , \quad f'''(0) = -6\frac{\Theta_{\rm r}}{T} + 3\left(\frac{\Theta_{\rm r}}{T}\right)^2 - \frac{1}{8}\left(\frac{\Theta_{\rm r}}{T}\right)^3 \, ,$$

from which, using (5.1.14), we obtain the expansion (5.1.13).

From (5.1.12) and (5.1.13), we find for the logarithm of the partition function after expanding:

---

[3] Whittaker, Watson, *A Modern Course of Analysis*, Cambridge at the Clarendon Press; V.I. Smirnow, *A Course of Higher Mathematics*, Pergamon Press, Oxford 1964: Vol. III, Part 2, p. 290.

[4] See e.g. A. Sommerfeld, *Thermodynamics and Statistical Physics*, Academic Press, NY 1950

$$Z_{\rm rot} = \frac{4\pi I^2}{(2\pi\hbar)^2} \int d\omega_1 \int d\omega_2\, {\rm e}^{-\frac{\beta I}{2}(\omega_1^2 + \omega_2^2)} = \frac{2IkT}{\hbar^2} \, .$$

$$\log Z_{\text{rot}} = \begin{cases} 3\,\mathrm{e}^{-\Theta_r/T} - \dfrac{9}{2}\,\mathrm{e}^{-2\Theta_r/T} + \mathcal{O}\!\left(\mathrm{e}^{-3\Theta_r/T}\right) & T \ll \Theta_r \\[2mm] \log\!\left(\dfrac{2T}{\Theta_r}\right) + \dfrac{\Theta_r}{6T} + \dfrac{1}{360}\!\left(\dfrac{\Theta_r}{T}\right)^2 + \mathcal{O}\!\left(\left(\dfrac{\Theta_r}{T}\right)^3\right) & T \gg \Theta_r\,. \end{cases}$$

(5.1.16a)

From this result, the contribution of the rotational degrees of freedom to the internal energy can be calculated:

$$E_{\text{rot}} = NkT^2 \frac{\partial}{\partial T}\log Z_{\text{rot}}$$

$$= \begin{cases} 3Nk\,\Theta_r\!\left(\mathrm{e}^{-\Theta_r/T} - 3\,\mathrm{e}^{-2\Theta_r/T} + \ldots\right) & T \ll \Theta_r \\[2mm] NkT\!\left(1 - \dfrac{\Theta_r}{6T} - \dfrac{1}{180}\!\left(\dfrac{\Theta_r}{T}\right)^2 + \ldots\right) & T \gg \Theta_r\,. \end{cases}$$

(5.1.16b)

The contribution to the heat capacity at constant volume is then

$$C_V^{\text{rot}} = Nk \begin{cases} 3\!\left(\dfrac{\Theta_r}{T}\right)^2 \mathrm{e}^{-\Theta_r/T}\!\left(1 - 6\,\mathrm{e}^{-\Theta_r/T} + \ldots\right) & T \ll \Theta_r \\[2mm] 1 + \dfrac{1}{180}\!\left(\dfrac{\Theta_r}{T}\right)^2 + \ldots & T \gg \Theta_r\,. \end{cases}$$

(5.1.16c)

In Fig. 5.1, we show the rotational contribution to the specific heat.

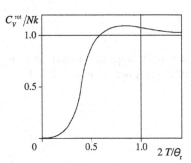

Fig. 5.1. The rotational contribution to the specific heat

At low temperatures, the rotational degrees of freedom are not thermally excited. Only at $T \approx \Theta_r/2$ do the rotational levels contribute. At high temperatures, i.e. in the classical region, the two rotational degrees of freedom make a contribution of $2kT/2$ to the internal energy. Only with the aid of quantum mechanics did it become possible to understand why, in contradiction to the equipartition theorem of classical physics, the specific heat per molecule can differ from the number of degrees of freedom multiplied by $k/2$. The rotational contribution to the specific heat has a maximum of 1.1 at the temperature $0.81\,\Theta_r/2$. For HCl, $\Theta_r/2$ is found to be $15.02\,\mathrm{K}$.

5.1.3 The Vibrational Contribution

We now come to the vibrational contribution, for which we introduce a characteristic temperature defined by

$$\hbar\omega = k\Theta_v \ . \tag{5.1.17}$$

We obtain the well-known partition function of a harmonic oscillator

$$Z_{\text{vib}} = \sum_{n=0}^{\infty} e^{-(n+\frac{1}{2})\frac{\Theta_v}{T}} = \frac{e^{-\Theta_v/2T}}{1 - e^{-\Theta_v/T}} \ , \tag{5.1.18}$$

whose logarithm is given by $\log Z_{\text{vib}} = -\frac{\Theta_v}{2T} - \log\left(1 - e^{-\Theta_v/T}\right)$. From it, we find for the internal energy:

$$E_{\text{vib}} = NkT^2 \frac{\partial}{\partial T} \log Z_{\text{vib}} = Nk\,\Theta_v \left[\frac{1}{2} + \frac{1}{e^{\Theta_v/T} - 1}\right] \ , \tag{5.1.19a}$$

and for the vibrational contribution to the heat capacity at constant volume

$$C_V^{\text{vib}} = Nk\frac{\Theta_v^2}{T^2}\frac{e^{\Theta_v/T}}{\left[e^{\Theta_v/T} - 1\right]^2} = Nk\frac{\Theta_v^2}{T^2}\frac{1}{[2\sinh\Theta_v/2T]^2} \ . \tag{5.1.19b}$$

At low and high temperatures, from (5.1.19b) we obtain the limiting cases

$$\frac{C_V^{\text{vib}}}{Nk} = \begin{cases} \left(\dfrac{\Theta_v}{T}\right)^2 e^{-\Theta_v/T} + \dots & T \ll \Theta_v \\[2ex] 1 - \dfrac{1}{12}\left(\dfrac{\Theta_v}{T}\right)^2 + \dots & T \gg \Theta_v \ . \end{cases} \tag{5.1.19c}$$

The excited vibrational energy levels are noticeably populated only at temperatures above Θ_v. The specific heat (5.1.19b) is shown in Fig. 5.2.

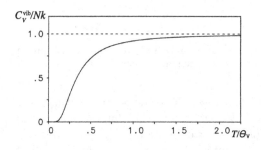

Fig. 5.2. The vibrational part of the specific heat (Eq. (5.1.19b))

The contribution of the *electronic energy* $\varepsilon_{\text{el}}^0$ to the partition function, free energy, internal energy, entropy, and to the chemical potential is, from (5.1.9):

$$Z_{el} = e^{-\varepsilon_{el}^0/kT} , \quad F_{el} = N\varepsilon_{el}^0 , \quad E_{el} = N\varepsilon_{el}^0 , \quad S_{el} = 0 , \quad \mu_{el} = \varepsilon_{el}^0 .$$

$$(5.1.20)$$

These contributions play a role in chemical reactions, where the (outer) electronic shells of the atoms undergo complete restructuring.

In a diatomic molecular gas, there are three degrees of freedom due to translation, two degrees of freedom of rotation, and one vibrational degree of freedom, which counts double ($E = \frac{p^2}{2m} + \frac{m}{2}\omega^2 x^2$; kinetic and potential energy each contribute $\frac{1}{2}kT$). The classical specific heat is therefore $7k/2$, as is observed experimentally at high temperatures. All together, this gives the temperature dependence of the specific heat as shown in Fig. 5.3. The curve is not continued down to a temperature of $T = 0$, since there the approximation of a classical ideal gas is certainly no longer valid.

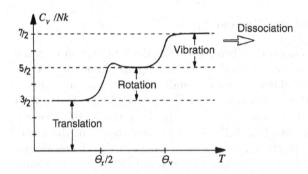

Fig. 5.3. The specific heat of a molecular gas at constant volume (schematic)

The rotational levels correspond to a wavelength of $\lambda = 0.1 - 1\,\mathrm{cm}$ and lie in the far infrared and microwave regions, while the vibrational levels at wavelengths of $\lambda = 2 \times 10^{-3} - 3 \times 10^{-3}\,\mathrm{cm}$ are in the infrared. The corresponding energies are $10^{-3} - 10^{-4}\,\mathrm{eV}$ and $0.06 - 0.04\,\mathrm{eV}$, resp. (Fig. 5.4). One electron volt corresponds to about $11000\,\mathrm{K}$ ($1\,\mathrm{K} \triangleq 0.86171 \times 10^{-4}\,\mathrm{eV}$). Some values of Θ_r and Θ_v are collected in Table 5.1.

In more complicated molecules, there are three rotational degrees of freedom and more vibrational degrees of freedom (for n atoms, in general $3n - 6$ vibrational degrees of freedom, and for linear molecules, $3n - 5$). In precise experiments, the coupling between the vibrational and rotational degrees of freedom and the anharmonicities in the vibrational degrees of freedom are also detected.

	H_2	HD	D_2	HCl	O_2
$\frac{1}{2}\Theta_r$ [K]	85	64	43	15	2
Θ_v [K]	6100	5300	4300	4100	2200

Table 5.1. The values of $\Theta_r/2$ and Θ_v for several molecules

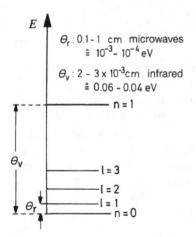

Fig. 5.4. The structure of the rotational and vibrational levels (schematic)

*5.1.4 The Influence of the Nuclear Spin

We emphasize from the outset that here, we make the assumption that the electronic ground state has zero orbital and spin angular momenta. For nuclei A and B, which have different nuclear spins S_A and S_B, one obtains an additional factor in the partition function, $(2S_A + 1)(2S_B + 1)$, i.e. $Z_i \to (2S_A + 1)(2S_B + 1)Z_i$. This leads to an additional term in the free energy per molecule of $-kT\log(2S_A + 1)(2S_B + 1)$, and to a contribution of $k\log(2S_A + 1)(2S_B + 1)$ to the entropy, i.e. a change of the chemical constants by $\log(2S_A + 1)(2S_B + 1)$ (see Eq. (3.9.29) and (5.2.5')). As a result, the internal energy and the specific heats remain unchanged.

For molecules such as H_2, D_2, O_2 which contain identical atoms, one must observe the Pauli principle. We consider the case of H_2, where the spin of the individual nuclei is $S_N = 1/2$.

Ortho hydrogen molecule: Nuclear spin triplet ($S_{tot} = 1$); the spatial wavefunction of the nuclei is antisymmetric ($l =$ odd (u))

Para hydrogen molecule: Nuclear spin singlet ($S_{tot} = 0$); the spatial wavefunction of the nuclei is symmetric ($l =$ even (g))

$$Z_u = \sum_{l \text{ odd(u)}} (2l + 1)\exp\left(-\frac{l(l+1)}{2}\frac{\Theta_r}{T}\right) \qquad (5.1.21a)$$

$$Z_g = \sum_{l \text{ even(g)}} (2l + 1)\exp\left(-\frac{l(l+1)}{2}\frac{\Theta_r}{T}\right). \qquad (5.1.21b)$$

In complete equilibrium, we have

$$Z = 3Z_u + Z_g .$$

At $T = 0$, the equilibrium state is the ground state $l = 0$, i.e. a para state. In fact, owing to the slowness of the transition between the two spin states at $T = 0$, a mixture of ortho and para hydrogen will be obtained. At high temperatures, $Z_u \approx Z_g \approx \frac{1}{2} Z_{\text{rot}} = \frac{T}{\Theta_r}$ holds and the mixing ratio of ortho to para hydrogen is 3:1. If we start from this state and cool the sample, then, leaving ortho-para conversion out of consideration, H_2 consists of a mixture of two types of molecules: $\frac{3}{4}N$ ortho and $\frac{1}{4}N$ para hydrogen, and the partition function of this (metastable) non-equilibrium state is

$$Z = (Z_u)^{3/4}(Z_g)^{1/4} . \qquad (5.1.22)$$

Then for the specific heat, we obtain

$$C_V^{\text{rot}} = \frac{3}{4} C_{Vo}^{\text{rot}} + \frac{1}{4} C_{Vp}^{\text{rot}} . \qquad (5.1.23)$$

In Fig. 5.5, the rotational parts of the specific heat in the metastable state ($\frac{3}{4}$ ortho and $\frac{1}{4}$ para), as well as for the case of complete equilibrium, are shown. The establishment of equilibrium can be accelerated by using catalysts.

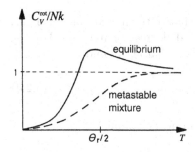

Fig. 5.5. The rotational part of the specific heat of diatomic molecules such as H_2: equilibrium (solid curve), metastable mixture (dashed)

In deuterium molecules, D_2, the nuclear spin per atom is $S = 1$,[5] which can couple in the molecule to ortho deuterium with a total spin of 2 or 0 and para deuterium with a total spin of 1. The degeneracy of these states is 6 and 3. The associated orbital angular momenta are even (g) and odd (u). The partition function, corresponding to Eq. (5.1.21a-b), is given by $Z = 6Z_g + 3Z_u$.

---

[5] QM I, page 187

*5.2 Mixtures of Ideal Molecular Gases

In this section, we investigate the thermodynamic properties of mixtures of molecular gases. The different types of particles (elements), of which there are supposed to be n, are enumerated by the index j. Then N_j refers to the particle number, $\lambda_j = \frac{h}{(2\pi m_j kT)^{1/2}}$ is the thermal wavelength, $c_j = \frac{N_j}{N}$ the concentration, $\varepsilon^0_{\mathrm{el},j}$ the electronic ground state energy, Z_j the overall partition function (see (5.1.2)), and $Z_{\mathrm{i},j}$ the partition function for the internal degrees of freedom of the particles of type j. Here, in contrast to (5.1.9), the electronic part is separated out. The total number of particles is $N = \sum_j N_j$.

The overall partition function of this non-interacting system is

$$Z = \prod_{j=1}^{n} Z_j \,, \tag{5.2.1}$$

and from it we find the free energy

$$F = -kT \sum_j N_j \left[1 + \log \frac{V Z_{\mathrm{i},j}}{N_j \lambda_j^3} \right] + \sum_j \varepsilon^0_{\mathrm{el},j} N_j \,. \tag{5.2.2}$$

From (5.2.2), we obtain the pressure, $P = -\left(\frac{\partial F}{\partial V} \right)_{T,\{N_j\}}$,

$$P = \frac{kT}{V} \sum_j N_j = \frac{kTN}{V} \,. \tag{5.2.3}$$

The equation of state (5.2.3) is identical to that of the monatomic ideal gas, since the pressure is due to the translational degrees of freedom. For the chemical potential μ_j of the component j (Sect. 3.9.1), we find

$$\mu_j = \left(\frac{\partial F}{\partial N_j} \right)_{T,V} = -kT \log \frac{V Z_{\mathrm{i},j}}{N_j \lambda_j^3} + \varepsilon^0_{\mathrm{el},j} \,; \tag{5.2.4}$$

or, if we use the pressure from (5.2.3) instead of the volume,

$$\mu_j = -kT \log \frac{kT Z_{\mathrm{i},j}}{c_j P \lambda_j^3} + \varepsilon^0_{\mathrm{el},j} \,. \tag{5.2.4'}$$

We now assume that the rotational degrees of freedom are completely unfrozen, but not the vibrational degrees of freedom ($\Theta_{\mathrm{r}} \ll T \ll \Theta_{\mathrm{v}}$). Then inserting $Z_{\mathrm{i},j} = Z_{\mathrm{rot},j} = \frac{2T}{\Theta_{\mathrm{r},j}}$ (see Eq. (5.1.13)) into (5.2.4') yields

$$\mu_j = \varepsilon^0_{\mathrm{el},j} - \frac{7}{2} kT \log kT - kT \log \frac{m_j^{3/2}}{2^{1/2} \pi^{3/2} \hbar^3 k \Theta_{\mathrm{r},j}} + kT \log c_j P \,. \tag{5.2.5}$$

We have taken the fact that the masses and the characteristic temperatures depend on the type of particle j into consideration here. The pressure enters the chemical potential of the component j in the combination $c_j P = P_j$ (partial pressure). The chemical potential (5.2.5) is a special case of the general form

$$\mu_j = \varepsilon_{\text{el},j}^0 - c_{P,j} T \log kT - kT\zeta_j + kT \log c_j P . \qquad (5.2.5')$$

For diatomic molecules in the temperature range mentioned above, $c_{P,j} = 7k/2$. The ζ_j are called chemical constants; they enter into the law of mass action (see Chap. 3.9.3). For the entropy, we find

$$
\begin{aligned}
S &= -\sum_j N_j \left(\frac{\partial \mu_j}{\partial T}\right)_{P,\{N_i\}} \\
&= \sum_j \left(c_{P,j} \log kT + c_{P,j} + k\zeta_j - k \log c_j P\right) N_j ,
\end{aligned}
\qquad (5.2.6)
$$

from which one can see that the coefficient $c_{P,j}$ is the specific heat at constant pressure of the component j.

Remarks to Sections 5.1 and 5.2: In the preceding sections, we have described the essential effects of the internal degrees of freedom of molecular gases. We now add some supplementary remarks about additional effects which depend upon the particular atomic structure.

(i) We first consider monatomic gases. The only internal degrees of freedom are electronic. In the noble gases, the electronic ground state has $L = S = 0$ and is thus not degenerate. The excited levels lie about 20 eV above the ground state, corresponding to a temperature of 200.000 K higher; in practice, they are therefore not thermally populated, and all the atoms remain in their ground state. One can also say that the electronic degrees of freedom are "frozen out". The nuclear spin S_N leads to a degeneracy factor $(2S_N + 1)$. Relative to pointlike classical particles, the partition function contains an additional factor $(2S_N + 1)e^{-\varepsilon_0/kT}$, which gives rise to a contribution to the free energy of $\varepsilon_0 - kT \log(2S_N + 1)$. This leads to an additional term of $k \log(2S_N + 1)$ in the entropy, but not to a change in the specific heat.

(ii) The excitation energies of other atoms are not as high as in the case of the noble gases, e.g. 2.1 eV for Na, or 24.000 K, but still, the excited states are not thermally populated. When the electronic shell of the atom has a nonzero S, but still $L = 0$, this leads together with the nuclear spin to a degeneracy factor of $(2S+1)(2S_N+1)$. The free energy then contains the additional term $\varepsilon_0 - kT \log((2S_N + 1)(2S + 1))$ with the consequences discussed above. Here, to be sure, one must consider the magnetic interaction between the nuclear and the electronic moments, which leads to the hyperfine interaction. This is e.g. in hydrogen of the order of 6×10^{-6} eV, leading to the well-known 21 cm line. The corresponding characteristic temperature is 0.07 K. The hyperfine splitting can therefore be completely neglected in the gas phase.

(iii) In the case that both the spin S and the orbital angular momentum L are nonzero, the ground state is $(2S + 1)(2L + 1)$-fold degenerate; this degeneracy is partially lifted by the spin-orbit coupling. The energy eigenvalues depend on the total angular momentum J, which takes on values between $S + L$ and $|S - L|$. For example, monatomic halogens in their ground state have $S = \frac{1}{2}$ and $L = 1$, according to Hund's first two rules. Because of the spin-orbit coupling, in the ground state $J = \frac{3}{2}$, and the levels with $J = \frac{1}{2}$ have a higher energy. For e.g. chlorine, the doubly-degenerate $^2P_{1/2}$ level lies $\delta\varepsilon = 0.11\,\text{eV}$ above the 4-fold degenerate $^2P_{3/2}$ ground state level. This corresponds to a temperature of $\frac{\delta\varepsilon}{k} = 1270\,\text{K}$. The partition function now contains a factor $Z_{\text{el}} = 4\,e^{-\varepsilon_0/kT} + 2\,e^{-(\varepsilon_0+\delta\varepsilon)/kT}$ due to the internal fine-structure degrees of freedom, which leads to an additional term in the free energy of $-kT\log Z_{\text{el}} = \varepsilon_0 - kT\log\left(4 + 2\,e^{-\frac{\delta\varepsilon}{kT}}\right)$. This yields the following electronic contribution to the specific heat:

$$C_V^{\text{el}} = Nk\,\frac{2\left(\frac{\delta\varepsilon}{kT}\right)^2 e^{\frac{\delta\varepsilon}{kT}}}{\left(2\,e^{\frac{\delta\varepsilon}{kT}}+1\right)^2}\,.$$

For $T \ll \delta\varepsilon/k$, $Z_{\text{el}} = 4$, only the four lowest levels are populated, and $C_V^{\text{el}} = 0$. For $T \gg \delta\varepsilon/k$, $Z_{\text{el}} = 6$, and all six levels are equally occupied, so that $C_V^{\text{el}} = 0$. For temperatures between these extremes, C_V^{el} passes through a maximum at about $\delta\varepsilon/k$. Both at low and at high temperatures, the fine structure levels express themselves only in the degeneracy factors, but do not contribute to the specific heat. One should note however that monatomic Cl is present only at very high temperatures, and otherwise bonds to give Cl_2.

(iv) In diatomic molecules, in many cases the lowest electronic state is not degenerate and the excited electronic levels are far from ε_0. The internal partition function contains only the factor $e^{-\varepsilon_0/kT}$ due to the electrons. There are, however, molecules which have a finite orbital angular momentum Λ or spin. This is the case in NO, for example. Since the orbital angular momentum has two possible orientations relative to the molecular axis, a factor of 2 in the partition function results. A finite electronic spin leads to a factor $(2S+1)$. For $S \neq 0$ and $\Lambda \neq 0$, there are again fine-structure effects which can be of the right order of magnitude to influence the thermodynamic properties. The resulting expressions take the same form as those in Remark (iii). A special case is that of the oxygen molecule, O_2. Its ground state $^3\Sigma$ has zero orbital angular momentum and spin $S = 1$; it is thus a triplet without fine structure. The first excited level $^1\Delta$ is doubly degenerate and lies relatively near at $\delta\varepsilon = 0.97\,\text{eV} \triangleq 11300\,\text{K}$, so that it can be populated at high temperatures. These electronic configurations lead to a factor of $e^{\frac{-\varepsilon_0}{kT}}\left(3+2\,e^{\frac{-\delta\varepsilon}{kT}}\right)$ in the partition function, with the consequences discussed in Remark (iii).

5.3 The Virial Expansion

5.3.1 Derivation

We now investigate a real gas, in which the particles interact with each other. In this case, the partition function can no longer be exactly calculated. For its evaluation, as a first step we will describe the virial expansion, an expansion in terms of the density. The grand partition function Z_G can be decomposed into the contributions for 0,1,2, etc. particles

$$Z_G = \text{Tr}\,e^{-(H-\mu N)/kT} = 1+Z(T,V,1)\,e^{\mu/kT}+Z(T,V,2)\,e^{2\mu/kT}+\ldots\,, \quad (5.3.1)$$

where $Z_N \equiv Z(T,V,N)$ represents the partition function for N particles.

From it, we obtain the grand potential, making use of the Taylor series expansion of the logarithm

$$\Phi = -kT \log Z_G = -kT \left[Z_1 e^{\mu/kT} + \left(Z_2 - \frac{1}{2} Z_1^2 \right) e^{2\mu/kT} + \dots \right], \quad (5.3.2)$$

where the logarithm has been expanded in powers of the fugacity $z = e^{\mu/kT}$. Taking the derivatives of (5.3.2) with respect to the chemical potential, we obtain the mean particle number

$$\bar{N} = -\left(\frac{\partial \Phi}{\partial \mu} \right)_{T,V} = Z_1 e^{\mu/kT} + 2 \left(Z_2 - \frac{1}{2} Z_1^2 \right) e^{2\mu/kT} + \dots . \quad (5.3.3)$$

Eq. (5.3.3) can be solved iteratively for $e^{\mu/kT}$, with the result

$$e^{\mu/kT} = \frac{\bar{N}}{Z_1} - \frac{2 \left(Z_2 - \frac{1}{2} Z_1^2 \right)}{Z_1} \left(\frac{\bar{N}}{Z_1} \right)^2 + \dots . \quad (5.3.4)$$

Eq. (5.3.4) represents a series expansion of $e^{\mu/kT}$ in terms of the density, since $Z_1 \sim V$. Inserting (5.3.4) into Φ has the effect that Φ is given in terms of $T, V, \bar{N}$ instead of its natural variables T, V, μ, which is favorable for constructing the equation of state:

$$\Phi = -kT \left[\bar{N} - \left(Z_2 - \frac{1}{2} Z_1^2 \right) \frac{\bar{N}^2}{Z_1^2} + \dots \right]. \quad (5.3.5)$$

These are the first terms of the so called *virial expansion*. By application of the Gibbs–Duhem relation $\Phi = -PV$, one can go from it directly to the expansion of the equation of state in terms of the particle number density $\rho = \bar{N}/V$

$$P = kT\rho \left[1 + B(T)\rho + C(T)\rho^2 + \dots \right]. \quad (5.3.6)$$

The coefficient of ρ^n in square brackets is called the $(n+1)$th virial coefficient. The leading correction to the equation of state of an ideal gas is determined by the *second virial coefficient*

$$B = -\left(Z_2 - \frac{1}{2} Z_1^2 \right) V/Z_1^2 . \quad (5.3.7)$$

This expression holds both in classical and in quantum mechanics.

Note: in the classical limit the integrations over momentum can be carried out, and (5.3.1) is simplified as follows:

$$Z_G(T,V,\mu) = \sum_{N=0}^{\infty} \frac{e^{\beta \mu N}}{N! \lambda^{3N}} Q(T,V,N) . \quad (5.3.8)$$

Here, $Q(T, V, N)$ is the configurational part of the partition function $Z(T, V, N)$

$$Q(T, V, N) = \int_V d^{3N}x \; e^{-\beta \sum_{i<j} v_{ij}} = \int_V d^{3N}x \prod_{i<j}(1 + f_{ij}) =$$

$$= \int_V d^{3N}x \; [1 + (f_{12} + f_{13} + \ldots) + (f_{12}f_{13} + \ldots) + \ldots]$$

(5.3.9)

with $f_{ij} = e^{-\beta v_{ij}} - 1$. In this expression, $\sum_{i<j} \equiv \frac{1}{2}\sum_i \sum_{j \neq i}$ refers to the sum over all pairs of particles. One can see from this that the virial expansion represents an expansion in terms of r_0^3/v, where r_0 is the range of the potential. The classical expansion is valid for $\lambda \ll r_0 \ll v^{1/3}$; see Eqs. (B.39a) and (B.39b) in Appendix B. Equation (5.3.9) can be used as the basis of a systematic graph-theoretical expansion (Ursell and Mayer 1939).

5.3.2 The Classical Approximation for the Second Virial Coefficient

In the case of a classical gas, one finds for the partition function for N particles

$$Z_N = \frac{1}{N! h^{3N}} \int d^3 p_1 \ldots d^3 p_N \int d^3 x_1 \ldots d^3 x_N \; e^{(\sum_i \mathbf{p}_i^2 / 2m + v(\mathbf{x}_1, \ldots, \mathbf{x}_N))/kT} \; ;$$

(5.3.10a)

after integrating over the $3N$ momenta, this becomes

$$Z_N = \frac{1}{\lambda^{3N} N!} \int d^3 x_1 \ldots d^3 x_N \; e^{-v(\mathbf{x}_1, \ldots, \mathbf{x}_N)/kT} \; ,$$

(5.3.10b)

where $v(\mathbf{x}_1, \ldots, \mathbf{x}_N)$ is the total potential of the N particles. The integrals over $\mathbf{x}_i$ are restricted to the volume V. If no external potential is present, and the system is translationally invariant, so that the two-particle interaction depends only upon $\mathbf{x}_1 - \mathbf{x}_2$, we find from (5.3.10b)

$$Z_1 = \frac{1}{\lambda^3} \int d^3 x_1 \; e^0 = \frac{V}{\lambda^3}$$

(5.3.11a)

and

$$Z_2 = \frac{1}{2\lambda^6} \int d^3 x_1 \, d^3 x_2 \; e^{-v(\mathbf{x}_1 - \mathbf{x}_2)/kT} = \frac{V}{2\lambda^6} \int d^3 y \; e^{-v(\mathbf{y})/kT} \; .$$

(5.3.11b)

This gives for the second virial coefficient (5.3.7):

$$B = -\frac{1}{2} \int d^3 y \; f(\mathbf{y}) = -\frac{1}{2} \int d^3 y \left(e^{-v(\mathbf{y})/kT} - 1 \right)$$

(5.3.12)

with $f(\mathbf{y}) = e^{-v(\mathbf{y})/kT} - 1$. To proceed, we now require the two-particle potential $v(\mathbf{y})$, also known as the pair potential. In Fig. 5.6, as an example, the Lennard–Jones potential is shown; it finds applications in theoretical models for the description of gases and liquids and it is defined in Eq. (5.3.16).

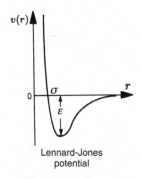

Lennard-Jones
potential

Fig. 5.6. The Lennard–Jones potential as an example of a pair potential $v(\mathbf{y})$, Eq. (5.3.16)

5.3.2.1 A Qualitative Estimate of $B(T)$

A typical characteristic of realistic potentials is the strong increase for overlapping atomic shells and the attractive interaction at larger distances. A typical shape is shown in Fig. 5.7. Up to the so called 'hard-core' radius σ, the potential is infinite, and outside this radius it is weakly negative. Thus the shape of $f(r)$ as shown in Fig. 5.7 results.

If we can now assume that in the region of the negative potential, $\left|\frac{v(\mathbf{x})}{kT}\right| \ll 1$, then we find for the function in (5.3.12)

$$f(\mathbf{x}) = \begin{cases} -1 & |\mathbf{x}| < \sigma \\ -\dfrac{v(\mathbf{x})}{kT} & |\mathbf{x}| \geq \sigma \end{cases} . \qquad (5.3.13)$$

From this, we obtain the second virial coefficient:

$$B(T) \approx -\frac{1}{2}\left[-\frac{4\pi}{3}\sigma^3 + 4\pi \int_\sigma^\infty dr\, r^2(-v(r))/kT\right] = b - \frac{a}{kT}, \qquad (5.3.14)$$

where

$$b = \frac{2\pi}{3}\sigma^3 = 4\,\frac{4\pi}{3}r_0^3 \qquad (5.3.15a)$$

denotes the fourfold molecular volume. For hard spheres of radius r_0, $\sigma = 2r_0$ and

$$a = -2\pi \int_\sigma^\infty dr\, r^2 v(r) = -\frac{1}{2}\int d^3x\, v(\mathbf{x})\Theta(r - \sigma) . \qquad (5.3.15b)$$

The result (5.3.14) for $B(T)$ is drawn in Fig. 5.8. In fact, $B(T)$ decreases again at higher temperatures, since the potential in Nature, unlike the artificial case of infinitely hard spheres, is not infinitely high (see Fig. 5.9).

Remark: From the experimental determination of the temperature dependence of the virial coefficients, we can gain information about the potential.

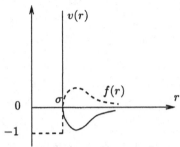
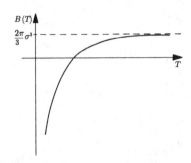

Fig. 5.7. A typical pair potential $v(r)$ (solid curve) and the associated $f(r)$ (dashed).

Fig. 5.8. The second virial coefficient from the approximate relation (5.3.14)

Examples:

Lennard–Jones potential ((12-6)-potential):

$$v(r) = 4\varepsilon \left[\left(\frac{\sigma}{r} \right)^{12} - \left(\frac{\sigma}{r} \right)^{6} \right] . \tag{5.3.16}$$

exp-6-Potential :

$$v(r) = \varepsilon \left[\exp\left(\frac{a - r}{\sigma_1} \right) - \left(\frac{\sigma_2}{r} \right)^{6} \right] . \tag{5.3.17}$$

The exp-6-potential is a special case of the so called Buckingham potential, which also contains a term $\propto -r^{-8}$.

5.3.2.2 The Lennard–Jones Potential

We will now discuss the second virial coefficient in the case of a Lennard–Jones potential

$$v(r) = 4\varepsilon \left[\left(\frac{\sigma}{r} \right)^{12} - \left(\frac{\sigma}{r} \right)^{6} \right] .$$

It proves expedient to introduce the dimensionless variables $r^* = r/\sigma$ and $T^* = kT/\varepsilon$. Integrating (5.3.12) by parts yields

$$B(T) = \frac{2\pi}{3} \sigma^3 \frac{4}{T^*} \int dr^* \, r^{*2} \left[\frac{12}{r^{*12}} - \frac{6}{r^{*6}} \right] e^{-\frac{4}{T^*}\left[\frac{1}{r^{*12}} - \frac{1}{r^{*6}} \right]} . \tag{5.3.18}$$

Expansion of the factor $\exp\left(\frac{4}{T^* r^{*6}} \right)$ in terms of $\frac{4}{T^* r^{*6}}$ leads to

$$B(T) = -\frac{2\pi}{3} \sigma^3 \sum_{j=0}^{\infty} \frac{2^{j-3/2}}{j!} \Gamma\left(\frac{2j-1}{4} \right) T^{*-(2j+1)/4}$$

$$= \frac{2\pi}{3} \sigma^3 \left[\frac{1.73}{T^{*1/4}} - \frac{2.56}{T^{*3/4}} - \frac{0.87}{T^{*5/4}} - \cdots \right] \tag{5.3.19}$$

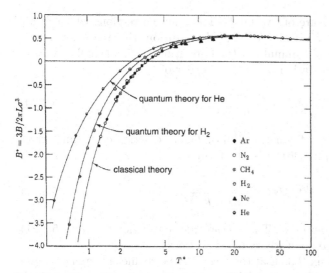

Fig. 5.9. The reduced second virial coefficient $B^* = 3B/2\pi L\sigma^3$ for the Lennard–Jones potential. L denotes the Loschmidt number (Avagadro's number, $L = 6.0221367 \cdot 10^{23}\,\mathrm{mol}^{-1}$); after Hirschfelder *et al.*[6] and R. J. Lunbeck, Dissertation, Amsterdam 1950

(see Hirschfelder *et al.*[6] Eq. (3.63)); the series converges quickly at large T^*. In Fig. 5.9, the reduced second virial coefficient is shown as a function of T^*.

Remarks:

(i) The agreement for the noble gases Ne, Ar, Kr, Xe after adjustment of σ and ε is good.
(ii) At $T^* > 100$, the decrease in $B(T)$ is experimentally somewhat greater than predicted by the Lennard–Jones interaction (i.e. the repulsion is weaker).
(iii) An improved fit to the experimental values is obtained with the exp-6-potential (5.3.17).
(iv) The possibility of representing the second virial coefficients for classical gases in a unified form by introducing dimensionless quantities is an expression of the so called law of corresponding states (see Sect. 5.4.3).

5.3.3 Quantum Corrections to the Virial Coefficients

The quantum-mechanical expression for the second virial coefficient $B(T)$ is given by (5.3.7), where the partition functions occurring there are to be computed quantum mechanically. The quantum corrections to $B(T)$ and the

---

[6] T. O. Hirschfelder, Ch. F. Curtiss and R. B. Bird, *Molecular Theory of Gases and Liquids*, John Wiley and Sons, Inc., New York 1954

other virial coefficients are of two kinds: There are corrections which result from statistics (Bose or Fermi statistics). In addition, there are corrections which arise from the non-commutativity of quantum mechanical observables. The corrections due to statistics are of the order of

$$B = \mp \frac{\lambda^3}{2^{5/2}} \propto \hbar^3 \qquad \text{for} \qquad \begin{array}{c} \text{bosons} \\ \text{fermions} \end{array} , \qquad (5.3.20)$$

as one can see from Sect. 4.2 or Eq. (B.43). The interaction quantum corrections, according to Eq. (B.46), take the form

$$B_{\text{qm}} = \int d^3y \, e^{-v(\mathbf{y})/kT} (\boldsymbol{\nabla} v(\mathbf{y}))^2 \frac{\hbar^2}{24m(kT)^2} , \qquad (5.3.21)$$

and are thus of the order of $\hbar^2$. The lowest-order correction given in (5.3.21) results from the non-commutativity of $\mathbf{p}^2$ and $v(\mathbf{x})$.

We show in Appendix B.33 that the second virial coefficient can be related to the time which the colliding particles spend within their mutual potential. The shorter this time, the more closely the gas obeys the classical equation of state for an ideal gas.

5.4 The Van der Waals Equation of State

5.4.1 Derivation

We now turn to the derivation of the equation of state of a *classical*, real (i.e. interacting) gas. We assume that the interactions of the gas atoms (molecules) consist only of a two-particle potential, which can be decomposed into a hard-core (H.C.) part, $v_{\text{H.C.}}(\mathbf{y})$ for $|\mathbf{y}| \leq \sigma$, and an attractive part, $w(\mathbf{y})$ (see Fig. 5.7):

$$v(\mathbf{y}) = v_{\text{H.C.}}(\mathbf{y}) + w(\mathbf{y}) . \qquad (5.4.1)$$

The expression "hard core" means that the gas molecules repel each other at short distances like impenetrable hard spheres, which is in fact approximately the case in Nature.

Our task is now to determine the partition function, for which after carrying out the integrations over momenta we obtain

$$Z(T, V, N) = \frac{1}{\lambda^{3N} N!} \int d^3x_1 \dots \int d^3x_N \, e^{-\sum_{i<j} v(\mathbf{x}_i - \mathbf{x}_j)/kT} . \qquad (5.4.2)$$

We still have to compute the configurational part. This can of course not be carried out exactly, but instead contains some intuitive approximations. Let us first ignore the attractive interaction and consider only the hard-core potential. This yields in the partition function for many particles:

$$\int d^3x_1 \ldots d^3x_N \; e^{-\sum_{i<j} v_{\mathrm{H.C.}}(\mathbf{x}_{ij})/kT} \approx (V - V_0)^N \; . \tag{5.4.3}$$

This result can be made plausible as follows: if the hard-core radius were zero, $\sigma = 0$, then the integration in (5.4.3) would give simply V^N; for a finite σ, each particle has only $V - V_0$ available, where V_0 is the volume occupied by the other $N - 1$ particles. This is not exact, since the size of the free volume $(V - V_0)$ depends on the configuration, as can be seen from Fig. 5.10. In (5.4.3), V_0 is to be understood the occupied volume for typical configurations which have a large statistical weight. Then, one can imagine carrying out the integrations in (5.4.3) successively, obtaining a factor $V - V_0$ for each particle.

Referring to Fig. 5.10, we can find the following bounds for V_0 with a particle number N: the smallest V_0 is obtained for spherical closest packing, $V_0^{\mathrm{min}} = 4\sqrt{2}\,r_0^3 N = 5.65\,r_0^3 N$. The largest V_0 is found when the spheres of radius $2r_0$ do not overlap, i.e. $V_0^{\mathrm{max}} = 8\frac{4\pi}{3}\,r_0^3 N = 33.51\,r_0^3 N$. The actual V_0 will lie between these extremes and can be determined as below from the comparison with the virial expansion, namely $V_0 = bN = 4\frac{4\pi}{3}\,r_0^3 N = 16.75\,r_0^3 N$.

Using (5.4.3), we can cast the partition function (5.4.2) in the form

$$Z(T, V, N) = \frac{(V - V_0)^N}{\lambda^{3N} N!} \frac{\int d^3x_1 \ldots \int d^3x_N \; e^{-\mathrm{H.C.}} e^{-\sum_{i<j} w(\mathbf{x}_i - \mathbf{x}_j)/kT}}{\int d^3x_1 \ldots \int d^3x_N \; e^{-\mathrm{H.C.}}} \; .$$

$$\tag{5.4.4}$$

Here, H.C. stands for the sum of all contributions from the hard-core potential divided by kT. The second fraction can be interpreted as the average of $\exp\left\{-\sum_{i<j} w(\mathbf{x}_i - \mathbf{x}_j)/kT\right\}$ in a gas which experiences only hard-core interactions. Before we treat this in more detail, we want to consider the second exponent more closely. For potentials whose range is much greater than σ and the distance between particles, it follows approximately that the potential acting on j due to the other particles,

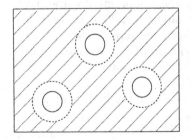 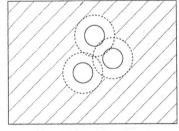

Fig. 5.10. Two configurations of three atoms within the volume V. In the first configuration, V_0 is larger than in the second. The center of gravity of an additional atom must be located outside the dashed circles. In the second configuration (closer packing), there will be more space for an additional atom (spheres of radius r_0 are represented by solid circles, spheres of radius $\sigma = 2r_0$ by dashed circles)

$\sum_{i \neq j} w(\mathbf{x}_i - \mathbf{x}_j) \approx (N-1) \int \frac{d^3 x}{V} w(\mathbf{x})$, i.e. the sum over all pairs

$$\sum_{i<j} w(\mathbf{x}_i - \mathbf{x}_j) \equiv \frac{1}{2} \sum_i \sum_{i \neq j} w(\mathbf{x}_i - \mathbf{x}_j) \approx \frac{1}{2} N(N-1)\bar{w} \approx \frac{1}{2} N^2 \bar{w} \quad (5.4.5a)$$

with

$$\bar{w} = \frac{1}{V} \int d^3 x \, w(\mathbf{x}) \equiv -\frac{2a}{V} . \quad (5.4.5b)$$

Thus we find for the partition function

$$Z(T, V, N) = \frac{(V - V_0)^N}{\lambda^{3N} N!} e^{-\frac{N(N-1)}{2} \frac{\bar{w}}{kT}} = \frac{(V - V_0)^N}{\lambda^{3N} N!} e^{\frac{N^2 a}{V kT}} . \quad (5.4.6)$$

In this calculation, the attractive part of the potential was replaced by its average value. Here, as in the molecular field theory for ferromagnetism which will be treated in the next chapter, we are using an "average-potential approximation".

Before we discuss the thermodynamic consequences of (5.4.6), we return once more to (5.4.4) and the note which followed it. The last factor can be written using a cumulant expansion, Eq. (1.2.16'), in the form

$$\left\langle e^{-\sum_{i<j} w(\mathbf{x}_i - \mathbf{x}_j)/kT} \right\rangle_{\mathrm{H.C.}} = \exp\left\{ -\left\langle \sum_{i<j} w(\mathbf{x}_i - \mathbf{x}_j)/kT \right\rangle_{\mathrm{H.C.}} \right.$$
$$\left. + \frac{1}{2} \left(\left\langle \left(\sum_{i<j} w(\mathbf{x}_i - \mathbf{x}_j)/kT \right)^2 \right\rangle_{\mathrm{H.C.}} - \left\langle \sum_{i<j} w(\mathbf{x}_i - \mathbf{x}_j)/kT \right\rangle_{\mathrm{H.C.}}^2 \right) + \ldots \right\} . $$
$$(5.4.7)$$

The average values $\langle \ \rangle_{\mathrm{H.C.}}$ are taken with respect to the canonical distribution function of the total hard-core potential. Therefore, $\left\langle \sum_{i<j} w(\mathbf{x}_i - \mathbf{x}_j) \right\rangle_{\mathrm{H.C.}}$ refers to the average of the attractive potential in the "free" volume allowed by the interaction of hard spheres. Under the assumption made earlier that the range is much greater than the hard-core radius σ and the particle distance, we again find (5.4.5a,b) and (5.4.6). The second term in the cumulant series (5.4.7) represents the mean square deviation of the attractive interactions. The higher the temperature, the more dominant the term $\bar{w}/kT$ becomes.

From (5.4.6), using $N! \simeq N^N e^{-N} \sqrt{2\pi N}$, we obtain the free energy,

$$F = -kTN \log \frac{e(V - V_0)}{\lambda^3 N} - \frac{N^2 a}{V} , \quad (5.4.8)$$

the pressure (the thermal equation of state),

$$P = -\left(\frac{\partial F}{\partial V} \right)_{T,N} = \frac{kTN}{V - V_0} - \frac{N^2 a}{V^2} , \quad (5.4.9)$$

and, with $E = -T^2 \left(\frac{\partial}{\partial T}\frac{F}{T}\right)_{V,N}$, the internal energy (caloric equation of state),

$$E = \frac{3}{2}NkT - \frac{N^2 a}{V} \,. \tag{5.4.10}$$

Finally, we can relate V_0 to the second virial coefficient. To do this, we expand (5.4.9) in terms of $1/V$ and identify the result with the virial expansion (5.3.6) and (5.3.14):

$$P = \frac{kTN}{V}\left[1 + \frac{V_0}{V} - \frac{aN}{kTV} + \cdots\right] \equiv \frac{kTN}{V}\left[1 + \left(b - \frac{a}{kT}\right)\frac{N}{V} + \cdots\right].$$

From this, we obtain

$$V_0 = Nb \,, \tag{5.4.11}$$

where b is the contribution to the second virial coefficient which results from the repulsive part of the potential. Inserting in (5.4.9), we find

$$P = \frac{kT}{v - b} - \frac{a}{v^2} \,, \tag{5.4.12}$$

where on the right-hand side, the specific volume $v = V/N$ was introduced. Equation (5.4.12) or equivalently (5.4.9) is the *van der Waals equation of state* for real gases,[7] and (5.4.10) is the associated caloric equation of state.

Remarks:

(i) The van der Waals equation (5.4.12) has, in comparison to the ideal gas equation $P = kT/v$, the following properties: the volume v is replaced by $v - b$, the free volume. For $v = b$, the pressure would become infinite. This modification with respect to the ideal gas is caused by the repulsive part of the potential.

(ii) The attractive interaction causes a reduction in the pressure via the term $-a/v^2$. This reduction becomes relatively more important as the temperature is lowered.

(iii) We make another comparison of the van der Waals equation to the ideal gas equation by writing (5.4.12) in the form

$$\left(P + \frac{a}{v^2}\right)(v - b) = kT \,.$$

Compared to $Pv = kT$, the specific volume v has been decreased by b, because the molecules are not pointlike, but instead occupy their own finite volumes. The mutual attraction of the molecules leads at a given pressure to a reduction of the volume; it thus acts like an additional pressure term. One can also readily understand the proportionality of this term to $1/v^2$. If one considers the surface layer of a liquid, it experiences a kind of attractive force from the deeper-lying layers, which must be proportional to the square of the density, since if the density were increased, the number of molecules in each layer would increase in proportion to the density, and the attractive force per unit area would thus increase proportionally to $1/v^2$.

[7] Johannes Dietrich van der Waals, 1837-1923: equation of state formulated 1873, Nobel prize 1910

The combined action of the two terms in the van der Waals equation results in qualitatively different shapes for the isotherms at low (T_1, T_2) and at high (T_3, T_4) temperatures. The family of van der Waals isotherms is shown in Fig. 5.11. For $T > T_c$, the isotherms are monotonic, while for $T < T_c$, they are S-shaped; the significance of this will be discussed below.

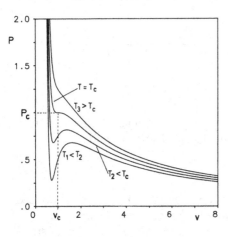

Fig. 5.11. The van der Waals isotherms in dimensionless units P/P_c and v/v_c

We see immediately that on the so called critical isotherm, there is a critical point, at which the first and second derivatives vanish, i.e. a horizontal point of inflection. The critical point T_c, P_c, V_c thus follows from $\frac{\partial P}{\partial V} = \frac{\partial^2 P}{\partial V^2} = 0$. This leads to the two conditions $-\frac{kT}{(v-b)^2} + \frac{2a}{v^3} = 0$, $\frac{kT}{(v-b)^3} - \frac{3a}{v^4} = 0$, from which the values

$$v_c = 3b\,, \quad kT_c = \frac{8}{27}\frac{a}{b}\,, \quad P_c = \frac{a}{27b^2} \tag{5.4.13}$$

are obtained. The dimensionless ratio

$$\frac{kT_c}{P_c v_c} = \frac{8}{3} = 2.\dot{6} \tag{5.4.14}$$

follows from this. The experimental value is found to be somewhat larger.

Note: It is apparent even from the derivation that the van der Waals equation can have only approximate validity. This is true of both the reduction of the repulsion effects to an effective molecular volume b, and of the replacement of the attractive (negative) part of the potential by its average value. The latter approximation improves as the range of the interactions increases. In the derivation, correlation effects were neglected, which is questionable especially in the neighborhood of the critical point, where strong density fluctuations will occur (see below). However, the van der Waals equation, in part with empirically modified van der Waals constants a and b, is able to give a qualitative description of condensation and of the behavior in the neighborhood of the critical point. There are numerous variations on the van der Waals equation; e.g. Clausius suggested the equation

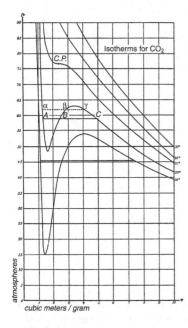

Fig. 5.12. Isotherms for carbonic acid obtained from Clausius' equation of state. From M. Planck, *Thermodynamik*, Veit & Comp, Leipzig, 1897, page 14

$$P = \frac{kT}{v - a} - \frac{c}{T(v + b)^2} \, .$$

The plot of its isotherms shown in Fig. 5.12 is similar to that obtained from the van der Waals theory.

5.4.2 The Maxwell Construction

At temperatures below T_c, the van der Waals isotherms have a typical S-shape (Fig. 5.12). The regions in which $(\partial P / \partial V)_T > 0$, i.e. the free energy is not convex and therefore the stability criterion (3.6.48b) is not obeyed, are particularly disturbing. The equation of state definitely requires modification in these regions. We now wish to consider the free energy within the van der Waals theory. As we finally shall see, an inhomogeneous state containing liquid and gaseous phases has a lower free energy. In Fig. 5.13, a van der Waals isotherm and below it the associated free energy $f(T, v) = F(T, V)/N$ are plotted. Although the lower figure can be directly read off from Eq. (5.4.8), it is instructive and useful for further discussion to determine the typical shape of the specific free energy from the isotherms $P(T, v)$ by integration of $P = -\left(\frac{\partial f}{\partial v}\right)_T$ over volume:

$$f(T, v) = f(T, v_a) - \int_{v_a}^{v} dv' \, P(T, v') \, . \tag{5.4.15}$$

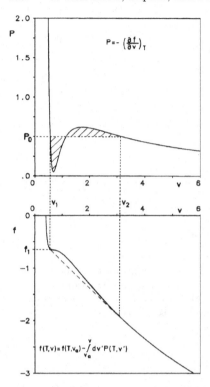

Fig. 5.13. A van der Waals isotherm and the corresponding free energy in the dimensionless units P/P_c, v/v_c and f/kT_c. The free energy of the heterogeneous state (dashed) is lower than the van der Waals free energy (solid curve)

The integration is carried out from an arbitrary initial value v_a of the specific volume up to v. We now draw in a horizontal line intersecting the van der Waals isotherm in such a way that the two shaded areas are equal. The pressure which corresponds to this line is denoted by P_0. This construction yields the two volume values v_1 and v_2. The values of the free energy at the volumes $v_{1,2}$ will be denoted by $f_{1,2} = f(T, v_{1,2})$. At the volumes v_1 and v_2, the pressure assumes the value P_0 and therefore the slope of $f(T, v)$ at these points has the value $-P_0$. As a reference for the graphical determination of the free energy, we draw a straight line through (v_1, f_1) with its slope equal to $-P_0$ (shown as a dashed line). If the pressure had the value P_0 throughout the whole interval between v_1 and v_2, then the free energy would be $f_1 - P_0(v - v_1)$. We can now readily see that the free energy which is shown in Fig. 5.13 follows from $P(T, v)$, since the van der Waals isotherm to the right of v_1 initially falls below the horizontal line $P = P_0$. Thus the negative integral, i.e. the free energy which corresponds to the van der Waals isotherm, lies above the dashed line. Only when the volume v_2 has been reached is $f_2 \equiv f(T, v_2) = f_1 - P_0(v_2 - v_1)$, owing to the equal areas which were presupposed in drawing the horizontal line, and the two curves meet again. Due to $P_0 = -\frac{\partial f}{\partial v}\big|_{v_1} = -\frac{\partial f}{\partial v}\big|_{v_2}$, the (dashed) line with slope $-P_0$ is precisely the double tangent to the curve $f(T, v)$. Since $P > P_0$ for $v < v_1$

and $P < P_0$ for $v > v_2$, f in these regions also lies above the double tangent. In Fig. 5.13 we can see that the free energy calculated in the van der Waals theory is not convex everywhere, $\left(0 > \frac{\partial^2 f}{\partial v^2} = -\frac{\partial P}{\partial v} = \frac{1}{\kappa_T}\right)$; this violates the thermodynamic inequality (3.3.5).

For comparison, we next consider a two-phase, heterogeneous system, whose entire material content is divided into a fraction $c_1 = \frac{v_2-v}{v_2-v_1}$ in the state (v_1, T) and a fraction $c_2 = \frac{v-v_1}{v_2-v_1}$ in the state (v_2, T). These states have the same pressure and temperature and can exist in mutual equilibrium. Since the free energy of this inhomogeneous state is given by the linear combination $c_1 f_1 + c_2 f_2$ of f_1 and f_2, it lies on the dashed line.[8] Thus, the free energy of this inhomogeneous state is lower than that from the van der Waals theory. In the interval $[v_1, v_2]$ (two-phase region), the substance divides into two phases, the liquid phase with temperature and volume (T, v_1), and the gas phase with (T, v_2). The pressure in this interval is P_0. The *real isotherm* is obtained from the van der Waals isotherm by replacing the S-shaped portion by the horizontal line at $P = P_0$, which divides the area equally. Outside the interval $[v_1, v_2]$, the van der Waals isotherm is unchanged. This construction of the equation of state from the van der Waals theory is called the *Maxwell construction*. The values of v_1 and v_2 depend on the temperature of the isotherm considered, i.e. $v_1 = v_1(T)$ and $v_2 = v_2(T)$. As T approaches T_c, the interval $[v_1(T), v_2(T)]$ becomes smaller and smaller; as the temperature decreases below T_c, the interval becomes larger. Correspondingly, the pressure $P_0(T)$ increases or decreases. In Fig. 5.14, the Maxwell construction for a family of van der Waals isotherms is shown. The points $(P_0(T), v_1(T))$ and

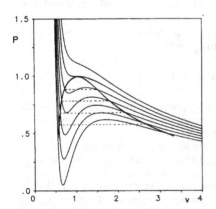
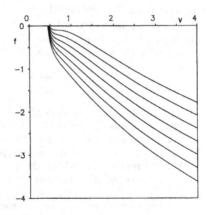

Fig. 5.14. Van der Waals isotherms, showing the Maxwell construction and the resulting coexistence curve (heavy curve) in the dimensionless units P/P_c and v/v_c, as well as the free energy f

---

[8] $c_1 + c_2 = 1$, $v_1 c_1 + v_2 c_2 = v$, $c_1 f_1 + c_2 f_2 = c_1 f_1 + c_2 (f_1 - P_0(v_2 - v_1)) = f_1 - P_0(v - v_1)$.

$(P_0(T), v_2(T))$ form the liquid branch and the gas branch of the coexistence curve (heavy curves in Fig. 5.14). The region within the coexistence curve is called the coexistence region or two-phase region. In this region the isotherms are horizontal, the state is heterogeneous, and it consists of both the liquid and gaseous phases from the two limiting points of the coexistence region.

Remarks:

(i) In Fig. 5.15, the PVT-surface which follows from the Maxwell construction is shown schematically. The van der Waals equation of state and the conclusions which can be drawn from it are in accord with the general considerations concerning the liquid-gas phase transition in the framework of thermodynamics which we gave in Sect. 3.8.1.

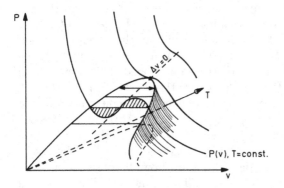

Fig. 5.15. The surface of the equation of state from the van der Waals theory with the Maxwell equal-area construction (schematic). Along with three isotherms at temperatures $T_1 < T_c < T_2$, the coexistence curve (surface) and its projection on the T-V plane are shown

(ii) The chemical potentials $\mu = f + Pv$ of the two coexisting liquid and gaseous phases are equal.

(iii) Kac, Uhlenbeck and Hemmer[9] calculated the partition function exactly for a one-dimensional model with an infinite-range potential

$$v(x) = \begin{cases} \infty & |x| < x_0 \\ -\kappa e^{-\kappa|x|} & |x| > x_0 \end{cases} \quad \text{and } \kappa \to 0 \,.$$

The result is an equation of state which is qualitatively the same as in the van der Waals theory. In the coexistence region, instead of the S-shaped curve, horizontal isotherms are found immediately.

(iv) A derivation of the van der Waals equation for long-range potentials akin to L. S. Ornstein's, in which the volume is divided up into cells and the most probable occupation number in each cell is calculated, was given by van Kampen[10]. The homogeneous and heterogeneous stable states were found. Within the coexistence region, the heterogeneous states – which are described by the horizontal line in the Maxwell construction – are absolutely stable. The two homogeneous states, represented by the S-shaped van der Waals isotherms, are metastable, as long as $\frac{\partial P}{\partial v} < 0$, and describe the superheated liquid and the supercooled vapor.

[9] M. Kac, G. E. Uhlenbeck and P. C. Hemmer, J. Math. Phys. **4**, 216 (1963)
[10] N. G. van Kampen, Phys. Rev. **135**, A362 (1964)

5.4.3 The Law of Corresponding States

If one divides the van der Waals equation by $P_c = \frac{a}{27b^2}$ and uses the reduced variables $P^* = \frac{P}{P_c}$, $V^* = \frac{v}{v_c}$, $T^* = \frac{T}{T_c}$, then a dimensionless form of the equation is obtained:

$$P^* = \frac{8T^*}{3V^* - 1} - \frac{3}{V^{*2}} \, . \tag{5.4.16}$$

In these units, the equation of state is the same for all substances. Substances with the same P^*, V^* and thus T^* are in corresponding states. Eq. (5.4.16) is called the "law of corresponding states"; it can also be cast in the form

$$\frac{P^*V^*}{T^*} = \frac{8}{3 - \frac{P^*}{T^*} \cdot \frac{T^*}{P^*V^*}} - \frac{3P^*}{T^{*2}} \frac{T^*}{P^*V^*} \, .$$

This means that P^*V^*/T^* as a function of P^* yields a family of curves with the parameter T^*. All the data from a variety of liquids at fixed T^* lie on a single curve (Fig. 5.16). This holds even beyond the validity range of the van der Waals equation. Experiments show that liquids behave similarly when P, V and T are measured in units of P_c, V_c and T_c. This is illustrated for a series of different substances in Fig. 5.16.

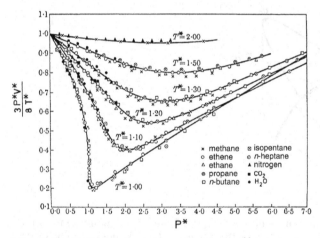

Fig. 5.16. The law of corresponding states.[11]

5.4.4 The Vicinity of the Critical Point

We now want to discuss the van der Waals equation in the vicinity of its critical point. To do this, we write the results in a form which makes the

---

[11] G. J. Su, Ind. Engng. Chem. analyt. Edn. **38**, 803 (1946)

analogy to other phase transitions transparent. The usefulness of this form
will become completely clear in connection with the treatment of ferromagnets in the next chapter. The *equation of state* in the *neighborhood of the
critical point* can be obtained by introducing the variables

$$\Delta P = P - P_c \,, \qquad \Delta v = v - v_c \,, \qquad \Delta T = T - T_c \tag{5.4.17}$$

and expanding the van der Waals equation (5.4.12) in terms of Δv and ΔT:

$$
\begin{aligned}
P &= \frac{k(T_c + \Delta T)}{2b + \Delta v} - \frac{a}{(3b + \Delta v)^2} \\
&= \frac{k(T_c + \Delta T)}{2b}\left(1 - \frac{\Delta v}{2b} + \left(\frac{\Delta v}{2b}\right)^2 - \left(\frac{\Delta v}{2b}\right)^3 + \left(\frac{\Delta v}{2b}\right)^4 \mp \cdots\right) \\
&\quad - \frac{a}{9b^2}\left(1 - 2\frac{\Delta v}{3b} + 3\left(\frac{\Delta v}{3b}\right)^2 - 4\left(\frac{\Delta v}{3b}\right)^3 + 5\left(\frac{\Delta v}{3b}\right)^4 \mp \cdots\right) .
\end{aligned}
$$

From this expansion, we find the equation of state in the immediate neighborhood of its critical point[12]

$$\Delta P^* = 4\,\Delta T^* - 6\,\Delta T^* \Delta v^* - \frac{3}{2}\,(\Delta v^*)^3 + \ldots \,; \tag{5.4.18}$$

it is in this approximation antisymmetric with respect to Δv^*, see Fig. 5.17.

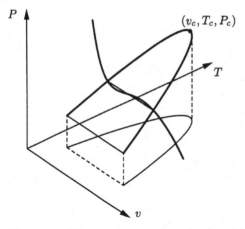

Fig. 5.17. The coexistence curve in the vicinity of the critical point. Due to the
term $4\,\Delta T^*$ in the equation of state (5.4.18), the coexistence region is inclined with
respect to the V-T plane. The isotherm shown is already so far from the critical
point that it is no longer strictly antisymmetric

---

[12] The term $\Delta T (\Delta v)^2$ and especially higher-order terms can be neglected in the
leading calculation of the coexistence curve, since it is effectively of order $(\Delta T)^2$
in comparison to $\sim (\Delta T)^{3/2}$ for the terms which were taken into account. The
corrections to the leading critical behavior will be summarized at the end of
this section. In Eq. (5.4.18), for clarity we use the reduced variables defined just
before Eq. (5.4.16): $\Delta P^* = \Delta P/P_c$ etc.

The Vapor-Pressure Curve: We obtain the vapor-pressure curve by projecting the coexistence region onto the P-T plane. Owing to the antisymmetry of the van der Waals isotherms with respect to Δv^* in the neighborhood of T_c (Eq. 5.4.18), we can easily determine the location of the two-phase region by setting $\Delta v^* = 0$ (cf. Fig. 5.17),

$$\Delta P^* = 4\,\Delta T^* \ . \tag{5.4.19}$$

The Coexistence Curve:
The coexistence curve is the projection of the coexistence region onto the V-T plane. Inserting (5.4.19) into (5.4.18), we obtain the equation $0 = 6\,\Delta T^* \Delta v^* + 3/2\,(\Delta v^*)^3$ with the solutions

$$\Delta v_G^* = -\Delta v_L^* = \sqrt{4(-\Delta T^*)} + \mathcal{O}(\Delta T^*) \tag{5.4.20}$$

for $T < T_c$. For $T < T_c$, the substance can no longer occur with a single density, but instead splits up into a less dense gaseous phase and a denser liquid phase (cf. Sect. 3.8). Δv_G^* and Δv_L^* represent the two values of the order parameter for this phase transition (see Chap. 7).

The Specific Heat:
$T > T_c$: From Eq. (5.4.10), the internal energy is found to be $E = \frac{3}{2}NkT - \frac{aN^2}{V}$.
Therefore, the specific heat at constant volume outside the coexistence region is

$$C_V = \frac{3}{2}Nk \ , \tag{5.4.21a}$$

as for an ideal gas. We now imagine that we can cool a system with precisely the critical density. Above T_c it has the homogeneous density $1/v_c$, while below T_c, it divides into the two fractions (as in (5.4.20)) $c_G = \frac{v_c - v_L}{v_G - v_L}$ and $c_L = \frac{v_G - v_c}{v_G - v_L}$ with a gaseous phase and a liquid phase. $T < T_c$: below T_c, the internal energy is given by

$$\frac{E}{N} = \frac{3}{2}kT - a\left(\frac{c_G}{v_G} + \frac{c_L}{v_L}\right) = \frac{3}{2}kT - a\frac{v_c + \Delta v_G + \Delta v_L}{(v_c + \Delta v_G)(v_c + \Delta v_L)} \ . \tag{5.4.21b}$$

If we insert (5.4.20), or, anticipating later results, (5.4.29),[13] we obtain

$$E = N\left(\frac{3}{2}kT - \frac{a}{v_c} + \frac{9}{2}k(T - T_c) + \frac{56}{25}\frac{a}{v_c}\left(\frac{T - T_c}{T_c}\right)^2 + \mathcal{O}\big((\Delta T)^{5/2}\big)\right) \ .$$

---

[13] With (5.4.20), one finds only the jump in the specific heat; in order to determine the linear term in (5.4.21b) as well, one must continue the expansion of v_G and Δv_L, Eq. (5.4.27). Including these higher terms, the coexistence curve is not symmetric.

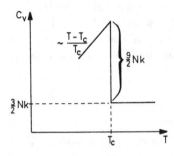

Fig. 5.18. The specific heat in the neighborhood of the critical point of the van der Waals liquid

The specific heat

$$C_V = \frac{3}{2}Nk + \frac{9}{2}Nk\left(1 + \frac{28}{25}\frac{T - T_c}{T_c} + \dots\right) \qquad \text{for } T < T_c \qquad (5.4.21c)$$

exhibits a discontinuity (see Fig. 5.18).

The Critical Isotherm:
In order to determine the critical isotherm, we set $\Delta T^* = 0$ in (5.4.18). The critical isotherm

$$\Delta P^* = -\frac{3}{2}(\Delta v^*)^3 \qquad (5.4.22)$$

is a parabola of third order; it passes through the critical point horizontally, which implies divergence of the isothermal compressibility.

The Compressibility:
To calculate the isothermal compressibility $\kappa_T = -\frac{1}{V}\left(\frac{\partial V}{\partial P}\right)_T$, we determine

$$N\left(\frac{\partial P^*}{\partial V^*}\right)_T = -6\,\Delta T^* - \frac{9}{2}(\Delta v^*)^2 \qquad (5.4.23)$$

from the van der Waals equation (5.4.18). For $T > T_c$, we find along the critical isochores ($\Delta v^* = 0$)

$$\kappa_T = \frac{1}{6P_c}\frac{1}{\Delta T^*} = \frac{T_c}{6P_c}\frac{1}{\Delta T} \,. \qquad (5.4.24a)$$

For $T < T_c$, along the coexistence curve (i.e. $\Delta v^* = \Delta v_G^* = -\Delta v_L^*$), using Eq. (5.4.20), we obtain the result $N\left(\frac{\partial P^*}{\partial V^*}\right)_T = -6\,\Delta T^* - \frac{9}{2}(\Delta v_G^*)^2 = 24\,\Delta T^*$, that is

$$\kappa_T = \frac{T_c}{12P_c}\frac{1}{(-\Delta T)} \,. \qquad (5.4.24b)$$

The isothermal compressibility diverges in the van der Waals theory above and below the critical temperature as $(T - T_c)^{-1}$. The accompanying long-range density fluctuations lead to an increase in light scattering in the forward direction (critical opalescence; see (9.4.51)).

Summary:
Comparison with experiments shows that liquids in the neighborhood of their critical points exhibit singular behavior, similar to the results described above. The coexistence line obeys a power law; however the exponent is not $1/2$, but instead $\beta \approx 0.326$; the specific heat is in fact divergent, and is characterized by a critical exponent α. The critical isotherm obeys $\Delta P \sim \Delta v^\delta$ and the isothermal compressibility is $\kappa_T \sim |T - T_c|^{-\gamma}$. Table 5.2 contains a summary of the results of the van der Waals theory and the power laws which are in general observed in Nature. The exponents β, α, δ, and γ are called critical exponents. The specific heat shows a discontinuity according to the van der Waals theory, as shown in Fig. 5.18. It is thus of the order of $(T - T_c)^0$ just to the left and to the right of the transition. The index d of the exponent 0 in Table 5.2 refers to this discontinuity. Compare Eq. (7.1.1).

Table 5.2. Critical Behavior according to the van der Waals Theory

Physical quantity	van der Waals	Critical behavior	Temperature range
$\Delta v_G = -\Delta v_L$	$\sim (T_c - T)^{\frac{1}{2}}$	$(T_c - T)^\beta$	$T < T_c$
c_V	$\sim (T - T_c)^{0d}$	$|T_c - T|^{-\alpha}$	$T \gtrless T_c$
ΔP	$\sim (\Delta v)^3$	$(\Delta v)^\delta$	$T = T_c$
κ_T	$\sim |T - T_c|^{-1}$	$|T - T_c|^{-\gamma}$	$T \gtrless T_c$

The Latent Heat Finally, we will determine the latent heat just below the critical temperature. The latent heat can be written using the Clausius–Clapeyron equation (3.8.8) in the form:

$$q = T(s_G - s_L) = T\frac{\partial P_0}{\partial T}(v_G - v_L) = T\frac{\partial P_0}{\partial T}(\Delta v_G - \Delta v_L).$$

Here, s_G and s_L refer to the entropies per particle of the gas and liquid phases and $\frac{\partial P_0}{\partial T}$ is the slope of the vaporization curve at the corresponding point. In the vicinity of the critical point, to leading order we can set $T\frac{\partial P_0}{\partial T} \approx T_c\frac{\partial P_0}{\partial T}\big|_{\text{c.p.}}$, where $(\partial P_0/\partial T)_{\text{c.p.}}$ is the slope of the evaporation curve at the critical point.

$$q = 2T_c\left(\frac{\partial P}{\partial T}\right)_{\text{c.p.}}\Delta v_G. \tag{5.4.25}$$

The slope of the vapor-pressure curve at T_c is finite (cf. Fig. 5.17 and Eq. (5.4.19)). Thus the latent heat decreases on approaching T_c according to the same power law as the order parameter, i.e. $q \propto (T_c - T)^\beta$; in the van der Waals theory, $\beta = \frac{1}{2}$.

By means of the thermodynamic relation (3.2.24) $C_P - C_V = -T \left(\frac{\partial P}{\partial T}\right)_V^2 / \left(\frac{\partial P}{\partial V}\right)_T$, we can also determine the critical behavior of the specific heat at constant pressure. Since $\left(\frac{\partial P}{\partial T}\right)_V$ is finite, the right-hand side behaves like the isothermal compressibility κ_T, and because C_V is only discontinuous or at most weakly singular, it follows in general that

$$C_P \sim \kappa_T \propto (T - T_c)^{-\gamma} ; \tag{5.4.26}$$

for a van der Waals liquid, $\gamma = 1$.

*Higher-Order Corrections to Eq. (5.4.18)

For clarity, we use the reduced quantities defined in (5.4.16). Then the van der Waals equation becomes

$$\Delta P^* = 4\Delta T^* - 6\Delta T^* \Delta v^* + 9\Delta T^* (\Delta v^*)^2 - \left(\frac{3}{2} + \frac{27}{2}\Delta T^*\right)(\Delta v^*)^3$$
$$+ \left(\frac{21}{4} + \frac{81}{4}\Delta T^*\right)(\Delta v^*)^4 + \left(\frac{99}{8} + \frac{243}{8}\Delta T^*\right)(\Delta v^*)^5 + \mathcal{O}\left((\Delta v^*)^6\right) . \tag{5.4.27}$$

The coexistence curve $\Delta v_{G/L}^*$ and the vapor-pressure curve, which we denote here by $\Delta P_0^*(\Delta T^*)$, are found from the van der Waals equation:

$$\Delta P^*(\Delta T^*, \Delta v_G^*) = \Delta P^*(\Delta T^*, \Delta v_L^*) = 0$$

with the Maxwell construction

$$\int_{\Delta v_L^*}^{\Delta v_G^*} d(\Delta v^*) \left(\Delta P^* - \Delta P_0^*(\Delta T^*)\right) = 0 .$$

For the *vapor-pressure curve* in the van der Waals theory, we obtain

$$\Delta P_0^* = 4\Delta T^* + \frac{24}{5}(-\Delta T^*)^2 + \mathcal{O}\left((-\Delta T^*)^{5/2}\right) , \tag{5.4.28}$$

and for the *coexistence curve*:

$$\Delta v_G^* = 2\sqrt{-\Delta T^*} + \frac{18}{5}(-\Delta T^*) + X(-\Delta T^*)^{3/2} + \mathcal{O}((\Delta T^*)^2)$$
$$\Delta v_L^* = -2\sqrt{-\Delta T^*} + \frac{18}{5}(-\Delta T^*) + Y(-\Delta T^*)^{3/2} + \mathcal{O}((\Delta T^*)^2) \tag{5.4.29}$$

(with $X - Y = \frac{294}{25}$, see problem 5.6). In contrast to the ferromagnetic phase transition, the order parameter is not exactly symmetric; instead, it is symmetric only near T_c, compare Eq. (5.4.20).

The *internal energy* is:

$$\frac{E}{N} = \frac{3}{2}kT - \frac{a}{v_c}\left(1 - 4\Delta T^* - \frac{56}{25}(\Delta T^*)^2 + \mathcal{O}\left(|\Delta T^*|^{5/2}\right)\right) \tag{5.4.30}$$

and the heat capacity is:

$$C_V = \frac{3}{2}Nk + \frac{9}{2}Nk\left(1 - \frac{28}{25}|\Delta T^*| + \mathcal{O}\left(|\Delta T^*|^{3/2}\right)\right). \tag{5.4.31}$$

For the calculation of the specific heat, only the difference $X - Y = 294/25$ enters. The vapor-pressure curve is no longer linear in ΔT^*, and the coexistence curve is no longer symmetric with respect to the critical volume.

5.5 Dilute Solutions

5.5.1 The Partition Function and the Chemical Potentials

We consider a solution where the solvent consists of N particles and the solute of N' atoms (molecules), so that the concentration is given by

$$c = \frac{N'}{N} \ll 1.$$

We shall calculate the properties of such a solution by employing the grand partition function[14]

$$Z_G(T, V, \mu, \mu') = \sum_{n'=0}^{\infty} Z_{n'}(T, V, \mu)z'^{n'}$$

$$= Z_0(T, V, \mu) + z'Z_1(T, V, \mu) + \mathcal{O}(z'^2). \tag{5.5.1}$$

It depends upon the chemical potentials of the solvent, μ, and of the solute, μ'. Since the solute is present only at a very low concentration, we have $\mu' \ll 0$ and therefore the fugacity $z' = e^{\mu'/kT} \ll 1$. In (5.5.1), $Z_0(T, V, \mu)$ means the grand partition function of the pure solvent and $Z_1(T, V, \mu)$ that of the solvent and a dissolved molecule.

From these expressions we find for the total pressure

$$-P = \frac{\Phi}{V} = -\frac{kT}{V}\log Z_G = \varphi_0(T, \mu) + z'\varphi_1(T, \mu) + \mathcal{O}(z'^2), \tag{5.5.2}$$

where $\varphi_0 = -\frac{kT}{V}\log Z_0$ and $\varphi_1 = -\frac{kT}{V}\frac{Z_1}{Z_0}$. In (5.5.2), $\varphi_0(T, \mu)$ is the contribution of the pure solvent and the second term is the correction due to

[14] Here, $Z_{n'}(T, V, \mu) = \sum_{n=0}^{\infty} \text{Tr}_n \text{Tr}_{n'} e^{-\beta(H_n + H'_{n'} + W_{n'n} - \mu n)}$, where Tr_n and $\text{Tr}_{n'}$ refer to the traces over n- and n'-particle states of the solvent and the solute, respectively. The Hamiltonians of these subsystems and their interactions are denoted by H_n, $H'_{n'}$ and $W_{n'n}$.

the dissolved solute. Here, Z_1 and therefore φ_1 depend on the interactions of the dissolved molecules with the solvent, but not however on the mutual interactions of the dissolved molecules. We shall now express the chemical potential μ in terms of the pressure. To this end, we use the inverse function φ_0^{-1} at fixed T, i.e. $\varphi_0^{-1}(T, \varphi_0(T, \mu)) = \mu$, obtaining

$$\mu = \varphi_0^{-1}(T, -P - z'\varphi_1(T, \mu))$$
$$= \varphi_0^{-1}(T, -P) - z' \frac{\varphi_1(T, \varphi_0^{-1}(T, -P))}{\frac{\partial \varphi_0}{\partial \mu}\big|_{\mu=\varphi_0^{-1}(T,-P)}} + \mathcal{O}(z'^2) . \tag{5.5.3}$$

The (mean) particle numbers are

$$N = -\frac{\partial \Phi}{\partial \mu} = -V\frac{\partial \varphi_0(T,\mu)}{\partial \mu} + \mathcal{O}(z') \tag{5.5.4a}$$

$$N' = -\frac{\partial \Phi}{\partial \mu'} = -\frac{z'V}{kT}\varphi_1(T,\mu) + \mathcal{O}(z'^2) . \tag{5.5.4b}$$

Inserting this into (5.5.3), we finally obtain

$$\mu(T, P, c) = \mu_0(T, P) - kTc + \mathcal{O}(c^2) , \tag{5.5.5}$$

where $\mu_0(T, P) \equiv \varphi_0^{-1}(T, -P)$ is the chemical potential of the pure solvent as a function of T and P. From (5.5.4b) and (5.5.4a), we find for the chemical potential of the solute:

$$\mu' = kT \log z' = kT \log\left(\frac{-N'kT}{V\varphi_1(T,\mu)}\right) + \mathcal{O}(z'^2)$$
$$= kT \log \frac{N'kT\frac{\partial \varphi_0(T,\mu)}{\partial \mu}}{N\varphi_1(T,\mu)} + \mathcal{O}(z') ; \tag{5.5.6}$$

and finally, using (5.5.5),

$$\mu'(T, P, c) = kT \log c + g(T, P) + \mathcal{O}(c) . \tag{5.5.7}$$

In the function $g(T, P) = kT \log(kT/v_0(T, P)\varphi_1(T, \mu_0(T, P)))$, which depends only on the thermodynamic variables T and P, the interactions of the dissolved molecules with the solvent also enter.

The simple dependences of the chemical potentials on the concentration are valid so long as one chooses T and P as independent variables. From (5.5.5), we can calculate the pressure as a function of T and μ. To do this, we use $P_0(T, \mu)$, the inverse function of $\mu_0(T, P)$, and rewrite (5.5.5) as follows:

$$\mu = \mu_0(T, P_0(T, \mu) + (P - P_0(T, \mu))) - kTc ;$$

we then expand in terms of $P - P_0(T, \mu)$ and use the fact that $\mu_0(T, P_0(T, \mu)) = \mu$ holds for the pure solvent:

$$\mu = \mu + \left(\frac{\partial \mu_0}{\partial P}\right)_T (P - P_0(T,\mu)) - kTc .$$

From the Gibbs-Duhem relation, we know that $\left(\frac{\partial \mu_0}{\partial P}\right)_T = v_0(P,T) = v + \mathcal{O}(c^2)$, from which it follows that

$$P = P_0(T,\mu) + \frac{c}{v}kT + \mathcal{O}(c^2) , \tag{5.5.8}$$

where v is the specific volume of the solvent. The interactions of the dissolved atoms with the solvent do not enter into $P(T,\mu,c)$ and $\mu(T,P,c)$ to the order we are considering, although we have not made any constraining assumptions about the nature of the interactions.

*An Alternate Derivation of (5.5.6) and (5.5.7) in the Canonical Ensemble

We again consider a system with two types of particles which are present in the amounts (particle numbers) N and N', where the concentration of the latter type, $c = \frac{N'}{N} \ll 1$, is very small. The mutual interactions of the dissolved atoms can be neglected in dilute solutions. The interaction of the solvent with the solute is denoted by $W_{N'N}$. Furthermore, the solute is treated classically. We initially make no assumptions regarding the solvent; in particular, it can be in any phase (solid, liquid, gaseous).

The partition function of the overall system then takes the form

$$
\begin{aligned}
Z &= \mathrm{Tr}\, e^{-H_N/kT} \int \frac{d\Gamma_{N'}}{N'! h^{3N'}} e^{-(H'_{N'} + W_{N'N})/kT} \\
&= \left(\mathrm{Tr}\, e^{-H_N/kT}\right) \frac{1}{N'! \lambda'^{3N}} \left\langle \int d^3x_1 \ldots d^3x_{N'}\, e^{-(V_{N'} + W_{N'N})/kT} \right\rangle ,
\end{aligned}
\tag{5.5.9a}
$$

where λ' is the thermal wavelength of the dissolved substance. H_N and $H'_{N'}$ are the Hamiltonians for the solvent and the solute molecules, $V_{N'}$ denotes the interactions of the solute molecules, and $W_{N'N}$ is the interaction of the solvent with the solute. A configurational contribution also enters into (5.5.9a):

$$
\begin{aligned}
Z_{\mathrm{conf}} &= \int d^3x_1 \ldots d^3x_{N'} \left\langle e^{-(V_{N'} + W_{N'N})/kT} \right\rangle \\
&\equiv \frac{\int d^3x_1 \ldots d^3x_{N'}\, \mathrm{Tr}\, e^{-H_N/kT}\, e^{-(V_{N'} + W_{N'N})/kT}}{\mathrm{Tr}\, e^{-H_N/kT}} .
\end{aligned}
\tag{5.5.9b}
$$

The trace runs over all the degrees of freedom of the solvent. When the latter must be treated quantum-mechanically, $W_{N'N}$ also contains an additional contribution due to the nonvanishing commutator of H_N and the interactions. $V_{N'}$ depends on the $\{x'\}$ and $W_{N'N}$ on the $\{x'\}$ and $\{x\}$ (coordinates of the solute molecules and the solvent). We assume that the interactions are short-ranged; then $V_{N'}$ can be neglected for all the typical configurations of the dissolved solute molecules:

$$
\begin{aligned}
\left\langle e^{-(V_{N'} + W_{N'N})/kT} \right\rangle &\approx \left\langle e^{-W_{N'N}/kT} \right\rangle \\
&= e^{-\langle W_{N'N}\rangle/kT + \frac{1}{2}\langle (W_{N'N}^2 - \langle W_{N'N}\rangle^2)\rangle/(kT)^2 \pm \cdots} \\
&= e^{-\sum_{n'=0}^{N'} \left(\frac{\langle W_{n'N}\rangle}{kT} - \frac{1}{2(kT)^2}\langle (W_{n'N}^2 - \langle W_{n'N}\rangle^2)\rangle \pm \cdots\right)} \\
&= e^{-N'\psi(T,V/N)} .
\end{aligned}
\tag{5.5.9c}
$$

Here, $W_{n'N}$ denotes the interaction of molecule n' with the N molecules of the solvent. In Eq. (5.5.9c), a cumulant expansion was carried out and we have taken into account that the overlap of the interactions of different molecules vanishes for all of the typical configurations. Owing to translational invariance, the expectation values $\langle W_{n'N} \rangle$ etc. are furthermore independent of $\mathbf{x}'$ and are the same for all n'. We thus find for each of the dissolved molecules a factor $e^{-\psi(T, V/N)}$, where ψ depends on the temperature and the specific volume of the solvent. It follows from (5.5.9c) that the partition function (5.5.9a) is

$$Z = \left(\operatorname{Tr} e^{-H_N/kT} \right) \frac{1}{N'!} \left(\frac{V}{\lambda'^3} \psi(T, V/N) \right)^{N'} . \qquad (5.5.10)$$

This result has the following physical meaning: *the dissolved molecules behave like an ideal gas.* They are subject at every point to the same potential from the surrounding solvent atoms, i.e. they are moving in a position-independent effective potential $kT\psi(T, V/N)$, whose value depends on the interactions, the temperature, and the density. The free energy therefore assumes the form

$$F(T, V, N, N') = F_0(T, V, N) - kTN' \log \frac{eV}{N'\lambda'^3} - N'\gamma(T, V/N) , \qquad (5.5.11)$$

where $F_0(T, V) = -kT \log \operatorname{Tr} e^{-H_N/kT}$ is the free energy of the pure solvent and $\gamma(T, V/N) = kT \log \psi(T, V/N)$ is due to the interactions of the dissolved atoms with the solvent. From (5.5.11), we find for the pressure

$$
\begin{aligned}
P = -\left(\frac{\partial F}{\partial V} \right)_{T,N,N'} &= P_0(T, V/N) + \frac{kTN'}{V} + N'\left(\frac{\partial}{\partial V}\gamma \right)_{T,N} \\
&= P_0(T, v) + \frac{kTc}{v} + c\left(\frac{\partial}{\partial v}\gamma(T, v) \right)_T ,
\end{aligned}
\qquad (5.5.12)
$$

where $c = \frac{N'}{N}$ and $v = \frac{V}{N}$ were employed.

We could calculate the chemical potentials from (5.5.11) as functions of T and v. In practice, however, one is usually dealing with physical conditions which fix the pressure instead of the specific volume. In order to obtain the chemical potentials as functions of the pressure, it is expedient to use the free enthalpy (Gibbs free energy). It is found from (5.5.11) and (5.5.12) to be

$$G = F + PV = G_0(T, P, N) - kTN' \left(\log \frac{eV}{N'\lambda'^3} - 1 \right) - N'\left(\gamma - V\frac{\partial\gamma}{\partial V} \right) , \qquad (5.5.13)$$

where $P_0(T, v)$ and $G_0(T, P, N)$ are the corresponding quantities for the pure solvent. From Equation (5.5.12), one can compute v as a function of P, T and c,

$$v = v_0(T, P) + \mathcal{O}\left(N'/N \right) .$$

If we insert this in (5.5.13), we find an expression for the free enthalpy of the form

$$G(T, P, N, N') = G_0(T, P, N) - kTN'\left(\log \frac{N}{N'} - 1 \right) + N'g(T, P) + \mathcal{O}\left(\frac{N'^2}{N} \right) , \qquad (5.5.14)$$

where $g(T, P) = \left. \left(-kT \log \frac{v}{\lambda'^3} - \left(\gamma - V\frac{\partial\gamma}{\partial V} \right) \right) \right|_{v=v_0(T,P)}$. Now we can compute the two chemical potentials as functions of T, P and c. For the chemical potential of the solvent, $\mu(T, P, c) = \left(\frac{\partial G}{\partial N} \right)_{T,P,N'}$, the result to leading order in the concentration is

$$\mu(T, P, c) = \mu_0(T, P) - kTc + \mathcal{O}(c^2) \ . \tag{5.5.15}$$

For the chemical potential of the solute, we find from (5.5.14)

$$\mu'(T, P, c) = \left(\frac{\partial G}{\partial N'}\right)_{N, P, T} = -kT \log \frac{1}{c} + g(T, P) + \mathcal{O}(c) \ . \tag{5.5.16}$$

The results (5.5.15) and (5.5.16) agree with those found in the framework of the grand canonical ensemble (5.5.5) and (5.5.7).

5.5.2 Osmotic Pressure

We let two solutions of the same substances (e.g. salt in water) be separated by a semipermeable membrane (Fig. 5.19). An example of a semipermeable membrane is a cell membrane.

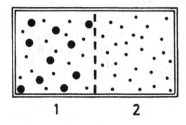

1   2

Fig. 5.19. A membrane which allows only the solvent to pass through ($=$ semipermeable) separates the two solutions. $\cdot$ = solvent, $\bullet$ = solute; concentrations c_1 and c_2

The semipermeable membrane allows only the solvent to pass through. Therefore, in chambers 1 and 2, there will be different concentrations c_1 and c_2. In equilibrium, the chemical potentials of the solvent on both sides of the membrane are equal, but not those of the solute. The *osmotic pressure* is defined by the pressure difference

$$\Delta P = P_1 - P_2 \ .$$

From (5.5.8), we can calculate the pressure on both sides of the membrane, and since in equilibrium, the chemical potentials of the solvent are equal, $\mu_1 = \mu_2$, it follows that the pressure difference is

$$\Delta P = \frac{c_1 - c_2}{v} kT \ . \tag{5.5.17}$$

The *van't Hoff formula* is obtained as a special case for $c_2 = 0$, $c_1 = c$, when only the pure solvent is present on one side of the membrane:

$$\Delta P = \frac{c}{v} kT = \frac{N'}{V} kT \ . \tag{5.5.17'}$$

Here, N' refers to the number of dissolved molecules in chamber 1 and V to its volume.

Notes:

(i) Equation (5.5.17′) holds for small concentrations independently of the nature of the solvent and the solute. We point out the formal similarity between the van't Hoff formula (5.5.17′) and the ideal gas equation. The osmotic pressure of a dilute solution of n moles of the dissolved substance is equal to the pressure that n moles of an ideal gas would exert on the walls of the overall volume V of solution and solvent.

(ii) One can gain a physical understanding of the origin of the osmotic pressure as follows: the concentrated part of the solution has a tendency to expand into the less concentrated region, and thus to equalize the concentrations.

(iii) For an aqueous solution of concentration $c = 0.01$, the osmotic pressure at room temperature amounts to $\Delta P = 13.3$ bar.

*5.5.3 Solutions of Hydrogen in Metals (Nb, Pd,...)

We now apply the results of Sect. 5.5.1 to an important practical example, the solution of hydrogen in metals such as Nb, Pd,... (Fig. 5.20). In the gas phase, hydrogen occurs in molecular form as H_2, while in metals, it dissociates. We thus have a case of *chemical equilibrium*, see Sect. 3.9.3.

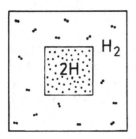

Fig. 5.20. Solution of hydrogen in metals: atomic hydrogen in a metal is represented by a dot, while molecular hydrogen in the surrounding gas phase is represented by a pair of dots.

The chemical potential of molecular hydrogen gas is

$$\mu_{H_2} = -kT \left[\log \frac{V}{N\lambda_{H_2}^3} + \log Z_i \right] = -kT \left[\log \frac{kT}{P\lambda_{H_2}^3} + \log Z_i \right] , \quad (5.5.18)$$

where Z_i also contains the electronic contribution to the partition function (Eq. (5.1.5c)). The chemical potential of atomic hydrogen dissolved in a metal is, according to Eq. (5.5.7), given by

$$\mu_H = kT \log c + g(T, P) . \quad (5.5.19)$$

The metals mentioned can be used for hydrogen storage. The condition for chemical equilibrium (3.9.26) is in this case $2\mu_H = \mu_{H_2}$; this yields the equilibrium concentration:

$$c = e^{(\mu_{H_2}/2 - g(T,P))/kT} = \left(\frac{P\lambda_{H_2}^3}{kT}\right)^{\frac{1}{2}} Z_i^{-\frac{1}{2}} \exp\left(\frac{-2g(T,P) + \varepsilon_{el}}{2kT}\right) . \quad (5.5.20)$$

Since $g(T, P)$ depends only weakly on P, the concentration of undissolved hydrogen is $c \sim P^{\frac{1}{2}}$. This dependence is known as *Sievert's law*.

5.5.4 Freezing-Point Depression, Boiling-Point Elevation, and Vapor-Pressure Reduction

Before we turn to a quantitative treatment of freezing-point depression, boiling-point elevation, and vapor-pressure reduction, we begin with a qualitative discussion of these phenomena. The free enthalpy of the liquid phase of a solution is lowered, according to Eq. (5.5.5), relative to its value in the pure solvent, an effect which can be interpreted in terms of an increase in entropy. The free enthalpies of the solid and gaseous phases remain unchanged. In Fig. 5.21, $G(T, P)$ is shown qualitatively as a function of the temperature and the pressure, keeping in mind its convexity, and assuming that the dissolved substance is soluble only in the liquid phase. The solid curve describes the pure solvent, while the change due to the dissolved substance is described by the chain curve. As a rule, the concentration of the solute in the liquid phase is largest and the associated entropy increase leads to a reduction of the free enthalpy. From these two diagrams, the depression of the freezing point, the elevation of the boiling point, and the reduction in the vapor pressure can be read off.

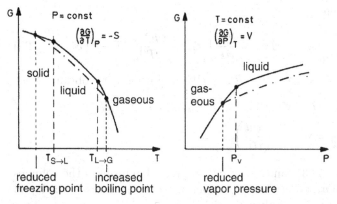

Fig. 5.21. The change in the free enthalpy on solution of a substance which dissolves to a notable extent only in the liquid phase. The solid curve is for the pure solvent, the chain curve for the solution. We can recognize the freezing-point depression, the boiling-point elevation, and the vapor-pressure reduction

Next we turn to the analytic treatment of these phenomena. We first consider the melting process. The concentrations of the dissolved substance in the liquid and solid phases are c_L and c_S.[15] The chemical potentials of the solvent in the liquid and the solid phase are denoted by μ^L and μ^S, and correspondingly in the pure system by μ_0^L and μ_0^S. From Eq. (5.5.5), we find that

$$\mu^L = \mu_0^L(P, T) - kTc_L$$

and

$$\mu^S = \mu_0^S(P, T) - kTc_S .$$

In equilibrium, the chemical potentials of the solvent must be equal, $\mu^L = \mu^S$, from which it follows that[16]

$$\mu_0^L(P, T) - kTc_L = \mu_0^S(P, T) - kTc_S . \tag{5.5.21}$$

For the pure solute, we obtain the melting curve, i.e. the relation between the melting pressure P_0 and the melting temperature T_0, from

$$\mu_0^L(P_0, T_0) = \mu_0^S(P_0, T_0) . \tag{5.5.22}$$

Let (P_0, T_0) be a point on the melting curve of the pure solvent. Then consider a point (P, T) on the melting curve which obeys (5.5.21), and which is shifted relative to (P_0, T_0) by ΔP and ΔT, that is

$$P = P_0 + \Delta P , \qquad T = T_0 + \Delta T .$$

If we expand Eq. (5.5.21) in terms of ΔP and ΔT, and use (5.5.22), we find the following relation

$$\frac{\partial \mu_0^L}{\partial P}\bigg|_0 \Delta P + \frac{\partial \mu_0^L}{\partial T}\bigg|_0 \Delta T - kTc_L = \frac{\partial \mu_0^S}{\partial P}\bigg|_0 \Delta P + \frac{\partial \mu_0^S}{\partial T}\bigg|_0 \Delta T - kTc_S . \tag{5.5.23}$$

We now recall that $G = \mu N = E - TS + PV$, and using it we obtain

$$dG = -SdT + VdP + \mu dN = d(\mu N) = \mu dN + Nd\mu ,$$

$$\left(\frac{\partial \mu}{\partial P}\right)_{T,N} = \frac{V}{N} = v, \qquad \left(\frac{\partial \mu}{\partial T}\right)_{P,N} = -\frac{S}{N} = -s .$$

The derivatives in (5.5.23) can therefore be expressed in terms of the volumes per molecule v_L and v_S, and the entropies per molecule s_L and s_S in the liquid and solid phases of the pure solvent,

---

[15] Since two phases and two components are present, the number of degrees of freedom is two (Gibbs' phase rule). One can for example fix the temperature and one concentration; then the other concentration and the pressure are determined.

[16] The chemical potentials of the solute must of course also be equal. From this fact, we can for example express the concentration in the solid phase, c_S, in terms of T and c_L. We shall, however, not need the exact value of c_S, since $c_S \ll c_L$ is negligible.

$$-(s_\mathrm{S} - s_\mathrm{L})\Delta T + (v_\mathrm{S} - v_\mathrm{L})\Delta P = (c_\mathrm{S} - c_\mathrm{L})kT \ . \tag{5.5.24}$$

Finally, we introduce the heat of melting $q = T(s_\mathrm{L} - s_\mathrm{S})$, thus obtaining

$$\frac{q}{T}\Delta T + (v_\mathrm{S} - v_\mathrm{L})\Delta P = (c_\mathrm{S} - c_\mathrm{L})kT \ . \tag{5.5.25}$$

The change in the transition temperature ΔT at a given pressure is obtained from (5.5.25), by setting $P = P_0$ or $\Delta P = 0$:

$$\Delta T = \frac{kT^2}{q}(c_\mathrm{S} - c_\mathrm{L}) \ . \tag{5.5.26}$$

As a rule, the concentration in the solid phase is much lower than that in the liquid phase, i.e. $c_\mathrm{S} \ll c_\mathrm{L}$; then (5.5.26) simplifies to

$$\Delta T = -\frac{kT^2}{q}c_\mathrm{L} < 0 \ . \tag{5.5.26'}$$

Since the entropy of the liquid is larger, or on melting, heat is absorbed, it follows that $q > 0$. As a result, the dissolution of a substance gives rise to a *freezing-point depression*.

Note: On solidification of a liquid, at first (5.5.26') holds, with the initial concentration c_L. Since however pure solvent precipitates out in solid form, the concentration c_L increases, so that it requires further cooling to allow the freezing process to continue. Freezing of a solution thus occurs over a finite temperature interval.

The above results can be transferred directly to the *evaporation process*. To do this, we make the replacements

$$\mathrm{L} \to \mathrm{G} \ , \ \mathrm{S} \to \mathrm{L}$$

and obtain from (5.5.25) for the liquid phase (L) and the gas phase (G) the relation

$$\frac{q}{T}\Delta T + (v_\mathrm{L} - v_\mathrm{G})\Delta P = (c_\mathrm{L} - c_\mathrm{G})kT \ . \tag{5.5.27}$$

Setting $\Delta P = 0$ in (5.5.27), we find

$$\Delta T = \frac{kT^2}{q}(c_\mathrm{L} - c_\mathrm{G}) \approx \frac{kT^2}{q}c_\mathrm{L} > 0 \ , \tag{5.5.28}$$

a *boiling-point elevation*. In the last equation, $c_\mathrm{L} \gg c_\mathrm{G}$ was assumed (this no longer holds near the critical point).

Setting $\Delta T = 0$ in (5.5.27), we find

$$\Delta P = \frac{c_\mathrm{L} - c_\mathrm{G}}{v_\mathrm{L} - v_\mathrm{G}}kT \approx -\frac{c_\mathrm{L} - c_\mathrm{G}}{v_\mathrm{G}}kT \ , \tag{5.5.29}$$

a *vapor-pressure reduction.* When the gas phase contains only the vapor of the pure solvent, (5.5.29) simplifies to

$$\Delta P = -\frac{c_{\mathrm{L}}}{v_{\mathrm{G}}}kT . \tag{5.5.30}$$

Inserting the ideal gas equation, $Pv_{\mathrm{G}} = kT$, we have

$$\Delta P = -c_{\mathrm{L}}P = -c_{\mathrm{L}}(P_0 + \Delta P) .$$

Rearrangement of the last equation yields the relative pressure change:

$$\frac{\Delta P}{P_0} = -\frac{c_{\mathrm{L}}}{1 + c_{\mathrm{L}}} \approx -c_{\mathrm{L}} , \tag{5.5.31}$$

known as *Raoult's law.* The relative vapor-pressure reduction increases linearly with the concentration of the dissolved substance. The results derived here are in agreement with the qualitative considerations given at the beginning of this subsection.

Problems for Chapter 5

5.1 The rotational motion of a diatomic molecule is described by the angular variables ϑ and φ and the canonically conjugate momenta p_ϑ and p_φ with the Hamilton function $H = \frac{p_\vartheta^2}{2I} + \frac{1}{2I \sin^2 \vartheta} p_\varphi^2$. Calculate the classical partition function for the canonical ensemble.
Result: $Z_{\mathrm{rot}} = \frac{2T}{\Theta_{\mathrm{r}}}$ (see footnote 4 to Eq. (5.1.15)).

5.2 Confirm the formulas (5.4.13) for the critical pressure, the critical volume, and the critical temperature of a van der Waals gas and the expansion (5.4.18) of $P(T,V)$ around the critical point up to the third order in Δv.

5.3 The expansion of the van der Waals equation in the vicinity of the critical point:
(a) Why is it permissible in the determination of the leading order to leave off the term $\Delta T(\Delta V)^2$ in comparison to $(\Delta V)^3$?
(b) Calculate the correction $\mathcal{O}(\Delta T)$ to the coexistence curve.
(c) Calculate the correction $\mathcal{O}((T - T_c)^2)$ to the internal energy.

5.4 The equation of state for a van der Waals gas is given in terms of reduced variables in Eq. (5.4.16).
Calculate the position of the inversion points (Chap. 3) in the p^*, T^* diagram. Where is the maximum of the curve?

5.5 Calculate the jump in the specific heat c_v for a van der Waals gas at a specific volume of $v \neq v_c$.

5.6 Show in general and for the van der Waals equation that κ_s and c_v exhibit the same behavior for $T \to T_c$.

5.7 Consider two metals 1 and 2 (with melting points T_1, T_2 and temperature independent heats of melting q_1, q_2), which form ideal mixtures in the liquid phase (i.e. as for small concentrations over the whole concentration range). In the solid phase these metals are not miscible. Calculate the eutectic point T_E (see also Sect. 3.9.2).

Hint: Set up the equilibrium conditions between pure solid phase 1 or 2 and the liquid phase. From these, the concentrations are determined:

$$c_i = e^{\lambda_i} , \quad \text{where } \lambda_i = \frac{q_i}{kT_i}\left(1 - \frac{T_i}{T}\right) ; \qquad i = 1, 2$$

using

$$\frac{\partial(G/T)}{\partial T} = -H/T^2 , \quad q_i = \Delta H_i , \quad G = \mu N .$$

5.8 Apply the van't Hoff formula (5.5.17′) to the following simple example: the concentration of the dissolved substance is taken to be $c = 0.01$, the solvent is water (at $20°C$); use $\rho_{H_2O} = 1\,g/cm^3$ ($20°C$). Find the osmotic pressure ΔP.

6. Magnetism

In this chapter, we will deal with the fundamental phenomenon of magnetism. We begin the first section by setting up the density matrix, starting from the Hamiltonian, and using it to derive the thermodynamic relations for magnetic systems. Then we continue with the treatment of diamagnetic and paramagnetic substances (Curie and Pauli paramagnetism). Finally, in Sect. 6.5.1, we investigate ferromagnetism. The basic properties of magnetic phase transitions will be studied in the molecular-field approximation (Curie–Weiss law, Ornstein-Zernike correlation function, etc.). The results obtained will form the starting point for the renormalization group theory of critical phenomena which is dealt with in the following chapter.

6.1 The Density Matrix and Thermodynamics

6.1.1 The Hamiltonian and the Canonical Density Matrix

We first summarize some facts about magnetic properties as known from electrodynamics and quantum mechanics. The Hamiltonian for N electrons in a magnetic field $\mathbf{H} = \mathrm{curl}\,\mathbf{A}$ is:

$$\mathcal{H} = \sum_{i=1}^{N} \frac{1}{2m}\left(\mathbf{p}_i - \frac{e}{c}\mathbf{A}\left(\mathbf{x}_i\right)\right)^2 - \boldsymbol{\mu}_i^{\mathrm{spin}}\cdot\mathbf{H}\left(\mathbf{x}_i\right) + W_{\mathrm{Coul}}. \tag{6.1.1}$$

The index i enumerates the electrons. The canonical momentum of the ith electron is $\mathbf{p}_i$ and the kinetic momentum is $m\mathbf{v}_i = \mathbf{p}_i - \frac{e}{c}\mathbf{A}\left(\mathbf{x}_i\right)$. The charge and the magnetic moment are given by[1]

$$e = -e_0\,, \quad \boldsymbol{\mu}_i^{\mathrm{spin}} = -\frac{g_e\mu_{\mathrm{B}}}{\hbar}\mathbf{S}_i\,, \tag{6.1.2a}$$

where along with the elementary charge e_0, the Bohr magneton

$$\mu_{\mathrm{B}} = \frac{e_0\hbar}{2mc} = 0.927\cdot 10^{-20}\,\frac{\mathrm{erg}}{\mathrm{Gauss}} = 0.927\cdot 10^{-23}\,\frac{\mathrm{J}}{\mathrm{T}} \tag{6.1.2b}$$

---

[1] QM I, p. 186

as well as the Landé-g-factor or the spectroscopic splitting factor of the electron

$$g_e = 2.0023 \tag{6.1.2c}$$

were introduced. The quantity $\gamma = \frac{eg_e}{2mc} = -\frac{g_e\mu_B}{\hbar}$ is called the magnetomechanical ratio or gyromagnetic ratio. The last term in (6.1.1) stands for the Coulomb interaction of the electrons with each other and with the nuclei.

The dipole-dipole interaction of the spins is neglected here. Its consequences, such as the demagnetizing field, will be considered in Sect. 6.6; see also remark (ii) at the end of Sect. 6.6.3. We assume that the magnetic field **H** is produced by some external sources. In vacuum, **B** = **H** holds. We use here the magnetic field **H**, corresponding to the more customary practice in the literature on magnetism. The current-density operator is thus given by[2]

$$\mathbf{j}(\mathbf{x}) \equiv -c\frac{\delta \mathcal{H}}{\delta \mathbf{A}(\mathbf{x})} = \sum_{i=1}^{N}\left\{\frac{e}{2m}\left[\left(\mathbf{p}_i - \frac{e}{c}\mathbf{A}(\mathbf{x}_i)\right), \delta(\mathbf{x}-\mathbf{x}_i)\right]_+ \right.$$
$$\left. + c\operatorname{curl}\left(\boldsymbol{\mu}_i^{\text{spin}}\,\delta(\mathbf{x}-\mathbf{x}_i)\right)\right\} \tag{6.1.3}$$

with $[A, B]_+ = AB + BA$. The current density contains a contribution from the electronic orbital motion and a spin contribution.

For the *total magnetic moment*, one obtains[3],[4]:

$$\boldsymbol{\mu} \equiv \frac{1}{2c}\int d^3x\,\mathbf{x}\times\mathbf{j}(\mathbf{x}) = \sum_{i=1}^{N}\left\{\frac{e}{2mc}\mathbf{x}_i\times\left(\mathbf{p}_i - \frac{e}{c}\mathbf{A}(\mathbf{x}_i)\right) + \boldsymbol{\mu}_i^{\text{spin}}\right\}\,. \tag{6.1.4}$$

When **H** is uniform, Eq. (6.1.4) can also be written in the form

$$\boldsymbol{\mu} = -\frac{\partial \mathcal{H}}{\partial \mathbf{H}}\,. \tag{6.1.5}$$

The magnetic moment of the ith electron for a uniform magnetic field (see Remark (iv) in Sect. 6.1.3) is – according to Eq. (6.1.4) – given by

$$\boldsymbol{\mu}_i = \boldsymbol{\mu}_i^{\text{spin}} + \frac{e}{2mc}\mathbf{L}_i - \frac{e^2}{2mc^2}\mathbf{x}_i\times\mathbf{A}(\mathbf{x}_i)$$
$$= \frac{e}{2mc}(\mathbf{L}_i + g_e\mathbf{S}_i) - \frac{e^2}{4mc^2}\left(\mathbf{H}\,\mathbf{x}_i^2 - \mathbf{x}_i(\mathbf{x}_i\cdot\mathbf{H})\right)\,. \tag{6.1.6}$$

---

[2] The intermediate steps which lead to (6.1.3)–(6.1.5) will be given at the end of this section.

[3] J. D. Jackson, *Classical Electrodynamics*, 2nd edition, John Wiley and sons, New York, 1975, p. 18.

[4] Magnetic moments are denoted throughout by μ, except for the spin magnetic moments of elementary particles which are termed μ^{spin}.

If $\mathbf{H} = H\mathbf{e}_z$, then (for a single particle) it follows that

$$\boldsymbol{\mu}_{iz} = \frac{e}{2mc}\left(\mathbf{L}_i + g_e\mathbf{S}_i\right)_z - \frac{e^2H}{4mc^2}\left(x_i^2 + y_i^2\right) = -\frac{\partial \mathcal{H}}{\partial H},$$

and the Hamiltonian is[5]

$$\mathcal{H} = \sum_{i=1}^{N}\left\{\frac{\mathbf{p}_i^2}{2m} - \frac{e}{2mc}\left(\mathbf{L}_i + 2\mathbf{S}_i\right)_z H + \frac{e^2H^2}{8mc^2}\left(x_i^2 + y_i^2\right)\right\} + W_{Coul}. \quad (6.1.7)$$

Here, we have used $g_e = 2$.

We now wish to set up the density matrices for magnetic systems; we can follow the steps in Chap. 2 to do this. An isolated magnetic system is described by a *microcanonical* ensemble,

$$\rho_{MC} = \delta\left(\mathcal{H} - E\right)/\Omega\left(E, \mathbf{H}\right) \quad \text{with} \quad \Omega\left(E, \mathbf{H}\right) = \mathrm{Tr}\,\delta(\mathcal{H} - E),$$

where, for the Hamiltonian, (6.1.1) is to be inserted. If the magnetic system is in contact with a heat bath, with which it can exchange energy, then one finds for the magnetic subsystem, just as in Chap. 2, the *canonical density matrix*

$$\rho = \frac{1}{Z}\,e^{-\mathcal{H}/kT}. \quad (6.1.8)$$

The normalization factor is given by the partition function

$$Z = \mathrm{Tr}\; e^{-\mathcal{H}/kT}. \quad (6.1.9a)$$

The canonical parameters (natural variables) are here the temperature, whose reciprocal is defined as in Chap. 2 in the microcanonical ensemble as the derivative of the entropy of the heat bath with respect to its energy, and the external magnetic field $\mathbf{H}$.[6] Correspondingly, the *canonical free energy*,

$$F(T, \mathbf{H}) = -kT \log Z, \quad (6.1.9b)$$

is a function of T and $\mathbf{H}$. The *entropy* S and the *internal energy* E are, by definition, calculated from

$$S = -k\langle\log\rho\rangle = \frac{1}{T}\left(E - F\right), \quad (6.1.10)$$

---

[5] See e.g. QM I, Sect. 7.2.

[6] In this chapter we limit our considerations to magnetic effects. Therefore, the particle number and the volume are treated as fixed. For phenomena such as magnetostriction, it is necessary to consider also the dependence of the free energy on the volume and more generally on the deformation tensor of the solid (see also the remark in 6.1.2.4).

and

$$E = \langle \mathcal{H} \rangle \, . \tag{6.1.11}$$

The *magnetic moment of the entire body* is defined as the thermal average of the total quantum-mechanical magnetic moment

$$\mathcal{M} \equiv \langle \boldsymbol{\mu} \rangle = -\left\langle \frac{\partial \mathcal{H}}{\partial \mathbf{H}} \right\rangle \, . \tag{6.1.12}$$

The *magnetization* $\mathbf{M}$ is defined as the magnetic moment per unit volume, i.e. for a uniformly magnetized body

$$\mathbf{M} = \frac{1}{V} \mathcal{M} \tag{6.1.13a}$$

and, in general,

$$\mathcal{M} = \int d^3x \, \mathbf{M}(\mathbf{x}) \, . \tag{6.1.13b}$$

For the differential of F, we find from (6.1.9)–(6.1.10)

$$dF = (F - E) \frac{dT}{T} - \mathcal{M} \cdot d\mathbf{H} \equiv -SdT - \mathcal{M} \cdot d\mathbf{H} \, , \tag{6.1.14a}$$

that is

$$\left(\frac{\partial F}{\partial T} \right)_{\mathbf{H}} = -S \quad \text{and} \quad \left(\frac{\partial F}{\partial \mathbf{H}} \right)_{T} = -\mathcal{M}. \tag{6.1.14b}$$

Using equation (6.1.10), one can express the internal energy E in terms of F and S and obtain from (6.1.14a) the First Law for magnetic systems:

$$dE = TdS - \mathcal{M}d\mathbf{H} \, . \tag{6.1.15}$$

The internal energy E contains the interaction of the magnetic moments with the magnetic field (see (6.1.7)). Compared to a gas, we have to make the following formal replacements in the First Law: $V \rightarrow \mathbf{H}$, $P \rightarrow \mathcal{M}$. Along with the (canonical) free energy $F(T, \mathbf{H})$, we introduce also the *Helmholtz free energy*[7]

$$A(T, \mathcal{M}) = F(T, \mathbf{H}) + \mathcal{M} \cdot \mathbf{H} \, . \tag{6.1.16}$$

Its differential is

$$dA = -SdT + \mathbf{H}d\mathcal{M} \, , \tag{6.1.17a}$$

i.e.

$$\left(\frac{\partial A}{\partial T} \right)_{\mathcal{M}} = -S \quad \text{and} \quad \left(\frac{\partial A}{\partial \mathcal{M}} \right)_{T} = \mathbf{H} \, . \tag{6.1.17b}$$

---

[7] The notation of the magnetic potentials is not uniform in the literature. This is true not only of the choice of symbols; even the potential $F(T, \mathbf{H})$, which depends on $\mathbf{H}$, is sometimes referred to as the Helmholtz free energy.

6.1.2 Thermodynamic Relations

*6.1.2.1 Thermodynamic Potentials

At this point, we summarize the definitions of the two potentials introduced in the preceding subsection. The following compilation, which indicates the systematic structure of the material, can be skipped over in a first reading:

$$F = F(T, \mathbf{H}) = E - TS , \qquad\qquad dF = -SdT - \mathbf{M}\,d\mathbf{H} \quad (6.1.18\text{a})$$
$$A = A(T, \mathbf{M}) = E - TS + \mathbf{M}\cdot\mathbf{H} , \qquad dA = -SdT + \mathbf{H}\,d\mathbf{M} . \quad (6.1.18\text{b})$$

In comparison to liquids, the thermodynamic variables here are $T, \mathbf{H}$ and $\mathbf{M}$ instead of T, P and V. The thermodynamic relations listed can be read off from the corresponding relations for liquids by making the substitutions $V \rightarrow -\mathbf{M}$ and $P \rightarrow \mathbf{H}$. There is also another analogy between magnetic systems and liquids: the density matrix of the grand potential contains the term $-\mu N$, which in a magnetic system corresponds to $-\mathbf{H}\cdot\mathbf{M}$. Particularly in the low-temperature region, where the properties of a magnetic system can be described in terms of spin waves (magnons), this analogy is useful. There, the value of the magnetization is determined by the number of thermally-excited spin waves. Therefore, we find the correspondence $M \leftrightarrow N$ and $H \leftrightarrow \mu$. Of course the *Maxwell relations* follow from (6.1.15) and (6.1.18a,b)

$$\left(\frac{\partial T}{\partial \mathbf{H}}\right)_S = -\left(\frac{\partial \mathbf{M}}{\partial S}\right)_{\mathbf{H}} , \qquad \left(\frac{\partial S}{\partial \mathbf{H}}\right)_T = \left(\frac{\partial \mathbf{M}}{\partial T}\right)_{\mathbf{H}} . \qquad (6.1.19)$$

*6.1.2.2 Magnetic Response Functions, Specific Heats, and Susceptibilities

Analogously to the specific heats of liquids, we define here the specific heats C_M and C_H (at constant M and H) as[8]

$$C_M \equiv T\left(\frac{\partial S}{\partial T}\right)_M = -T\left(\frac{\partial^2 A}{\partial T^2}\right)_M \qquad\qquad (6.1.20\text{a})$$

$$C_H \equiv T\left(\frac{\partial S}{\partial T}\right)_H = \left(\frac{\partial E}{\partial T}\right)_H = -T\left(\frac{\partial^2 F}{\partial T^2}\right)_H . \qquad (6.1.20\text{b})$$

Instead of the compressibilities as for liquids, in the magnetic case one has the isothermal susceptibility

$$\chi_T \equiv \left(\frac{\partial M}{\partial H}\right)_T = -\frac{1}{V}\left(\frac{\partial^2 F}{\partial H^2}\right)_T \qquad\qquad (6.1.21\text{a})$$

---

[8] To keep the notation simple, we will often write $\mathbf{H}$ and $\mathbf{M}$ as H and M, making the assumption that $\mathbf{M}$ is parallel to $\mathbf{H}$ and that H and M are the components in the direction of $\mathbf{H}$.

and the adiabatic susceptibility

$$\chi_S \equiv \left(\frac{\partial M}{\partial H}\right)_S = \frac{1}{V}\left(\frac{\partial^2 E}{\partial H^2}\right)_S .$$ (6.1.21b)

In analogy to Chap. 3, one finds that

$$C_H - C_M = TV\alpha_H^2/\chi_T ,$$ (6.1.22a)

$$\chi_T - \chi_S = TV\alpha_H^2/C_H$$ (6.1.22b)

and

$$\frac{C_H}{C_M} = \frac{\chi_T}{\chi_S} .$$ (6.1.22c)

Here, we have defined

$$\alpha_H \equiv \left(\frac{\partial M}{\partial T}\right)_H .$$ (6.1.23)

Eq. (6.1.22a) can also be rewritten as

$$C_H - C_M = TV\alpha_M^2\, \chi_T ,$$ (6.1.22d)

where

$$\alpha_M = \left(\frac{\partial H}{\partial T}\right)_M = -\frac{\alpha_H}{\chi_T}$$ (6.1.22e)

was used.

*6.1.2.3 Stability Criteria and the Convexity of the Free Energy

One can also derive inequalities of the type (3.3.5) and (3.3.6) for the magnetic susceptibilities and the specific heats:

$$\chi_T \geq 0 , \qquad C_H \geq 0 \qquad \text{and} \qquad C_M \geq 0 .$$ (6.1.24a,b,c)

To derive these inequalities on a statistical-mechanical basis, we assume that the Hamiltonian has the form

$$\mathcal{H} = \mathcal{H}_0 - \boldsymbol{\mu} \cdot \mathbf{H} ,$$ (6.1.25)

where $\mathbf{H}$ thus enters only linearly and $\boldsymbol{\mu}$ commutes with $\mathcal{H}$. It then follows that

$$\chi_T = \frac{1}{V}\left(\frac{\partial \langle \mu \rangle}{\partial H}\right)_T = \frac{1}{V}\left(\frac{\partial}{\partial H}\frac{\mathrm{Tr}\, e^{-\beta\mathcal{H}}\mu}{\mathrm{Tr}\, e^{-\beta\mathcal{H}}}\right)_T = \frac{\beta}{V}\left\langle (\mu - \langle\mu\rangle)^2 \right\rangle \geq 0 \quad (6.1.26a)$$

and

$$C_H = \left(\frac{\partial}{\partial T}\langle\mathcal{H}\rangle\right)_H = \left(\frac{\partial}{\partial T}\frac{\mathrm{Tr}\,e^{-\beta\mathcal{H}}\mathcal{H}}{\mathrm{Tr}\,e^{-\beta\mathcal{H}}}\right)_H = \frac{1}{kT^2}\left\langle(\mathcal{H}-\langle\mathcal{H}\rangle)^2\right\rangle \geq 0 \,,$$

$$(6.1.26b)$$

with which we have demonstrated (6.1.24a) and (6.1.24b). Eq. (6.1.24c) can be shown by taking the second derivative of

$$A(T,\mathcal{M}) = F(T,H) + H\mathcal{M}$$

with respect to the temperature at constant M (problem 6.1). As a result, $F(T,H)$ is concave[9] in T and in H, while $A(T,\mathcal{M})$ is concave in T and convex in $\mathcal{M}$: $\left(\left(\frac{\partial^2 A}{\partial T^2}\right)_H = -\frac{C_M}{T} \leq 0 \quad , \quad \left(\frac{\partial^2 A}{\partial \mathcal{M}^2}\right)_T = \left(\frac{\partial H}{\partial \mathcal{M}}\right)_T = 1/\chi_T \geq 0\right).$

In this derivation, we have used the fact that the Hamiltonian $\mathcal{H}$ has the general form (6.1.25), and therefore, diamagnetic effects (proportional to H^2) are negligible.

Remark: In analogy to the extremal properties treated in Sect. 3.6.4, the canonical free energy F for fixed T and H in magnetic systems strives towards a minimal value, as does the Helmholtz free energy A for fixed T and M. At these minima, the stationarity conditions $\delta F = 0$ and $\delta A = 0$ hold, i.e.:
$dF < 0$ when T and H are fixed, and $dA < 0$ when T and M are fixed.

6.1.2.4 Internal Energy

$E \equiv \langle\mathcal{H}\rangle$ is the internal energy, which is found in a natural manner from statistical mechanics. It contains the energy of the material including the effects of the electromagnetic field, but not the field energy itself. It is usual to introduce a second internal energy, also, which we denote by U and which is defined as

$$U = E + \mathbf{M}\cdot\mathbf{H} \; ; \tag{6.1.27a}$$

it thus has the complete differential

$$dU = TdS + \mathbf{H}d\mathbf{M} \, . \tag{6.1.27b}$$

From this, we derive

$$T = \left(\frac{\partial U}{\partial S}\right)_{\mathbf{M}} \quad , \quad \mathbf{H} = \left(\frac{\partial U}{\partial \mathbf{M}}\right)_S \tag{6.1.27c}$$

and the Maxwell relation

$$\left(\frac{\partial \mathbf{H}}{\partial S}\right)_{\mathbf{M}} = \left(\frac{\partial T}{\partial \mathbf{M}}\right)_S. \tag{6.1.28}$$

---

[9] See also R. B. Griffiths, J. Math. Phys. **5**, 1215 (1964). In fact, it is sufficient for the proof of (6.1.24a) to show that μ enters $\mathcal{H}$ linearly. Cf. M. E. Fisher, Rep. Progr. Phys. **XXX**, 615 (1967), p. 644.

Remarks:

(i) As was emphasized in footnote 5, throughout this chapter the particle number and the volume are treated as fixed. In the case of variable volume and variable particle number, the generalization of the First Law takes on the form

$$dU = TdS - PdV + \mu dN + \mathbf{H}d\mathbf{M} \tag{6.1.29}$$

and, correspondingly,

$$dE = TdS - PdV + \mu dN - \mathbf{M}d\mathbf{H} . \tag{6.1.30}$$

The grand potential

$$\Phi(T, V, \mu, \mathbf{H}) = -kT \log \mathrm{Tr}\, e^{-\beta(\mathcal{H} - \mu N)} \tag{6.1.31a}$$

then has the differential

$$d\Phi = -SdT - PdV - \mu dN - \mathbf{M}d\mathbf{H} , \tag{6.1.31b}$$

where the chemical potential μ is not to be confused with the microscopic magnetic moment $\boldsymbol{\mu}$.

(ii) We note that the free energies of the crystalline solid are not rotationally invariant, but instead are invariant only with respect to rotations of the corresponding point group. Therefore, the susceptibility $\chi_{ij} = \frac{\partial M_i}{\partial H_j}$ is a second-rank tensor. In this textbook, we present the essential statistical methods, but we forgo a discussion of the details of solid-state physics or element specific aspects. The methods presented here should permit the reader to master the complications which arise in treating real, individual problems.

6.1.3 Supplementary Remarks

(i) *The Bohr–van Leeuwen Theorem.*
The content of the *Bohr–van Leeuwen theorem* is the nonexistence of magnetism in classical statistics.
The classical partition function for a charged particle in the electromagnetic field is given by

$$Z_{cl} = \frac{\int d^{3N}p \int d^{3N}x}{(2\pi\hbar)^{3N} N!} e^{-\mathcal{H}(\{\mathbf{p}_i - \frac{e}{c}\mathbf{A}(\mathbf{x}_i)\}, \{\mathbf{x}_i\})/kT} . \tag{6.1.32}$$

Making the substitution $\mathbf{p}'_i = \mathbf{p}_i - \frac{e}{c}\mathbf{A}(\mathbf{x}_i)$, we can see that Z_{cl} becomes independent of $\mathbf{A}$ and thus also of $\mathbf{H}$. Then we have $\mathbf{M} = -\frac{\partial F}{\partial \mathbf{H}} = 0$, and $\chi = -\frac{1}{V}\frac{\partial^2 F}{\partial H^2} = 0$. Since the spin is also a quantum-mechanical phenomenon, dia-, para-, and ferromagnetism are likewise quantum phenomena. One might

ask how this statement can be reconciled with the 'classical' Langevin paramagnetism which will be discussed below. In the latter, a large but fixed
value of the angular momentum is assumed, so that a non-classical feature
is introduced into the theory. In classical physics, angular momenta, atomic
radii, etc. vary continuously and without limits.[10]

(ii) Here, we append the simple intermediate computations leading to (6.1.3)–
(6.1.5). In (6.1.3), we need to evaluate $-c\frac{\delta\mathcal{H}}{\delta A(x)}$. The first term in (6.1.1) evidently
leads to the first term in (6.1.3). In the component of the current j_α, taking the
derivative of the second term leads to

$$c\frac{\delta}{\delta A_\alpha(\mathbf{x})}\sum_{i=1}^N \boldsymbol{\mu}_i^{\text{spin}}\cdot\text{curl}\,\mathbf{A}(\mathbf{x}_i)=c\frac{\delta}{\delta A_\alpha(\mathbf{x})}\sum_{i=1}^N \mu_{i\beta}^{\text{spin}}\epsilon_{\beta\gamma\delta}\frac{\partial}{\partial x_{i\gamma}}A_\delta(\mathbf{x}_i)=$$

$$=c\sum_{i=1}^N \mu_{i\beta}^{\text{spin}}\epsilon_{\beta\gamma\delta}\frac{\partial}{\partial x_{i\gamma}}\delta(\mathbf{x}-\mathbf{x}_i)\delta_{\alpha\delta}=c\left(\sum_{i=1}^N \text{rot}\left[\boldsymbol{\mu}_i^{\text{spin}}\delta(\mathbf{x}-\mathbf{x}_i)\right]\right)_\alpha.$$

Pairs of Greek indices imply a summation. Since the derivative of the third term in
(6.1.1) yields zero, we have demonstrated (6.1.3).

(iii) In (6.1.4), the first term is obtained in a readily-apparent manner from the
first term in (6.1.3). For the second term, we carry out an integration by parts and
use $\partial_\delta x_\beta = \delta_{\delta\beta}$, obtaining

$$\frac{1}{2}\sum_{i=1}^N\left(\int d^3x\,\mathbf{x}\times\text{curl}\left[\boldsymbol{\mu}_i^{\text{spin}}\delta(\mathbf{x}-\mathbf{x}_i)\right]\right)_\alpha=$$

$$=\frac{1}{2}\sum_{i=1}^N\int d^3x\,\epsilon_{\alpha\beta\gamma}\,x_\beta\,\epsilon_{\gamma\delta\rho}\,\partial_\delta\left[\mu_{i\rho}^{\text{spin}}\delta(\mathbf{x}-\mathbf{x}_i)\right]=$$

$$=-\frac{1}{2}\sum_{i=1}^N\int d^3x\,\epsilon_{\alpha\beta\gamma}\,\epsilon_{\gamma\delta\rho}\,\delta_{\delta\beta}\mu_{i\rho}^{\text{spin}}\delta(\mathbf{x}-\mathbf{x}_i)=$$

$$=-\frac{1}{2}\sum_{i=1}^N\int d^3x\,(-2\delta_{\alpha\rho})\,\mu_{i\rho}^{\text{spin}}\delta(\mathbf{x}-\mathbf{x}_i)=\sum_{i=1}^N \mu_{i\alpha}^{\text{spin}},$$

with which we have demonstrated (6.1.4).

(iv) Finally, we show the validity of (6.1.5).
We can write the vector potential of a uniform magnetic field in the form
$\mathbf{A}=\frac{1}{2}\mathbf{H}\times\mathbf{x}$, since $\text{curl}\,\mathbf{A}=\frac{1}{2}\left(\mathbf{H}(\nabla\cdot\mathbf{x})-(\mathbf{H}\cdot\nabla)\mathbf{x}\right)$ yields $\mathbf{H}$. To obtain the
derivative, we use $\frac{1}{2}\epsilon_{\sigma\alpha\tau}\,x_{i\tau}$ for the derivative with respect to H_α after the second
equals sign below, finding

$$-\frac{\partial\mathcal{H}}{\partial H_\alpha}=-\sum_{i=1}^N\frac{2}{2m}\left(\mathbf{p}_i-\frac{e}{c}\mathbf{A}\right)_\sigma\left(-\frac{e}{c}\right)\frac{\partial}{\partial H_\alpha}\frac{1}{2}\epsilon_{\sigma\rho\tau}\,H_\rho x_{i\tau}+\mu_{i\alpha}^{\text{spin}}=$$

$$=\frac{e}{2mc}\sum_{i=1}^N\left(\mathbf{x}_i\times\left(\mathbf{p}_i-\frac{e}{c}\mathbf{A}\right)\right)_\alpha+\mu_{i\alpha}^{\text{spin}},$$

(6.1.33)

which is in fact the right-hand side of (6.1.4).

---

[10] A detailed discussion of this theorem and the original literature citations are to
be found in J. H. van Vleck, *The Theory of Electric and Magnetic Susceptibility*,
Oxford, University Press, 1932.

In the Hamiltonian (6.1.1), W_{Coul} contains the mutual Coulomb interaction of the electrons and their interactions with the nuclei. The thermodynamic relations derived in Sect. (6.1.2) are thus generally valid; in particular, they apply to ferromagnets, since there the decisive exchange interaction is merely a consequence of the Coulomb interactions together with *Fermi–Dirac* statistics.

In addition to the interactions included in (6.1.1), there are also the magnetic dipole interaction between magnetic moments and the spin-orbit interaction,[11] which lead among other things to *anisotropy effects*. The derived thermodynamic relations also hold for these more general cases, whereby the susceptibilities and specific heats become shape-dependent owing to the long-range dipole interactions. In Sect. 6.6, we will take up the effects of the dipole interactions in more details. For elliptical samples, the internal magnetic field is uniform, $\mathbf{H}_i = \mathbf{H} - D\mathbf{M}$, where D is the demagnetizing tensor (or simply the appropriate demagnetizing factor, if the field is applied along one of the principal axes). We will see that instead of the susceptibility with respect to the external field $\mathbf{H}$, one can employ the susceptibility with respect to the macroscopic internal field, and that this susceptibility is shape-independent.[12] In the following four sections, which deal with basic statistical-mechanical aspects, we leave the dipole interactions out of consideration; this is indeed quantitatively justified in many situations. In the next two sections 6.2 and 6.3, we deal with the magnetic properties of non-interacting atoms and ions; these can be situated within solids. The angular momentum quantum numbers of individual atoms in their ground states are determined by *Hund's rules*.[13]

6.2 The Diamagnetism of Atoms

We consider atoms or ions with closed electronic shells, such as for example helium and the other noble gases or the alkali halides. In this case, the quantum numbers of the orbital angular momentum and the total spin in the ground state are zero, $S = 0$ and $L = 0$, and as a result the total angular momentum $\mathbf{J} = \mathbf{L} + \mathbf{S}$ is also $J = 0$.[14] Therefore, we have

---

[11] The spin-orbit interaction $\propto \mathbf{L} \cdot \mathbf{S}$ leads in effective spin models to anisotropic interactions. The orbital angular momentum is influenced by the crystal field of the lattice, transferring the anisotropy of the lattice to the spin.

[12] For non-elliptical samples, the magnetization is not uniform. In this case, $\frac{\partial M}{\partial H}$ depends on position within the sample and has only a local significance. It is then expedient to introduce a total susceptibility $\chi_{T,S}^{\text{tot}} = \left(\frac{\partial M}{\partial H}\right)_{T,S}$, which differs from (6.1.33) in the homogeneous case only by a factor of V.

[13] See e.g. QM I, Chap. 13 and Table I.12

[14] The diamagnetic contribution is also present in other atoms, but in the magnetic fields which are available in the laboratory, it is negligible compared to the paramagnetic contribution.

$\mathbf{L}\,|0\rangle = \mathbf{S}\,|0\rangle = \mathbf{J}\,|0\rangle = 0$, where $|0\rangle$ designates the ground state. The paramagnetic contribution to the Hamiltonian (6.1.7) thus vanishes in every order
of perturbation theory. It suffices to treat the remaining diamagnetic term
in (6.1.7) in first-order perturbation theory, since all the excited states lie at
much higher energies. Owing to the rotational symmetry of the wavefunc-
tions of closed shells, we find $\langle 0|\sum_i \left(x_i^2 + y_i^2\right)|0\rangle = \frac{2}{3}\langle 0|\sum_i r_i^2 |0\rangle$ and, for
the energy shift of the ground state,

$$E_1 = \frac{e^2 H^2}{12mc^2}\langle 0|\sum_i r_i^2 |0\rangle . \tag{6.2.1}$$

From this it follows for the magnetic moment and the susceptibility of a single
atom:

$$\langle \mu_z \rangle = -\frac{\partial E_1}{\partial H} = -\frac{e^2 \langle 0|\sum_i r_i^2 |0\rangle}{6mc^2}H, \quad \chi = \frac{\partial \langle \mu_z \rangle}{\partial H} = -\frac{e^2 \langle 0|\sum_i r_i^2 |0\rangle}{6mc^2}, \tag{6.2.2}$$

where the sums run over all the electrons in the atom. The magnetic moment
is directed oppositely to the applied field and the susceptibility is negative.
We can estimate the magnitude of this so called *Langevin diamagnetism* using
the Bohr radius:

$$\chi = -\frac{25 \times 10^{-20} \times 10^{-16}}{6 \times 10^{-27} \times 10^{21}}\,\mathrm{cm}^3 \approx -5 \times 10^{-30}\mathrm{cm}^3,$$

$$\chi \text{ per mole} = -5 \times 10^{-30} \times 6 \times 10^{23}\,\frac{\mathrm{cm}^3}{\mathrm{mole}} \approx -3 \times 10^{-6}\,\frac{\mathrm{cm}^3}{\mathrm{mole}} .$$

The experimental values of the molar susceptibility of the noble gases are
collected in Table 6.1.

Table 6.1. Molar susceptibilities of the noble gases

	He	Ne	Ar	Kr	Xe
χ in $10^{-6}\,\mathrm{cm}^3/\mathrm{mole}$	-1.9	-7.2	-15.4	-28.0	-43.0

An intuitively apparent interpretation of this diamagnetic susceptibility runs
as follows: the field $\mathbf{H}$ induces an additional current $\Delta j = -er\Delta\omega$, whereby the
orbital frequency of the electronic motion increases by the Larmor frequency $\Delta\omega = \frac{eH}{2mc}$. The sign of this change corresponds to Lenz's law, so that both the magnetic
moment μ_z and the induced magnetic field are opposite to the applied field $\mathbf{H}$:

$$\mu_z \sim \frac{r\Delta j}{2c} \sim -\frac{r^2 \Delta\omega e}{2c} \sim -\frac{e^2 r^2 H}{4mc^2} .$$

We also note that the result (6.2.2) is proportional to the square of the Bohr radius
and therefore to the fourth power of $\hbar$, confirming the quantum nature of magnetic
phenomena.

6.3 The Paramagnetism of Non-coupled Magnetic Moments

Atoms and ions with an odd number of electrons, e.g. Na, as well as atoms and ions with partially filled inner shells, e.g. Mn^{2+}, Gd^{3+}, or U^{4+} (transition elements, ions which are isoelectronic with transition elements, rare-earth and actinide elements) have nonvanishing magnetic moments even when $\mathbf{H} = 0$,

$$\boldsymbol{\mu} = \frac{e}{2mc}\left(\mathbf{L} + g_e\,\mathbf{S}\right) = \frac{e}{2mc}\left(\mathbf{J} + \mathbf{S}\right) \quad (g_e = 2) \ . \tag{6.3.1}$$

Here, $\mathbf{J} = \mathbf{L} + \mathbf{S}$ is the total angular momentum operator. For relatively low external magnetic fields (i.e. $e\hbar H/mc \ll$ spin-orbit coupling)) with $\mathbf{H}$ applied along the z-axis, the theory of the *Zeeman effect*[15] gives the energy-level shifts

$$\Delta E_{M_J} = g\mu_B M_J H \ , \tag{6.3.2}$$

where M_J runs over the values $\mathrm{M}_J = -J,\ldots,J$[16] and the Landé factor

$$g = 1 + \frac{J(J+1) + S(S+1) - L(L+1)}{2J(J+1)} \tag{6.3.3}$$

was used. Familiar special cases are $L = 0 : g = 2, M_J \equiv M_S = \pm\frac{1}{2}$ and $S = 0 : g = 1, M_J \equiv M_L = -L,\ldots,L$.

The Landé factor can be made plausible in the classical picture where $\mathbf{L}$ and $\mathbf{S}$ precess independently around the spatially fixed direction of the constant of the motion $\mathbf{J}$. Then we find:

$$(\mathbf{L} + 2\mathbf{S})_z = \frac{\mathbf{J}\cdot(\mathbf{L} + 2\mathbf{S})}{|\mathbf{J}|}\frac{J_z}{|\mathbf{J}|} = J_z\,\frac{\mathbf{J}^2 + \mathbf{J}\cdot\mathbf{S}}{\mathbf{J}^2}$$

$$= J_z\left(1 + \frac{\mathbf{S}^2 + \frac{1}{2}\left(\mathbf{J}^2 - \mathbf{L}^2 - \mathbf{S}^2\right)}{\mathbf{J}^2}\right) \ .$$

The partition function then becomes

$$Z = \left(\sum_{m=-J}^{J} e^{-\eta m}\right)^N = \left(\frac{\sinh\eta\,(2J+1)\,/2}{\sinh\eta/2}\right)^N , \tag{6.3.4}$$

with the abbreviation

$$\eta = \frac{g\mu_B H}{kT} \ . \tag{6.3.5}$$

Here, we have used the fact that

---

[15] Cf. e.g. QM I, Sect. 14.2
[16] $\mathbf{J} = \mathbf{L} + \mathbf{S}$, $J_z\,|m_j\rangle = \hbar m_j\,|m_j\rangle$

$$\sum_{m=-J}^{J} e^{-\eta m} = e^{-\eta J} \sum_{r=0}^{2J} e^{\eta r} = e^{-\eta J} \frac{e^{\eta(2J+1)} - 1}{e^{\eta} - 1} = \left(\frac{\sinh \eta \, (2J+1)\,/2}{\sinh \eta/2} \right).$$

For the free energy, we find from (6.3.4)

$$F(T,H) = -kTN \log \left\{ \frac{\sinh \eta \, (2J+1)\,/2}{\sinh \eta/2} \right\}, \tag{6.3.6}$$

from which we obtain the magnetization

$$M = -\frac{1}{V} \frac{\partial F}{\partial H} = n g \mu_\mathrm{B} J B_J (\eta) \tag{6.3.7}$$

($n = \frac{N}{V}$). The magnetization is oriented parallel to the magnetic field $\mathbf{H}$. In Eq. (6.3.7) we have introduced the *Brillouin function* B_J, which is defined as

$$B_J(\eta) = \frac{1}{J} \left\{ (J + \frac{1}{2}) \coth \eta (J + \frac{1}{2}) - \frac{1}{2} \coth \frac{\eta}{2} \right\} \tag{6.3.8}$$

(Fig. 6.1). We now consider the asymptotic limiting cases:

$$\eta \to 0: \qquad \coth \eta = \frac{1}{\eta} + \frac{\eta}{3} + \mathcal{O}\left(\eta^3\right), \; B_J(\eta) = \frac{J+1}{3}\eta + \mathcal{O}\left(\eta^3\right) \tag{6.3.9a}$$

and

$$\eta \to \infty: \qquad B_J(\infty) = 1. \tag{6.3.9b}$$

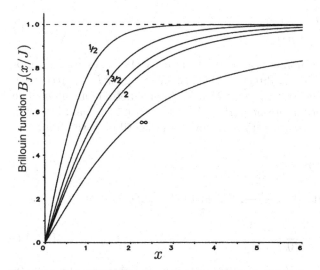

Fig. 6.1. The Brillouin function for $J = 1/2, 1, 3/2, 2, \infty$ as a function of $x = \frac{g\mu_\mathrm{B} J H}{kT} = \eta J$. For classical moments B_∞ is identical to the Langevin function

Inserting (6.3.9a) into (6.3.7), we obtain for low applied fields ($H \ll kT/Jg\mu_B$)

$$M = n \left(g\mu_B\right)^2 \frac{J(J+1)H}{3kT} ,\tag{6.3.10a}$$

while from (6.3.9b), for high fields ($H \gg kT/Jg\mu_B,$), we find

$$M = ng\mu_B J \tag{6.3.10b}$$

This signifies complete alignment (saturation) of the magnetic moments. An important special case is represented by spin-$\frac{1}{2}$ systems. Setting $J = \frac{1}{2}$ in (6.3.8), we find

$$B_{\frac{1}{2}}(\eta) = 2\coth\eta - \coth\frac{\eta}{2} = \frac{(\cosh^2\frac{\eta}{2} + \sinh^2\frac{\eta}{2})}{\sinh\frac{\eta}{2}\cosh\frac{\eta}{2}} - \frac{\cosh\frac{\eta}{2}}{\sinh\frac{\eta}{2}} = \tanh\frac{\eta}{2} .$$
$$\tag{6.3.11}$$

This result can be more directly obtained by using the fact that for spin $S = 1/2$, the partition function of a spin is given by $Z = 2\cosh\eta/2$ and the average value of the magnetization by $M = ng\mu_B Z^{-1}\sinh\eta/2$. Letting $J = \infty$, while at the same time $g\mu_B \to 0$, so that $\mu = g\mu_B J$ remains finite, we find

$$B_\infty(\eta) = \coth\eta J - \frac{1}{\eta J} = \coth\frac{\mu H}{kT} - \frac{kT}{\mu H} .\tag{6.3.12a}$$

$B_\infty(\eta)$ is called the *Langevin function* for classical magnetic moments μ; together with (6.3.7), it determines the magnetization

$$M = n\mu\left(\coth\frac{\mu H}{kT} - \frac{kT}{\mu H}\right)\tag{6.3.12b}$$

of "classical" magnetic moments of magnitude μ. A classical magnetic moment μ can be oriented in any direction in space; its energy is $E = -\mu H\cos\vartheta$, where ϑ is the angle between the field **H** and the magnetic moment μ. The classical partition function for one particle is $Z = \int d\Omega\, e^{-E/kT}$ and leads via (6.1.9b) once again to (6.3.12b). Finally, for the susceptibility we obtain

$$\chi = n\left(g\mu_B\right)^2\frac{J}{kT}B_J'(\eta) .\tag{6.3.13}$$

In small magnetic fields $H \ll \frac{kT}{Jg\mu_B}$, this gives the *Curie law*

$$\chi_{Curie} = n\left(g\mu_B\right)^2\frac{J(J+1)}{3kT} .\tag{6.3.14}$$

The magnetic behavior of non-coupled moments characterized by (6.3.7), (6.3.13), and (6.3.14) is termed *paramagnetism*. The Curie law is typical of

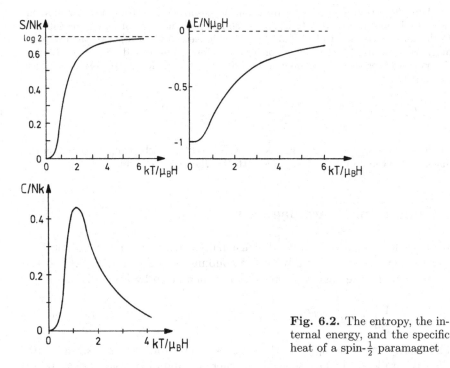

Fig. 6.2. The entropy, the internal energy, and the specific heat of a spin-$\frac{1}{2}$ paramagnet

preexisting elementary magnetic moments which need only be oriented by the applied field, in contrast to the polarization of harmonic oscillators, whose moments are induced by the field (cf. problem 6.4).

We include a remark about the magnitudes. The diamagnetic susceptibility per mole, from the estimate which follows Eq. (6.2.2), is equal to about $\chi^{mole} \approx -10^{-5}\mathrm{cm}^3/\mathrm{mole}$. The paramagnetic susceptibility at room temperature is roughly 500 times larger, i.e. $\chi^{mole} \approx 10^{-2}$–$10^{-3}\mathrm{cm}^3/\mathrm{mole}$. The entropy of a paramagnet is

$$S = -\left(\frac{\partial F}{\partial T}\right)_H = Nk\left(\log\left(\frac{\sinh \frac{\eta(2J+1)}{2}}{\sinh \frac{\eta}{2}}\right) - \eta J B_J(\eta)\right) . \tag{6.3.15}$$

For spin $\frac{1}{2}$, (6.3.15) simplifies to

$$S = Nk\left(\log\left(2\cosh \frac{\mu_\mathrm{B}H}{kT}\right) - \frac{\mu_\mathrm{B}H}{kT}\tanh \frac{\mu_\mathrm{B}H}{kT}\right) \tag{6.3.16}$$

with the limiting case

$$S = Nk\log 2 \quad \text{for} \quad H \to 0 . \tag{6.3.16'}$$

The entropy, the internal energy, and the specific heat of the paramagnet are reproduced in Figs. 6.2a,b,c. The bump in the specific heat is typical of 2-level systems and is called a Schottky anomaly in connection with defects.

Van Vleck paramagnetism: The quantum number of the total angular momentum also becomes zero, $J = 0$, when a shell has just one electron less than half full. In this case, according to Eq. (6.3.2) we have indeed $\langle 0| \mathbf{J} + \mathbf{S} |0\rangle = 0$, but the paramagnetic term in (6.1.7) yields a nonzero contribution in second order perturbation theory. Together with the diamagnetic term, one obtains for the energy shift of the ground state

$$\Delta E_0 = -\sum_n \frac{|\langle 0| (\mathbf{L} + 2\mathbf{S}) \cdot \mathbf{H} |n\rangle|^2}{E_n - E_0} + \frac{e^2 H^2}{8mc^2} \langle 0| \sum_i (x_i^2 + y_i^2) |0\rangle . \qquad (6.3.17)$$

The first, paramagnetic term, named for *van Vleck*[17], which also plays a role in the magnetism of molecules[18], competes with the diamagnetic term.

6.4 Pauli Spin Paramagnetism

We consider now a free, three-dimensional electron gas in a magnetic field and restrict ourselves initially to the coupling of the magnetic field to the electron spins. The energy eigenvalues are then given by Eq. (6.1.7):

$$\epsilon_{\mathbf{p}\pm} = \frac{\mathbf{p}^2}{2m} \pm \frac{1}{2} g_e \mu_B H . \qquad (6.4.1)$$

The energy levels are split by the magnetic field. Electrons whose spins are aligned parallel to the field have higher energies, and these states are therefore less occupied (see Fig. 6.3).

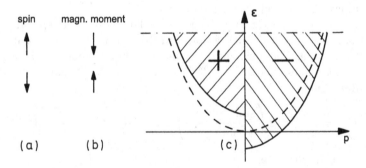

Fig. 6.3. Orientation **(a)** of the spins, and **(b)** of the magnetic moments. **(c)** The energy as a function of **p** (on the left for positive spins and on the right for negative spins)

---

[17] J. H. van Vleck, *The Theory of Magnetic and Electric Susceptiblities*, Oxford University Press, 1932.
[18] Ch. Kittel, *Introduction to Solid State Physics*, Third edition, John Wiley, New York, 1967

The number of electrons in the two states is found to be

$$N_\pm = \frac{V}{(2\pi\hbar)^3} \int d^3p\, n\left(\frac{p^2}{2m} \pm \frac{1}{2}g_e\mu_B H\right) = \int_0^\infty d\epsilon\, \frac{1}{2}\nu(\epsilon)n\left(\epsilon \pm \frac{1}{2}g_e\mu_B H\right) ,$$

$$(6.4.2)$$

where the density of states has been introduced:

$$\nu(\epsilon) = \frac{gV}{(2\pi\hbar)^3} \int d^3p\, \delta(\epsilon - \epsilon_p) = N\frac{3}{2}\frac{\epsilon^{1/2}}{\epsilon_F^{3/2}} ; \qquad (6.4.3)$$

it fulfills the normalization condition $\int_0^{\epsilon_F} d\epsilon\, \nu(\epsilon) = N$. In the case that $g_e\mu_B H \ll \mu \approx \epsilon_F$, we can expand in terms of H:

$$N_\pm = \int_0^\infty d\epsilon\, \frac{1}{2}\nu(\epsilon)\left[n(\epsilon) \pm n'(\epsilon)\frac{1}{2}g_e\mu_B H + \mathcal{O}\left(H^2\right)\right]. \qquad (6.4.4)$$

For the magnetization, using the above result we obtain:

$$M = -\mu_B(N_+ - N_-)/V = -\frac{\mu_B^2 H}{V}\int_0^\infty d\epsilon\, \nu(\epsilon)n'(\epsilon) + \mathcal{O}\left(H^3\right) , \qquad (6.4.5)$$

where we have set $g_e = 2$. For $T \to 0$, we find from (6.4.5) the magnetization

$$M = \mu_B^2\nu(\epsilon_F)H/V + \mathcal{O}\left(H^3\right) = \frac{3}{2}\mu_B^2\frac{NH}{V\epsilon_F} + \mathcal{O}\left(H^3\right) \qquad (6.4.6)$$

and the magnetic susceptibility

$$\chi_P = \frac{3}{2}\mu_B^2\frac{N}{V\epsilon_F} + \mathcal{O}\left(H^2\right) . \qquad (6.4.7)$$

This result describes the phenomenon of *Pauli spin paramagnetism*.

Supplementary remarks:

(i) For $T \neq 0$, we must take the change of the chemical potential into account, making use of the Sommerfeld expansion:

$$N = \int_0^\infty d\epsilon\, \nu(\epsilon)n(\epsilon) + \mathcal{O}\left(H^2\right) = \int_0^\mu d\epsilon\, \nu(\epsilon) + \frac{\pi^2(kT)^2}{6}\nu'(\mu) + \mathcal{O}\left(H^2, T^4\right)$$

$$= \int_0^{\epsilon_F} d\epsilon\, \nu(\epsilon) + (\mu - \epsilon_F)\nu(\epsilon_F) + \frac{\pi^2(kT)^2}{6}\nu'(\epsilon_F) + \mathcal{O}\left(H^2, T^4\right) . \qquad (6.4.8)$$

Since the first term on the right-hand side is equal to N, we write

$$\mu - \epsilon_F = -\frac{\pi^2(kT)^2}{6}\frac{\nu'(\epsilon_F)}{\nu(\epsilon_F)} + \mathcal{O}\left(H^2, T^4\right) . \qquad (6.4.9)$$

Integrating by parts, we obtain from (6.4.5) and (6.4.9)

$$
\begin{aligned}
M &= \frac{\mu_B^2 H}{V} \int\limits_0^\infty d\epsilon\, \nu'(\epsilon) n(\epsilon) + \mathcal{O}\left(H^3\right) \\
&= \frac{\mu_B^2 H}{V} \left[\nu(\mu) + \frac{\pi^2 (kT)^2}{6} \nu''(\mu) + \mathcal{O}\left(H^3, T^4\right)\right] \\
&= \frac{\mu_B^2 H}{V} \left[\nu(\epsilon_F) - \frac{\pi^2 (kT)^2}{6} \left(\frac{\nu'(\epsilon_F)^2}{\nu(\epsilon_F)} - \nu''(\epsilon_F)\right)\right] + \mathcal{O}\left(H^3, T^4\right) \ .
\end{aligned}
$$

(6.4.10)

(ii) The Pauli susceptibility (6.4.7) can be interpreted similarly to the linear specific heat of a Fermi gas (see Sect. 4.3.2):

$$
\chi_P = \chi_{\text{Curie}} \frac{\nu(\epsilon_F)}{N} kT = \mu_B^2\, \nu(\epsilon_F)/V \ .
\tag{6.4.11}
$$

Naively, one might expect that the susceptibility of N electrons would be equal to the Curie susceptibility χ_{Curie} from Eq. (6.3.14) and therefore would diverge as $1/T$. It was Pauli's accomplishment to realize that not all of the electrons contribute, but instead only those near the Fermi energy. The number of thermally excitable electrons is $kT\nu(\epsilon_F)$.

(iii) The Landau quasiparticle interaction (see Sect. 4.3.3e, Eq. 4.3.29c) yields

$$
\chi_P = \frac{\mu_B^2\, \nu(\epsilon_F)}{V(1 + F_a)} \ .
\tag{6.4.12}
$$

In this expression, F_a is an antisymmetric combination of the interaction parameters.[19]

(iv) In addition to Pauli spin paramagnetism, the electronic orbital motions give rise to Landau diamagnetism [20]

$$
\chi_L = -\frac{e^2 k_F}{12\pi^2 m c^2} \ .
\tag{6.4.13}
$$

For a free electron gas, $\chi_L = -\frac{1}{3}\chi_P$. The lattice effects in a crystal have differing consequences for χ_L and χ_P. Eq. (6.4.13) holds for free electrons neglecting the Zeeman term. The magnetic susceptibility for free spin-$\frac{1}{2}$ fermions is composed of three parts: it is the sum

$$
\chi = \chi_P + \chi_L + \chi_{\text{Osc}} \ .
$$

---

[19] D. Pines and Ph. Nozières, *The Theory of Quantum Liquids* Vol. I: Normal Fermi Liquids, W. A. Benjamin, New York 1966, p. 25

[20] See e.g. D. Wagner, *Introduction to the Theory of Magnetism*, Pergamon Press, Oxford, 1972.

χ_{Osc} is an oscillatory part, which becomes important at high magnetic fields H and is responsible for de Haas–van Alphen oscillations.

(v) Fig. 6.3c can also be read differently from the description given above. If one introduces the densities of states for spin $\pm\hbar/2$

$$
\begin{aligned}
\nu_\pm(\epsilon) &= \frac{V}{(2\pi\hbar)^3} \int d^3p\, \delta(\epsilon - \epsilon_{p\pm}) \\
&= \frac{V}{(2\pi\hbar)^3} \int d^3p\, \delta\left(\epsilon - \left(\frac{p^2}{2m} \pm \frac{1}{2}g_e\mu_{\text{B}}H\right)\right) \\
&= \frac{mV}{2\pi^2\hbar^3} \int_0^\infty dp\, p\, \Theta\left(\epsilon \mp \frac{1}{2}g_e\mu_{\text{B}}H\right) \delta\left(p - \sqrt{2m\left(\epsilon \mp \frac{1}{2}g_e\mu_{\text{B}}H\right)}\right) \\
&= N\frac{3}{4\epsilon_F^{3/2}} \Theta\left(\epsilon \mp \frac{1}{2}g_e\mu_{\text{B}}H\right)\left(\epsilon \mp \frac{1}{2}g_e\mu_{\text{B}}H\right)^{1/2} ,
\end{aligned}
$$

then the solid curves which are drawn on the left and the right also refer to $\nu_+(\epsilon)$ and $\nu_-(\epsilon)$.

6.5 Ferromagnetism

6.5.1 The Exchange Interaction

Ferromagnetism and antiferromagnetism are based on variations of the exchange interaction, which is a consequence of the Pauli principle and the Coulomb interaction (cf. the remark following Eq. (6.1.33)). In the simplest case of the exchange interaction of two electrons, two atoms or two molecules with the spins S_1 and S_2, the interaction has the form $\pm J\, S_1 \cdot S_2$, where J is a positive constant which depends on the distance between the spins. The exchange constant $\pm J$ is determined by the overlap integrals, containing the Coulomb interaction.[21] When the exchange energy is negative,

$$
E = -J\, S_1 \cdot S_2 , \tag{6.5.1a}
$$

then a parallel spin orientation is favored. This leads in a solid to ferromagnetism (Fig. 6.4b); then below the Curie temperature T_c, a spontaneous magnetization occurs within the solid. When the exchange energy is positive,

$$
E = J\, S_1 \cdot S_2 , \tag{6.5.1b}
$$

then an antiparallel spin orientation is preferred. In a suitable lattice structure, this can lead to an antiferromagnetic state: below the Néel temperature T_N, an alternating (staggered) magnetic order occurs (Fig. 6.4c). Above the

---

[21] See Chaps. 13 and 15, QM I

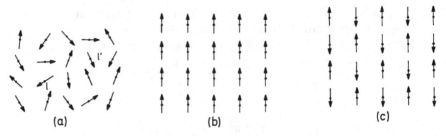

Fig. 6.4. A crystal lattice of magnetic ions. The spin S_l is located at the position x_l, and l denumerates the lattice sites. **(a)** the paramagnetic state; **(b)** the ferromagnetic state; **(c)** the antiferromagnetic state.

respective transition temperature (T_C or T_N), a paramagnetic state occurs (Fig. 6.4a). The exchange interaction is, to be sure, short-ranged; but owing to its electrostatic origin it is in general considerably stronger than the dipole-dipole interaction. Examples of ferromagnetic materials are Fe, Ni, EuO; and typical antiferromagnetic materials are MnF_2 and $RbMnF_3$.

In the rest of this section, we turn to the situation described by equation (6.5.1a), i.e. to ferromagnetism, and return to (6.5.1b) only in the discussion of phase transitions. We now imagine that the magnetic ions are located on a simple cubic lattice with lattice constant a, and that a negative exchange interaction ($J > 0$) acts between them (Fig. 6.4a). The lattice sites are enumerated by the index l. The position of the lth ion is denoted by x_l and its spin is S_l. All the pairwise interaction energies of the form (6.5.1a) contribute to the total Hamiltonian [22]:

$$\mathcal{H} = -\frac{1}{2} \sum_{l,l'} J_{ll'} \, \mathbf{S}_l \cdot \mathbf{S}_{l'} \, . \tag{6.5.2}$$

Here, we have denoted the exchange interaction between the spins at the lattice sites l and l' by $J_{ll'}$. The sum runs over all l and l', whereby the factor $1/2$ guarantees that each pair of spins is counted only once in (6.5.2). The exchange interaction obeys $J_{ll'} = J_{l'l}$, and we set $J_{ll} = 0$ so that we do not need to exclude the occurrence of the same l-values in the sum. The Hamiltonian (6.5.2) represents the *Heisenberg model* [23]. Since only scalar products of spin vectors occur, it has the following important property: $\mathcal{H}$

---

[22] In fact, there are also interactions within a solid between more than just two spins, which we however neglect here.

[23] The *direct exchange* described above occurs only when the moments are near enough so that their wavefunctions overlap. More frequently, one finds an *indirect exchange*, which couples more distant moments. The latter acts via an intermediate link, which can be a quasi-free electron in a metal or a bound electron in an insulator. The resulting interaction is called in the first case the RKKY (Rudermann, Kittel, Kasuya, Yosida) interaction and in the second, it is referred

is invariant with respect to a common rotation of all the spin vectors. No direction is especially distinguished and therefore the ferromagnetic order which can occur may point in any arbitrary direction. Which direction is in fact chosen by the system is determined by small anisotropy energies or by an external magnetic field. In many substances, this rotational invariance is nearly ideally realized, e.g. in EuO, EuS, Fe and in the antiferromagnet RbMnF$_3$. In other cases, the anisotropy of the crystal structure may have the effect that the magnetic moments orient in only two directions, e.g. along the positive and negative z-axis, instead of in an arbitrary spatial direction. This situation can be described by the *Ising model*

$$\mathcal{H} = -\frac{1}{2} \sum_{l,l'} J_{ll'} S_l^z S_{l'}^z \; . \tag{6.5.3}$$

This model is considerably simpler than the Heisenberg model (6.5.2), since the Hamiltonian is diagonal in the spin eigenstates of S_l^z. But even for (6.5.3), the evaluation of the partition function is in general not trivial. As we shall see, the one-dimensional Ising model can be solved exactly in an elementary way for an interaction restricted to the nearest neighbors. The solution of the two-dimensional model, i.e. the calculation of the partition function, requires special algebraic or graph-theoretical methods, and in three dimensions the model has yet to be solved exactly. When the lattice contains N sites, then the partition function $Z = \text{Tr } e^{-\beta \mathcal{H}}$ has contributions from all together 2^N configurations (every spin can take on the two values $\pm \hbar/2$ independently of all the others). A naive summation over all these configurations is possible even for the Ising model only in one dimension. In order to understand the essential physical effects which accompany ferromagnetism, in the next section we will apply the molecular field approximation. It can be carried out for all problems related to ordering. We will demonstrate it using the Ising model as an example.

6.5.2 The Molecular Field Approximation for the Ising Model

We consider the Hamiltonian of the Ising model in an external magnetic field

$$\mathcal{H} = -\frac{1}{2} \sum_{l,l'} J(l - l') \sigma_l \sigma_{l'} - h \sum_l \sigma_l \; . \tag{6.5.4}$$

---

to as superexchange (see e.g. C. M. Hurd, Contemp. Phys. **23**, 469 (1982)). Also in cases where direct exchange is not predominant and even for itinerant magnets (with 3d and 4s electrons which are not localized, but instead form bands), the magnetic phenomena, in particular their behavior near the phase transition, can be described using an effective Heisenberg model. A derivation of the Heisenberg model from the Hubbard model can be found in D. C. Mattis, *The Theory of Magnetism*, Harper and Row, New York, 1965.

In comparison to (6.5.3), equation (6.5.4) contains some changes of notation. Instead of the spin operators S_l^z, we have introduced the Pauli spin matrices σ_l^z and use the eigenstates of the σ_l^z as basis functions; their eigenvalues are

$$\sigma_l = \pm 1 \quad \text{for every } l .$$

The Hamiltonian becomes simply a function of (commuting) numbers. By writing the exchange interaction in the form $J(l - l')$ ($J(l - l') = J(l' - l) = J_{ll'} \hbar^2 / 4, J(0) = 0$), we express the fact that the system is translationally invariant, i.e. $J(l - l')$ depends only on the distance between the lattice sites. The effect of an applied magnetic field is represented by the term $-h \sum_l \sigma_l$. The factor $-\frac{1}{2} g \mu_B$ has been combined with the magnetic field H into $h = -\frac{1}{2} g \mu_B H$; the sign convention for h is chosen so that the σ_l are aligned parallel to it.

Due to the translational invariance of the Hamiltonian, it proves to be expedient for later use to introduce the Fourier transform of the exchange coupling,

$$\tilde{J}(\mathbf{k}) = \sum_l J(l) e^{-i \mathbf{k} \cdot \mathbf{x}_l} . \tag{6.5.5}$$

Frequently, we will require $\tilde{J}(\mathbf{k})$ for small wavenumbers $\mathbf{k}$. Due to the finite range of $J(l - l')$, we can expand the exponential functions in (6.5.5)

$$\tilde{J}(\mathbf{k}) = \sum_l J(l) - \frac{1}{2} \sum_l (\mathbf{k} \cdot \mathbf{x}_l)^2 J(l) + \dots . \tag{6.5.5'}$$

For cubic and square lattices, and in general when reflection symmetry is present, the linear term in $\mathbf{k}$ makes no contribution.

We can interpret the Hamiltonian (6.5.4) in the following manner: for some configuration of all the spins $\sigma_{l'}$, a local field

$$h_l = h + \sum_{l'} J(l - l') \sigma_{l'} \tag{6.5.6}$$

acts on an arbitrarily chosen spin σ_l. If h_l were a fixed applied field, we could immediately write down the partition function for the spin σ_l. Here, however, the field h_l depends on the configuration of the spins and the value of σ_l itself enters into the local fields which act upon its neighbors. In order to avoid this difficulty by means of an approximation, we replace the local field (6.5.6) by its average value, i.e. by the *mean field*

$$\langle h_l \rangle = h + \sum_{l'} J(l - l') \langle \sigma_{l'} \rangle = h + \tilde{J}(0) m . \tag{6.5.7}$$

In the second part of this equation, we have introduced the average value

$$m = \langle \sigma_l \rangle , \tag{6.5.8}$$

which is position-independent, owing to the translational invariance of the Hamiltonian; thus m refers to the average magnetization per lattice site (per spin). Furthermore, we use the abbreviation

$$\tilde{J} \equiv \tilde{J}(0) \equiv \sum_l J(l) \qquad (6.5.9)$$

for the Fourier transform at $\mathbf{k} = 0$ (see (6.5.5')). Eq. (6.5.7) contains, in addition to the external field, the *molecular field* $\tilde{J}m$. The density matrix then has the simplified form

$$\rho \propto \prod_l e^{\sigma_l (h + \tilde{J}m)/kT} \ .$$

Formally, we have reduced the problem to that of a paramagnet, where the molecular field must still be determined self-consistently from the magnetization (6.5.8).

We still want to derive the molecular field approximation, justified above with intuitive arguments, in a more formal manner. We start with an arbitrary interaction term in (6.5.4), $-J(l - l')\sigma_l \sigma_{l'}$, and rewrite it up to a prefactor as follows:

$$
\begin{aligned}
\sigma_l \sigma_{l'} &= \big((\langle \sigma_l \rangle + \sigma_l - \langle \sigma_l \rangle)\big)\big((\langle \sigma_{l'} \rangle + \sigma_{l'} - \langle \sigma_{l'} \rangle)\big) \\
&= \langle \sigma_l \rangle \langle \sigma_{l'} \rangle + \langle \sigma_l \rangle (\sigma_{l'} - \langle \sigma_{l'} \rangle) \\
&\quad + \langle \sigma_{l'} \rangle (\sigma_l - \langle \sigma_l \rangle) + (\sigma_l - \langle \sigma_l \rangle)(\sigma_{l'} - \langle \sigma_{l'} \rangle) \ .
\end{aligned}
\qquad (6.5.10)
$$

Here, we have ordered the terms in powers of the deviation from the mean value. We now neglect terms which are nonlinear in these fluctuations. This yields the following approximate replacements:

$$\sigma_l \sigma_{l'} \ \rightarrow \ -\langle \sigma_l \rangle \langle \sigma_{l'} \rangle + \langle \sigma_l \rangle \sigma_{l'} + \langle \sigma_{l'} \rangle \sigma_l \ , \qquad (6.5.10')$$

which lead from (6.5.4) to the Hamiltonian in the *molecular field approximation*

$$\mathcal{H}_{\mathrm{MFT}} = \frac{1}{2} m^2 \, N \, \tilde{J}(0) - \sum_l \sigma_l \big(h + \tilde{J}(0)m\big) \ . \qquad (6.5.11)$$

We refer to the Remarks for comments about the validity and admissibility of this approximation. With the simplified Hamiltonian (6.5.11), we obtain the density matrix

$$\rho_{\mathrm{MFT}} = Z_{\mathrm{MFT}}^{-1} \, e^{\beta \left[\sum_l \sigma_l (h + \tilde{J}m) - \frac{1}{2} m^2 \tilde{J} N \right]} \qquad (6.5.12)$$

and the partition function

$$Z_{\text{MFT}} = \text{Tr}\, e^{\beta\left[\sum_l \sigma_l(h+\tilde{J}m)-\frac{1}{2}m^2\tilde{J}N\right]} = \prod_l \left(\sum_{\sigma_l=\pm1} e^{\beta\sigma_l(h+\tilde{J}m)}\right) e^{-\frac{1}{2}\beta m^2 \tilde{J}N}$$

$$(6.5.13)$$

in the molecular field approximation, where $\text{Tr} \equiv \sum_{\{\sigma_l=\pm1\}}$. We thus find for (6.5.13)

$$Z_{\text{MFT}} = \left(e^{-\frac{1}{2}\beta m^2 \tilde{J}} 2\cosh \beta(h + \tilde{J}m)\right)^N .$$

$$(6.5.13')$$

Using $m = \frac{1}{N}kT\frac{\partial}{\partial h}\log Z_{\text{MFT}}$, we obtain the equation of state in the molecular field approximation:

$$m = \tanh\left(\beta(\tilde{J}m + h)\right) ,$$

$$(6.5.14)$$

which is an implicit equation for m. Compared to the equation of state of a paramagnet, the field h is amplified by the internal molecular field $\tilde{J}m$. As we shall see later, (6.5.14) can be solved analytically for h. It is however instructive to solve (6.5.14) first for limiting cases. To do this, it will prove expedient to introduce the following abbreviations:

$$T_c = \frac{\tilde{J}}{k} \quad \text{and} \quad \tau = \frac{T-T_c}{T_c} .$$

$$(6.5.15)$$

We will immediately see that T_c has the significance of the transition temperature, the *Curie temperature*. Above T_c, the magnetization is zero in the absence of an applied field; below this temperature, it increases continuously with decreasing temperature to a finite value. We first determine the behavior in the neighborhood of T_c, where we can expand in terms of τ, h and m.

a) $h = 0$:
For zero applied field and in the vicinity of T_c, (6.5.14) can be expanded in a Taylor series,

$$m = \tanh \beta\tilde{J}m = \frac{T_c}{T}m - \frac{1}{3}\left(\frac{T_c}{T}m\right)^3 + \dots$$

$$(6.5.16)$$

which can be cut off at the third order so as to retain the leading term of the solution. The solutions of (6.5.16) are

$$m = 0 \quad \text{for} \quad T > T_c$$

$$(6.5.17a)$$

and

$$m = \pm m_0 , \quad m_0 = \sqrt{3}(-\tau)^{1/2} \quad \text{for} \quad T < T_c .$$

$$(6.5.17b)$$

The first solution, $m = 0$, is found for all temperatures, the second only for $T \leq T_c$, i.e. $\tau \leq 0$. Since the free energy of the second solution is smaller (see below and in Fig. 6.9), it is the stable solution below T_c. From these considerations we find the temperature ranges given in (6.5.17). For $T \leq T_c$, the *spontaneous magnetization*, denoted as m_0, is observed (6.5.17b); it follows a square-root law (Fig. 6.5). This quantity is called the *order parameter* of the ferromagnetic phase transition.

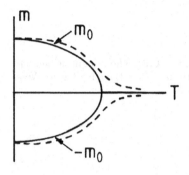

Fig. 6.5. The spontaneous magnetization (solid curve), and the magnetization in an applied field (dashed). The spontaneous magnetization in the Ising model has two possible orientations, $+m_0$ or $-m_0$

b) h and τ nonzero:

for small h and τ and thus small m, the expansion of (6.5.14)

$$m \left(1 - \frac{T_c}{T} \right) = \frac{h}{kT} - \frac{1}{3} \left(\frac{h}{kT} + \frac{T_c}{T} m \right)^3 + \ldots ,$$

leads to the *magnetic equation of state*

$$\frac{h}{kT_c} = \tau m + \frac{1}{3} m^3 \qquad (6.5.18)$$

in the neighborhood of T_c. An applied magnetic field produces a finite magnetization even above T_c and leads qualitatively to the dashed curve in Fig. 6.5.

c) $\tau = 0$:

exactly at T_c, we find from (6.5.18) the critical isotherm:

$$m = \left(\frac{3h}{kT_c} \right)^{1/3} , \qquad h \sim m^3 . \qquad (6.5.19)$$

d) Susceptibility for small τ:

we now compute the isothermal magnetic susceptibility $\chi = \left(\frac{\partial m}{\partial h} \right)_T$, by differentiating the equation of state (6.5.18) with respect to h

$$\frac{1}{kT_c} = \tau \chi + m^2 \chi . \qquad (6.5.20)$$

In the limit $h \to 0$, we can insert the spontaneous magnetization (6.5.17) into (6.5.20) and obtain for the isothermal magnetic susceptibility

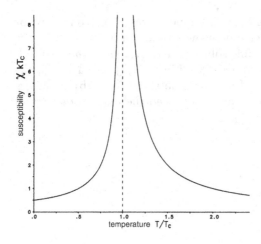

Fig. 6.6. The magnetic susceptibility (6.5.21): the Curie–Weiss
law

$$\chi = \frac{1/kT_c}{\tau + m^2} = \begin{cases} \dfrac{1/k}{(T - T_c)} & T > T_c \\[3mm] \dfrac{1/k}{2(T_c - T)} & T < T_c \end{cases} \quad ; \tag{6.5.21}$$

this is the *Curie–Weiss law* shown in Fig. 6.6.

Remark: We can understand the divergent susceptibility at T_c by starting from the Curie law for paramagnetic spins (6.3.10a), adding the internal molecular field $\tilde{J}m$ to the field h, and then determining the magnetization from it:

$$m = \frac{1}{kT}(h + \tilde{J}m) \rightarrow \frac{m}{h} = \frac{1/k}{T - T_c} \; . \tag{6.5.22}$$

Following these limiting cases, we solve (6.5.14) generally. We first discuss the graphical solution of this equation, referring to Fig. 6.7.

e) A graphical solution of the equation $m = \tanh\big(\beta(h + \tilde{J}m)\big)$

To find a graphical solution, it is expedient to introduce the auxiliary variable $y = m + \frac{h}{kT_c}$. Then one finds m as a function of h by determining the intersection of the line $y - \frac{h}{kT_c}$ with $\tanh\frac{T_c}{T}y$:

$$m = y - \frac{h}{kT_c} = \tanh\frac{T_c}{T}y \; .$$

For $T \geq T_c$, Fig. 6.7a exhibits exactly one intersection for each value of h. This yields the monotonically varying curve for $T \geq T_c$ in Fig. 6.8. For $T < T_c$, from Fig. 6.7b the slope of $\tanh\frac{T_c}{T}y$ at $y = 0$ is greater than 1 and therefore we find three intersections for small absolute values of h, while the solution for high fields remains unique. This leads to the function for $T < T_c$ which is shown in Fig. 6.8.

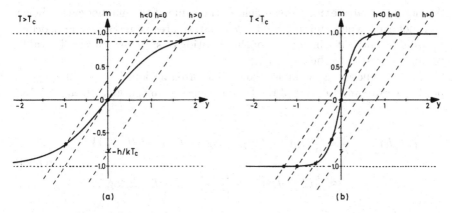

Fig. 6.7. The graphical solution of Eq. (6.5.14).

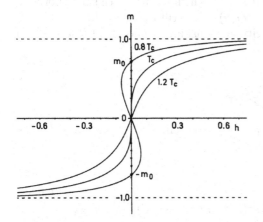

Fig. 6.8. The magnetic equation of state in the molecular field approximation (6.5.23). The dotted vertical line on the m-axis represents the inhomogeneous state (6.5.28)

For small h, $m(h)$ is not uniquely determined. Particularly noticeable is the fact that the S-shaped curve $m(h)$ contains a section with negative slope, i.e. negative susceptibility. In order to clarify the stability of the solution, we need to consider the free energy. We first note that for large h, the magnetization approaches its saturation value (Fig. 6.7).

In fact, one can immediately compute the function $h(m)$ from Eq. (6.5.14) analytically, since from

$$\beta(\tilde{J}m + h) = \operatorname{arctanh} m \equiv \frac{1}{2}\log\frac{1+m}{1-m}$$

the equation of state

$$h = -kT_c m + \frac{kT}{2}\log\frac{1+m}{1-m} \tag{6.5.23}$$

follows. Its shape is shown in Fig. 6.8 for $T \lessgtr T_c$ at the two values $T = 0.8\,T_c$ and $1.2\,T_c$ taken as examples, in agreement with the graphical construction.

As mentioned above, for a given field h, the value of the magnetization below T_c is not everywhere unique; e.g. for $h = 0$, the three values 0 and $\pm m_0$ occur. In order to find out which parts of the equation of state are physically stable, we must investigate the free energy.

The free energy in the molecular field approximation, $F = -kT \log Z_{\mathrm{MFT}}$, per lattice site and in units of the Boltzmann constant, is given from (6.5.13′) by

$$f(T,h) = \frac{F}{Nk} = \frac{1}{2}T_c m^2 - T \log \left\{ 2\cosh\left((T_c m + h/k)/T\right)\right\}$$
$$\approx \frac{1}{2}(T - T_c)m^2 + \frac{T_c}{12}m^4 - mh/k - T\log 2 \ . \tag{6.5.24}$$

We give here in the first line the complete expression, and in the second line the expansion in terms of m, h and $T - T_c$, which applies in the neighborhood of the phase transition. Here, $m = m(h)$ must still be inserted.

From (6.5.24), the heat capacity at vanishing applied field (for $T \approx T_c$) can be found:

$$c_{h=0} = -NkT \left.\frac{\partial^2 f}{\partial T^2}\right|_{h=0} = \begin{cases} 0 & T > T_c \\ \frac{3}{2}Nk\frac{T}{T_c} & T < T_c \end{cases} \ ;$$

here, a jump of magnitude $\Delta c_{h=0} = \frac{3}{2}Nk$ is seen. We calculate directly the *Helmholtz* free energy

$$a(T,m) = f + mh/k$$
$$= \frac{1}{2}T_c m^2 - T \log \left\{ 2\cosh\left((T_c m + h/k)/T\right)\right\} + mh/k \ , \tag{6.5.25}$$

in which $h = h(m)$ is to be inserted. From the determining equation for m (6.5.14), it follows that

$$T \log \left\{ 2\cosh\left((T_c m + h/k)/T\right)\right\} =$$
$$= T \log 2 + T \log \left(\frac{1}{1 - \tanh^2\left((T_c m + h/k)/T\right)} \right)^{1/2}$$
$$= T \log 2 - \frac{T}{2}\log(1 - m^2) \ .$$

Combining this with (6.5.23) and inserting into (6.5.25), we obtain

$$a(T,m) = -\frac{1}{2}T_c m^2 - T\log 2 + \frac{1}{2}T\log(1 - m^2) + \frac{Tm}{2}\log\frac{1+m}{1-m}$$
$$\approx -T\log 2 + \frac{1}{2}(T_c - T)m^2 + \frac{T_c}{12}m^4 \ ; \tag{6.5.26}$$

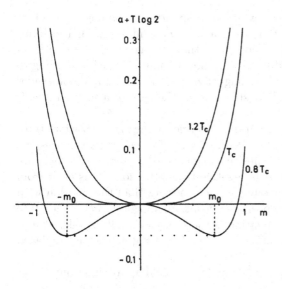

Fig. 6.9. The Helmholtz free energy in the molecular field approximation above and below T_c, for $T = 0.8\,T_c$ and $T = 1.2\,T_c$.

here, the second line holds near T_c. The Helmholtz free energy above and below T_c is shown in Fig. 6.9.

We first wish to point out the similarity of the free energy for $T < T_c$ with that of the van der Waals gas. For temperatures $T < T_c$, there is a region in $a(T, m)$ which violates the stability criterion (6.1.24a). The magnetization can be read off from Fig. 6.9 using

$$h = k \left(\frac{\partial a}{\partial m} \right)_T , \qquad (6.5.27)$$

by drawing a tangent with the slope h to the function $a(T, m)$. Above T_c, this construction gives a unique answer; below T_c, however, it is unique only for a sufficiently strong applied field. We continue the discussion of the low-temperature phase and determine the reorientation of the magnetization on changing the direction of the applied magnetic field, starting with a magnetic field h for which only a single value of the magnetization results from the tangent construction. Lowering the field causes m to decrease until at $h = 0$, the value m_0 is obtained. Exactly the same tangent, namely that with slope zero, applies to the point $-m_0$. Regions of magnetization m_0 and $-m_0$ can therefore be present in equilibrium with each another. When a fraction c of the body has the magnetization $-m_0$ and a fraction $1 - c$ has the magnetization m_0, then for $0 \leq c \leq 1$ the average magnetization is

$$m = -cm_0 + (1 - c)m_0 = (1 - 2c)m_0 \qquad (6.5.28)$$

in the interval between $-m_0$ and m_0.

The free energy of this inhomogeneously magnetized object is $a(m_0)$ (dotted line in Fig. 6.9), and is thus lower than the part of the molecular-field solution which arches upwards and which corresponds to a homogeneous state in the coexistence region of the two states $+m_0$ and $-m_0$. In the interval $[-m_0, m_0]$, the system does not enter the homogeneous state with its higher free energy, but instead breaks up into domains[24] which according to Eq. (6.5.28) yield all together the magnetization m. We remind the reader of the analogy to the Maxwell construction in the case of a van der Waals liquid.

For completeness, we compare the free energies of the magnetization states belonging to a small but nonzero h. Without loss of generality we can assume that h is positive. Along with the positive magnetization, for small h there are also two solutions of (6.5.27) with negative magnetizations. It is clear from Fig. 6.9 that the latter two have higher free energies than the solution with positive magnetization. For a positive (negative) magnetic field, the state with positive (negative) magnetization is thermodynamically stable. The S-shaped part of the equation of state (for $T < T_c$) in Fig. 6.8 is thus replaced by the dotted vertical line.

Finally, we give the entropy in the molecular field approximation:

$$s = \frac{S}{Nk} = -\left(\frac{\partial a}{\partial T}\right)_m = -\left[\frac{1+m}{2}\log\frac{1+m}{2} + \frac{1-m}{2}\log\frac{1-m}{2}\right] ; \quad (6.5.29)$$

it depends only on the average magnetization m.

The internal energy is given by

$$e = \frac{E}{Nk} = a - mh/k + Ts = -\frac{1}{2} T_c m^2 - mh/k . \qquad (6.5.30)$$

This can be more readily seen from (6.5.11) by taking an average value $\langle \mathcal{H} \rangle$ with the density matrix (6.5.12). From $h = k\frac{\partial a(T,m)}{\partial m}$ it again follows that $m = \tanh\frac{T_c m + h/k}{T}$, i.e. we recover Eq. (6.5.14).

Remarks:

(i) The molecular field approximation can also be applied to other models, for example the Heisenberg model, and also for quite different cooperative phenomena. The results are completely analogous.

(ii) The effect of the remaining spins on an arbitrarily chosen spin is replaced in molecular field theory be a mean field. In the case of a short-range interaction, the real field configuration will deviate considerably from this mean value. The more long-ranged the interaction, the more spins contribute to the local field, and the more closely it thus approaches the average field. The

---

[24] The number of domains can be greater than just two. When there are only a few domains, the interface energy is negligible in comparison to the gain in volume energy; see problem 7.6. In reality, the dipole interaction, anisotropies and inhomogeneities in the crystal play a role in the formation of domains. They form in such a way that the energy including that of the magnetic field is minimized.

molecular field approximation is therefore exact in the limit of long-range interactions (see also problem 6.13, the Weiss model). We note here the analogy between the molecular field theory and the Hartree-Fock theory of atoms and other many-body systems.

(iii) We want to point out another aspect of the molecular field approximation: its results do not depend at all on the dimensionality. This contradicts intuition and also exact calculations. In the case of short-range interactions, one-dimensional systems in fact do not undergo a phase transition; there are too few neighbors to lead to a cooperative ordering phenomenon.

(iv) In the next chapter, we shall turn to a detailed comparison of the gas-liquid transition and the ferromagnetic transition. We point out here in anticipation that the van der Waals liquid and the ferromagnet show quite similar behavior in the immediate vicinity of their critical points in the molecular field approximation; e.g. $(\rho_G - \rho_c) \sim (T_c - T)^{1/2}$ and $M_0 \sim (T_c - T)^{1/2}$, and likewise, the isothermal compressibility and the magnetic susceptibility both diverge as $(T_c - T)^{-1}$. This similarity is not surprising; in both cases, the interactions with the other gas atoms or spins is replaced by a mean field which is determined self-consistently from the ensuing equation of state.

(v) If one compares the critical power laws (6.5.17), (6.5.19), and (6.5.21) with experiments, with the exact solution of the two-dimensional Ising model, and with numerical results from computer simulations or series expansions, it is found that in fact qualitatively similar power laws hold, but the critical exponents are different from those found in the molecular field theory. The lower the dimensionality, the greater the deviations found. Instead of (6.5.17), (6.5.19), and (6.5.21), one finds generalized power laws:

$$m_0 \sim |\tau|^\beta \qquad\qquad T < T_c, \qquad\qquad (6.5.31\text{a})$$
$$m \sim h^{1/\delta} \qquad\qquad T = T_c, \qquad\qquad (6.5.31\text{b})$$
$$\chi \sim |\tau|^{-\gamma} \qquad\qquad T \gtrless T_c, \qquad\qquad (6.5.31\text{c})$$
$$c_h \sim |\tau|^{-\alpha} \qquad\qquad T \gtrless T_c. \qquad\qquad (6.5.31\text{d})$$

The *critical exponents* β, δ, γ and α which occur in these expressions in general differ from their molecular field values $1/2, 3, 1$ and 0 (corresponding to the jump). For instance, in the *two-dimensional Ising model*, $\beta = 1/8, \delta = 15, \gamma = 7/4$, and $\alpha = 0$ (logarithmic).

Remarkably, the values of the critical exponents do not depend on the lattice structure, but only on the dimensionality of the system. All Ising systems with short-range forces have the same critical exponents in d dimensions. Here, we have an example of the so called *universality*. The critical behavior depends on only a very few quantities, such as the dimensionality of the system, the number of components of the order parameter and the symmetry of the Hamiltonian. Heisenberg ferromagnets have different critical exponents from Ising ferromagnets, but within these groups, they are all the same. With these remarks about the actual behavior in the neighborhood of a critical point, we will close the discussion. In particular, we postpone the description of additional analogies between phase transitions to the next chapter. We now return to the molecular field approximation and use it to compute the magnetic susceptibility and the position-dependent spin correlation function.

6.5.3 Correlation Functions and Susceptibility

In this subsection, we shall consider the Ising model in the presence of a spatially varying applied magnetic field h_l. The Hamiltonian is then given by

$$\mathcal{H} = \mathcal{H}_0 - \sum_l h_l \sigma_l = -\frac{1}{2} \sum_{l,l'} J(l - l') \sigma_l \sigma_{l'} - \sum_l h_l \sigma_l . \tag{6.5.32}$$

The magnetization per spin at position l now depends on the lattice site l:

$$m_l = \langle \sigma_l \rangle \equiv \mathrm{Tr} \left[e^{-\beta \mathcal{H}} \sigma_l \right] / \mathrm{Tr}\, e^{-\beta \mathcal{H}} . \tag{6.5.33}$$

We first define the *susceptibility*

$$\chi(\mathbf{x}_l, \mathbf{x}_{l'}) = \frac{\partial m_l}{\partial h_{l'}} , \tag{6.5.34}$$

which describes the response at the site l to a change in the field at the site l'. The *correlation function* is defined as

$$G(\mathbf{x}_l, \mathbf{x}_{l'}) \equiv \langle \sigma_l \sigma_{l'} \rangle - \langle \sigma_l \rangle \langle \sigma_{l'} \rangle$$
$$= \langle (\sigma_l - \langle \sigma_l \rangle)(\sigma_{l'} - \langle \sigma_{l'} \rangle) \rangle . \tag{6.5.35}$$

The correlation function (6.5.35) is a measure of how strongly the deviations from the mean values at the sites l and l' are correlated with each other. Susceptibility and correlation function are related through the important *fluctuation-response theorem*

$$\chi(\mathbf{x}_l, \mathbf{x}_{l'}) = \frac{1}{kT} G(\mathbf{x}_l, \mathbf{x}_{l'}) . \tag{6.5.36}$$

This theorem (6.5.36) can be derived by taking the derivative of (6.5.33) with respect to $h_{l'}$.

For a translationally invariant system, we have

$$\chi(\mathbf{x}_l, \mathbf{x}_{l'})|_{\{h_l = 0\}} = \chi(\mathbf{x}_l - \mathbf{x}_{l'}) \quad \text{and} \quad G(\mathbf{x}_l, \mathbf{x}_{l'})|_{\{h_l = 0\}} = G(\mathbf{x}_l - \mathbf{x}_{l'}) . \tag{6.5.37}$$

At small fields h_l, we find $(m_l' \equiv m_l - m)$

$$m_l' = \sum_{l'} \chi(\mathbf{x}_l - \mathbf{x}_{l'}) h_{l'} . \tag{6.5.38}$$

A periodic field

$$h_l = h_{\mathbf{q}} e^{i\mathbf{q}\mathbf{x}_l} \tag{6.5.39}$$

therefore gives rise to a magnetization of the form

$$m_l' = e^{i\mathbf{q}\mathbf{x}_l} \sum_{l'} \chi\left(\mathbf{x}_l - \mathbf{x}_{l'}\right) e^{-i\mathbf{q}(\mathbf{x}_l - \mathbf{x}_{l'})} h_{\mathbf{q}} = \chi\left(\mathbf{q}\right) e^{i\mathbf{q}\mathbf{x}_l} h_{\mathbf{q}} , \tag{6.5.40}$$

where

$$\chi\left(\mathbf{q}\right) = \sum_{l'} \chi\left(\mathbf{x}_l - \mathbf{x}_{l'}\right) e^{-i\mathbf{q}(\mathbf{x}_l - \mathbf{x}_{l'})} = \frac{1}{kT} \sum_l G\left(\mathbf{x}_l\right) e^{-i\mathbf{q}\mathbf{x}_l} \tag{6.5.41}$$

is the Fourier transform of the susceptibility, and following the equals sign (6.5.36) has been inserted. In particular for $\mathbf{q} = 0$, we find the following relation between the uniform susceptibility and the correlation function:

$$\chi \equiv \chi\left(0\right) = \frac{1}{kT} \sum_l G\left(\mathbf{x}_l\right) . \tag{6.5.42}$$

Since the correlation function (6.5.35) can never be greater than 1, ($|\sigma_l| = 1$), and is in no case divergent, the divergence of the uniform susceptibility, Eq. (6.5.21) (i.e. the susceptibility referred to a spatially uniform field) can only be due to the fact that the correlations at T_c attain an infinitely long range.

6.5.4 The Ornstein–Zernike Correlation Function

We now want to calculate the correlation function introduced in the previous section within the molecular field approximation. As before, we denote the field by h_l, so that the mean value $m_l = \langle \sigma_l \rangle$ is also site dependent. In the molecular-field approximation, the density matrix is given by

$$\rho_{\text{MFT}} = Z^{-1} \exp\left[\beta \sum_l \sigma_l(h_l + \sum_{l'} J(l - l')\langle \sigma_{l'} \rangle)\right]. \tag{6.5.43}$$

The Fourier transform of the exchange coupling, which we take to be short-ranged, can be written for small wavenumbers as

$$\tilde{J}(\mathbf{k}) \equiv \sum_l J(l) e^{-i\mathbf{k}\mathbf{x}_l} \approx \tilde{J} - \mathbf{k}^2 \frac{1}{6} \sum_l \mathbf{x}_l^2 J(l) \equiv \tilde{J} - k^2 J . \tag{6.5.44}$$

Here, we have replaced the exponential function by its Taylor series. Due to the mirror symmetry of a cubic lattice, $J\left(-l\right) = J\left(l\right)$, and therefore there is no linear term in $\mathbf{k}$. Furthermore, we have $\sum_l \left(\mathbf{k} \cdot \mathbf{x}_l\right)^2 J\left(l\right) = \frac{1}{3}k^2 \sum_l \mathbf{x}_l^2 J\left(l\right)$.

The constant J is defined by

$$J = \frac{1}{6} \sum_l x_l^2 J(l) .$$

(6.5.45)

Using the density matrix (6.5.43), we obtain for the mean value of σ_l, in analogy to (6.5.14) in Sect. 6.5.2, the result

$$\langle \sigma_l \rangle = \tanh \left[\beta(h_l + \sum_{l'} J(l - l') \langle \sigma_{l'} \rangle) \right] .$$

(6.5.46)

We now take the derivative $\frac{\partial}{\partial h_{l'}}$ of the last equation (6.5.46), and finally set all the $h_{l'} = 0$, obtaining for the susceptibility:

$$\chi(\mathbf{x}_l - \mathbf{x}_{l'}) = \frac{1}{\cosh^2 \left[\beta \sum_{l''} J(l - l'')m \right]} \times$$
$$\times \left(\beta \delta_{ll'} + \beta \sum_{l''} J(l - l'') \chi(\mathbf{x}_{l''} - \mathbf{x}_{l'}) \right)$$

(6.5.47)

The Fourier-transformed susceptibility (6.5.41) is obtained from (6.5.47), recalling the convolution theorem:

$$\chi(\mathbf{q}) = \frac{1}{\cosh^2 \beta \tilde{J} m} \left(\beta + \beta \tilde{J}(\mathbf{q}) \chi(\mathbf{q}) \right) .$$

(6.5.48)

Furthermore, using $\cosh^2 \beta \tilde{J} m = \frac{1}{1 - \tanh^2 \beta \tilde{J} m} = \frac{1}{1 - m^2}$, where we have inserted the determining equation for m, Eq. (6.5.16), we obtain the general result

$$\chi(\mathbf{q}) = \frac{\beta}{\frac{1}{1 - m^2} - \beta \tilde{J}(\mathbf{q})} .$$

(6.5.49)

From this last equation, together with (6.5.15) and (6.5.44), we find in the neighborhood of T_c:

$$\chi(\mathbf{q}) = \frac{\beta}{1 - \frac{T_c}{T} + m_0^2 + \frac{Jq^2}{kT}} \qquad \text{for} \quad T \approx T_c$$

(6.5.50)

or also

$$\chi(\mathbf{q}) = \frac{1}{J(q^2 + \xi^{-2})} ,$$

(6.5.50')

where the correlation length

$$\xi = \left(\frac{J}{kT_c} \right)^{\frac{1}{2}} \begin{cases} \tau^{-1/2} & T > T_c \\ (-2\tau)^{-1/2} & T < T_c \end{cases}$$

(6.5.51)

has been introduced, with $\tau = (T - T_c)/T_c$. The susceptibility in real space is obtained by inverting the Fourier transform:

$$\chi(\mathbf{x}_l - \mathbf{x}_{l'}) = \frac{1}{N} \sum_{\mathbf{q}} \chi(\mathbf{q}) e^{i\mathbf{q}(\mathbf{x}_l - \mathbf{x}_{l'})} = \frac{V}{N(2\pi)^3} \int d^3q\, \chi(\mathbf{q})\, e^{i\mathbf{q}(\mathbf{x}_l - \mathbf{x}_{l'})} .$$

$$(6.5.52)$$

For the second equals sign it was assumed that the system is macroscopic, so that the sum over $\mathbf{q}$ can be replaced by an integral (cf. (4.1.2b) and (4.1.14a) with $\mathbf{p}/\hbar \to \mathbf{q}$) .

To compute the susceptibility for large distances it suffices to make use of the result for $\chi(\mathbf{q})$ at small values of $\mathbf{q}$ (Eq. (6.5.50′)); then with the lattice constant a we find

$$\chi(\mathbf{x}_l - \mathbf{x}_{l'}) = \frac{a^3}{(2\pi)^3} \int d^3q\, \frac{e^{i\mathbf{q}(\mathbf{x}_l - \mathbf{x}_{l'})}}{J(q^2 + \xi^{-2})} = \frac{a^3 e^{-|\mathbf{x}_l - \mathbf{x}_{l'}|/\xi}}{4\pi J |\mathbf{x}_l - \mathbf{x}_{l'}|} . \qquad (6.5.53)$$

From χ calculated in this way, we find the correlation function via (6.5.37):

$$G(\mathbf{x}) = kT\chi(\mathbf{x}) = \frac{kT a^3 e^{-|\mathbf{x}|/\xi}}{4\pi J |\mathbf{x}|} , \qquad (6.5.53')$$

which in this context is called the *Ornstein–Zernike correlation function*. The Ornstein–Zernike correlation function and its Fourier transform are shown in Fig. 6.10 and Fig. 6.11 for the temperatures $T = 1.01\, T_c$ and $T = T_c$. In these figures, the correlation length ξ at $T = 1.01\, T_c$ is also indicated. The quantity ξ_0 is defined by $\xi_0 = (J/kT_c)^{1/2}$, according to (6.5.48). At large distances $\chi(\mathbf{x})$ decreases exponentially as $\frac{1}{|\mathbf{x}|} e^{-|\mathbf{x}|/\xi}$. The correlation length ξ characterizes

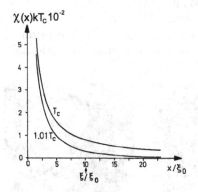

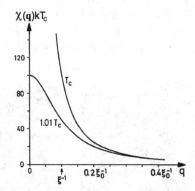

Fig. 6.10. The Ornstein–Zernike correlation function for $T = 1.01\, T_c$ and for $T = T_c$. Distances are measured in units of $\xi_0 = (J/kT_c)^{1/2}$.

Fig. 6.11. The Fourier transform of the Ornstein–Zernike susceptibility for $T = 1.01\, T_c$ and for $T = T_c$. The reciprocal of the correlation length for $T = 1.01\, T_c$ is indicated by the arrow.

the typical length over which the spin fluctuations are correlated. For $|\mathbf{x}| \gg \xi$, $G(\mathbf{x})$ is practically zero. At T_c, $\xi = \infty$, and $G(\mathbf{x})$ obeys the power law

$$G(\mathbf{x}) = \frac{kT_c v}{4\pi J |\mathbf{x}|} \tag{6.5.54}$$

with the volume of the unit cell $v = a^3$. $\chi(\mathbf{q})$ varies as $1/q^2$ for $\xi^{-1} \ll q$ and for $q = 0$, it is identical with the Curie–Weiss susceptibility. On approaching T_c, $\chi(0)$ becomes larger and larger. We note further that the continuum theory and thus $(6.5.50')$ and $(6.5.52)$ apply only to the case when $|\mathbf{x}| \gg a$. An important experimental tool for the investigation of magnetic phenomena is neutron scattering. The magnetic moment of the neutron interacts with the field produced by the magnetic moments in the solid and is therefore sensitive to magnetic structure and to static and dynamic fluctuations. The elastic scattering cross-section is proportional to the static susceptibility $\chi(\mathbf{q})$. Here, $\mathbf{q}$ is the momentum transfer, $\mathbf{q} = \mathbf{k}_{in} - \mathbf{k}_{out}$, where $\mathbf{k}_{in(out)}$ are the wave numbers of the incident and scattered neutrons. The increase of $\chi(\mathbf{q})$ at small $\mathbf{q}$ for $T \to T_c$ leads to intense forward scattering. This is termed *critical opalescence* near the Curie temperature, in analogy to the corresponding phenomenon in light scattering near the critical point of the gas-liquid transition.

The correlation length ξ diverges at the critical point; the correlations become more and more long-ranged as T_c is approached. Therefore, statistical fluctuations of the magnetic moments are correlated with each other over larger and larger regions. Furthermore, a field acting at the position $\mathbf{x}$ induces a polarization not only at that position, but also up to a distance ξ, as a result of $(6.5.37)$. The increase of the correlations can also be recognized in the spin configurations illustrated in Fig. 6.12. Here, 'snapshots' from a computer simulation of the Ising model are shown. White pixels represent $\sigma = +1$ and black pixels are for $\sigma = -1$. At twice the transition temperature, the spins are correlated only over very short distances (of a few lattice constants). At $T = 1.1 T_c$, the increase of the correlation length is clearly recognizable.

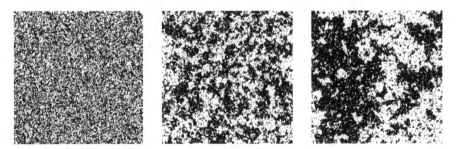

Fig. 6.12. A 'snapshot' of the spin configuration of a two-dimensional Ising model at $T = 2T_c, T = 1.1 T_c$ and $T = T_c$. White pixels represent $\sigma = +1$, and black pixels refer to $\sigma = -1$.

Along with very small clusters, both the black and the white clusters can be made out up to the correlation length ξ $(T = 1.1\,T_c)$. At $T = T_c$, $\xi = \infty$. In the figure, one sees two large white and black clusters. If the area viewed were to be enlarged, it would become clear that these are themselves located within an even larger cluster, which itself is only a member of a still larger cluster. There are thus correlated regions on all *length scales*. We observe here a *scale invariance* to which we shall return later. When we enlarge the unit of length, the larger clusters become smaller clusters, but since there are clusters up to infinitely large dimensions, the picture remains the same.

The Ornstein–Zernike theory (6.5.51) and (6.5.53$'$) reproduces the correct behavior qualitatively. The correlation length diverges however in reality as $\xi = \xi_0 \tau^{-\nu}$, where in general $\nu \neq \frac{1}{2}$, and also the shape of $G(\mathbf{x})$ differs from (6.5.53$'$) (see Chap. 7).

*6.5.5 Continuum Representation

6.5.5.1 Correlation Functions and Susceptibilities

It is instructive to derive the results obtained in the preceding sections in a continuum representation. The formulas which occur in this derivation will also allow a direct comparison with the Ginzburg–Landau theory, which we will treat later (in Chap. 7). Critical anomalies occur at large wavelengths. In order to describe this region, it is sufficient and expedient to go to a continuum formulation:

$$h_l \rightarrow h\left(\mathbf{x}\right) , \; \sigma_l \rightarrow \sigma\left(\mathbf{x}\right) , \; m_l \rightarrow m(\mathbf{x}) ,$$

$$\sum_l h_l \sigma_l \rightarrow \int \frac{d^3 x}{v} h\left(\mathbf{x}\right) \sigma\left(\mathbf{x}\right) . \tag{6.5.55}$$

Here, a is the lattice constant and $v = a^3$ is the volume of the unit cell. The sum over l becomes an integral over $\mathbf{x}$ in the limit $v \rightarrow \infty$. The partial derivative becomes a functional derivative[25] $(v \rightarrow 0)$

$$\frac{\delta m\left(\mathbf{x}\right)}{\delta h\left(\mathbf{x}'\right)} = \frac{1}{v}\frac{\partial m_l}{\partial h_{l'}} \quad \text{etc., e.g.} \quad \frac{\delta h\left(\mathbf{x}\right)}{\delta h\left(\mathbf{x}'\right)} = \delta\left(\mathbf{x} - \mathbf{x}'\right) . \tag{6.5.56}$$

For the susceptibility and correlation function we thus obtain from (6.5.34)

$$\chi\left(\mathbf{x} - \mathbf{x}'\right) = v\,\frac{\delta m\left(\mathbf{x}\right)}{\delta h\left(\mathbf{x}'\right)} = \frac{\partial m_l}{\partial h_{l'}} = \frac{1}{kT}\,G\left(\mathbf{x} - \mathbf{x}'\right) . \tag{6.5.57}$$

For small $h\left(\mathbf{x}\right)$, we find

$$m\left(\mathbf{x}\right) = \int \frac{d^3 x'}{v}\chi\left(\mathbf{x} - \mathbf{x}'\right) h\left(\mathbf{x}'\right) . \tag{6.5.58}$$

---

[25] The general definition of the functional derivative is to be found in W.I. Smirnov, *A Course of Higher Mathematics*, Vol. V, Pergamon Press, Oxford 1964 or in QM I, Sect. 13.3.1

A periodic field

$$h\left(\mathbf{x}'\right) = h_{\mathbf{q}}\, e^{i\mathbf{q}\mathbf{x}'} \tag{6.5.59}$$

induces a magnetization of the form

$$m\left(\mathbf{x}\right) = e^{i\mathbf{q}\mathbf{x}} \int \frac{d^3 x'}{v} \chi\left(\mathbf{x} - \mathbf{x}'\right) e^{-i\mathbf{q}\left(\mathbf{x}-\mathbf{x}'\right)} h_{\mathbf{q}} = \chi\left(\mathbf{q}\right) e^{i\mathbf{q}\mathbf{x}} h_{\mathbf{q}}\,, \tag{6.5.60}$$

where

$$\chi\left(\mathbf{q}\right) = \int \frac{d^3 y}{v} \chi\left(\mathbf{y}\right) e^{-i\mathbf{q}\mathbf{y}} = \frac{1}{kTv} \int d^3 y\, e^{-i\mathbf{q}\mathbf{y}}\, G\left(\mathbf{y}\right) \tag{6.5.61}$$

is the Fourier transform of the susceptibility, and after the second equals sign, we have made use of (6.5.37). In particular, for $\mathbf{q} = 0$, we find the following relation between the uniform susceptibility and the correlation function:

$$\chi \equiv \chi\left(0\right) = \frac{1}{kTv} \int d^3 y\, G\left(\mathbf{y}\right)\,. \tag{6.5.62}$$

6.5.5.2 The Ornstein–Zernike Correlation Function

As before, the field $h\left(\mathbf{x}\right)$ and with it also the mean value $\langle\sigma(\mathbf{x})\rangle$ are position dependent. The density matrix in the molecular field approximation and in the continuum representation is given by:

$$\rho_{\mathrm{MFT}} = Z^{-1} \exp\left[\beta \int \frac{d^3 x}{v}\sigma\left(\mathbf{x}\right) \left(h\left(\mathbf{x}\right) + \int \frac{d^3 x'}{v} J\left(\mathbf{x} - \mathbf{x}'\right) \langle\sigma\left(\mathbf{x}'\right)\rangle\right)\right]\,. \tag{6.5.63}$$

The Fourier transform of the exchange coupling for small wavenumbers assumes the form

$$\tilde{J}\left(\mathbf{k}\right) = \int \frac{d^3 x}{v} J\left(\mathbf{x}\right) e^{-i\mathbf{k}\cdot\mathbf{x}} \approx \tilde{J} - \frac{1}{6}k^2 \int \frac{d^3 x}{v}\mathbf{x}^2 J(\mathbf{x}) \equiv \tilde{J} - k^2 J\,, \tag{6.5.64}$$

where the exponential function has been replaced by its Taylor expansion. Owing to the spherical symmetry of the exchange interaction $J(\mathbf{x}) \equiv J(|\mathbf{x}|)$, there is no linear term in $\mathbf{k}$ and we find $\int d^3 x\,(\mathbf{k}\mathbf{x})^2\, J\left(\mathbf{x}\right) = \frac{1}{3}k^2 \int d^3 x\,\mathbf{x}^2\, J\left(\mathbf{x}\right)$. The constant J is defined by $J = \frac{1}{6v} \int d^3 x\,\mathbf{x}^2\, J\left(\mathbf{x}\right)$. The inverse transform of (6.5.64) yields

$$J\left(\mathbf{x}\right) = v\left(\tilde{J} + J\boldsymbol{\nabla}^2\right) \delta\left(\mathbf{x}\right)\,. \tag{6.5.65}$$

For phenomena at small $\mathbf{k}$ or large distances, the real position dependence of the exchange interaction can be replaced by (6.5.65). We insert this into (6.5.63) and obtain the mean value of $\sigma\left(x\right)$, analogously to (6.5.14) in Sect. 6.5.2:

$$\langle\sigma\left(\mathbf{x}\right)\rangle = \tanh\left[\beta\left(h\left(\mathbf{x}\right) + \tilde{J}\left\langle\sigma\left(\mathbf{x}\right)\right\rangle + J\boldsymbol{\nabla}^2 \langle\sigma(\mathbf{x})\rangle\right)\right]\,. \tag{6.5.66}$$

In the neighborhood of T_c, we can carry out an expansion similar to that in (6.5.16), $m\left(\mathbf{x}\right) \equiv \langle\sigma\left(\mathbf{x}\right)\rangle$,

$$\tau m\left(\mathbf{x}\right) - \frac{J}{kT_c}\boldsymbol{\nabla}^2 m\left(\mathbf{x}\right) + \frac{1}{3}m\left(\mathbf{x}\right)^3 = \frac{h\left(\mathbf{x}\right)}{kT_c}\,, \tag{6.5.67}$$

with $\tau = (T - T_c)/T_c$, where the second term on the left-hand side occurs due to the spatial inhomogeneity of the magnetization. The equations of the continuum limit can be obtained from the corresponding equations of the discrete representation at any step, e.g. (6.5.67) follows from (6.5.46), by carrying out the substitutions $\langle \sigma_l \rangle = m_l \rightarrow m(\mathbf{x})$, $J(l) \rightarrow J(\mathbf{x}) = \left(\tilde{J} + J\nabla^2 \right) \delta(\mathbf{x})$.

Now we take the functional derivative $\frac{\delta}{\delta h(\mathbf{x}')}$ of the last equation, (6.5.67),

$$\left[\tau - \frac{J}{kT_c} \nabla^2 + m_0^2 \right] \chi(\mathbf{x} - \mathbf{x}') = v\delta(\mathbf{x} - \mathbf{x}')/kT_c . \tag{6.5.68}$$

Since the susceptibility is calculated in the limit $h \rightarrow 0$, the spontaneous magnetization m_0, which is given by the molecular-field expressions (6.5.17a,b), appears on the left side. The solution of this differential equation, which also occurs in connection with the Yukawa potential, is given in three dimensions by

$$\chi(\mathbf{x} - \mathbf{x}') = \frac{v e^{-|\mathbf{x} - \mathbf{x}'|/\xi}}{4\pi J |\mathbf{x} - \mathbf{x}'|} . \tag{6.5.69}$$

The Fourier transform is

$$\chi(\mathbf{q}) = \frac{1}{J(q^2 + \xi^{-2})} . \tag{6.5.70}$$

In this expression, we have introduced the *correlation length*:

$$\xi = \left(\frac{J}{kT_c} \right)^{1/2} \begin{cases} \tau^{-1/2} & T > T_c \\ (-2\tau)^{-1/2} & T < T_c . \end{cases} \tag{6.5.71}$$

The results thus obtained agree with those of the previous section; for their discussion, we refer to that section.

*6.6 The Dipole Interaction, Shape Dependence, Internal and External Fields

6.6.1 The Hamiltonian

In this section, we investigate the influence of the dipole interaction. The total Hamiltonian for the magnetic moments $\boldsymbol{\mu}_l$ is given by

$$\mathcal{H} \equiv \mathcal{H}_0(\{\boldsymbol{\mu}_l\}) + \mathcal{H}_d(\{\boldsymbol{\mu}_l\}) - \sum_l \boldsymbol{\mu}_l \mathbf{H}_a . \tag{6.6.1}$$

$\mathcal{H}_0$ contains the exchange interaction between the magnetic moments and $\mathcal{H}_d$ represents the dipole interaction

$$\mathcal{H}_d = \frac{1}{2} \sum_{l,l'} A_{ll'}^{\alpha\beta} \mu_l^\alpha \mu_{l'}^\beta$$

$$= \frac{1}{2} \sum_{l,l'} \left(\frac{\delta_{\alpha\beta}}{|\mathbf{x}_l - \mathbf{x}_{l'}|^3} - \frac{3(\mathbf{x}_l - \mathbf{x}_{l'})_\alpha (\mathbf{x}_l - \mathbf{x}_{l'})_\beta}{|\mathbf{x}_l - \mathbf{x}_{l'}|^5} \right) \mu_l^\alpha \mu_{l'}^\beta , \tag{6.6.2}$$

and $\mathbf{H}_a$ is the externally applied magnetic field. The dipole interaction is long-ranged, in contrast to the exchange interaction; it decreases as the third power of the distance. Although the dipole interaction is in general considerably weaker than the exchange interaction – its interaction energy corresponds to a temperature of about[26] 1 K – it plays an important role for some phenomena due to its long range and also due to its anisotropy.

The goal of this section is to obtain predictions about the free energy and its derivatives for the Hamiltonian (6.6.1),

$$F(T, H_a) = -kT \log \text{Tr} \, e^{-\mathcal{H}/kT} \tag{6.6.3}$$

and to analyze the modifications which result from including the dipole interaction. Before we turn to the microscopic theory, we wish to derive some elementary consequences of classical magnetostatics for thermodynamics; their justification within the framework of statistical mechanics will be given at the end of this section.

6.6.2 Thermodynamics and Magnetostatics

6.6.2.1 The Demagnetizing Field

It is well known from electrodynamics[27] (magnetostatics) that in a magnetized body, in addition to the externally applied field $\mathbf{H}_a$, there is a demagnetizing field $\mathbf{H}_d$ which results from the dipole fields of the individual magnetic moments, so that the effective field in the interior of the magnet, $\mathbf{H}_i$,

$$\mathbf{H}_i = \mathbf{H}_a + \mathbf{H}_d \,, \tag{6.6.4a}$$

is in general different from $\mathbf{H}_a$. The field $\mathbf{H}_d$ is uniform only in *ellipsoids* and their limiting shapes, and we will thus limit ourselves as usual to *this type of bodies*. For ellipsoids, the demagnetizing field has the form $\mathbf{H}_d = -D\mathbf{M}$ and thus the (macroscopic) field in the interior of the body is

$$\mathbf{H}_i = \mathbf{H}_a - D\mathbf{M} \,. \tag{6.6.4b}$$

Here, D is the demagnetizing tensor and $\mathbf{M}$ is the magnetization (per unit volume). When $\mathbf{H}_a$ is applied along one of the principal axes, D can be interpreted as the appropriate demagnetizing factor in Eq. (6.6.4b). For $\mathbf{H}_a$ and therefore $\mathbf{M}$ parallel to the axis of a long cylindrical body, $D = 0$; for $\mathbf{H}_a$ and $\mathbf{M}$ perpendicular to an infinitely extended thin sheet, $D = 4\pi$; and for a sphere, $D = \frac{4\pi}{3}$. The value of the internal field thus depends on the shape of the sample and the direction of the applied field.

---

[26] See e.g. the estimate in N. W. Ashcroft and N. D. Mermin, *Solid State Physics*, Holt, Rinehart and Winston, New York, 1976, p. 673.

[27] A. Sommerfeld, *Electrodynamics*, Academic Press, New York 1952; R. Becker and F. Sauter, *Theorie der Elektrizität*, Vol. 1, 21st Edition, p. 52, Teubner, Stuttgart, 1973; R. Becker, *Electromagnetic Fields and Interactions*, Blaisdell, 1964; J. D. Jackson, *Classical Electrodynamics*, 2nd edition, John Wiley, New York, 1975.

6.6.2.2 Magnetic Susceptibilities

We now need to distinguish between the susceptibility relative to the applied field, $\chi_a(H_a) = \frac{\partial M}{\partial H_a}$, and the susceptibility relative to the internal field, $\chi_i(H_i) = \frac{\partial M}{\partial H_i}$. We consider for the moment only fields in the direction of the principal axes, so that we do not need to take the tensor character of the susceptibilities into account. We emphasize that the usual definition in electrodynamics is the second one. This is due to the fact that $\chi_i(H_i)$ is a pure materials property[28], and that owing to curl $\mathbf{H}_i = \frac{4\pi}{c}\mathbf{j}$, the field $\mathbf{H}_i$ can be controlled in the core of a coil by varying the current density $\mathbf{j}$.

Taking the derivative of Eq. (6.6.4b) with respect to M, one obtains the relation between the two susceptibilities:

$$\frac{1}{\chi_i(H_i)} = \frac{1}{\chi_a(H_a)} - D \; . \tag{6.6.5a}$$

It is physically clear that the susceptibility $\chi_i(H_i)$ relative to the internal field H_i acting in the interior of the body is a specific materials parameter which is independent of the shape, and that therefore the shape dependence of $\chi_a(H_a)$

$$\chi_a(H_a) = \frac{\chi_i(H_i)}{1 + D\chi_i(H_i)} \tag{6.6.5b}$$

results form the occurrence of D in (6.6.5b) and (6.6.4b).[29]

If the field is not applied along one of the principal axes of the ellipsoid, one can derive the tensor relation by taking the derivative of the component α of (6.6.4b) with respect to M_β:

$$\left(\chi_i^{-1}\right)_{\alpha\beta} = \left(\chi_a^{-1}\right)_{\alpha\beta} - D_{\alpha\beta} \; . \tag{6.6.5c}$$

Relations of the type (6.6.5a–c) can be found in the classical thermodynamic literature.[30]

---

[28] In the literature on magnetism, $\chi_i(H_i)$ is called the true susceptibility and $\chi_a(H_a)$ the apparent susceptibility. E. Kneller, *Ferromagnetismus*, Springer, Berlin, 1962, p. 97.

[29] When $\chi_i \lesssim 10^{-4}$, as in many practical situations, the demagnetization correction can be neglected. On the other hand, there are also cases in which the shape of the object can become important. In paramagnetic salts, χ_i increases at low temperatures according to Curie's law, and it can become of the order of 1; in superconductors, $4\pi\chi_i = -1$ (perfect diamagnetism or Meissner effect).

[30] R. Becker and W. Döring, *Ferromagnetismus*, Springer, Berlin, 1939, p. 8; A. B. Pippard, *Elements of Classical Thermodynamics*, Cambridge at the University Press 1964, p. 66.

6.6.2.3 Free Energies and Specific Heats

Starting from the free energy $F(T, H_a)$ with the differential

$$dF = -SdT - VMdH_a \, , \tag{6.6.6}$$

we can define a new free energy by means of a Legendre transformation

$$\hat{F}(T, \mathbf{H}_i) = F(T, \mathbf{H}_a) + \frac{V}{2} M^\alpha D^{\alpha\beta} M^\beta \, . \tag{6.6.7a}$$

The differential of this free energy is, using (6.6.4b), given by

$$d\hat{F}(T, \mathbf{H}_i) = -SdT - VMdH_i \, . \tag{6.6.7b}$$

Since the entropy $S(T, \mathbf{H}_i)$ and the magnetization $\mathbf{M}(T, \mathbf{H}_i)$ as functions of the internal field must be independent of the shape of the sample, all the derivatives of $\hat{F}(T, \mathbf{H}_i)$ are shape independent. Therefore, the free energy $\hat{F}(T, \mathbf{H}_i)$ is itself shape independent. From (6.6.6) and (6.6.7b), it follows that

$$S = - \left(\frac{\partial F}{\partial T} \right)_{H_a} = - \left(\frac{\partial \hat{F}}{\partial T} \right)_{H_i} \tag{6.6.8}$$

and

$$M = -\frac{1}{V} \left(\frac{\partial F}{\partial H_a} \right)_T = -\frac{1}{V} \left(\frac{\partial \hat{F}}{\partial H_i} \right)_T \, . \tag{6.6.9}$$

The specific heat can also be defined for a constant internal field

$$C_{H_i} = \frac{T}{V} \left(\frac{\partial S}{\partial T} \right)_{H_i} \tag{6.6.10a}$$

and for a constant applied (external) field

$$C_{H_a} = \frac{T}{V} \left(\frac{\partial S}{\partial T} \right)_{H_a} \, . \tag{6.6.10b}$$

Using the Jacobian as in Sect. 3.2.4, one can readily obtain the following relations

$$C_{H_a} = C_{H_i} \frac{1}{1 + D\chi_{i_T}} \tag{6.6.11a}$$

$$C_{H_i} = C_{H_a} + T \frac{\left(\frac{\partial M}{\partial T} \right)_{H_a} D \left(\frac{\partial M}{\partial T} \right)_{H_a}}{1 - D\chi_{a_T}} \tag{6.6.11b}$$

and

$$C_{H_a} = C_{H_i} - T \frac{\left(\frac{\partial M}{\partial T} \right)_{H_i} D \left(\frac{\partial M}{\partial T} \right)_{H_i}}{1 + D\chi_{i_T}} \, , \tag{6.6.11c}$$

where the index T indicates the isothermal susceptibility. The shape independence of $\chi_i(H_i)$ and $\hat{F}(T, H_i)$, which is plausible for the physical reasons given above, has also been derived using perturbation-theoretical methods.[31] For a vanishingly small field, the shape-independence could be proven without resorting to perturbation theory.[32]

---

[31] P. M. Levy, Phys. Rev. **170**, 595 (1968); H. Horner, Phys. Rev. **172**, 535 (1968)
[32] R. B. Griffiths, Phys. Rev.**176**, 655 (1968)

6.6.2.4 The Local Field

Along with the internal field, one occasionally also requires the local field H_{loc}. It is the field present at the position of a magnetic moment. One obtains it by imagining a sphere to be centered on the lattice site under consideration, which is large compared to the unit cell but small compared to the overall ellipsoid (see Fig. 6.13). We obtain for the local field:[27]

$$H_{\text{loc}} = H_a + \phi M \qquad (6.6.12a)$$

$$\text{with} \quad \phi = \phi_0 + \frac{4\pi}{3} - D . \qquad (6.6.12b)$$

Here, ϕ_0 is the sum of the dipole fields of the average moments within the fictitious sphere. The medium outside the imaginary sphere can be treated as a continuum, and its contribution is that of a solid polarized ellipsoid $(-D)$, minus that of a polarized sphere $\left(\frac{4\pi}{3}\right)$. For a cubic lattice, ϕ_0 vanishes for reasons of symmetry.[27] One can also introduce a free energy

$$\hat{F}(T, H_{\text{loc}}) = F(T, H_a) - \frac{1}{2}VM\phi M \qquad (6.6.13a)$$

with the differential

$$d\hat{F} = -SdT - VMdH_{\text{loc}} . \qquad (6.6.13b)$$

Since, owing to (6.6.12a,b), (6.6.7a), and (6.6.13a), it follows that

$$\hat{F}(T, H_{\text{loc}}) = \hat{F}(T, H_i) + \frac{1}{2}VM\left(\phi_0 + \frac{4\pi}{3}\right)M , \qquad (6.6.14)$$

so that $\hat{F}$ differs from $\hat{F}$ only by a term which is independent of the external shape and is itself therefore shape-independent. One can naturally also

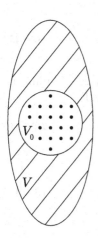

Fig. 6.13. The definition of the local field. An ellipsoid of volume V and a fictitious sphere of volume V_0 (schematic, not to scale)

define susceptibilities at constant H_{loc} and derive relations corresponding to equations (6.6.13a-c) and (6.6.11a-c), in which essentially H_i is replaced by H_{loc} and D by ϕ.

6.6.3 Statistical–Mechanical Justification

In this subsection, we will give a microscopic justification of the thermodynamic results obtained in the preceding section and derive Hamiltonians for the calculation of the shape-independent free energies $\hat{F}(T, H_i)$ and $\hat{F}(T, H_{\text{loc}})$ of equations (6.6.7a) and (6.6.13a). The magnetic moments will be represented by the their mean values and fluctuations. The dipole interaction will be decomposed into a short-range and a long-range part. For the interactions of the fluctuations, the long-range part can be neglected. The starting point will be the Hamiltonian (6.6.1), in which we introduced the fluctuations around (deviations from) the mean value $\langle \mu_l^\alpha \rangle$

$$\delta\mu_l^\alpha \equiv \mu_l^\alpha - \langle \mu_l^\alpha \rangle : \tag{6.6.15}$$

$$
\begin{aligned}
\mathcal{H} &= \mathcal{H}_0(\{\boldsymbol{\mu}_l\}) + \frac{1}{2}\sum_{l,l'} A_{ll'}^{\alpha\beta}\,\delta\mu_l^\alpha\,\delta\mu_{l'}^\beta + \frac{1}{2}\sum_{l,l'} A_{ll'}^{\alpha\beta}\langle\mu_l^\alpha\rangle\langle\mu_{l'}^\beta\rangle \\
&\quad + \sum_{l,l'} A_{ll'}^{\alpha\beta}\delta\mu_l^\alpha\langle\mu_{l'}^\beta\rangle - \sum_l \mu_l^\alpha H_a^\alpha \\
&= \mathcal{H}_0(\{\boldsymbol{\mu}_l\}) + \frac{1}{2}\sum_{l,l'} A_{ll'}^{\alpha\beta}\,\delta\mu_l^\alpha\,\delta\mu_{l'}^\beta - \frac{1}{2}\sum_{l,l'} A_{ll'}^{\alpha\beta}\langle\mu_l^\alpha\rangle\langle\mu_{l'}^\beta\rangle \\
&\quad - \sum_l \mu_l^\alpha(H_a^\alpha + H_{d,l}^\alpha)
\end{aligned}
\tag{6.6.16}
$$

with the thermal average of the field at the lattice point l due to the remaining dipoles:

$$H_{d,l}^\alpha = -\sum_{l'} A_{ll'}^{\alpha\beta}\langle\mu_{l'}^\beta\rangle . \tag{6.6.17}$$

For ellipsoids in an external magnetic field, the magnetization is uniform, $(\langle\mu_{l'}^\beta\rangle = \frac{V}{N}M^\beta)$; likewise the dipole field (demagnetizing field):

$$H_{d,l}^\alpha = H_{\text{loc}}^\alpha \equiv (\phi_0 + D_0 - D)_{\alpha\beta}M^\beta . \tag{6.6.18}$$

In going from (6.6.17) to (6.6.18), the dipole sum

$$\phi_{\alpha\beta} = -\frac{V}{N}\sum_{l'} A_{ll'}^{\alpha\beta} = (\phi_0 + D_0 - D)_{\alpha\beta} \tag{6.6.19}$$

was decomposed into a discrete sum over the subvolume V_0 (the Lorentz sphere) and the region $V - V_0$, in which a continuum approximation can be applied:

$$(D_0 - D)_{\alpha\beta} = -\int_{V-V_0} d^3x \, \frac{\partial}{\partial x_\alpha} \frac{\partial}{\partial x_\beta} \frac{1}{|\mathbf{x}|}$$

$$= \delta_{\alpha\beta} \left(\int_{S_1} df_\alpha \frac{\partial}{\partial x_\beta} \frac{1}{|\mathbf{x}|} - \int_{S_2} df_\alpha \frac{\partial}{\partial x_\beta} \frac{1}{|\mathbf{x}|} \right) . \tag{6.6.20}$$

The first surface integral extends over the surface of the Lorentz sphere and the second over the (external) surface of the ellipsoid (sample).

With this, we can write the Hamiltonian in the form

$$\mathcal{H} = \mathcal{H}_0(\{\boldsymbol{\mu}_l\}) + \frac{1}{2} \sum_{l,l'} A_{ll'}^{\alpha\beta} \, \delta\mu_l^\alpha \, \delta\mu_{l'}^\beta - \sum_l \mu_l^\alpha H_{\text{loc}}^\alpha + \frac{1}{2} V M_\alpha \phi_{\alpha\beta} M_\beta . $$

$$\tag{6.6.21}$$

Since the long-range property of the dipole interaction plays no role in the interaction between the fluctuations $\delta\mu_l$, the first two terms in the Hamiltonian are shape-independent. The sample shape enters only in the local field H_{loc} and in the fourth term on the right-hand side. Comparison with (6.6.13a) shows that the free energy $\hat{F}(T, H_{\text{loc}})$, which, apart from its dependence on H_{loc}, is shape independent, can be determined by computation of the partition function with the first three terms of (6.6.21).

If the dipole interactions between the fluctuations is completely neglected,[33] one obtains the approximate effective Hamiltonian

$$\hat{\mathcal{H}} = \mathcal{H}_0(\{\boldsymbol{\mu}_l\}) - \sum_l \boldsymbol{\mu}_l \mathbf{H}_{\text{loc}} , \tag{6.6.22}$$

in which the dipole interaction expresses itself only in the demagnetizing field.

The exact treatment of the second term, $\frac{1}{2}\sum_{l,l'} A_{ll'}^{\alpha\beta} \, \delta\mu_l^\alpha \, \delta\mu_{l'}^\beta$ in (6.6.21) is carried out as follows: since the expectation value based on approximate application of the Ornstein–Zernike theory decreases as $\langle \delta\mu_l \delta\mu_{l'} \rangle \approx \frac{e^{-r_{ll'}/\xi}}{r}$, and $A_{ll'} \sim \frac{1}{r_{ll'}^3}$, the interaction of the fluctuations is negligible at large distances. The shape of the sample thus plays no role in this term in the limit $V \to \infty$ with the shape kept unchanged. One can thus replace $A_{ll'}^{\alpha\beta}$ by

$$^\sigma A_{ll'}^{\alpha\beta} = \frac{\partial}{\partial x^\alpha} \frac{\partial}{\partial x^\beta} \frac{e^{-\sigma|\mathbf{x}|}}{|\mathbf{x}|} , \tag{6.6.23}$$

with the cutoff length σ^{-1}, or more precisely

$$\frac{1}{2} \sum_{l,l'} A_{ll'}^{\alpha\beta} \delta\mu_l^\alpha \delta\mu_{l'}^\beta = \lim_{\sigma\to 0} \lim_{V\to\infty} \frac{1}{2} \sum_{l,l'} {}^\sigma A_{ll'}^{\alpha\beta} \delta\mu_l^\alpha \delta\mu_{l'}^\beta . \tag{6.6.24}$$

---

[33] J.H. van Vleck, J. Chem. Phys. **5**, 320, (1937), Eq. (36).

Inserting $\delta\mu_l = \mu_l - \langle\mu_l\rangle$, we obtain for the right-hand side of (6.6.24)

$$\lim_{\sigma\to 0}\lim_{V\to\infty}\frac{1}{2}\sum_{l,l'}{}^\sigma A_{ll'}^{\alpha\beta}\left(\mu_l^\alpha\mu_{l'}^\beta - 2\mu_l^\alpha\langle\mu_{l'}^\beta\rangle + \langle\mu_l^\alpha\rangle\langle\mu_{l'}^\beta\rangle\right)$$

$$= \lim_{\sigma\to 0}\lim_{V\to\infty}\frac{1}{2}\sum_{l,l'}{}^\sigma A_{ll'}^{\alpha\beta}\mu_l^\alpha\mu_{l'}^\beta + \qquad\qquad (6.6.25)$$

$$+ \sum_l(\phi_0 + D_0)_{\alpha\beta}M^\beta\mu_l^\alpha - \frac{V}{2}(\phi_0 + D_0)M^2\,.$$

In the order: first the thermodynamic limit $V\to\infty$, then $\sigma\to 0$, the first term in (6.6.25) is shape-independent. Since in the second and third terms, the sum over l' is cut off by $e^{-|\mathbf{x}_l-\mathbf{x}_{l'}|\sigma}$, the contribution $-D$ due to the external boundary of the ellipsoid does not appear here. Inserting (6.6.24) and (6.6.25) into (6.6.21), we find the *Hamiltonian in final form*[34]

$$\mathcal{H} = \hat{\mathcal{H}} - \frac{V}{2}MDM \qquad\qquad (6.6.26\text{a})$$

with

$$\hat{\mathcal{H}} = \mathcal{H}_0(\{\mu_l\}) + \int\frac{d^3q}{(2\pi)^3}v_a A_\mathbf{q}^{\alpha\beta}\mu_\mathbf{q}^\alpha\mu_{-\mathbf{q}}^\beta - \sum_l\mu_l^\alpha H_i^\alpha\,. \qquad (6.6.26\text{b})$$

Here, the Fourier transforms

$$\mu_\mathbf{q}^\alpha = \frac{1}{\sqrt{N}}\sum_l e^{-i\mathbf{q}\mathbf{x}_l}\mu_l^\alpha\,, \qquad\qquad (6.6.27\text{a})$$

$$A_\mathbf{q}^{\alpha\beta} = \sum_{l\neq 0}e^{-i\mathbf{q}(\mathbf{x}_{l'}-\mathbf{x}_1)}A_{l0}^{\alpha\beta} \qquad\qquad (6.6.27\text{b})$$

and the internal field $\mathbf{H}_i = \mathbf{H}_a - DM$ have been introduced. The Fourier transform (6.6.27b) can be evaluated using the *Ewald method* [35]; for cubic lattices, it yields[36]

$$A_\mathbf{q}^{\alpha\beta} = \frac{1}{v_a}\left(\frac{4\pi}{3}\left(\delta^{\alpha\beta} - \frac{3q^\alpha q^\beta}{q^2}\right) + \alpha_1 q^\alpha q^\beta + \left(\alpha_2 q^2 - \alpha_3(q^\alpha)^2\right)\delta^{\alpha\beta}\right.$$

$$\left. + \mathcal{O}\left(q^4, (q^\alpha)^4, (q^\alpha)^2(q^\beta)^2\right)\right)\,, \qquad\qquad (6.6.27\text{b}')$$

where v_a is the volume of the primitive unit cell and the α_i are constants which depend on the lattice structure. The first two terms in $\hat{\mathcal{H}}$, Eq. (6.6.26b),

[34] See also W. Finger, Physica **90 B**, 251 (1977).

[35] P. P. Ewald, Ann. Phys. **54**, 57 (1917); *ibid.*, **54**, 519 (1917); *ibid.*, **64**, 253 (1921)

[36] M. H. Cohen and F. Keffer, Phys. Rev. **99**, 1135 (1955); A. Aharony and M. E. Fisher, Phys. Rev. B **8**, 3323 (1973)

are shape-independent. The sample shape enters only into the internal field H_i and in the last term of (6.6.26a). Comparison of Eq. (6.6.26a) with Eq. (6.6.7a) shows that the shape-independent free energy $\hat{F}(T, H_i)$ can be calculated from the partition function derived from the Hamiltonian $\hat{\mathcal{H}}$, Eq. (6.6.26b). We note in particular the nonanalytic behavior of the term $q_\alpha q_\beta/q^2$ in the limit $q \to 0$; it is caused by the $1/r^3$-dependence of the dipole interaction. Due to this term, the longitudinal and transverse wavenumber-dependent susceptibilities (with respect to the wavevector) are different from each other.[37] We recall that the short-ranged exchange interaction can be expanded as a Taylor series in $\mathbf{q}$:

$$\mathcal{H}_0 = -\frac{1}{2} \int d^3q \, \tilde{J}(\mathbf{q}) \mu_{\mathbf{q}} \mu_{-\mathbf{q}}$$
$$\tilde{J}(\mathbf{q}) = \tilde{J} - J\mathbf{q}^2 + \mathcal{O}(q^4) \,. \tag{6.6.28}$$

In addition to the effects of the demagnetizing field and the resulting shape dependence, which we have treated in detail, the dipole interaction, even though it is in general much weaker than the exchange interaction, has a number of important consequences owing to its long range and its anisotropic character:[37] (i) It changes the values of the critical exponents in the neighborhood of ferromagnetic phase transitions; (ii) it can stabilize magnetic order in systems of low dimensionality, which otherwise would not occur due to the large thermal fluctuations; (iii) the total magnetic moment $\boldsymbol{\mu} = \sum_l \boldsymbol{\mu}_l$ is no longer conserved. This has important consequences for the dynamics; and (iv) the dipole interaction is important in nuclear magnetism, where it is larger than or comparable to the indirect exchange interaction.

We can now include the dipole interactions in the results of Sects. 6.1 to 6.5 in the following manner:

(i) If we neglect the dipole interaction between the fluctuations of the magnetic moments $\delta\mu_l = \mu_l - \langle\mu_l\rangle$ as an approximation, we can take the spatially uniform part of the dipole fields into account by replacing the field $\mathbf{H}$ by the local field $\mathbf{H}_{\text{loc}}$.

(ii) If, in addition to the exchange interactions possibly present, we also include the dipole interaction between the fluctuations, then according to (6.6.26), the *complete Hamiltonian* contains the *internal field* $\mathbf{H}_i$. The field $\mathbf{H}$ must therefore be replaced by $\mathbf{H}_i$; furthermore, the shape-dependent term $-\frac{V}{2}MDM$ enters into the Hamiltonian $\mathcal{H}$, Eq. (6.6.26a), and, via the term $\hat{\mathcal{H}}$, also the shape-independent part of the dipole interaction, i.e. Eq. (6.6.27b').

---

[37] E. Frey and F. Schwabl, Advances in Physics **43**, 577 (1994)

6.6.4 Domains

The spontaneous magnetization per spin, $m_0(T)$, is shown in Fig. 6.5. The total magnetic moment of a uniformly magnetized sample without an external field would be $Nm_0(T)$, and its spontaneous magnetization per unit volume $M_0(T) = Nm_0(T)/V$, where N is the overall number of magnetic moments. In fact, as a rule the magnetic moment is smaller or even zero. This results from the fact that a sample in general breaks up into domains with different directions of magnetization. Within each domain, $|\mathbf{M}(\mathbf{x}, T)| = M_0(T)$. Only when an external field is applied do the domains which are oriented parallel to the field direction grow at the cost of the others, and reorientation occurs until finally $Nm_0(T)$ has been reached. The spontaneous magnetization is therefore also called the *saturation magnetization*. We want to illustrate domain formation, making use of two examples.

(i) One possible domain structure in a ferromagnetic bar below T_c is shown in Fig. 6.14. One readily sees that for the configuration with 45°-walls throughout the sample,

$$\operatorname{div} \mathbf{M} = 0 \, . \tag{6.6.29}$$

Then it follows from the basic equations of magnetostatics

$$\operatorname{div} \mathbf{H}_i = -4\pi \operatorname{div} \mathbf{M} \tag{6.6.30a}$$
$$\operatorname{curl} \mathbf{H}_i = 0 \tag{6.6.30b}$$

that, in the interior of the sample,

$$\mathbf{H}_i = 0 \, , \tag{6.6.31}$$

and thus also $\mathbf{B} = 4\pi\mathbf{M}$ in the interior. From the continuity conditions it follows that $\mathbf{B} = \mathbf{H} = 0$ outside the sample. The domain configuration is therefore energetically more favorable than a uniformly magnetized sample.
(ii) Domain structures also express themselves in a measurement of the total magnetic moment $\mathfrak{M}$ of a sphere. The calculated magnetization $M = \frac{\mathfrak{M}}{V}$ as a function of the applied field is indicated by the curves in Fig. 6.15.

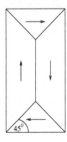

Fig. 6.14. The domain structure in a prism-shaped sample

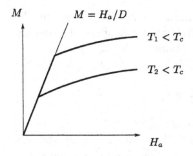

Fig. 6.15. The magnetization within a sphere as a function of the external field $H_a, T_1 < T_2 < T_c$; D is the demagnetizing factor.

Let the magnetization within a uniformly magnetized region as a function of the internal field $H_i = H_a - DM$ be given by the function $M = M(H_i)$. As long as the overall magnetization of the sphere is less than the saturation magnetization, the domains have a structure such that $H_i = 0$, and therefore, $M = \frac{1}{D}H_a$ must hold.[38] For $H_a = DM_{\text{spont}}$, the sample is finally uniformly magnetized, corresponding to the saturation magnetization. For $H_a > DM_{\text{spont}}$, M can be calculated from $M = M(H_a - DM)$.

6.7 Applications to Related Phenomena

In this section, we discuss consequences of the results of this chapter on magnetism for other areas of physics: polymer physics, negative temperatures and the melting curve of ^{3}He.

6.7.1 Polymers and Rubber-like Elasticity

Polymers are long chain molecules which are built up of similar links, the monomers. The number of monomers is typically $N \approx 100,000$. Examples of polymers are polyethylene, $(CH_2)_N$, polystyrene, $(C_8H_8)_N$, and rubber, $(C_5H_8)_N$, where the number of monomers is $N > 100,000$ (see Fig. 6.16).

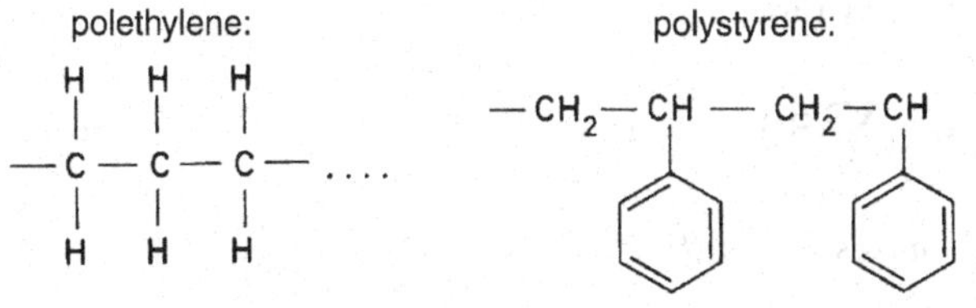

polethylene:

```
    H   H   H
    |   |   |
— C — C — C —  ....
    |   |   |
    H   H   H
```

polystyrene:

$$-CH_2-CH\ -\ CH_2-CH$$

Fig. 6.16. The structures of polyethylene and polystyrene

---

[38] S. Arajs and R. V Calvin, J. Appl. Phys. **35**, 2424 (1964).

To find a description of the mechanical and thermal properties we set up the following simple model (see Fig. 6.17): the starting point in space of monomer 1 is denoted by $\mathbf{X}_1$, and that of a general monomer i by $\mathbf{X}_i$. The position (orientation) of the ith monomer is then given by the vector $\mathbf{S}_i \equiv \mathbf{X}_{i+1} - \mathbf{X}_i$:

$$\mathbf{S}_1 = \mathbf{X}_2 - \mathbf{X}_1, \ldots, \mathbf{S}_i = \mathbf{X}_{i+1} - \mathbf{X}_i, \ldots, \mathbf{S}_N = \mathbf{X}_{N+1} - \mathbf{X}_N . \qquad (6.7.1)$$

We now assume that aside from the chain linkage of the monomers there are no interactions at all between them, and that they can freely assume any arbitrary orientation, i.e. $< \mathbf{S}_i \cdot \mathbf{S}_j >= 0$ for $i \neq j$. The length of a monomer is denoted by a, i.e. $\mathbf{S}_i^2 = a^2$.

Fig. 6.17. A polymer, composed of a chain of monomers

Since the line connecting the two ends of the polymer can be represented in the form

$$\mathbf{X}_{N+1} - \mathbf{X}_1 = \sum_{i=1}^{N} \mathbf{S}_i , \qquad (6.7.2)$$

it follows that

$$\langle \mathbf{X}_{N+1} - \mathbf{X}_1 \rangle = 0 . \qquad (6.7.3)$$

Here, we average independently over all possible orientations of the $\mathbf{S}_i$. The last equation means that the coiled polymer chain is oriented randomly in space, but makes no statement about its typical dimensions. A suitable measure of the mean square length is

$$\left\langle (\mathbf{X}_{N+1} - \mathbf{X}_1)^2 \right\rangle = \left\langle \left(\sum \mathbf{S}_i \right)^2 \right\rangle = a^2 N . \qquad (6.7.4)$$

We define the so called *radius of gyration*

$$R \equiv \sqrt{\left\langle (\mathbf{X}_{N+1} - \mathbf{X}_1)^2 \right\rangle} = a N^{\frac{1}{2}} , \qquad (6.7.5)$$

which characterizes the size of the polymer coil that grows as the square root of the number of monomers.

In order to study the *elastic properties*, we allow a force to act on the ends of the polymer, i.e. the force $\mathbf{F}$ acts on $\mathbf{X}_{N+1}$ and the force $-\mathbf{F}$ on $\mathbf{X}_1$ (see Fig. 6.17). Under the influence of this tensile force, the energy depends on the positions of the two ends:

$$
\begin{aligned}
\mathcal{H} &= -(\mathbf{X}_{N+1} - \mathbf{X}_1) \cdot \mathbf{F} \\
&= -\left[(\mathbf{X}_{N+1} - \mathbf{X}_N) + (\mathbf{X}_N - \mathbf{X}_{N-1}) + \ldots + (\mathbf{X}_2 - \mathbf{X}_1)\right] \cdot \mathbf{F} \\
&= -\mathbf{F} \cdot \sum_{i=1}^{N} \mathbf{S}_i \,.
\end{aligned}
\tag{6.7.6}
$$

Polymers under tension can therefore be mapped onto the problem of a paramagnet in a magnetic field, Sect. 6.3. The force corresponds to the applied magnetic field in the paramagnetic case, and the length of the polymer chain to the magnetization. Thus, the thermal average of the distance vector between the ends of the chain is

$$
\mathbf{L} = \left\langle \sum_{i=1}^{N} \mathbf{S}_i \right\rangle = Na \left(\coth \frac{aF}{kT} - \frac{kT}{aF} \right) \frac{\mathbf{F}}{F} \,.
\tag{6.7.7}
$$

We have used the Langevin function for classical moments in this expression, Eq. (6.3.12b), and multiplied by the unit vector in the direction of the force, $\mathbf{F}/F$. If aF is small compared to kT, we find (corresponding to Curie's law)

$$
\mathbf{L} = \frac{Na^2}{3kT} \mathbf{F} \,.
\tag{6.7.8}
$$

For the change in the length, we obtain from the previous equation

$$
\frac{\partial L}{\partial F} \sim \frac{1}{T}
\tag{6.7.9a}
$$

and

$$
\frac{\partial L}{\partial T} = -\frac{Na^2}{3kT^2} |\mathbf{F}| \,.
\tag{6.7.9b}
$$

The length change per unit force or the elastic constant decreases with increasing temperature according to (6.7.9a). A still more spectacular result is that for the expansion coefficient $\frac{\partial L}{\partial T}$: rubber contracts when its temperature is increased! This is in complete contrast to crystals, which as a rule expand with increasing temperature. The reason for the elastic behavior of rubber is easy to see: the higher the temperature, the more dominant is the entropy term in the free energy, $F = E - TS$, which strives towards a minimum. The entropy increases, i.e. the polymer becomes increasingly disordered or coiled and therefore pulls together. The general dependence of the length on $a|\mathbf{F}|/kT$ is shown in Fig. 6.18.

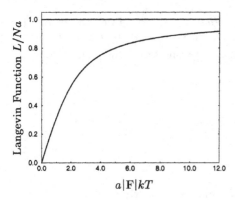

Fig. 6.18. The length of a polymer under the influence of a tensile force **F**.

Remark: In the model considered here, we have not taken into account that a monomer has a limited freedom of orientation, since each position can be occupied by at most one monomer. In a theory which takes this effect into account, the dependence $R = aN^{1/2}$ in Eq. (6.7.5) is replaced by $R = aN^\nu$. The exponent ν has a significance analogous to that of the exponent of the correlation length in phase transitions, and the degree of polymerization (chain length) N corresponds to the reciprocal distance from the critical point, τ^{-1}. The properties of polymers, in which the volume already occupied is excluded, correspond to a random motion in which the path cannot lead to a point already passed through (self-avoiding random walk). The properties of both these phenomena follow from the n-component ϕ^4 model (see Sect. 7.4.5) in the limit $n \to 0$.[39] An approximate formula for ν is due to Flory: $\nu_{\text{Flory}} = 3/(d+2)$.

6.7.2 Negative Temperatures

In isolated systems whose energy levels are bounded above and below, thermodynamic states with negative absolute temperatures can be established. Examples of such systems with energy levels that are bounded towards higher energies are two-level systems or paramagnets in an external magnetic field h.

We consider a paramagnet consisting of N spins of quantum number $S = 1/2$ with an applied field along the z direction. Considering the quantum numbers of the Pauli spin matrices $\sigma_l = \pm 1$, the Hamiltonian has the following diagonal structure

$$\mathcal{H} = -h \sum_l \sigma_l . \tag{6.7.10}$$

The magnetization per lattice site is defined by $m = \langle \sigma \rangle$ and is independent of the lattice position l. The entropy is given by

---

[39] P.-G. de Gennes, *Scaling Concepts in Polymer Physics*, Cornell University Press, Ithaca, 1979.

$$S(m) = -kN \left[\frac{1+m}{2} \log \frac{1+m}{2} + \frac{1-m}{2} \log \frac{1-m}{2} \right]$$
$$= -k \left[N_+ \log \frac{N_+}{N} + N_- \log \frac{N_-}{N} \right] , \tag{6.7.11}$$

and the internal energy E depends on the magnetization via

$$E = -Nhm = -h(N_+ - N_-) , \tag{6.7.12}$$

with $N_\pm = N(1 \pm m)/2$. These expressions follow immediately from the treatment in the microcanonical ensemble (Sect. 2.5.2.2) and can also be obtained from Sect. 6.3 by elimination of T and B. For $m = 1$ (all spins parallel to the field h), the energy is $E = -Nh$; for $m = -1$ (all spins antiparallel to h), the energy is $E = Nh$. The entropy is given in Fig. 2.9 as a function of the energy. It is maximal for $E = 0$, i.e. in the state of complete disorder. The temperature is obtained by taking the derivative of the entropy with respect to the energy:

$$T = \frac{1}{\left(\frac{\partial S}{\partial E}\right)_h} = \frac{2h}{k} \left[\log \frac{1+m}{1-m} \right]^{-1} . \tag{6.7.13}$$

It is shown as a function of the energy in Fig. 2.10. In the interval $0 < m \le 1$, i.e. $-1 \le E/Nh < 0$, the temperature is positive, as usual. For $m < 0$, that is when the magnetization is oriented antiparallel to the magnetic field, the absolute temperature becomes negative, i.e. $T < 0$! With increasing energy, the temperature T goes from 0 to ∞, then through $-\infty$, and finally to -0. Negative temperatures thus belong to higher energies, and are therefore "hotter" than positive temperatures. In a state with a negative temperature, more spins are in the excited state than in the ground state. One can also see that negative temperatures are in fact hotter than positive by bringing two such systems into thermal contact. Take system 1 to have the positive temperature $T_1 > 0$ and system 2 the negative temperature $T_2 < 0$. We assume that the exchange of energy takes place quasistatically; then the total entropy is $S = S_1(E_1) + S_2(E_2)$ and the (constant) total energy is $E = E_1 + E_2$. From the increase of entropy, it follows with $\frac{dE_2}{dt} = -\frac{dE_1}{dt}$ that

$$0 < \frac{dS}{dt} = \frac{\partial S_1}{\partial E_1} \frac{dE_1}{dt} + \frac{\partial S_2}{\partial E_2} \frac{dE_2}{dt} = \left(\frac{1}{T_1} - \frac{1}{T_2} \right) \frac{dE_1}{dt} . \tag{6.7.14}$$

Since the factor in brackets, $\left(\frac{1}{T_1} + \frac{1}{|T_2|} \right)$, is positive, $\frac{dE_1}{dt} > 0$ must also hold; this means that energy flows from subsystem 2 at a negative temperature into subsystem 1.

We emphasize that the energy dependence of $S(E)$ represented in Fig. 2.9 and the negative temperatures which result from it are a direct consequence of the boundedness of the energy levels. If the energy levels were not bounded from above, then a finite energy input could not lead to an infinite temperature or even beyond it. We also note that the specific heat per lattice site of this spin system is given by

$$\frac{C}{Nk} = \left(\frac{2h}{kT}\right)^2 \frac{e^{2h/kT}}{\left(1 + e^{2h/kT}\right)^2} \tag{6.7.15}$$

and vanishes both at $T = \pm 0$ as well as at $T = \pm\infty$.

We now discuss two examples of negative temperatures:

(i) Nuclear spins in a magnetic field:
The first experiment of this kind was carried out by Purcell and Pound[40] in a nuclear magnetic resonance experiment using the nuclear spins of ^{7}Li in LiF. The spins were first oriented at the temperature T by the field **H**. Then the direction of **H** was so quickly reversed that the nuclear spins could not follow it, that is faster than a period of the nuclear spin precession. The spins are then in a state with the negative temperature $-T$. The mutual interaction of the spins is characterized by their spin-spin relaxation time of $10^{-5} - 10^{-6}$ sec. This interaction is important, since it allows the spin system to reach internal equilibrium; it is however negligible for the energy levels in comparison to the Zeeman energy. For nuclear spins, the interaction with the lattice in this material is so slow (the spin-lattice relaxation time is 1 to 10 min) that the spin system can be regarded as completely isolated for times in the range of seconds. The state of negative temperature is maintained for some minutes, until the magnetization reverses through interactions with the lattice and the temperature returns to its initial value of T. In dilute gases, a state of spin inversion with a lifetime of days can be established.

(ii) Lasers (pulsed lasers, ruby lasers):
By means of irradiation with light, the atoms of the laser medium are excited (Fig. 6.19). The excited electron drops into a metastable state. When more

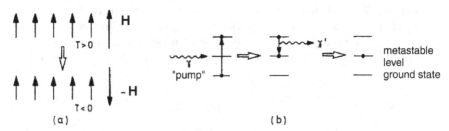

Fig. 6.19. Examples of negative temperatures: (a) nuclear spins in a magnetic field H, which is rotated by 180° (b) a ruby laser. The "pump" raises electrons into an excited state. The electron can fall into a metastable state by emission of a photon. When a population inversion is established, the temperature is negative

---

[40] E. M. Purcell and R. V. Pound, Phys. Rev. **81**, 279 (1951)

electrons are in this excited state than in the ground state, i.e. when a population inversion has been established, the state is described by a negative temperature.

*6.7.3 The Melting Curve of ^{3}He

The anomalous behavior of the melting curve of ^{3}He (Fig. 6.20) is related to the magnetic properties of solid ^{3}He.[41] As we already discussed in connection

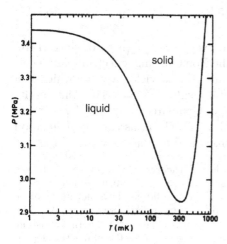

Fig. 6.20. The melting curve of ^{3}He at low temperatures

with the Clausius–Clapeyron equation,

$$\frac{dP}{dT} = \frac{S_S - S_L}{V_S - V_L} < 0 \, , \qquad (6.7.16)$$

the solid has a higher entropy than the liquid in the temperature range below $0.32\,\mathrm{K}$. The minimum in the melting curve occurs according to the Clausius–Clapeyron equation just when the entropies are equal. The magnetic effects in ^{3}He result from the nuclear spins and are therefore considerably weaker than in electronic magnetism. The exchange interaction J is so small that the antiferromagnetic order in solid ^{3}He sets in only at temperatures of $T_N \sim 10^{-3}\,\mathrm{K}$. Hence, the spins are disordered and make a contribution to the entropy (cf. (6.3.16$'$)) of:

$$S_S = Nk \left[\log 2 - \mathcal{O}\!\left(\left(\frac{J}{kT} \right)^2 \right) \right] . \qquad (6.7.17)$$

[41] J. Wilks, *The properties of Liquid and Solid Helium*, Clarendon Press, Oxford, 1967; C. M. Varma and N.,R. Werthamer, in K.-H. Bennemann and J. B. Ketterson, Eds., *The physics of liquid and solid He* Part I, p. 549, J. Wiley, New York, 1976; A. C. Anderson, W. Reese, J. C. Wheatley, Phys. Rev. **130**, 495 (1963); O. V. Lounasmaa, *Experimental Principles and Methods Below 1K*, Academic Press, London, 1974.

Since the lattice entropy at $0.3\,\mathrm{K}$ is negligibly small, this is practically the entire entropy in the solid. The entropy of the liquid is (from Eq. (4.3.19))

$$S_{\mathrm{L}} \approx kN \frac{\pi^2}{2} \frac{T}{T_F} \qquad T_F \approx 1\,\mathrm{K} . \tag{6.7.18}$$

According to the Clausius–Clapeyron equation (6.7.16), the melting curve has a minimum when $S_{\mathrm{L}} = S_{\mathrm{S}}$:

$$T_{\min} = \frac{2T_F}{\pi^2} \log 2 \sim \frac{2T_F}{\pi^2} 0.69 \sim 0.15\,\mathrm{K} . \tag{6.7.19}$$

Below $T_{\min}$, the slope of the melting curve $\frac{dP}{dT} = \frac{S_{\mathrm{L}}-S_{\mathrm{S}}}{V_{\mathrm{L}}-V_{\mathrm{S}}} < 0$, since there, $S_{\mathrm{L}} < S_{\mathrm{S}}$ and $V_{\mathrm{L}} > V_{\mathrm{S}}$. This leads to the Pomeranchuk effect, already mentioned in Sect. 3.8.2. The above estimate of $T_{\min}$ yields a value which is a factor of 2 smaller than the experimental result, $T_{\min}^{\exp} = 0.3\,\mathrm{K}$. This results from the value of S_{L}, which is too large. Compared to an ideal gas, there are correlations in an interacting Fermi liquid which, as can be understood intuitively, lead to a lowering of its entropy and to a larger value of $T_{\min}$.

Before the discovery of the two superfluid phases of $^3\mathrm{He}$, the existence of a maximum in the melting curve below $10^{-3}\mathrm{K}$ was theoretically discussed.[41] It was expected due to the T^3-dependence of the specific heat in the antiferromagnetically ordered phase and the linear specific heat of the Fermi liquid. This picture however changed with the discovery of the superfluid phases of $^3\mathrm{He}$ (see Fig. 4.10). The specific heat of the liquid behaves at low temperatures like $e^{-\Delta/kT}$, with a constant Δ (energy gap), and therefore the melting curve rises for $T \to 0$ and has the slope 0 at $T = 0$.

Literature

A.I. Akhiezer, V.G. Bar'yakhtar and S.V. Peletminskii, *Spin Waves*, North Holland, Amsterdam, 1968

N.W. Ashcroft and N.D. Mermin, *Solid State Physics*, Holt, Rinehart and Winston, New York, 1976

R. Becker u. W. Döring, *Ferromagnetismus*, Springer, Berlin, 1939

W.F. Brown, *Magnetostatic Principles in Ferromagnetism*, North Holland, Amsterdam, 1962

F. Keffer, *Spin Waves*, Encyclopedia of Physics, Vol. XVIII/2, p. 1. Ferromagnetism, ed. S. Flügge (Springer, Berlin, Heidelberg, New York 1966)

Ch. Kittel, *Introduction to Solid State Physics*, 3rd ed., John Wiley, 1967

Ch. Kittel, *Thermal Physics*, John Wiley, New York, 1969

D.C. Mattis, *The Theory of Magnetism*, Harper and Row, New York, 1965

A.B. Pippard, *Elements of Classical Thermodynamics*, Cambridge at the University Press, 1964

H.E. Stanley, *Introduction to Phase Transitions and Critical Phenomena*, Clarendon Press, Oxford, 1971

J.H. van Vleck, *The Theory of Magnetic and Electric Susceptibilites*, Oxford University Press, 1932

D. Wagner, *Introduction to the Theory of Magnetism*, Pergamon Press, Oxford, 1972

Problems for Chapter 6

6.1 Derive (6.1.24c) for the Hamiltonian of (6.1.25), by taking the second derivative of

$$A(T, M) = -kT \log \operatorname{Tr} e^{-\beta \mathcal{H}} + HM$$

with respect to T for fixed M.

6.2 The classical paramagnet: Consider a system of N non-interacting, classical magnetic moments, $\boldsymbol{\mu}_i$ ($\sqrt{\boldsymbol{\mu}_i^2} = m$) in a magnetic field $\mathbf{H}$, with the Hamiltonian $\mathcal{H} = -\sum_{i=1}^{N} \boldsymbol{\mu}_i \mathbf{H}$. Calculate the classical partition function, the free energy, the entropy, the magnetization, and the isothermal susceptibility. Refer to the suggestions following Eq. (6.3.12b).

6.3 The quantum-mechanical paramagnet, in analogy to the main text:
(a) Calculate the entropy and the internal energy of an ideal paramagnet as a function of T. Show that for $T \to \infty$,

$$S = Nk \ln (2J + 1) \,,$$

and discuss the temperature dependence in the vicinity of $T = 0$.
(b) Compute the heat capacities C_H and C_M for a non-interacting spin-$1/2$ system.

6.4 The susceptibility and mean square deviation of harmonic oscillators: Consider a quantum-mechanical harmonic oscillator with a charge e in an electric field E

$$\mathcal{H} = \frac{p^2}{2m} + \frac{m\omega^2}{2}x^2 - eEx \,.$$

Show that the dielectric susceptibility is given by

$$\chi = \frac{\partial \langle ex \rangle}{\partial E} = \frac{e^2}{m\omega^2}$$

and that the mean square deviation takes the form

$$\langle x^2 \rangle = \frac{\hbar}{2\omega m} \coth \frac{\beta \hbar \omega}{2} \,,$$

from which it follows that

$$\chi = \frac{2 \tanh \frac{\beta \hbar \omega}{2}}{\hbar \omega} \langle x^2 \rangle \,.$$

Compare these results with the paramagnetism of non-coupled magnetic moments! Take account of the difference between rigid moments and induced moments, and the resulting different temperature dependences of the susceptibility. Take the classical limit $\beta \hbar \omega \ll 1$.

6.5 Consider a solid with N degrees of freedom, which are each characterized by two energy levels at Δ and $-\Delta$. Show that

$$E = -N\Delta \tanh \frac{\Delta}{kT} \quad , \quad C = \frac{dE}{dT} = Nk \left(\frac{\Delta}{kT}\right)^2 \frac{1}{\cosh^2 \frac{\Delta}{kT}}$$

holds. How does the specific heat behave for $T \gg \Delta/k$ and for $T \ll \Delta/k$?

6.6 When the system described in 6.5 is disordered, so that all values of Δ within the interval $0 \leq \Delta \leq \Delta_0$ occur with equal probabilities, show that then the specific heat for $kT \ll \Delta_0$ is proportional to T.

Hint: The internal energy of this system can be found from problem 6.5 by averaging over all values of Δ. This serves as a model for the linear specific heat of glasses at low temperatures.

6.7 Demonstrate the validity of the fluctuation-response theorem, Eq. (6.5.35).

6.8 Two defects are introduced into a ferromagnet at the sites $\mathbf{x}_1$ and $\mathbf{x}_2$, and produce there the magnetic fields h_1 and h_2. Calculate the interaction energy of these defects for $|\mathbf{x}_1 - \mathbf{x}_2| > \xi$. For which signs of the h_i is there an attractive interaction of the defects?

Suggestions: The energy in the molecular field approximation is

$\bar{E} = \sum_{l,l'} \langle S_l \rangle \langle S_{l'} \rangle J(l - l')$.

For each individual defect, $\langle S_l \rangle_{1,2} = G(\mathbf{x}_l - \mathbf{x}_{1,2}) h_{1,2}$, where G is the Ornstein-Zernike correlation function. For two defects which are at a considerable distance apart, $\langle S_l \rangle$ can be approximated as a linear superposition of the single-defect averages. The interaction energy can be obtained by calculating $\bar{E}$ for this linear superposition and subtracting the energies of the single defects.

6.9 The one-dimensional Ising model: Calculate the partition function Z_N for a one-dimensional Ising model with N spins obeying the Hamiltonian

$$\mathcal{H} = -\sum_{i=1}^{N-1} J_i S_i S_{i+1} \ .$$

Hint: Prove the recursion relation $Z_{N+1} = 2Z_N \cosh(J_N/kT)$.

6.10 (a) Calculate the two-spin correlation function $G_{i,n} := \langle S_i S_{i+n} \rangle$ for the one-dimensional Ising model in problem 6.9.
Hint: The correlation function can be found by taking the appropriate derivatives of the partition function with respect to the interactions. Observe that $S_i^2 = 1$.
Result: $G_{i,n} = \tanh^n(J/kT)$ for $J_i = J$.
(b) Determine the behavior of the correlation length defined by $G_{i,n} = e^{-n/\xi}$ for $T \to 0$.
(c) Calculate the susceptibility from the fluctuation-response theorem:

$$\chi = \frac{(g\mu_B)^2}{kT} \sum_i^N \sum_j^N \langle S_i S_j \rangle \ .$$

Hint: Consider how many terms with $|i - j| = 0$, $|i - j| = 1$, $|i - j| = 2$ etc. occur in the double sum. Compute the geometric series which appear.

Result:

$$\chi = \frac{(g\mu_B)^2}{kT}\left\{N\left(\frac{1+\alpha}{1-\alpha}\right) - \frac{2\alpha\left(1-\alpha^N\right)}{(1-\alpha)^2}\right\} \; ; \; \alpha = \tanh\frac{J}{kT}.$$

(d) Show that in the thermodynamic limit, $(N \to \infty)$ $\chi \propto \xi$ for $T \to 0$, and thus $\gamma/\nu = 1$.
(e) Plot χ^{-1} in the thermodynamic limit as a function of temperature.
(f) How can one obtain from this the susceptibility of an antiferromagnetically-coupled linear chain? Plot and discuss χ as a function of temperature.

6.11 Show that in the molecular-field approximation for the Ising model, the internal energy E is given by

$$E = \left(-\frac{1}{2}kT_c\, m^2 - hm\right)N$$

and the entropy S by

$$S = kN\left[-\frac{T_c}{T}m^2 - \frac{1}{kT}hm + \log\left(2\cosh(kT_c m + h)/kT\right)\right].$$

Inserting the equation of state, show also that

$$S = -kN\left(\frac{1+m}{2}\log\frac{1+m}{2} + \frac{1-m}{2}\log\frac{1-m}{2}\right).$$

Finally, expand $a(T,m) = e - Ts + mh$ up to the 4th power in m.

6.12 An improvement of the molecular field theory for an Ising spin system can be obtained as follows (Bethe–Peierls approximation): the interaction of a spin σ_0 with its z neighbors is treated exactly. The remaining interactions are taken into account by means of a molecular field h', which acts only on the z neighbors. The Hamiltonian is then given by:

$$\mathcal{H} = -h'\sum_{j=1}^{z}\sigma_j - J\sum_{j=1}^{z}\sigma_0\sigma_j - h\sigma_0 \,.$$

The applied field h acts directly on the central spin and is likewise included in h'. H' is determined self-consistently from the condition $\langle\sigma_0\rangle = \langle\sigma_j\rangle$.
(a) Show that the partition function $Z(h',T)$ has the form

$$Z = \left[2\cosh\left(\frac{h'}{kT}+\frac{J}{kT}\right)\right]^z e^{-h/kT} + \left[2\cosh\left(\frac{h'}{kT}-\frac{J}{kT}\right)\right]^z e^{h/kT}$$

$$= Z_+ + Z_- \,.$$

(b) Calculate the average values $\langle\sigma_0\rangle$ and $\langle\sigma_j\rangle$ for simplicity with $h = 0$.
Result:

$$\langle\sigma_0\rangle = (Z_+ - Z_-)/Z \,,$$

$$\langle\sigma_j\rangle = \frac{1}{z}\sum_{j=1}^{z}\langle\sigma_j\rangle = \frac{1}{z}\frac{\partial}{\partial\left(\frac{h'}{kT}\right)}\log Z =$$

$$= \frac{1}{z}\left[Z_+\tanh\left(\frac{h'}{kT}-\frac{J}{kT}\right) + Z_-\tanh\left(\frac{h'}{kT}-\frac{J}{kT}\right)\right].$$

(c) The equation $\langle \sigma_0 \rangle = \langle \sigma_j \rangle$ has a nonzero solution below T_c:

$$\frac{h'}{kT(z-1)} = \frac{1}{2} \log \frac{\cosh\left(\frac{J}{kT} + \frac{h'}{kT}\right)}{\cosh\left(\frac{J}{kT} - \frac{h'}{kT}\right)} .$$

Determine T_c and h' by expanding the equation in terms of $\frac{h'}{kT}$.
Result:

$$\tanh \frac{J}{kT_c} = 1/(z-1)$$

$$\left(\frac{h'}{kT}\right)^2 = 3 \frac{\cosh^3(J/kT)}{\sinh(J/kT)} \left\{\tanh \frac{J}{kT} - \frac{1}{z-1} + \dots\right\} .$$

6.13 In the so called Weiss model, each of the N spins interacts equally strongly with every other spin:

$$\mathcal{H} = -\frac{1}{2} \sum_{l,l'} J \sigma_l \sigma_{l'} - h \sum_l \sigma_l .$$

Here, $J = \frac{\hat{J}}{N}$. This model can be solved exactly; show that it yields the result of molecular field theory.

6.14 Magnons (= spin waves) in ferromagnets. The *Heisenberg Hamiltonian*, which gives a satisfactory description of certain ferromagnets, is given by

$$\mathcal{H} = -\frac{1}{2} \sum_{l,l'} J(|\mathbf{x}_l - \mathbf{x}_{l'}|) \mathbf{S}_l \mathbf{S}_{l'} ,$$

where l and l' are nearest neighbors on a cubic lattice. By applying the *Holstein–Primakoff transformation*,

$$S_l^+ = \sqrt{2S}\, \varphi(n_l)\, a_l, \quad S_l^- = \sqrt{2S}\, a_l^+ \varphi(n_l), \quad S_l^z = S - n_l$$

$(S_l^\pm = S_l^x \pm i S_l^y)$ with $\varphi(n_l) = \sqrt{1 - n_l/2S}$, $n_l = a_l^\dagger a_l$ and $[a_l, a_{l'}^\dagger] = \delta_{ll'}$, as well as $[a_l, a_{l'}] = 0$ – the spin operators are transformed into Bose operators.
(a) Show that the commutation relations for the spin operators are fulfilled.
(b) Represent the Heisenberg Hamiltonian up to second order (harmonic approximation) in the Bose operators a_l by expanding the square roots in the above transformation in a Taylor series.
(c) Diagonalize $\mathcal{H}$ (by means of a Fourier transformation) and determine the magnon dispersion relations.

6.15 (a) Show that a magnon lowers the z-component of the total spin operator $S^z \equiv \sum_l S_l^z$ by $\hbar$.
(b) Calculate the temperature dependence of the magnetization.
(c) Show that in a one- and a two-dimensional spin lattice, there can be no ferromagnetic order at finite temperatures!

6.16 Assume a Heisenberg model in an external field **H**,

$$\mathcal{H} = -\frac{1}{2} \sum_{l,l'} J\left(l - l'\right) \mathbf{S}_l \mathbf{S}_{l'} - \boldsymbol{\mu} \cdot \mathbf{H} \,,$$

$$\boldsymbol{\mu} = -\frac{g\mu_B}{\hbar} \sum_l \mathbf{S}_l \,.$$

Show that the isothermal susceptibilities $\chi_{||}$ (parallel to **H**) and $\chi_\perp$ (perpendicular to **H**) are not negative.
Suggestions: Include an additional field $\Delta\mathbf{H}$ in the Hamiltonian and take the derivative with respect to this field. For $\chi_{||}$, i.e. $\Delta\mathbf{H} \,||\, \mathbf{H}$, the assertion follows as in Sect. 3.3 for the compressibility. For an arbitrarily oriented $\Delta\mathbf{H}$, it is expedient to use the expansion given in Appendix C.

6.17 Denote the specific heat at constant magnetization by c_M, and at constant field by c_H. Show that relation (6.1.22c) holds for the isothermal and the adiabatic susceptibility. Volume changes of the magnetic material are to be neglected here.

6.18 A paramagnetic material obeys the Curie law

$$M = c\frac{H}{T} \,,$$

where c is a constant. Show, keeping in mind $T\,dS = dE - H\,dM$, that

$$dT_{ad} = \frac{H}{c_H T} c\,dH$$

for an adiabatic change (keeping the volume constant). c_H is the specific heat at constant magnetic field.

6.19 A paramagnetic substance obeys the Curie law $M = \frac{c}{T}H$ (c const.) and its internal energy E is given by $E = aT^4$ ($a > 0$, const.).
(a) What quantity of heat δQ is released on isothermal magnetization if the magnetic field is increased from 0 to H_1?
(b) How does the temperature change if the field is now reduced adiabatically from H_1 to 0?

6.20 Prove the relationships between the shape-dependent and the shape-independent specific heat (6.6.11a), (6.6.11b) and (6.6.11c).

6.21 Polymers in a restricted geometry: Consider a polymer which is in a cone-shaped box (as shown). Why does the polymer move towards the larger opening? (no calculation necessary!)

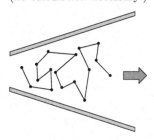

7. Phase Transitions, Scale Invariance, Renormalization Group Theory, and Percolation

This chapter builds upon the results of the two preceding chapters dealing with the ferromagnetic phase transition and the gas-liquid transition. We start with some general considerations on symmetry breaking and phase transitions. Then a variety of phase transitions and critical points are discussed, and analogous behavior is pointed out. Subsequently, we deal in detail with critical behavior and give its phenomenological description in terms of static scaling theory. In the section that follows, we discuss the essential ideas of renormalization group theory on the basis of a simple model, and use it to derive the scaling laws. Finally, we introduce the Ginzburg–Landau theory; it provides an important cornerstone for the various approximation methods in the theory of critical phenomena.

The first, introductory section of this chapter exhibits the richness and variety of phase-transition phenomena and tries to convey the fascination of this field to the reader. It represents a departure from the main thrust of this book, since it offers only phenomenological descriptions without statistical, theoretical treatment. All of these manifold phenomena connected with phase transitions can be described by a single unified theory, the renormalization group theory, whose theoretical efficacy is so great that it is also fundamental to the quantum field theory of elementary particles.

7.1 Phase Transitions and Critical Phenomena

7.1.1 Symmetry Breaking, the Ehrenfest Classification

The fundamental laws of Nature governing the properties of matter (Maxwell's electrodynamics, the Schrödinger equation of a many-body system) exhibit a number of distinct symmetry properties. They are invariant with respect to spatial and temporal translations, with respect to rotations and inversions. The states which exist in Nature do not, in general, display the full symmetry of the underlying natural principles. A solid is invariant only with respect to the discrete translations and rotations of its point group.

Matter can furthermore exist in different states of aggregation or phases, which differ in their symmetry and as a result in their thermal, mechanical, and electromagnetic properties. The external conditions (pressure P, temperature T, magnetic field $\mathbf{H}$, electric field $\mathbf{E}$, ...) determine in which of the possible phases a chemical substance with particular internal interactions will present itself. If the external forces or the temperature are changed, at particular values of these quantities the system can undergo a transition from one phase to another: a phase transition takes place.

The Ehrenfest Classification: as is clear from the examples of phase transitions already treated, the free energy (or some other suitable thermodynamic potential) is a non-analytic function of a control parameter at the phase transition. The following classification of phase transitions, due to *Ehrenfest*, is commonly used: a *phase transition of n-th order* is defined by the property that at least one of the n-th derivatives of its thermodynamic potential is discontinuous, while all the lower derivatives are continuous at the transition. When one of the first derivatives shows a discontinuity, we speak of a first-order phase transition; when the first derivatives vary continuously but the second derivatives exhibit discontinuities or singularities, we speak of a second-order phase transition (or critical point), or of a continuous phase transition.

The understanding of the question as to which phases will be adopted by a particular material under particular conditions certainly belongs among the most interesting topics of the physics of condensed matter. Due to the differing properties of different phases, this question is also of importance for materials applications. Furthermore, the behavior of matter in the vicinity of phase transitions is also of fundamental interest. Here, we wish to indicate two aspects in particular: why is it that despite the short range of the interactions, one observes long-range correlations of the fluctuations, in the vicinity of a critical point T_c and long-range order below T_c? And secondly, what is the influence of the internal symmetry of the order parameter? Fundamental questions of this type are of importance far beyond the field of condensed-matter physics. Renormalization group theory was originally developed in the framework of quantum field theory. In connection with critical phenomena, it was formulated by Wilson[1] in such a way that the underlying structure of nonlinear field theories became apparent, and that also allowed systematic and detailed calculations. This decisive breakthrough led not only to an enormous increase in the knowledge and deeper understanding of condensed matter, but also had important repercussions for the quantum-field theoretical applications of renormalization group theory in elementary particle physics.

*7.1.2 Examples of Phase Transitions and Analogies

We begin by describing the essential features of phase transitions, referring to Chaps. 5 and 6, where the analogy and the common features between

---

[1] K. G. Wilson, Phys. Rev. B **4**, 3174, 3184 (1971)

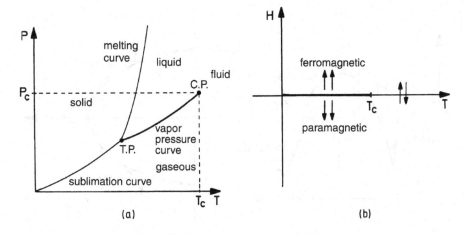

Fig. 7.1a,b. Phase diagrams of **(a)** a liquid $(P\text{-}T)$ and **(b)** a ferromagnet $(H\text{-}T)$. (Triple point = T.P., critical point = C.P.)

the liquid-gas transition and the ferromagnetic phase transition were already mentioned, and here we take up their analysis. In Fig. 7.1a,b, the phase diagrams of a liquid and a ferromagnet are shown. The two ferromagnetic ordering possibilities for an Ising ferromagnet (spin "up" and spin "down") correspond to the liquid and the gaseous phases. The critical point corresponds to the Curie temperature. As a result of the symmetry of the Hamiltonian for $H = 0$ with respect to the operation $\sigma_l \rightarrow -\sigma_l$ for all l, the phase boundary is situated symmetrically in the $H\text{-}T$ plane.

Ferromagnetic order is characterized by the order parameter m at $H = 0$. It is zero above T_c and $\pm m_0$ below T_c, as shown in the $M\text{-}T$ diagram in Fig. 7.1d. The corresponding quantity for the liquid can be seen in the $V\text{-}T$ diagram of Fig. 7.1c.

Here, the order parameter is $(\rho_L - \rho_c)$ or $(\rho_G - \rho_c)$. In everyday life, we usually observe the liquid-gas transition at constant pressure far below P_c. On heating, the density changes discontinuously as a function of the temperature. Therefore, the vaporization transition is usually considered to be a first-order phase transition and the critical point is the end point of the vaporization curve, at which the difference between the gas and the liquid ceases to exist. The analogy between the gas-liquid and the ferromagnetic transitions becomes clearer if one investigates the liquid in a so called Natterer tube[2]. This is a sealed tube in which the substance thus has a fixed, given density. If one chooses the amount of material so that the density is equal to the critical density ρ_c, then above T_c there is a fluid phase, while on cooling, this phase splits up into a denser liquid phase separated from the less dense gas phase by a meniscus. This corresponds to cooling a ferromagnet

---

[2] See the reference in Sect. 3.8.

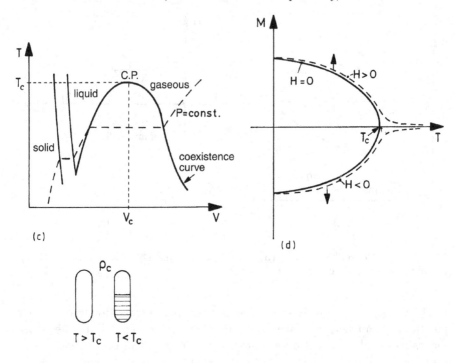

Fig. 7.1c,d. The order parameter for **(c)** the gas-liquid transition (below, two Natterer tubes are illustrated), and for **(d)** the ferromagnetic transition

at $H = 0$. Above T_c, the disordered paramagnetic state is present, while below it, the sample splits up into (at least two) negatively and positively oriented ferromagnetic phases.[3] Fig. 7.1e,f shows the isotherms in the P-V and M-H diagrams. The similarity of the isotherms becomes clear if the second picture is rotated by 90°. In ferromagnets, the particular symmetry again expresses itself. Since the phase boundary curve in the P-T diagram of the liquid is slanted, the horizontal sections of the isotherms in the P-V diagram are not congruent. Finally, Fig. 7.1g,h illustrates the surface of the equation of state.

The behavior in the immediate vicinity of a critical point is characterized by power laws with critical exponents which are summarized for ferromagnets and liquids in Table 7.1. As in Chaps. 5 and 6, $\tau = \frac{T - T_c}{T_c}$. The critical exponents $\beta, \gamma, \delta, \alpha$ for the order parameter, the susceptibility, the critical isotherm, and the specific heat are the goal of theory and experiment. Additional analogies will be seen later in connection with the correlation functions and the scattering phenomena which follow from them.

---

[3] In Ising systems there are two magnetization directions; in Heisenberg systems without an applied field, the magnetization can be oriented in any arbitrary direction, since the Hamiltonian (6.5.2) is rotationally invariant.

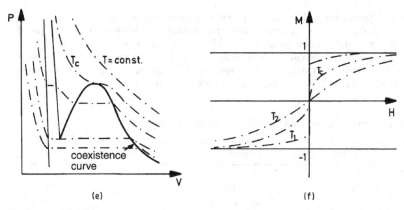

Fig. 7.1e,f. The isotherms **(e)** in the P-V and **(f)** in the M-H diagram

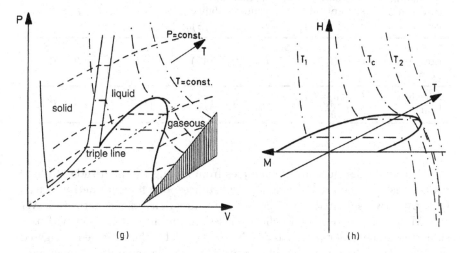

Fig. 7.1g,h. The surface of the equation of state for a liquid **(g)** and for a ferromagnet **(h)**

The general *definition* of the value of a critical exponent of a function $f(T - T_c)$, which is not *a priori* a pure power law is given by

$$\text{exponent} = \lim_{T \to T_c} \frac{d \log f(T - T_c)}{d \log(T - T_c)} \,. \tag{7.1.1}$$

When f has the form $f = a + (T - T_c)$, one finds:

$$\frac{d \log(a + T - T_c)}{d \log(T - T_c)} = \frac{1}{a + (T - T_c)} \cdot \frac{1}{\frac{d \log(T - T_c)}{d(T - T_c)}} = \frac{T - T_c}{a + (T - T_c)} \longrightarrow 0 \,.$$

When f is logarithmically divergent, the following expression holds:

$$\frac{d \log \log(T - T_c)}{d \log(T - T_c)} = \frac{1}{\log(T - T_c)} \longrightarrow 0 \,.$$

In these two cases, the value of the critical exponent is zero. The first case occurs for the specific heat in the molecular field approximation, the second for the specific heat of the two-dimensional Ising model. The reason for introducing critical exponents even for such cases can be seen from the scaling laws which will be treated in the next section. To distinguish between the different meanings of the exponent zero (discontinuity and logarithm), one can write 0_d and $0_{\log}$.

Table 7.1. Ferromagnets and Liquids: Critical Exponents

	Ferromagnet	Liquid	Critical behavior	
Order parameter	M	$(V_{G,L} - V_c)$ or $(\rho_{G,L} - \rho_c)$	$(-\tau)^{\beta}$	$T < T_c$
Isothermal susceptibility	Magnetic susceptibility $\chi_T = \left(\frac{\partial M}{\partial H}\right)_T$	Isothermal compressibility $\kappa_T = -\frac{1}{V}\left(\frac{\partial V}{\partial P}\right)_T$	$\propto |\tau|^{-\gamma}$	$T \gtrless T_c$
Critical isotherm $(T = T_c)$	$H = H(M)$	$P = P(V - V_c)$	$\sim M^{\delta}$ $\sim (V - V_c)^{\delta}$	$T = T_c$
Specific heat	$C_{M=0} = C_{H=0}$ $= T\left(\frac{\partial S}{\partial T}\right)_H$	$C_V = T\left(\frac{\partial S}{\partial T}\right)_V$	$\propto |\tau|^{-\alpha}$	$T \gtrless T_c$

We want to list just a few examples from among the multitude of phase transitions[4]. In the area of magnetic substances, one finds antiferromagnets (e.g. with two sublattices having opposite directions of magnetization $\mathbf{M}_1$ and $\mathbf{M}_2$), ferrimagnets, and helical phases. In an antiferromagnet with two sublattices, the order parameter is $\mathbf{N} = \mathbf{M}_1 - \mathbf{M}_2$, the so called staggered magnetization. In binary liquid mixtures, there are separation transitions, where the order parameter characterizes the concentration. In the case of structural phase transitions, the lattice structure changes at the transition, and the order parameter is given by the displacement field or the strain tensor. Examples are ferroelectrics[5] and distortive transitions, where the order parameter is given by e.g. the electric polarization $\mathbf{P}$ or the rotation angle φ of a molecular group. Finally, there are transitions into macroscopic quantum states, i.e. superfluidity and superconductivity. Here, the order parameter is a complex field ψ, the macroscopic wavefunction, and the broken symmetry is the gauge invariance with respect to the phase of ψ. In the liquid-solid transition, the translational symmetry is broken and the order parameter is

---

[4] We mention two review articles in which the literature up to 1966 is summarized: M. E. Fisher, *The Theory of Equilibrium Critical Phenomena*, p. 615; and P. Heller, *Experimental Investigations of Critical Phenomena*, p. 731, both in Reports on Progress in Physics XXX (1967).

[5] In a number of structural phase transitions, the order parameter jumps discontinuously to a finite value at the transition temperature. In this case, according to Ehrenfest's classification, we are dealing with a first-order phase transition.

a component of the Fourier-transformed density. This transition line does not end in a critical point.

Table 7.2 lists the order parameter and an example of a typical substance for some of these phase transitions.

Table 7.2. Phase transitions (critical points), order parameters, and substances

Phase transition	Order parameter		Substance
Paramagnet–ferromagnet (Curie temperature)	Magnetization	$\mathbf{M}$	Fe
Paramagnet–antiferromagnet (Néel temperature)	staggered magnetization	$\mathbf{N} = \mathbf{M}_1 - \mathbf{M}_2$	$RbMnF_3$
Gas-liquid (Critical point)	Density	$\rho - \rho_c$	CO_2
Separation of binary liquid mixtures	Concentration	$c - c_c$	Methanol-n-Hexane
Order–disorder transitions	Sublattice occupation	$N_A - N_B$	Cu-Zn
Paraelectric–ferroelectric	Polarization	$\mathbf{P}$	$BaTiO_3$
Distortive structural transitions	Rotation angle	φ	$SrTiO_3$
Elastic phase transitions	Strain	ϵ	KCN
He I–He II (Lambda point)	Bose condensate	Ψ	^{4}He
Normal conductor–superconductor	Cooper-pair amplitude	Δ	Nb_3Sn

In general, the *order parameter* is understood to be a quantity which is zero above the critical point and finite below it, and which characterizes the structural or other changes which occur in the transition, such as the expectation value of lattice displacements or a component of the total magnetic moment.

To clarify some concepts, we discuss at this point a generalized *anisotropic, ferromagnetic Heisenberg model*:

$$\mathcal{H} = -\frac{1}{2}\sum_{l,l'}\left\{J_{\|}(l-l')\sigma_l^z\sigma_{l'}^z + J_{\perp}(l-l')(\sigma_l^x\sigma_{l'}^x + \sigma_l^y\sigma_{l'}^y)\right\} - h\sum_l \sigma_l^z , \quad (7.1.2)$$

where $\boldsymbol{\sigma}_l = (\sigma_l^x, \sigma_l^y, \sigma_l^z)$ is the three-dimensional Pauli spin operator at lattice site $\mathbf{x}_l$ and N is the number of lattice sites. This Hamiltonian contains the *uniaxial ferromagnet* for $J_{\|}(l-l') > J_{\perp}(l-l') \geq 0$, and for $J_{\perp}(l-l') = J_{\|}(l-l')$, it describes the *isotropic Heisenberg model* (6.5.2). In the former case, the order parameter referred to the number of lattice sites ($h = 0$) is the single-component quantity $\langle \frac{1}{N} \sum_l \sigma_l^z \rangle$, i.e. the number of components n is $n = 1$. In the latter case, the order parameter is $\langle \frac{1}{N} \sum_l \boldsymbol{\sigma}_l \rangle$, which can point in any arbitrary direction ($h = 0$!); here, the number of components is $n = 3$. For $J_\perp(l - l') > J_\parallel(l - l') \geq 0$, we find the so called *planar ferromagnet*, in which the order parameter $\frac{1}{N} \sum_l \langle (\sigma_l^x, \sigma_l^y, 0) \rangle$ has two components, $n = 2$. A special case of the uniaxial ferromagnet is the Ising model (6.5.4), with $J_\perp(l - l') = 0$. The *uniaxial ferromagnet* has the following symmetry elements: all rotations around the z-axis, the discrete symmetry $(\sigma_l^x, \sigma_l^y, \sigma_l^z) \rightarrow (\sigma_l^x, \sigma_l^y, -\sigma_l^z)$ and products thereof. Below T_c, the invariance with respect to this discrete symmetry is broken. In the planar ferromagnet, the (continuous) rotational symmetry around the z-axis, and in the case of the isotropic Heisenberg model, the $O(3)$ symmetry – i.e. the rotational invariance around an arbitrary axis – is broken.

One could ask why e.g. for the the Ising Hamiltonian without an external field, $\frac{1}{N}\langle \sum \sigma_l \rangle$ can ever be nonzero, since from the invariance operation $\{\sigma_l^z\} \rightarrow \{-\sigma_l^z\}$, it follows that $\frac{1}{N}\langle \sum_l \sigma_l^z \rangle = -\frac{1}{N}\langle \sum_l \sigma_l^z \rangle$. In a finite system, $\frac{1}{N}\langle \sum \sigma_l^z \rangle_h$ is analytic in h for finite h, and

$$\lim_{h \to 0} \frac{1}{N} \left\langle \sum_l \sigma_l^z \right\rangle_h = 0 . \tag{7.1.3}$$

For finite N, configurations with spins oriented opposite to the field also contribute to the partition function, and their weight increases with decreasing values of h.

The mathematically precise definition of the order parameter is:

$$\langle \sigma \rangle = \lim_{h \to 0} \lim_{N \to \infty} \frac{1}{N} \left\langle \sum_l \sigma_l^z \right\rangle_h ; \tag{7.1.4}$$

first, the thermodynamic limit $N \to \infty$ is taken, and then $h \to 0$. This quantity can be nonzero below T_c. For $N \to \infty$, states with the 'wrong' orientation have vanishing weights in the partition function for arbitrarily small but finite fields.

7.1.3 Universality

In the vicinity of critical points, the topology of the phase diagrams of such diverse systems as a gas-liquid mixture and a ferromagnetic material are astonishingly similar; see Fig. 7.1. Furthermore, experiments and computer simulations show that the critical exponents for the corresponding phase transitions for broad classes of physical systems are the same and depend only on the number of components and the symmetry of the order parameter, the spatial dimension and the character of the interactions, i.e. whether short-ranged, or long-ranged (e.g. Coulomb, dipolar forces). This remarkable feature is termed *universality*. The microscopic details of these strongly interacting many-body systems express themselves only in the prefactors (amplitudes) of the power laws, and even the ratios of these amplitudes are universal numbers.

The reason for this remarkable result lies in the divergence of the correlation length $\xi = \xi_0 \left(\frac{T-T_c}{T_c}\right)^{-\nu}$. On approaching T_c, ξ becomes the only relevant length scale of the system, which at long distances dominates all of the microscopic scales. Although the phase transition is caused as a rule by short-range interactions of the microscopic constituents, due to the long-range fluctuations (see 6.12), the dependence on the microscopic details such as the lattice structure, the lattice constant, or the range of the interactions (as long as they are short-ranged) is secondary. In the critical region, the system behaves collectively, and only global features such as its spatial dimension and its symmetry play a role; this makes the universal behavior understandable.

The universality of critical phenomena is not limited to materials classes, but instead it extends beyond them. For example, the static critical behavior of the gas-liquid transition is the same as that of Ising ferromagnets. Planar ferromagnets behave just like ^{4}He at the lambda point. Even without making use of renormalization group theory, these relationships can be understood with the aid of the following transformations[6]: the grand partition function of a gas can be approximately mapped onto that of a lattice gas which is equivalent to a magnetic Ising model (occupied/unoccupied cells $\widehat{=}$ spin up/down). The Hamiltonian of a Bose liquid can be mapped onto that of a planar ferromagnet. The gauge invariance of the Bose Hamiltonian corresponds to the two-dimensional rotational invariance of the planar ferromagnet.

7.2 The Static Scaling Hypothesis[7]

7.2.1 Thermodynamic Quantities and Critical Exponents

In this section, we discuss the analytic structure of the thermodynamic quantities in the vicinity of the critical point and draw from it typical conclusions about the critical exponents. This generally-applicable procedure will be demonstrated using the terminology of ferromagnetism. In the neighborhood of T_c, the equation of state according to Eq. (6.5.16) takes on the form

---

[6] See e.g. M. E. Fisher, *op. cit.*, and problem 7.16.

[7] Although the so called scaling theory of critical phenomena can be derived microscopically through renormalization group theory (see Sect. 7.3.4), it is expedient for the following reasons to first introduce it on a phenomenological basis: (i) as a motivation for the procedures of renormalization group theory; (ii) as an illustration of the structure of scaling considerations for physical situations where field-theoretical treatments based on renormalization group theory are not yet available (e.g. for many nonequilibrium phenomena). Scaling treatments, starting from critical phenomena and high-energy scaling in elementary particle physics, have acquired a great influence in the most diverse fields.

$$\frac{h}{kT_c} = \tau m + \frac{1}{3}m^3 \tag{7.2.1}$$

which can be rearranged as follows: $\frac{1}{kT_c}\frac{h}{|\tau|^{3/2}} = \mathrm{sgn}(\tau)\frac{m}{|\tau|^{1/2}} + \frac{1}{3}\left(\frac{m}{|\tau|^{1/2}}\right)^3$.
Solving for m, we obtain the following dependence of m on τ and h:

$$m(\tau, h) = |\tau|^{1/2} m_{\pm}\left(\frac{h}{|\tau|^{3/2}}\right) \quad \text{for} \quad T \gtrless T_c. \tag{7.2.2}$$

The functions $m_{\pm}$ for $T \gtrless T_c$ are determined by (7.2.1). In the vicinity of the critical point, the magnetization depends on τ and h in a very special way: apart from the factor $|\tau|^{1/2}$, it depends only on the ratio $h/|\tau|^{3/2}$. The magnetization is a generalized homogeneous function of τ and h. This implies that (7.2.2) is invariant with respect to the scale transformation

$$h \to hb^3, \quad \tau \to \tau b^2, \quad \text{and} \quad m \to mb.$$

This scaling invariance of the physical properties expresses itself for example in the specific heat of $^4\mathrm{He}$ at the lambda point (Fig. 7.2).

We know from Chap. 6 and Table 7.1 that the real critical exponents differ from their molecular-field values in (7.2.2). It is therefore reasonable to extend the equation of state (7.2.2) to arbitrary critical exponents[8]:

$$m(\tau, h) = |\tau|^{\beta} m_{\pm}\left(\frac{h}{|\tau|^{\delta\beta}}\right); \tag{7.2.3}$$

in this expression, β and δ are critical exponents and the $m_{\pm}$ are called *scaling functions*.

At the present stage, (7.2.3) remains a hypothesis; it is, however, possible to prove this hypothesis using renormalization group theory, as we shall demonstrate later in Sect. 7.3, for example in Eq. (7.3.40′). For the present, we take (7.2.3) as given and ask what its general consequences are. The two scaling functions $m_{\pm}(y)$ must fulfill certain boundary conditions which follow from the critical properties listed in Eq. (6.5.31) and Table 7.1. The magnetization is always oriented parallel to h when the applied field is nonzero and remains finite in the limit $h \to 0$ below T_c, while above T_c, it goes to zero:

---

[8] In addition to being a natural generalization of molecular field theory, one can understand the scaling hypothesis (7.2.3) by starting with the fact that singularities are present only for $\tau = 0$ and $h = 0$. How strong the effects of the singularities will be depends on the distance from the critical point, τ, and on $h/|\tau|^{\beta\delta}$, i.e. the ratio between the applied field and the field-equivalent of τ; that is $h_\tau = m^\delta = |\tau|^{\beta\delta}$. As long as $h \ll h_\tau$, the system is effectively in the low-field limit and $m \approx |\tau|^\beta m_{\pm}(0)$. On the other hand, if τ becomes so small that $|\tau| \lesssim h^{1/\beta\delta}$, then the influence of the applied field predominates. Any additional reduction of τ produces no further change: m remains at the value which it had for $|\tau| = h^{1/\beta\delta}$, i.e. $h^{\frac{1}{\delta}} m_{\pm}(1)$. In the limit $\tau \to 0$, $m_{\pm}(y) \longrightarrow y^\beta$ must hold, so that the singular dependence on τ in $m(\tau, h)$ cancels out.

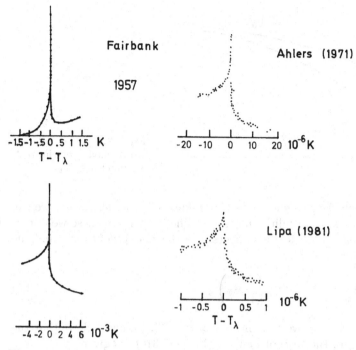

Fig. 7.2. The specific heat at constant pressure, c_P, at the lambda transition of ^{4}He. The shape of the specific heat stays the same on changing the temperature scale ($1\,\mathrm{K}$ to $10^{-6}\,\mathrm{K}$)

$$\lim_{y\to 0} m_-(y) = \mathrm{sgn}\, y\,, \quad m_+(0) = 0\,. \tag{7.2.4a}$$

The thermodynamic functions are non-analytic precisely at $\tau = 0, h = 0$. For nonzero h, the magnetization is finite over the whole range of temperatures and remains an analytic function of τ even for $\tau = 0$; the $|\tau|^{\beta}$ dependence of (7.2.3) must be compensated by the function $m_\pm(h/|\tau|^{\delta\beta})$. Therefore, the two functions $m_\pm$ must behave as

$$\lim_{y\to\infty} m_\pm(y) \propto y^{1/\delta} \tag{7.2.4b}$$

for large arguments. It follows from this that for $\tau = 0$, i.e. at the critical point, $m \sim h^{1/\delta}$. The scaling functions $m_\pm(y)$ are plotted in Fig. 7.3.

Eq. (7.2.3), like the molecular-field version of the scaling law given above, requires that the magnetization must be a generalized homogeneous function of τ and h and is therefore invariant with respect to *scale transformations*:

$$h \to h b^{\frac{\beta\delta}{\nu}}, \tau \to \tau b^{\frac{1}{\nu}}, \quad \text{and} \quad m \to m b^{\beta/\nu}\,.$$

The name *scaling law* is derived from this scale invariance. Equation (7.2.3) contains additional information about the thermodynamics; by integration,

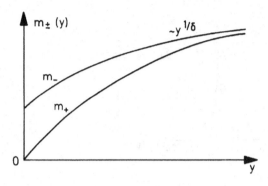

Fig. 7.3. The qualitative behavior of the scaling functions $m_\pm$

we can determine the free energy and by taking suitable derivatives we can find the magnetic susceptibility and the specific heat. From these we obtain relations between the critical exponents. For the *susceptibility*, we find the scaling law from Eq. (7.2.3):

$$\chi \equiv \left(\frac{\partial m}{\partial h} \right)_T = |\tau|^{\beta - \delta\beta} m'_\pm \left(\frac{h}{|\tau|^{\delta\beta}} \right) , \tag{7.2.5}$$

and in the limit $h \to 0$, we thus have $\chi \propto |\tau|^{\beta - \delta\beta}$. It then follows that the critical exponent of the susceptibility, γ (Eq. (6.5.31c)), is given by

$$\gamma = -\beta(1 - \delta) . \tag{7.2.6}$$

The *specific free energy* is found through integration of (7.2.3):

$$f - f_0 = - \int_{h_0}^{h} dh \, m(\tau, h) = -|\tau|^{\beta + \delta\beta} \int_{h_0/|\tau|^{\delta\beta}}^{h/|\tau|^{\delta\beta}} dx \, m_\pm(x) .$$

Here, h_0 must be sufficiently large so that the starting point for the integration lies outside the critical region. The free energy then takes on the following form:

$$f(\tau, h) = |\tau|^{\beta + \delta\beta} \hat{f}_\pm \left(\frac{h}{|\tau|^{\beta\delta}} \right) + f_{reg} . \tag{7.2.7}$$

In this expression, $\hat{f}$ is defined by the value resulting from the upper limit of the integral and f_{reg} is the non-singular part of the free energy. The specific heat at constant magnetic field is obtained by taking the second derivative of (7.2.7),

$$c_h = -\frac{\partial^2 f}{\partial \tau^2} \sim A_\pm |\tau|^{\beta(1+\delta) - 2} + B_\pm . \tag{7.2.8}$$

The $A_\pm$ in this expression are amplitudes and the $B_\pm$ come from the regular part. Comparison with the behavior of the specific heat as characterized by the critical exponent α (Eq. (6.5.31d)) yields

$$\alpha = 2 - \beta(1 + \delta) \ . \tag{7.2.9}$$

The relations between the critical exponents are termed *scaling relations*, since they follow from the scaling laws for the thermodynamic quantities. If we add (7.2.6) and (7.2.9), we obtain

$$\gamma + 2\beta = 2 - \alpha \ . \tag{7.2.10}$$

From (7.2.6) and (7.2.9), one can see that the remaining thermodynamic critical exponents are determined by β and δ.

7.2.2 The Scaling Hypothesis for the Correlation Function

In the molecular field approximation, we obtained the Ornstein–Zernike behavior in Eqns. (6.5.50) and (6.5.53′) for the wavevector-dependent susceptibility $\chi(\mathbf{q})$ and the correlation function $G(\mathbf{x})$:

$$\chi(\mathbf{q}) = \frac{1}{\tilde{J}q^2} \frac{(q\xi)^2}{1 + (q\xi)^2} \ , \quad G(\mathbf{x}) = \frac{kT_c\,v\,e^{-|\mathbf{x}|/\xi}}{4\pi\tilde{J}\,|\mathbf{x}|} \quad \text{with} \quad \xi = \xi_0 \tau^{-\frac{1}{2}} \ . \tag{7.2.11}$$

The generalization of this law is ($q \ll a^{-1}, |\mathbf{x}| \gg a, \xi \gg a$ with the lattice constant a):

$$\chi(\mathbf{q}) = \frac{1}{q^{2-\eta}} \, \hat{\chi}(q\xi) \ , \quad G(\mathbf{x}) = \frac{1}{|\mathbf{x}|^{1+\eta}} \, \hat{G}\left(|\mathbf{x}|/\xi\right) \ , \quad \xi = \xi_0\,\tau^{-\nu} \ ,$$

$$\tag{7.2.12a,b,c}$$

where the functions $\hat{\chi}(q\xi)$ and $\hat{G}(|\mathbf{x}|/\xi)$ are still to be determined. In (7.2.12c), we assumed that the correlation length ξ diverges at the critical point. This divergence is characterized by the critical exponent ν. Just at T_c, $\xi = \infty$ and therefore there is no longer any finite characteristic length; the correlation function $G(\mathbf{x})$ can thus only fall off according to a power law $G(\mathbf{x}) \sim \frac{1}{|\mathbf{x}|^{1+\eta}} \hat{G}(0)$. The possibility of deviations from the $1/|\mathbf{x}|$-behavior of the Ornstein–Zernike theory was taken into account by introducing the additional critical exponent η. In the immediate vicinity of T_c, ξ is the only relevant length and therefore the correlation function also contains the factor $\hat{G}(|\mathbf{x}|/\xi)$. Fourier transformation of $G(\mathbf{x})$ yields (7.2.12a) for the wavevector-dependent susceptibility, which for its part represents an evident generalization of the Ornstein–Zernike expression. We recall (from Sects. 5.4.4 and 6.5.5.2) that the increase of $\chi(\mathbf{q})$ for small q on approaching T_c leads to critical opalescence.

In (7.2.11) and (7.2.12b), a three-dimensional system was assumed. Phase transitions are of course highly interesting also in two dimensions, and furthermore it has proved fruitful in the theory of phase transitions to consider arbitrary dimensions (even non-integral dimensions). We therefore generalize the relations to arbitrary dimensions d:

$$G(\mathbf{x}) = \frac{1}{|\mathbf{x}|^{d-2+\eta}} \hat{G}(|\mathbf{x}|/\xi) \; . \tag{7.2.12b'}$$

Equations (7.2.12a) and (7.2.12c) remain valid also in d dimensions, whereby of course the exponents ν and η and the form of the functions $\hat{G}$ and $\hat{\chi}$ depend on the spatial dimension. From (7.2.12a) and (7.2.12b') at the critical point we obtain

$$G(\mathbf{x}) \propto \frac{1}{|\mathbf{x}|^{d-2+\eta}} \quad \text{and} \quad \chi \propto \frac{1}{q^{2-\eta}} \quad \text{for} \quad T = T_c \; . \tag{7.2.13}$$

Here, we have assumed that $\hat{G}(0)$ and $\hat{\chi}(\infty)$ are finite, which follows from the finite values of $G(\mathbf{x})$ at finite distances and of $\chi(\mathbf{q})$ at finite wavenumbers (and $\xi = \infty$).

We now consider the limiting case $\mathbf{q} \to 0$ for temperatures $T \neq T_c$. Then we find from (7.2.12a)

$$\chi = \lim_{q \to 0} \chi(\mathbf{q}) \propto \frac{(q\xi)^{2-\eta}}{q^{2-\eta}} = \xi^{2-\eta} \; . \tag{7.2.14}$$

This dependence is obtained on the basis of the following arguments: for finite ξ, the susceptibility remains finite even in the limit $\mathbf{q} \to 0$. Therefore, the factor $\frac{1}{q^{2-\eta}}$ in (7.2.12a) must be compensated by a corresponding dependence of $\hat{\chi}(q\xi)$, from which the relation (7.2.14) follows for the homogeneous susceptibility. Since its divergence is characterized by the critical exponent γ according to (6.5.31c), it follows from (7.2.14) together with (7.2.12c) that there is an additional scaling relation

$$\gamma = \nu(2 - \eta) \; . \tag{7.2.15}$$

Relations of the type (7.2.3), (7.2.7), and (7.2.12b') are called *scaling laws*, since they are invariant under the following scale transformations:

$$x \to x/b, \quad \xi \to \xi/b, \quad \tau \to \tau b^{1/\nu}, \quad h \to h b^{\beta\delta/\nu}$$
$$m \to m b^{\beta/\nu}, \quad f_s \to f_s b^{(2-\alpha)/\nu}, \quad G \to G b^{(d-2+\eta)/\nu} \; , \tag{7.2.16}$$

where f_s stands for the singular part of the (specific) free energy.

If we in addition assume that these scale transformations are based on a microscopic elimination procedure by means of which the original system with lattice constant a and N lattice sites is mapped onto a new system with the same lattice constant a but a reduced number Nb^{-d} of degrees of freedom, then we find

$$\frac{F_s(\tau, h)}{N} = b^{-d} \frac{F_s(\tau b^{1/\nu}, h b^{\beta\delta/\nu})}{Nb^{-d}} \; , \tag{7.2.17}$$

which implies the *hyperscaling relation*

$$2 - \alpha = d\nu \, , \tag{7.2.18}$$

which also contains the spatial dimension d. According to equations (7.2.6), (7.2.9), (7.2.15), and (7.2.18), all the critical exponents are determined by two independent ones.

For the *two-dimensional Ising model* one finds the exponents of the correlation function, $\nu = 1$ and $\eta = 1/4$, from the exponents quoted following Eq. (6.5.31d) and the scaling relations (7.2.15) and (7.2.18).

7.3 The Renormalization Group

7.3.1 Introductory Remarks

The term 'renormalization' of a theory refers to a certain reparametrization with the goal of making the renormalized theory more easily dealt with than the original version. Historically, renormalization was developed by Stückelberg and Feynman in order to remove the divergences from quantum-field theories such as quantum electrodynamics. Instead of the bare parameters (masses, coupling constants), the Lagrange function is expressed in terms of physical masses and coupling coefficients, so that ultraviolet divergences due to virtual transitions occur only within the connection between the bare and the physical quantities, leaving the renormalized theory finite. The renormalization procedure is not unique; the renormalized quantities can for example depend upon a cutoff length scale, up to which certain virtual processes are taken into account. *Renormalization group theory* studies the dependence on this length scale, which is also called the "flow parameter". The name "renormalization group" comes from the fact that two consecutive renormalization group transformations lead to a third such transformation.

In the field of critical phenomena, where one must explain the observed behavior at large distances (or in Fourier space at small wavenumbers), it is reasonable to carry out the renormalization procedure by a suitable elimination of the short-wavelength fluctuations. A partial evaluation of the partition function in this manner is easier to carry out than the calculation of the complete partition function, and can be done using approximation methods. As a result of the elimination step, the remaining degrees of freedom are subject to modified, *effective* interactions.

Quite generally, one can expect the following advantages from such a renormalization group transformation:

(i) The new coupling constants could be smaller. By repeated applications of the renormalization procedure, one could thus finally obtain a practically free theory, without interactions.

(ii) The successively iterated coupling coefficients, also called "parameter flow", could have a *fixed point*, at which the system no longer changes under additional renormalization group transformations. Since the elimination of degrees of freedom is accompanied by a change of the underlying lattice spacing, or length scale, one can anticipate that the fixed points are under certain circumstances related to critical points. Furthermore, it can be hoped that the flow in the vicinity of these fixed points can yield information about the universal physical quantities in the neighborhood of the critical points.

The scenario described under (i) will in fact be found for the one-dimensional Ising model, and that described under (ii) for the two-dimensional Ising model.

The renormalization group method brings to bear the scale invariance in the neighborhood of a critical point. In the case of so called *real-space transformations* (in contrast to transformation in Fourier space), one eliminates certain degrees of freedom which are defined on a lattice, and thus carries out a partial trace operation on the partition function. The lattice constant of the resulting system is then readjusted and the internal variables are renormalized in such a manner that the new Hamiltonian corresponds to the original one in its form. By comparison, one defines effective, scale-independent coupling constants, whose flow behavior is then investigated. We first study the one-dimensional Ising model and then the two-dimensional. Finally, the general structure of such transformations will be discussed with the derivation of scaling laws. A brief schematic treatment of continuous field-theoretical formulations will be undertaken following the Ginzburg–Landau theory.

7.3.2 The One-Dimensional Ising Model, Decimation Transformation

We will first illustrate the renormalization group method using the one-dimensional Ising model, with the ferromagnetic exchange constant J in zero applied field, as an example. The Hamiltonian is

$$\mathcal{H} = -J \sum_l \sigma_l \sigma_{l+1} , \qquad (7.3.1)$$

where l runs over all the sites in the one-dimensional chain; see Fig. 7.4. We introduce the abbreviation $K = J/kT$ into the partition function for N spins with periodic boundary conditions $\sigma_{N+1} = \sigma_1$,

$$Z_N = \mathrm{Tr}\, e^{-\mathcal{H}/kT} = \sum_{\{\sigma_l = \pm 1\}} e^{K \sum_l \sigma_l \sigma_{l+1}} . \qquad (7.3.2)$$

The decimation procedure consists in partially evaluating the partition function, by carrying out the sum over every second spin in the first step. In Fig. 7.4, the lattice sites for which the trace is taken are marked with a cross.

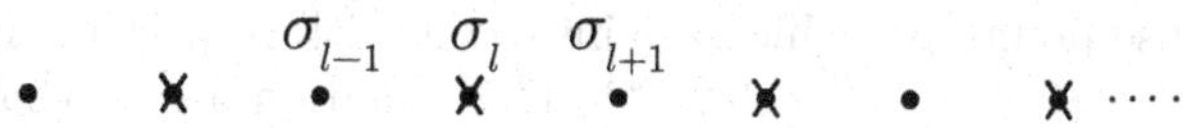

Fig. 7.4. An Ising chain; the trace is carried out over all the lattice points which are marked with a cross. The result is a lattice with its lattice constant doubled

A typical term in the partition function is then

$$\sum_{\sigma_l=\pm 1} e^{K\sigma_l(\sigma_{l-1}+\sigma_{l+1})} = 2\cosh K(\sigma_{l-1}+\sigma_{l+1}) = e^{2g+K'\sigma_{l-1}\sigma_{l+1}} , \quad (7.3.3)$$

with coefficients g and K' which are still to be determined. Here, we have taken the sum over $\sigma_l = \pm 1$ after the first equals sign. Since $\cosh K(\sigma_{l-1} + \sigma_{l+1})$ depends only on whether σ_{l-1} and σ_{l+1} are parallel or antiparallel, the result can in any case be brought into the form given after the second equals sign. The coefficients g and K' can be determined either by expansion of the exponential function or, still more simply, by comparing the two expressions for the possible orientations. If $\sigma_{l-1} = -\sigma_{l+1}$, we find

$$2 = e^{2g-K'} , \quad (7.3.4a)$$

and if $\sigma_{l-1} = \sigma_{l+1}$, the result is

$$2\cosh 2K = e^{2g+K'} . \quad (7.3.4b)$$

From the product of (7.3.4a) and (7.3.4b) we obtain $4\cosh 2K = e^{4g}$, and from the quotient, $\cosh 2K = e^{2K'}$; thus the recursion relations are:

$$K' = \frac{1}{2}\log\cosh 2K \quad (7.3.5a)$$

$$g = \frac{1}{2}(\log 2 + K') . \quad (7.3.5b)$$

Repeating this decimation procedure a total of k times, we obtain from (7.3.5a,b) for the kth step the following recursion relation:

$$K^{(k)} = \frac{1}{2}\log\left(\cosh 2K^{(k-1)}\right) \quad (7.3.6a)$$

$$g(K^{(k)}) = \frac{1}{2}\log 2 + \frac{1}{2}K^{(k)} . \quad (7.3.6b)$$

The decimation produces another Ising model with an interaction between nearest neighbors having a coupling constant $K^{(k)}$. Furthermore, a spin-independent contribution $g(K^{(k)})$ to the energy is generated; in the kth step, it is given by (7.3.6b).

In a transformation of this type, it is expedient to determine the *fixed points* which in the present context will prove to be physically relevant. Fixed points are those points K^* which are invariant with respect to the transformation, i.e. here $K^* = \frac{1}{2} \log(\cosh 2K^*)$. This equation has two solutions,

$$K^* = 0 \quad (T = \infty) \quad \text{and} \quad K^* = \infty \quad (T = 0) . \qquad (7.3.7)$$

The recursion relation (7.3.6a) is plotted in Fig. 7.5. Starting with the initial value K_0, one obtains $K'(K_0)$, and by a reflection in the line $K' = K$, $K'(K'(K_0))$, and so forth. One can see that the coupling constant decreases continually; the system moves towards the fixed point $K^* = 0$, i.e. a non-interacting system. Therefore, for a finite K_0, we never arrive at an ordered state: there is no phase transition. Only for $K = \infty$, i.e. for a finite exchange interaction J and $T = 0$, do the spins order.

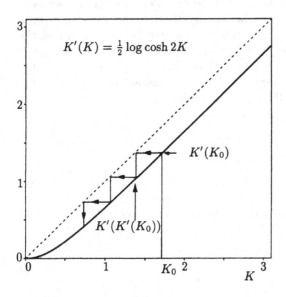

Fig. 7.5. The recursion relation for the one-dimensional Ising model with interactions between nearest neighbors (heavy solid curve), the line $K' = K$ (dashed), and the iteration steps (thin lines with arrows)

Making use of this renormalization group (RG) transformation, we can calculate the partition function and the free energy. The partition function for all together N spins with the coupling constant K, using (7.3.3), is

$$Z_N(K) = e^{Ng(K')} Z_{\frac{N}{2}}(K') = e^{Ng(K') + \frac{N}{2}g(K'')} Z_{\frac{N}{2^2}}(K'') , \qquad (7.3.8)$$

and, after the nth step,

$$Z_N(K) = \exp\left[N \sum_{k=1}^{n} \frac{1}{2^{k-1}} g(K^{(k)}) + \log Z_{\frac{N}{2^n}}(K^{(n)}) \right] . \qquad (7.3.9)$$

The reduced free energy per lattice site and kT is defined by

$$\tilde{f} = -\frac{1}{N} \log Z_N(K) . \qquad (7.3.10)$$

As we have seen, the interactions become weaker as a result of the renor-malization group transformation, which gives rise to the following possible application: after several steps the interactions have become so weak that perturbation-theory methods can be used, or the interaction can be altogether neglected. Setting $K^{(n)} \approx 0$, from (7.3.9) we obtain the approximation:

$$\tilde{f}^{(n)}(K) = -\sum_{k=1}^{n} \frac{1}{2^{k-1}} g(K^{(k)}) - \frac{1}{2^n} \log 2 , \qquad (7.3.11)$$

since the free energy per spin of a field-free spin-1/2 system without inter-actions is $-\log 2$. Fig. 7.6 shows $\tilde{f}^{(n)}(K)$ for $n = 1$ to 5. We can see how quickly this approximate solution approaches the exact reduced free energy $\tilde{f}(K) = -\log(2\cosh K)$. The one-dimensional Ising model can be exactly solved by elementary methods (see problem 6.9), as well as by using the transfer matrix method, cf. Appendix F.

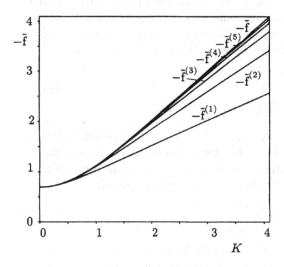

Fig. 7.6. The reduced free energy of the one-dimensional Ising model. $\tilde{f}$ is the exact free energy, $\tilde{f}^{(1)}, \tilde{f}^{(2)}, \dots$ are the approximations (7.3.11)

7.3.3 The Two-Dimensional Ising Model

The application of the decimation procedure to the two-dimensional Ising model is still more interesting, since this model exhibits a phase transition at a finite temperature $T_c > 0$. We consider the square lattice rotated by $45°$ which is illustrated in Fig. 7.7, with a lattice constant of one.

The Hamiltonian multiplied by β, $H = \beta\mathcal{H}$, is

$$H = -\sum_{n.n.} K\sigma_i\sigma_j , \qquad (7.3.12)$$

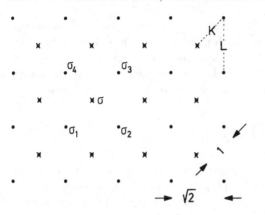

Fig. 7.7. A square spin lattice, rotated by 45°. The lattice sites are indicated by points. In the decimation transformation, the spins at the sites which are also marked by a cross are eliminated. K is the interaction between nearest neighbors and L is the interaction between next-nearest neighbors

where the sum runs over all pairs of nearest neighbors (n.n.) and $K = J/kT$. When in the partial evaluation of the partition function the trace is taken over the spins marked by crosses, we obtain a new square lattice of lattice constant $\sqrt{2}$. How do the coupling constants transform? We pick out one of the spins with a cross, σ, denote its neighbors as $\sigma_1, \sigma_2, \sigma_3$, and σ_4, and evaluate their contribution to the partition function:

$$\sum_{\sigma=\pm 1} e^{K(\sigma_1+\sigma_2+\sigma_3+\sigma_4)\sigma} = e^{\log(2\cosh K(\sigma_1+\sigma_2+\sigma_3+\sigma_4))}$$

$$= e^{A'+\frac{1}{2}K'(\sigma_1\sigma_2...+\sigma_3\sigma_4)+L'(\sigma_1\sigma_3+\sigma_2\sigma_4)+M'\sigma_1\sigma_2\sigma_3\sigma_4} .$$

(7.3.13)

This transformation (taking a partial trace) yields a modified interaction between nearest neighbors, K' (here, the elimination of two crossed spins contributes); in addition, new interactions between the next-nearest neighbors (such as σ_1 and σ_3) and a four-spin interaction are generated:

$$H' = \left(A' + K' \sum_{n.N.} \sigma_i\sigma_j + L' \sum_{"u.n.N."} \sigma_i\sigma_j + ...\right) .$$

(7.3.12')

The coefficients A', K', L' and M' can readily be found from (7.3.13) as functions of K, by using $\sigma_i^2 = 1$, $i = 1,\ldots,4$ (see problem 7.2):

$$A'(K) = \log 2 + \frac{1}{8}\{\log\cosh 4K + 4\log\cosh 2K\} ,$$

(7.3.14)

$$K'(K) = \frac{1}{4}\log\cosh 4K , \quad L'(K) = \frac{1}{2}K'(K)$$

(7.3.13')

$$M'(K) = \frac{1}{8}\{\log\cosh 4K - 4\log\cosh 2K\} .$$

Putting the critical value $K_c = J/kT_c = 0.4406$ (exact result[9]) into this relation as an estimate for the initial value K, we find $M' \ll L' \le K'$. In

---

[9] The partition function of the Ising model on a square lattice without an external field was evaluated exactly by L. Onsager, Phys. Rev. **65**, 117 (1944), using the transfer matrix method (see Appendix F.).

the first elimination step, the original Ising model is transformed into one with three interactions; in the next step we must take these into account and obtain still more interactions, and so on. In a quantitatively usable calculation it will thus be necessary to determine the recursion relations for an extended number of coupling constants. Here, we wish only to determine the essential structure of such recursion relations and to simplify them sufficiently so that an analytic solution can be found. Therefore, we neglect the coupling constant M' and all the others which are generated by the elimination procedure, and restrict ourselves to K' and L' as well as their initial values K and L. This is suggested by the smallness of M' which we mentioned above.

We now require the recursion relation including the coupling constant L, which acts between σ_1 and σ_4, etc. Thus, expanding (7.3.13') up to second order in K and taking note of the fact that an interaction L between next-nearest neighbors in the original Hamiltonian appears as a contribution to the interactions of the nearest neighbors in the primed Hamiltonian, we find the following recursion relations on elimination of the crossed spins (Fig. 7.7):

$$K' = 2K^2 + L \tag{7.3.15a}$$
$$L' = K^2 . \tag{7.3.15b}$$

These relations can be arrived at intuitively as follows: the spin σ mediates an interaction of the order of K times K, i.e. K^2 between σ_1 and σ_3, likewise the crossed spin just to the left of σ. This leads to $2K^2$ in K'. The interaction L between next-nearest neighbors in the original model makes a direct contribution to K'. Spin σ also mediates a diagonal interaction between σ_1 and σ_4, leading thus to the relation $L' = K^2$ in (7.3.15b).

However, it should be clear that in contrast to the one-dimensional case, new coupling constants are generated in every elimination step. One cannot expect that these recursion relations, which have been restricted as an approximation to a reduced parameter space (K, L), will yield quantitatively accurate results. They do contain all the typical features of this type of recursion relations.

In Fig. 7.8, we have shown the recursion relations (7.3.15a,b)[10]. Starting from values $(K, 0)$, the recursion relation is repeatedly applied, likewise for initial values $(0, L)$. The following picture emerges: for small initial values, the flux lines converge to $K = L = 0$, and for large initial values they converge to $K = L = \infty$. These two regions are separated by two lines, which meet at $K_c^* = \frac{1}{3}$ and $L_c^* = \frac{1}{9}$. Further on it will become clear that this fixed point is connected to the critical point.

We now want to investigate analytically the more important properties of the flow diagram which follows from the recursion relations (7.3.15a,b). As a

---

[10] For clarity we have drawn in only every other iteration step in Fig. 7.8. We will return to this point at the end of this section, after investigating the analytic behavior of the recursion relation.

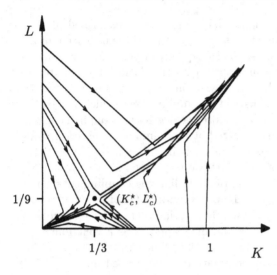

Fig. 7.8. A flow diagram of Eq. (7.3.15a,b) (only every other point is indicated.) Three fixed points can be recognized: $K^* = L^* = 0, K^* = L^* = \infty$ and $K_c^* = \frac{1}{3}, L_c^* = \frac{1}{9}$

first step, the *fixed points* must be determined from (7.3.15a,b), i.e. K^* and L^*, which obey $K^* = 2K^{*2} + L^*$ and $L^* = K^*$. These conditions give three fixed points

$$\text{(i)} \quad K^* = L^* = 0, \quad \text{(ii)} \quad K^* = L^* = \infty, \quad \text{and} \quad \text{(iii)} \quad K_c^* = \frac{1}{3}, \; L_c^* = \frac{1}{9}\,.$$

$$(7.3.16)$$

The high-temperature fixed point (i) corresponds to a temperature $T = \infty$ (disordered phase), while the low-temperature fixed point (ii) corresponds to $T = 0$ (ordered low-temperature phase). The critical behavior can be related only to the non-trivial fixed point (iii), $(K_c^*, L_c^*) = (\frac{1}{3}, \frac{1}{9})$.

That the initial values of K and L which lead to the fixed point (K_c^*, L_c^*) represent critical points can be seen in the following manner: the RG transformation leads to a lattice with its lattice constant increased by a factor of $\sqrt{2}$. The correlation length of the transformed system ξ' is thus smaller by a factor of $\sqrt{2}$:

$$\xi' = \xi/\sqrt{2}\,. \tag{7.3.17}$$

However, at the fixed point, the coupling constants K_c^*, L_c^* are invariant, so that for ξ of the fixed point, we have $\xi' = \xi$, i.e. at the fixed point, it follows that $\xi = \xi/\sqrt{2}$, thus

$$\xi = \begin{cases} \infty & \text{or} \\ 0 & . \end{cases} \tag{7.3.18}$$

The value 0 corresponds to the high-temperature and to the low-temperature fixed points. At finite K^*, L^*, ξ cannot be zero, but only ∞. Calculating back through the transformation shows that the correlation length at each point along the critical trajectory which leads to the fixed point is infinite. Therefore, all the points of the "critical trajectory", i.e. the trajectory leading to the fixed point, are critical points of Ising models with nearest-neighbor and next-nearest-neighbor interactions.

In order to determine the critical behavior, we examine the behavior of the coupling constants in the vicinity of the "non-trivial" fixed point; to this end, we linearize the transformation equations (7.3.15a,b) around (K_c^*, L_c^*) in the lth step:

$$\delta K_l = K_l - K_c^* \quad , \quad \delta L_l = L_l - L_c^* \, . \tag{7.3.19}$$

We thereby obtain the following linear recursion relation:

$$\begin{pmatrix} \delta K_l \\ \delta L_l \end{pmatrix} = \begin{pmatrix} 4K_c^* & 1 \\ 2K_c^* & 0 \end{pmatrix} \begin{pmatrix} \delta K_{l-1} \\ \delta L_{l-1} \end{pmatrix} = \begin{pmatrix} \frac{4}{3} & 1 \\ \frac{2}{3} & 0 \end{pmatrix} \begin{pmatrix} \delta K_{l-1} \\ \delta L_{l-1} \end{pmatrix} . \tag{7.3.20}$$

The eigenvalues of the transformation matrix can be determined from $\lambda^2 - \frac{4}{3}\lambda - \frac{2}{3} = 0$, i.e.

$$\lambda_{1,2} = \frac{1}{3}(2 \pm \sqrt{10}) = \begin{cases} 1.7208 \\ -0.3874 \, . \end{cases} \tag{7.3.21a}$$

The associated eigenvectors can be obtained from $\left(4 - (2 \pm \sqrt{10})\right)\delta K + 3\delta L = 0$, i.e.

$$\delta L = \pm \frac{\sqrt{10} - 2}{3} \delta K \quad \text{and thus}$$

$$\mathbf{e}_1 = \left(1, \frac{\sqrt{10} - 2}{3}\right) \quad \text{and} \quad \mathbf{e}_2 = \left(1, -\frac{\sqrt{10} + 2}{3}\right) \tag{7.3.21b}$$

with the scalar product $\mathbf{e}_1 \cdot \mathbf{e}_2 = \frac{1}{3}$.

We now start from an Ising model with coupling constants K_0 and L_0 (including the division by kT). We first expand the deviations of the initial coupling constants K_0 and L_0 from the fixed point in the basis of the eigenvectors (7.3.21):

$$\begin{pmatrix} K_0 \\ L_0 \end{pmatrix} = \begin{pmatrix} K_c^* \\ L_c^* \end{pmatrix} + c_1\mathbf{e}_1 + c_2\mathbf{e}_2 \, , \tag{7.3.22}$$

with expansion coefficients c_1 and c_2. The decimation procedure is repeated several times; after l transformation steps, we obtain the coupling constants K_l and L_l:

$$\begin{pmatrix} K_l \\ L_l \end{pmatrix} = \begin{pmatrix} K_c^* \\ L_c^* \end{pmatrix} + \lambda_1^l c_1\mathbf{e}_1 + \lambda_2^l c_2\mathbf{e}_2 \, . \tag{7.3.23}$$

If the Hamiltonian H differs from H^* only by an increment in the direction $\mathbf{e}_2$, the successive application of the renormalization group transformation leads to the fixed point, since $|\lambda_2| < 1$ (see Fig. 7.9).

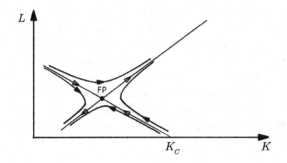

Fig. 7.9. Flow diagram based on the recursion relation (7.3.22), which is linearized around the nontrivial fixed point (FP)

Let us now consider the original nearest-neighbor Ising model with the coupling constant $K_0 \equiv \frac{J}{kT}$ and with $L_0 = 0$, and first determine the critical value K_c; this is the value of K_0 which leads to the fixed point. The condition for K_c, from the above considerations, is given by

$$\begin{pmatrix} K_c \\ 0 \end{pmatrix} = \begin{pmatrix} \frac{1}{3} \\ \frac{1}{9} \end{pmatrix} + 0 \cdot \mathbf{e}_1 + c_2 \begin{pmatrix} 1 \\ -\frac{\sqrt{10}+2}{3} \end{pmatrix} . \tag{7.3.24}$$

These two linear equations have the solution

$$c_2 = \frac{1}{3(\sqrt{10}+2)}, \quad \text{and therefore} \quad K_c = \frac{1}{3} + \frac{1}{3(\sqrt{10}+2)} = 0.3979 . \tag{7.3.25}$$

For $K_0 = K_c$, the linearized RG transformation leads to the fixed point, i.e. this is the critical point of the nearest-neighbor Ising model, $K_c = \frac{J}{kT_c}$. From the nonlinear recursion relation (7.3.15a,b), we find for the critical point the slighty smaller value $K_c^{n.l.} = 0.3921$. Both values differ from Onsager's exact solution, which gives $K_c = 0.4406$, but they are much closer than the value from molecular field theory, $K_c = 0.25$.

For $K_0 = K_c$, only $c_2 \neq 0$, and the transformation leads to the fixed point. For $K_0 \neq K_c$, we also have $c_1 \propto (K_0 - K_c) = -\frac{J}{kT_c^2}(T - T_c) \cdots \neq 0$. This increases with each application of the RG transformation, and thus leads away from the fixed point (K_c^*, L_c^*) (Fig. 7.9), so that the flow runs either to the low-temperature fixed point (for $T < T_c$) or to the high-temperature fixed point (for $T > T_c$).

Now we may determine the critical exponent ν for the correlation length, beginning with the recursion relation

$$(K - K_c)' = \lambda_1 (K - K_c) \tag{7.3.26}$$

and writing λ_1 as a power of the new length scale

$$\lambda_1 = (\sqrt{2})^{y_1} . \tag{7.3.27}$$

For the exponent y_1 defined here, we find the value

$$y_1 = 2\frac{\log \lambda_1}{\log 2} = 1.566 . \tag{7.3.28}$$

From $\xi' = \xi/\sqrt{2}$ (Eq. (7.3.17)), it follows that $(K'-K_c)^{-\nu} = (K-K_c)^{-\nu}/\sqrt{2}$, i.e.

$$(K' - K_c) = (\sqrt{2})^{\frac{1}{\nu}} (K - K_c) . \tag{7.3.29}$$

Comparing this with the first relation (7.3.26), we obtain

$$\nu = \frac{1}{y_1} = 0.638 . \tag{7.3.30}$$

This is, to be sure, quite a ways from 1, the known exact value of the two-dimensional Ising model, but nevertheless it is larger than 0.5, the value from the molecular-field approximation. A considerable improvement can be obtained by extending the recursion relation to several coupling coefficients.

Let us now consider the effect of a finite magnetic field h (including the factor β). The recursion relation can again be established intuitively. The field h acts directly on the remaining spins, as well as a (somewhat underestimated) additional field Kh which is due to the orienting action of the field on the eliminated neighboring spins, so that all together we have

$$h' = h + Kh . \tag{7.3.31}$$

The fixed point value of this recursion relation is $h^* = 0$. Linearization around the fixed point yields

$$h' = (1 + K^*)h = \frac{4}{3}h ; \tag{7.3.32}$$

thus the associated eigenvalue is

$$\lambda_h = \frac{4}{3} . \tag{7.3.33}$$

$K_0 - K_c$ (or $T - T_c$) and h are called the *relevant "fields"*, since the eigenvalues λ_1 and λ_h are larger than 1, and they therefore increase as a result of the renormalization group transformation and lead away from the fixed point. In contrast, c_2 is an *"irrelevant field"*, since $|\lambda_2| < 1$, and therefore c_2 becomes increasingly smaller with repeated RG transformations. Here, "fields" refers to fields in the usual sense, but also to coupling constants in the Hamiltonian. The structure found here is typical of models which describe critical points, and remains the same even when one takes arbitrarily many coupling constants into account in the transformation: there are *two relevant fields* ($T - T_c$ and h, the conjugate field to the order parameter), and *all the other fields are irrelevant*.

We add a remark concerning the flow diagram 7.9. There, owing to the negative sign of λ_2, only every other point is shown. This corresponds to a twofold application of the transformation and an increase of the lattice constant by a factor of 2, as well as $\lambda_1 \rightarrow \lambda_1^2, \lambda_2 \rightarrow \lambda_2^2$. Then the second eigenvalue λ_2^2 is also positive, since otherwise the trajectory would move along an oscillatory path towards the fixed point.

7.3.4 Scaling Laws

Although the decimation procedure described in Sect. 7.3.3 with only a few parameters does not give quantitatively satisfactory results and is also unsuitable for the calculation of correlation functions, it does demonstrate the general structure of RG transformations, which we shall now use as a starting point for deriving the scaling laws.

A general RG transformation $\mathcal{R}$ maps the original Hamiltonian $\mathcal{H}$ onto a new one,

$$\mathcal{H}' = \mathcal{R}\mathcal{H} . \tag{7.3.34}$$

This transformation also implies the rescaling of all the lengths in the problem, and that $N' = Nb^{-d}$ holds for the number of degrees of freedom N in d dimensions (here, $b = \sqrt{2}$ for the decimation transformation of 7.3.1).

The fixed-point Hamiltonian is determined by

$$\mathcal{R}(\mathcal{H}^*) = \mathcal{H}^* . \tag{7.3.35}$$

For small deviations from the fixed-point Hamiltonian,

$$\mathcal{R}(\mathcal{H}^* + \delta\mathcal{H}) = \mathcal{H}^* + \mathcal{L}\,\delta\mathcal{H} ,$$

we can expand in terms of the deviation $\delta\mathcal{H}$. From the expansion, we obtain the linearized recursion relation

$$\mathcal{L}\delta\mathcal{H} = \delta\mathcal{H}' . \tag{7.3.36a}$$

The eigenoperators $\delta\mathcal{H}_1, \delta\mathcal{H}_2, \ldots$ of this linear transformation are determined by the eigenvalue equation

$$\mathcal{L}\delta\mathcal{H}_i = \lambda_i\delta\mathcal{H}_i . \tag{7.3.36b}$$

A given Hamiltonian $\mathcal{H}$, which differs only slightly from $\mathcal{H}^*$, can be represented by $\mathcal{H}^*$ and the deviations from it:

$$\mathcal{H} = \mathcal{H}^* + \tau\delta\mathcal{H}_\tau + h\delta\mathcal{H}_h + \sum_{i\geq 3} c_i\delta\mathcal{H}_i , \tag{7.3.37}$$

where $\delta\mathcal{H}_\tau$ and $\delta\mathcal{H}_h$ denote the two *relevant* perturbations with

$$|\lambda_\tau| = b^{y_\tau} > 1 , \quad |\lambda_h| = b^{y_h} > 1 ; \tag{7.3.38}$$

they are related to the temperature variable $\tau = \frac{T-T_c}{T_c}$ and the external field h, while $|\lambda_j| = b^{y_j} < 1$ and thus $y_j < 0$ for $j \geq 3$ are connected with the *irrelevant* perturbations.[11] The coefficients τ, h, and c_j are called *scaling*

---

[11] Compare the discussion following Eq. (7.3.33). The (only) irrelevant field there is denoted by c_2. In the following, we assume that $\lambda_i \geq 0$.

fields. For the Ising model, $\delta\mathcal{H}_h = \sum_l \sigma_l$. Denoting the initial values of the fields by c_i, we find that the free energy transforms after l steps to

$$F_N(c_i) = F_{N/b^{dl}}(c_i \lambda_i^l) . \tag{7.3.39a}$$

For the free energy per spin,

$$f(c_i) = \frac{1}{N} F_N(c_i) , \tag{7.3.39b}$$

we then find in the linear approximation

$$f(\tau, h, c_3, \ldots) = b^{-dl} f\left(\tau b^{y_\tau l}, h b^{y_h l}, c_3 b^{y_3 l}, \ldots\right) . \tag{7.3.40}$$

Here, we have left off an additive term which has no influence on the following derivation of the scaling law; it is, however, important for the calculation of the free energy. The scaling parameter l can now be chosen in such a way that $|\tau| b^{y_\tau l} = 1$, which makes the first argument of f equal to ± 1. Then we find

$$f(\tau, h, c_3, \ldots) = |\tau|^{d/y_\tau} \hat{f}_\pm\left(h|\tau|^{-y_h/y_\tau}, c_3|\tau|^{|y_3|/y_\tau}, \ldots\right) , \tag{7.3.40'}$$

where $\hat{f}_\pm(x, y, \ldots) = f(\pm 1, x, y, \ldots)$ and $y_\tau, y_h > 0, y_3, \ldots < 0$. Close to T_c, the dependence on the irrelevant fields $c_3, \ldots$ can be neglected, and Eq. (7.3.40') then takes on precisely the scaling form (Eq. 7.2.7), with the conventional exponents

$$\beta\delta = y_h/y_\tau \tag{7.3.41a}$$

and

$$2 - \alpha = \frac{d}{y_\tau} . \tag{7.3.41b}$$

Taking the derivative with respect to h yields

$$\beta = \frac{d - y_h}{y_\tau} \quad \text{and} \quad \gamma = \frac{d - 2y_h}{y_\tau} . \tag{7.3.41c,d}$$

We have thus derived the scaling law, Eq. (7.2.7), within the RG theory for fixed points with just one relevant field, along with the applied magnetic field and the irrelevant operators. Furthermore, the dependence on the irrelevant fields $c_3, \ldots$ gives rise to corrections to the scaling laws, which must be taken into account for temperatures outside the asymptotic region.

In order to make the connection between y_τ and the exponent ν, we recall that l iterations reduce the correlation length to $\xi' = b^{-l}\xi$, which implies that $(\tau b^{y_\tau l})^{-\nu} = b^{-l}\tau^{-\nu}$ and, as a result,

$$\nu = \frac{1}{y_\tau} \tag{7.3.41e}$$

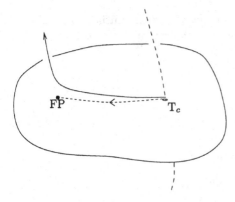

Fig. 7.10. The critical hypersurface. A trajectory within the critical hypersurface is shown as a dashed curve. The full curve is a trajectory near the critical hypersurface. The coupling coefficients of a particular physical system as a function of the temperature are indicated by the long-dashed curve

(cf. Eq. (7.3.30) for the two-dimensional Ising model). From the existence of a fixed-point Hamiltonian with two relevant operators, the scaling form of the free energy can be derived, and it is also possible to calculate the critical exponents. Even the form of the scaling functions $\hat{f}$ and $\hat{m}$ can be computed with perturbation-theoretical methods, since the arguments are finite. A similar procedure can be applied to the correlation function, Eq. (7.2.12b'). At this point it is important to renormalize the spin variable, $\sigma' = b^{\zeta}\sigma$, whereby it is found that setting the value

$$\zeta = (d - 2 + \eta)/2 \tag{7.3.41f}$$

guarantees the validity of (7.2.13) at the critical point.

We add a few remarks about the generic structure of the flow diagram in the vicinity of a critical fixed point (Fig. 7.10). In the multidimensional space of the coupling coefficients, there is a direction (the relevant direction) which leads away from the fixed point (we assume that $h = 0$). The other eigenvectors of the linearized RG transformation span the critical hypersurface. Further away from the fixed point, this hypersurface is no longer a plane, but instead is curved. The trajectories from each point on the critical hypersurface lead to the critical fixed point. When the initial point is close to but not precisely on the critical hypersurface, the trajectory at first runs parallel to the hypersurface until the relevant portion has become sufficiently large so that finally the trajectory leaves the neighborhood of the critical hypersurface and heads off to either the high-temperature or the low-temperature fixed point. For a given physical system (ferromagnet, liquid, ...), the parameters $\tau, c_3, \ldots$ depend on the temperature (the long-dashed curve in Fig. 7.10). The temperature at which this curve intersects the critical hypersurface is the transition temperature T_c.

From this discussion, the *universality properties* should be apparent. All systems which belong to a particular part of the parameter space, i.e. to the region of attraction of a given fixed point, are described by the same power laws in the vicinity of the critical hypersurface of the fixed point.

*7.3.5 General RG Transformations in Real Space

A general RG transformation in real space maps a particular spin system $\{\sigma\}$ with the Hamiltonian $\mathcal{H}\{\sigma\}$, defined on a lattice, onto a new spin system with fewer degrees of freedom (by $N'/N = b^{-d}$) and a new Hamiltonian $\mathcal{H}'\{\sigma'\}$. It can be represented by a transformation $T\{\sigma',\sigma\}$, such that

$$e^{-G-\mathcal{H}'\{\sigma'\}} = \sum_{\{\sigma\}} T\{\sigma',\sigma\} e^{-\mathcal{H}\{\sigma\}} \tag{7.3.42}$$

with the conditions

$$\sum_{\{\sigma'\}} \mathcal{H}'\{\sigma'\} = 0 \tag{7.3.43a}$$

and

$$\sum_{\{\sigma'\}} T\{\sigma',\sigma\} = 1 \;, \tag{7.3.43b}$$

which guarantee that

$$e^{-G} \operatorname{Tr}_{\{\sigma'\}} e^{-\mathcal{H}'\{\sigma'\}} = \operatorname{Tr}_{\{\sigma\}} e^{-\mathcal{H}\{\sigma\}} \tag{7.3.44a}$$

is fulfilled ($\operatorname{Tr}_{\{\sigma\}} \equiv \sum_{\{\sigma\}}$). This yields a relation between the free energy F of the original lattice and the free energy F' of the primed lattice:

$$F' + G = F \;. \tag{7.3.44b}$$

The constant G is independent of the configuration of the $\{\sigma'\}$ and is determined by equation (7.3.43a).

Important examples of such transformations are decimation transformations, as well as linear and nonlinear block-spin transformations. The simplest realization consists of

$$T\{\sigma',\sigma\} = \Pi_{i' \in \Omega'} \frac{1}{2} \left(1 + \sigma'_{i'} \, t_{i'}(\sigma) \right) \;, \tag{7.3.45}$$

where Ω denotes the lattice sites of the initial lattice and Ω' those of the new lattice, and the function $t_{i'}(\sigma)$ determines the nature of the transformation.

α) Decimation Transformation (Fig. 7.11)

$$t_{i'}\{\sigma\} = \zeta \sigma_{i'}$$
$$b = \sqrt{2} \;, \quad \zeta = b^{(d-2+\eta)/2} \;, \tag{7.3.46a}$$

where ζ rescales the amplitude of the remaining spins.

Then,

$$\langle \sigma_x \sigma_0 \rangle = \zeta^2 \langle \sigma'_{x'} \, \sigma'_0 \rangle \;.$$

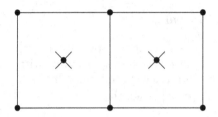

Fig. 7.11. A decimation transformation

β) Linear Block-Spin Transformation (on a triangular lattice, Fig.7.12)

$$t_{i'}\{\sigma\} = p(\sigma_{i'}^1 + \sigma_{i'}^2 + \sigma_{i'}^3)$$

$$b = \sqrt{3} \quad , \quad p = \frac{1}{3}(\sqrt{3})^{\eta/2} = 3^{-1+\eta/4} \;. \tag{7.3.46b}$$

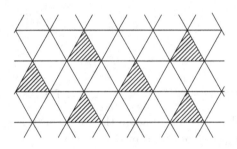

Fig. 7.12. A block-spin transformation

γ) Nonlinear Block-Spin Transformation

$$t_{i'}\{\sigma\} = p(\sigma_{i'}^1 + \sigma_{i'}^2 + \sigma_{i'}^3) + q\sigma_{i'}^1\sigma_{i'}^2\sigma_{i'}^3 \;. \tag{7.3.46c}$$

An important special case

$$p = -q = \frac{1}{2}, \quad \sigma_{i'}' = \text{sgn}(\sigma_{i'}^1 + \sigma_{i'}^2 + \sigma_{i'}^3) \;.$$

These so called real-space renormalization procedures were introduced by Niemeijer and van Leeuwen[12]. The simplified variant given in Sect. 7.3.3 is from[13]. The block-spin transformation for a square Ising lattice is described in[14]. For a detailed discussion with additional references, we refer to the article by Niemeijer and van Leeuwen[15].

[12] Th. Niemeijer and J. M. J. van Leeuwen, Phys. Rev. Lett. **31**, 1411 (1973).
[13] K. G. Wilson, Rev. Mod. Phys. **47**, 773 (1975).
[14] M. Nauenberg and B. Nienhuis, Phys. Rev. Lett. **33**, 344 (1974).
[15] Th. Niemeijer and J. M. J. van Leeuwen, in *Phase Transitions and Critical Phenomena* Vol. 6, Eds. C. Domb and M. S. Green, p. 425, Academic Press, London 1976.

*7.4 The Ginzburg–Landau Theory

7.4.1 Ginzburg–Landau Functionals

The Ginzburg–Landau theory is a continuum description of phase transitions. Experience and the preceding theoretical considerations in this chapter show that the microscopic details such as the lattice structure, the precise form of the interactions, etc. are unimportant for the critical behavior, which manifests itself at distances which are much greater than the lattice constant. Since we are interested only in the behavior at small wavenumbers, we can go to a macroscopic continuum description, roughly analogous to the transition from microscopic electrodynamics to continuum electrodynamics. In setting up the Ginzburg–Landau functional, we will make use of an intuitive justification; a microscopic derivation is given in Appendix E. (see also problem 7.15).

We start with a ferromagnetic system consisting of Ising spins ($n = 1$) on a d-dimensional lattice. The generalization to arbitrary dimensions is interesting for several reasons. First, it contains the physically relevant dimensions, three and two. Second, it may be seen that certain approximation methods are exact above four dimensions; this gives us the possibility of carrying out perturbation expansions around the dimension four (Sect. 7.4.5).

Instead of the spins S_l on the lattice, we introduce a continuum magnetization

$$m(\mathbf{x}) = \frac{1}{\tilde{N}a_0^d} \sum_l g(\mathbf{x} - \mathbf{x}_l) S_l \ . \tag{7.4.1}$$

Here, $g(\mathbf{x} - \mathbf{x}_l)$ is a weighting function, which is equal to one within a cell with N spins and is zero outside it. The linear dimension of this cell, a_c, is supposed to be much larger than the lattice constant a_0 but much smaller than the length L of the crystal, i.e. $a_0 \ll a_c \ll L$. The function $g(\mathbf{x} - \mathbf{x}_l)$ is assumed to vary continuously from the value 1 to 0, so that $m(\mathbf{x})$ is a continuous function of x; see Fig. 7.13.

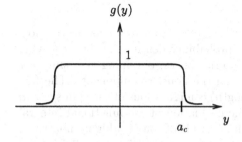

Fig. 7.13. The weighting function $g(y)$ along one of the d cartesian coordinates

Making use of

$$\int d^d x \, g(\mathbf{x} - \mathbf{x}_l) = \tilde{N}a_0^d$$

and of the definition (7.4.1), we can rewrite the Zeeman term as follows:

$$\sum_l h S_l = h \sum_l \frac{1}{\tilde{N} a_0^d} \int d^d x\, g(\mathbf{x} - \mathbf{x}_l) S_l = \int d^d x\, h m(\mathbf{x}) . \qquad (7.4.2)$$

From the canonical density matrix for the spins, we obtain the probability density for the configurations $m(\mathbf{x})$. Generally, we have

$$\mathcal{P}[m(\mathbf{x})] = \left\langle \delta\left(m(\mathbf{x}) - \frac{1}{\tilde{N} a_0^d} \sum_l g(\mathbf{x} - \mathbf{x}_l) S_l \right) \right\rangle . \qquad (7.4.3)$$

For $\mathcal{P}[m(\mathbf{x})]$, we write

$$\mathcal{P}[m(\mathbf{x})] \propto e^{-\mathcal{F}[m(\mathbf{x})]/kT} , \qquad (7.4.4)$$

in which the Ginzburg–Landau functional $\mathcal{F}[m(\mathbf{x})]$ enters; it is a kind of Hamiltonian for the magnetization $m(\mathbf{x})$. The tendency towards ferromagnetic ordering due to the exchange interaction must express itself in the form of the functional $\mathcal{F}[m(\mathbf{x})]$

$$\mathcal{F}[m(\mathbf{x})] = \int d^d x \left(a m^2(\mathbf{x}) + \frac{b}{2} m^4(\mathbf{x}) + c\left(\nabla m(\mathbf{x})\right)^2 - h m(\mathbf{x}) \right) . \qquad (7.4.5)$$

In the vicinity of T_c, only configurations of $m(\mathbf{x})$ with small absolute values should be important, and therefore the Taylor expansion (7.4.5) should be allowed. Before we turn to the coefficients in (7.4.5), we make a few remarks about the significance of this functional.

Due to the averaging (7.4.1), short-wavelength variations of S_l do not contribute to $m(\mathbf{x})$. The long-wavelength variations, however, with wavelengths larger than a_z, are reflected fully in $m(\mathbf{x})$. The partition function of the magnetic system therefore has the form

$$Z = Z_0(T) \int \mathcal{D}[m(\mathbf{x})] e^{-\mathcal{F}[m(\mathbf{x})]/kT} . \qquad (7.4.6)$$

Here, the functional integral $\int \mathcal{D}[m(\mathbf{x})] \ldots$ refers to a sum over all the possible configurations of $m(\mathbf{x})$ with the probability density $e^{-\mathcal{F}[m(\mathbf{x})]/kT}$. One can represent $m(\mathbf{x})$ by means of a Fourier series, obtaining the sum over all configurations by integration over all the Fourier components. The factor $Z_0(T)$ is due to the (short-wavelength) configurations of the spin system, which do not contribute to $m(\mathbf{x})$. The evaluation of the functional integral which occurs in the partition function (7.4.6) is of course a highly nontrivial problem and will be carried out in the following Sections 7.4.2 and 7.4.5 using approximation methods. The *free energy* is

$$F = -kT \log Z . \qquad (7.4.7)$$

We now come to the coefficients in the expansion (7.4.5). First of all, this expansion took into account the fact that $\mathcal{F}[m(\mathbf{x})]$ has the same symmetry as the microscopic spin Hamiltonian, i.e. aside from the Zeeman term, $\mathcal{F}[m(\mathbf{x})]$ is an even function of $m(\mathbf{x})$. Owing to (7.4.2), the field h expresses itself only in the Zeeman term, $-\int d^d x\, h\, m(\mathbf{x})$, and the coefficients a, b, c are independent of h. For reasons of stability, large values of $m(\mathbf{x})$ must have a small statistical weight, which requires that $b > 0$. If for some system $b \leq 0$, the expansion must be extended to higher orders in $m(\mathbf{x})$. These circumstances occur in first-order phase transitions and at tricritical points. The ferromagnetic exchange interaction has a tendency to orient the spins uniformly. This leads to the term $c\nabla m\nabla m$ with $c > 0$, which suppresses inhomogeneities in the magnetization.

Finally, we come to the values of a. For $h = 0$ and a uniform $m(x) = m$, the probability weight $e^{-\beta\mathcal{F}}$ is shown in Fig. 7.14.

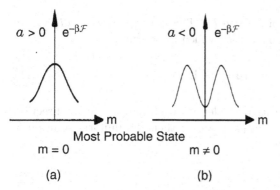

Fig. 7.14. The probability density $e^{-\beta\mathcal{F}}$ as a function of a uniform magnetization. (a) For $a > 0$ $(T > T_c^0)$ and (b) for $a < 0$ $(T < T_c^0)$

When $a > 0$, then the most probable configuration is $m = 0$; when $a < 0$, then the most probable configuration is $m \neq 0$. Thus, a must change its sign,

$$a = a'(T - T_c^0)\,, \tag{7.4.8}$$

in order for the phase transition to occur. Due to the nonlinear terms and to fluctuations, the real T_c will differ from T_c^0. The coefficients b and c are finite at T_c^0.

If one starts from a Heisenberg model instead of from an Ising model, the replacements

$$S_l \to \mathbf{S}_l \quad \text{and} \quad m(\mathbf{x}) \to \mathbf{m}(\mathbf{x})$$

$$m^4(\mathbf{x}) \to \left(\mathbf{m}(\mathbf{x})^2\right)^2 \quad, \quad (\nabla m)^2 \to \nabla_\alpha \mathbf{m}\nabla_\alpha \mathbf{m}\,. \tag{7.4.9}$$

must be made, leading to Eq. (7.4.10). Ginzburg–Landau functionals can be introduced for every type of phase transition. It is also not necessary to attempt a microscopic derivation: the form is determined in most cases from knowledge of the symmetry of the order parameter. Thus, the Ginzburg–Landau theory was first applied to the case of superconductivity long before the advent of the microscopic BCS theory. The Ginzburg–Landau theory was also particularly successful in treating superconductivity, because here simple approximations (see Sect. 7.4.2) are valid even close to the transition (see also Sect. 7.4.4).

7.4.2 The Ginzburg–Landau Approximation

We start with the Ginzburg–Landau functional for an order parameter with n components, $\mathbf{m}(\mathbf{x})$, $n = 1, 2, \ldots,$:

$$\mathcal{F}[\mathbf{m}(\mathbf{x})] = \int d^d x \left[a\mathbf{m}^2(\mathbf{x}) + \frac{1}{2}b(\mathbf{m}(\mathbf{x})^2)^2 + c(\nabla \mathbf{m})^2 - \mathbf{h}(\mathbf{x})\mathbf{m}(\mathbf{x}) \right]. \quad (7.4.10)$$

The integration extends over a volume L^d. The *most probable configuration* of $\mathbf{m}(\mathbf{x})$ is given by the stationary state which is determined by

$$\frac{\delta \mathcal{F}}{\delta \mathbf{m}(\mathbf{x})} = 2(a + b\mathbf{m}(\mathbf{x})^2 - c\nabla^2)\mathbf{m}(\mathbf{x}) - \mathbf{h}(\mathbf{x}) = 0. \quad (7.4.11)$$

Let $\mathbf{h}$ be independent of position and let us take $\mathbf{h}$ to lie in the x_1-direction without loss of generality, $\mathbf{h} = h\mathbf{e}_1, (h \gtrsim 0)$; then the uniform solution is found from

$$2(a + b\mathbf{m}^2)\mathbf{m} - h\mathbf{e}_1 = 0. \quad (7.4.12)$$

We discuss *special cases*:

(i) $\mathbf{h} \to 0$: *spontaneous magnetization* and *specific heat*
When there is no applied field, (7.4.12) has the following solutions:

$$\mathbf{m} = 0 \quad \text{for} \quad a > 0$$

$$(\mathbf{m} = 0) \quad \text{and} \quad \mathbf{m} = \pm\mathbf{e}_1 m_0, \quad m_0 = \sqrt{\frac{-a}{b}} \quad \text{for} \quad a < 0. \quad (7.4.13)$$

The (Gibbs) *free energy* for the configurations (7.4.13) is[16]

$$F(T, h = 0) = F[0] = 0 \qquad \text{for} \quad T > T_c^0 \qquad (7.4.14a)$$

$$F(T, h = 0) = F[m_0] = -\frac{1}{2}\frac{a^2}{b}L^d \qquad \text{for} \quad T < T_c^0. \qquad (7.4.14b)$$

---

[16] Instead of really computing the functional integral $\int \mathcal{D}[\mathbf{m}(\mathbf{x})]e^{-\mathcal{F}[\mathbf{m}(\mathbf{x})]/kT}$ as is required by (7.4.6) and (7.4.7) for the determination of the free energy, $\mathbf{m}(\mathbf{x})$ was replaced everywhere by its most probable value.

We will always leave off the regular term $F_{reg} = -kT \log Z_0$. The state $\mathbf{m} = 0$ would have a higher free energy for $T < T_c^0$ than the state m_0; therefore, $\mathbf{m} = 0$ was already put in parentheses in (7.4.13). For $T < T_c^0$, we thus find a finite *spontaneous magnetization*. The onset of this magnetization is characterized by the critical exponent β, which here takes on the value $\beta = \frac{1}{2}$ (Fig. 7.15).

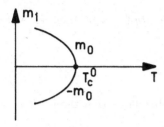

Fig. 7.15. The spontaneous magnetization in the Ginzburg–Landau approximation

Specific Heat
From (7.4.14a,b), we immediately find the specific heat

$$L^d c_{h=0} = T\left(\frac{\partial S}{\partial T}\right)_{h=0} = -T\left(\frac{\partial^2 F}{\partial T^2}\right)_{h=0} = \begin{cases} 0 & T > T_c^0 \\ T\frac{a'^2}{b}L^d & T < T_c^0 \end{cases} , \quad (7.4.15)$$

with a' from (7.4.8). The specific heat exhibits a jump

$$\Delta c_{h=0} = T_c^0 \frac{a'^2}{b} , \quad (7.4.16)$$

and the critical exponent α is therefore zero (see Eq. (7.1.1)), $\alpha = 0$.

(ii) *The equation of state for $h > 0$ and the susceptibility*
We decompose $\mathbf{m}$ into a longitudinal part, $\mathbf{e}_1 m_1$, and a transverse part, $\mathbf{m}_\perp = (0, m_2, ..., m_n)$. Evidently, Eq. (7.4.12) gives

$$\mathbf{m}_\perp = 0 \quad (7.4.17)$$

and the magnetic equation of state

$$h = 2(a + bm_1^2)m_1 . \quad (7.4.18)$$

We can simplify this in limiting cases:
α) $T = T_c^0$

$$h = 2bm_1^3 \quad \text{i.e.} \quad \delta = 3 . \quad (7.4.19)$$

β) $T > T_c^0$

$$m_1 = \frac{h}{2a} + \mathcal{O}(h^3) . \quad (7.4.20)$$

$\gamma)\ T < T_c^0$

$m_1 = m_0 \operatorname{sgn}(h) + \Delta m$ \quad yields

$$m_1 = m_0 \operatorname{sgn}(h) + \frac{h}{4bm_0^2} + \mathcal{O}(h^2\operatorname{sgn}(h))$$

$$= m_0 \operatorname{sgn}(h) + \frac{h}{-4a} + \mathcal{O}(h^2\operatorname{sgn}(h)) . \tag{7.4.21}$$

We can now also calculate the *magnetic susceptibility* for $h = 0$, either by differentiating the equation of state (7.4.18)

$$2(a + 3bm_1^2)\frac{\partial m_1}{\partial h} = 1$$

or directly, by inspection of (7.4.20) and (7.4.21). It follows that the isothermal susceptibility is given by

$$\chi_T = \left(\frac{\partial m_1}{\partial h}\right)_T = \begin{cases} \frac{1}{2a} & T > T_c^0 \\ \frac{1}{4|a|} & T < T_c^0 \end{cases} . \tag{7.4.22}$$

The critical exponent γ has, as in molecular field theory, a value of $\gamma = 1$.

7.4.3 Fluctuations in the Gaussian Approximation

7.4.3.1 Gaussian Approximation

Next we want to investigate the influence of fluctuations of the magnetization. To this end, we first expand the Ginzburg–Landau functional in terms of the deviations from the most probable state up to second order

$$\mathbf{m}(\mathbf{x}) = m_1\mathbf{e}_1 + \mathbf{m}'(\mathbf{x}) , \tag{7.4.23}$$

where

$$\mathbf{m}'(\mathbf{x}) = L^{-d/2} \sum_{k \in B} \mathbf{m}_k e^{ikx} \tag{7.4.24}$$

characterizes the deviation from the most probable value. Because of the underlying cell structure, the summation over k is restricted to the Brillouin zone B: $-\frac{\pi}{a_c} < k_i < \frac{\pi}{a_c}$. The condition that $\mathbf{m}(\mathbf{x})$ be real yields

$$\mathbf{m}_k^* = \mathbf{m}_{-k} . \tag{7.4.25}$$

A) $T > T_c^0$ and $h = 0$:

In this region, $m_1 = 0$, and the Fourier series (7.4.24) diagonalizes the harmonic part $\mathcal{F}_h$ of the Ginzburg–Landau functional

$$\mathcal{F}_h = \int d^dx \left(a\mathbf{m}'^2 + c(\nabla\mathbf{m}')^2\right) = \sum_k (a + ck^2)\mathbf{m}_k\mathbf{m}_{-k} . \tag{7.4.26}$$

We can now readily calculate the partition function (7.4.6) in the Gaussian approximation above T_c^0:

$$Z_G = Z_0 \int \prod_k d\mathbf{m}_k e^{-\beta\mathcal{F}_h} . \tag{7.4.27}$$

We decompose $\mathbf{m}_k$ into real and imaginary parts, finding for each k and each of the n components of $\mathbf{m}_k$ a Gaussian integral, so that

$$Z_G = Z_0 \prod_k \left(\sqrt{\frac{\pi}{\beta(a + ck^2)}}\right)^n \tag{7.4.28}$$

results, and thus the free energy (the stationary solution $m_1 = 0$ makes no contribution) is

$$F(T,0) = F_0 - kT\frac{n}{2}\sum_k \log\frac{\pi}{\beta(a + ck^2)} . \tag{7.4.29}$$

The specific heat, using $\sum_{\mathbf{k}} \cdots = \frac{V}{(2\pi)^d}\int d^dk\ldots$ and Eq. (7.4.8), is then

$$c_{h=0} = -T\frac{\partial^2 F/L^d}{\partial T^2} = k\frac{n}{2}(Ta')^2 \int \frac{d^dk}{(2\pi)^d}\frac{1}{(a + ck^2)^2} + \cdots . \tag{7.4.30}$$

The dots stand for less singular terms. We define the quantity

$$\xi = \sqrt{\frac{c}{a}} = \left(\frac{c}{a'}\right)^{1/2}(T - T_c^0)^{-1/2} , \tag{7.4.31}$$

which diverges in the limit $T \to T_c^0$ and will be found to represent the correlation length in the calculation of the correlation function (7.4.47). By introducing $q = \xi k$ into (7.4.30) as a new integration variable, we find the singular part of the specific heat

$$c_{h=0}^{sing.} = \tilde{A}_+\xi^{4-d} \tag{7.4.32}$$

with the amplitude

$$\tilde{A}_+ = k\frac{n}{2}\left(\frac{Ta'}{c}\right)^2 \int_{q<\Lambda\xi} \frac{d^dq}{(2\pi)^d}\frac{1}{(1 + q^2)^2} . \tag{7.4.33}$$

Here, the radius Λ of the Brillouin sphere enters; it is introduced at the end of Appendix E. The amplitude $\tilde{A}_+$ characterizes the strength of the singularity above T_c. Here and in the following, d-dimensional integrals of the type

$$\int \frac{d^d k}{(2\pi)^d} f(k^2) = \int \frac{d\Omega_d}{(2\pi)^d} \int dk\, k^{d-1}\, f(k^2) \qquad (7.4.34a)$$

occur, where

$$K_d \equiv \int \frac{d\Omega_d}{(2\pi)^d} = \left(2^{d-1}\, \pi^{d/2}\, \Gamma\left(\frac{d}{2}\right)\right)^{-1} \qquad (7.4.34b)$$

is the surface of a d-dimensional unit sphere divided by $(2\pi)^d$. In the further evaluation of (7.4.32) and (7.4.33), the three cases $d < 4$, $d = 4$, $d > 4$ must be distinguished:

<u>$d < 4$</u>

$$\int_0^{\Lambda\xi} dq\, \frac{q^{d-1}}{(1+q^2)^2} = \int_0^\infty dq\, \frac{q^{d-1}}{(1+q^2)^2} - \underbrace{\int_{\Lambda\xi}^\infty dq\, q^{d-5}}_{(\Lambda\xi)^{d-4}} = \text{finite} + \mathcal{O}\left((\Lambda\xi)^{d-4}\right)$$

<u>$d = 4$</u>

$$\int_0^{\Lambda\xi} dq\, \frac{q^3}{(1+q^2)^2} \sim \int^{\Lambda\xi} \frac{dq}{q} \sim \log \Lambda\xi$$

<u>$d > 4$</u>

$$\int_0^{\Lambda\xi} dq \left(\frac{q^{d-1}}{(1+q^2)^2} - q^{d-5}\right) + \int_0^{\Lambda\xi} dq\, q^{d-5}$$

$$= -\int_0^{\Lambda\xi} dq\, \frac{q^{d-5} + 2q^{d-3}}{(1+q^2)^2} + \frac{1}{d-4}(\Lambda\xi)^{d-4} .$$

The overall result is summarized in (7.4.35):

$$c_{h=0}^{sing} = \begin{cases} A_+ (T - T_c^0)^{-\frac{4-d}{2}} & d < 4 \\ \sim \log(T - T_c^0) & d = 4 \\ A - B(T - T_c^0)^{\frac{d-4}{2}} & d > 4 . \end{cases} \qquad (7.4.35)$$

For $d \leq 4$, the specific heat diverges at T_c; for $d > 4$, it exhibits a cusp. The amplitude A_+ for $d < 4$ is given by

$$A_+ = \frac{n}{2} T^2 \left(\frac{a'}{c}\right)^{\frac{d}{2}} K_d \int_0^\infty dq\, \frac{q^{d-1}}{(1+q^2)^2} . \qquad (7.4.36)$$

Below $d = 4$, the critical exponent of the specific heat is $(c_{h=0} \sim (T - T_c)^{-\alpha})$

$$\alpha = \frac{1}{2}(4 - d) ; \qquad (7.4.37)$$

in particular, for $d = 3$ in the Gaussian approximation, $\alpha = \frac{1}{2}$. Comparison with exact results and experiments shows that the Gaussian approximation overestimates the fluctuations.

B) $T < T_c^0$

Now we turn to the region $T < T_c^0$ and distinguish between the longitudinal (m_1) and the transverse components (m_i)

$$m_1(x) = m_1 + m_1'(x), \quad m_i(x) = m_i'(x) \quad \text{for} \quad i \geq 2 \tag{7.4.38}$$

with the Fourier components m_{1k}' and m_{ik}', where the latter are present only for $n \geq 2$. In the present context, including non-integer values of d, vectors will be denoted by just x, etc. From (7.4.10), we find for the Ginzburg–Landau functional in second order in the fluctuations:

$$\mathcal{F}_h[\mathbf{m}] = \mathcal{F}[m_1] + \sum_k \left[\left(-2a + \frac{3h}{2m_1} + \mathcal{O}(h^2) + ck^2 \right) |m_{1k}'|^2 \right.$$
$$\left. + \left(\frac{h}{2m_1} + ck^2 \right) \sum_{i \geq 2} |m_{ik}|^2 \right]. \tag{7.4.39}$$

To arrive at this expression, the following ancillary calculation was used:

$$a \left(m_1^2 + 2m_1 m_1' + m_1'^2 + m_\perp^2 \right)$$
$$+ \frac{b}{2} \left(m_1^4 + 4m_1^3 m_1' + 6m_1^2 m_1'^2 + 2m_1^2 m_\perp^2 \right) - h(m_1 + m_1')$$
$$= am_1^2 + \frac{b}{2} m_1^4 - hm_1 + \left(a + 3bm_1^2 \right) m_1'^2 + \underbrace{\left(a + bm_1^2 \right)}_{\frac{h}{2m_1}} m_\perp^2 .$$

Analogously to the computation leading from (7.4.26) to (7.4.29), we find for the free energy of the low-temperature phase at $h = 0$

$$F(T, 0) = F_0(T, h) + F_{G.L.}(T, 0) -$$
$$- \frac{1}{2} kT \sum_k \left\{ \log \frac{\pi}{\beta(2|a| + ck^2)} + (n - 1) \log \frac{\pi}{\beta ck^2} \right\}. \tag{7.4.40}$$

The first term results from Z_0; the second from $\mathcal{F}[m_1]$, the stationary solution considered in the Ginzburg–Landau approximation; the third term from the longitudinal fluctuations; and the fourth from the transverse fluctuations. The latter do not contribute to the specific heat, since their energy is temperature independent for $h = 0$:

$$c_{h=0} = T\frac{a'^2}{b} + \tilde{A}_- \xi^{4-d} = T\frac{a'^2}{b} + A_- (T_c - T)^{-\frac{4-d}{2}}, \tag{7.4.41}$$

where the low-temperature correlation length

$$\xi = \left(\sqrt{\frac{2|a|}{c}}\right)^{-1} = \left(\frac{c}{2a'}\right)^{1/2} (T_c^0 - T)^{-1/2} , \qquad T < T_c^0 \qquad (7.4.42)$$

is to be inserted. The amplitudes in (7.4.23) and (7.4.41) obey the relations

$$\tilde{A}_- = \frac{4}{n}\tilde{A}_+ , \quad A_- = \frac{2^{d/2}}{n}A_+ . \qquad (7.4.43)$$

The ratio of the amplitudes of the singular contribution to the specific heat depends only on the number of components n and the spatial dimension d, and is in this sense universal. The transverse fluctuations do not contribute to the specific heat below T_c; therefore, the factor $\frac{1}{n}$ enters the amplitude ratio.

7.4.3.2 Correlation Functions

We now calculate the correlation functions in the Gaussian approximation. We start by considering $T > T_c^0$. In order to calculate this type of quantity, with which we shall meet up repeatedly later, we introduce the generating functional

$$Z[h] = \frac{1}{Z_G} \int \prod_k dm_k \, e^{-\beta \mathcal{F}_h + \sum h_k m_{-k}}$$

$$= \frac{1}{Z_G} \int \prod_k dm_k \, e^{-\beta \sum_k (a+ck^2)|m_k|^2 + h_k m_{-k}} . \qquad (7.4.44)$$

To evaluate the Gaussian integrals in (7.4.44), we introduce the substitution

$$\tilde{m}_k = m_k - \frac{1}{2\beta}(a + ck^2)^{-1}h_k , \qquad (7.4.45)$$

obtaining

$$Z[h] = \exp\left[\frac{1}{4\beta}\sum_k \frac{1}{a+ck^2}h_k h_{-k}\right] . \qquad (7.4.46)$$

Evidently,

$$\langle m_k m_{-k'}\rangle = \frac{\partial^2}{\partial h_{-k}\partial h_{k'}}Z[h]\bigg|_{h=0} ,$$

from which we find the correlation function by making use of (7.4.46):

$$\langle m_k m_{-k'}\rangle = \delta_{kk'}\frac{1}{2\beta(a+ck^2)} \equiv \delta_{k,k'}G(k) . \qquad (7.4.47)$$

Here, we have taken into account the fact that in the sum over k in (7.4.46), each term $h_k h_{-k} = h_{-k} h_k$ occurs twice. From the last equation, the meaning of the *correlation length* (7.4.31) becomes clear, since in *real space*, Eq. (7.4.47) gives

$$\langle m(x)m(x')\rangle = \frac{1}{L^d}\sum_k e^{ik(x-x')}\frac{1}{2\beta(a+ck^2)} = \int \frac{d^d k}{(2\pi)^d}\frac{e^{ik(x-x')}}{2\beta c(\xi^{-2}+k^2)}$$

$$= \frac{\xi^{2-d}}{2\beta c}\int_{q<\Lambda\xi}\frac{d^d q}{(2\pi)^d}\frac{e^{iq(x-x')/\xi}}{1+q^2} .$$

$$(7.4.48)$$

The correlation length is characterized by the critical exponent $\nu = \frac{1}{2}$. For $T = T_c^0$, one can see immediately from the second expression that

$$\langle m(x)m(x')\rangle \sim \frac{1}{|x-x'|^{d-2}} , \qquad (7.4.49)$$

i.e. the exponent η introduced in (7.2.13) is zero in this approximation: $\eta = 0$. In *three* dimensions, we find from (7.4.48) the *Ornstein–Zernike correlation function*:

$$\langle m(\mathbf{x})m(\mathbf{x}')\rangle = \frac{1}{8\pi\beta c}\frac{e^{-r/\xi}}{r} , \qquad r = |\mathbf{x}-\mathbf{x}'| . \qquad (7.4.50)$$

Remark: The correlation function (7.4.47) obeys

$$\lim_{k\to 0} G(k) = kT\chi_T , \qquad (7.4.51)$$

where χ_T is the isothermal susceptibility (7.4.22a).

For $T < T_c^0$ we distinguish for $n > 1$ between the longitudinal correlation function and the transverse ($i \geq 2$) correlation function:

$$G_\parallel(k) = \langle m'_{1k} m'_{1-k}\rangle \quad \text{and} \quad G_\perp(k) = \langle m_{ik} m_{i-k}\rangle . \qquad (7.4.52)$$

For $n = 1$, only $G_\parallel(k)$ is relevant. From (7.4.39), it follows in analogy to (7.4.47) that

$$G_\parallel(k) = \frac{1}{2\beta[-2a+\frac{3h}{2m_1}+ck^2]} \xrightarrow{h\to 0} \frac{1}{2\beta[2a'(T_c^0-T)+ck^2]} \qquad (7.4.53)$$

and

$$G_\perp(k) = \frac{1}{2\beta[\frac{h}{2m_1}+ck^2]} \xrightarrow{h\to 0} \frac{1}{2\beta ck^2} \qquad (7.4.54a)$$

$$G_\perp(0) = \frac{Tm_1}{h} . \qquad (7.4.54b)$$

The divergence of the transverse susceptibility (correlation function) (7.4.54a) at $h = 0$ is a result of rotational invariance, owing to which it costs no energy to rotate the magnetization.

We first want to summarize the results of the Gaussian approximation, then treat the limits of its validity, and finally, in Sect. 7.4.4.1, to discuss the form of the correlation functions below T_c^0 in a more general way.

In summary for the *critical exponents*, we have:

$$\alpha_{\text{Fluct}} = 2 - \frac{d}{2}, \ \beta = \frac{1}{2}, \ \gamma = 1, \ \delta = 3, \ \nu = \frac{1}{2}, \ \eta = 0 \qquad (7.4.55)$$

and for the *amplitude ratios* of the specific heat, the longitudinal correlation function and the isothermal susceptibility:

$$\frac{\tilde{A}_+}{\tilde{A}_-} = \frac{n}{4}, \quad \frac{\tilde{C}_+}{\tilde{C}_-} = 1, \quad \text{and} \quad \frac{C_+}{C_-} = 2. \qquad (7.4.56)$$

The amplitudes are defined in (7.4.32), (7.4.41), (7.4.57), and (7.4.58):

$$G(k) = \tilde{C}_\pm \frac{\xi^2}{1 + (\xi k)^2}, \quad \tilde{C}_\pm = \frac{1}{2\beta c}, \qquad (7.4.57)$$

$$\chi = C_\pm |T - T_c|^{-1}, \quad T \gtrless T_c. \qquad (7.4.58)$$

7.4.3.3 Range of Validity of the Gaussian Approximation

The range of validity of the Gaussian approximation and of more elaborate perturbation-theoretical calculations can be estimated by comparing the higher orders with lower orders. For example, the fourth order must be much smaller than the second, or the Gaussian contribution to the specific heat must be smaller than the stationary value. The Ginzburg–Landau approximation is permissible if the fluctuations are small compared to the stationary value, i.e. from Eqns. (7.4.16) and (7.4.41),

$$\Delta c \gg \xi^{4-d} \left(\frac{Ta'}{c} \right)^2 \mathcal{N} , \qquad (7.4.59)$$

where $\mathcal{N}$ is a numerical factor. Then we require that

$$\tau^{(4-d)/2} \gg \frac{\mathcal{N}}{\xi_0^d \Delta c} \qquad (7.4.60)$$

with $\tau = \frac{T - T_c^0}{T_c^0}$ and $\xi_0 = \sqrt{\frac{c}{a' T_c^0}}$.

For dimensions $d < 4$, the Ginzburg–Landau approximation fails near T_c^0. From (7.4.60), we find a characteristic temperature $\tau_{GL} = (\frac{\mathcal{N}}{\xi_0^d \Delta c})^{2/(4-d)}$,

Table 7.3. The correlation length and the critical region

Superconductors[17]	$\xi_0 \sim 10^3$ Å	$\tau_{GL} = 10^{-10} - 10^{-14}$
Magnets	$\xi_0 \sim$ Å	$\tau_{GL} \sim 10^{-2}$
λ−Transition	$\xi_0 \sim 4$ Å	$\tau_{GL} \sim 0.3$

the so called Ginzburg–Levanyuk temperature; it depends on the Ginzburg–Landau parameters (see Table 7.3).

In this connection, $d_c = 4$ appears as a limiting dimension (upper critical dimension). For $d < 4$, the Ginzburg–Landau approximation fails when $\tau < \tau_{GL}$. It is then no longer sufficient to add the fluctuation contribution; instead, one has to take interactions between the fluctuations into account. Above four dimensions, the corrections to the Gaussian approximation on approaching T_c^0 become smaller, so that there, the Gaussian approximation applies. For $d > 4$, the exponent of the fluctuation contribution is negative, from Eq. (7.4.35): $\alpha_{\text{Fluct}} < 0$. Then the ratio can be $\frac{c_{h=0}(T_c^0)}{\Delta c} \gtrless 1$.

7.4.4 Continuous Symmetry and Phase Transitions of First Order

7.4.4.1 Susceptibilities for $T < T_c$

A) Transverse Susceptibility

We found for the transverse correlation function(7.4.54a) that $G_\perp(k) = \frac{1}{2\beta[\frac{h}{2m_1}+ck^2]}$ and we now want to show that the relation $G_\perp(0) = \frac{Tm_1}{h}$ is a general result of rotational invariance. To this end, we imagine that an external field $\mathbf{h}$ acts on a ferromagnet. Now we investigate the influence of an additional infinitesimal, transverse field $\delta\mathbf{h}$ which is perpendicular to $\mathbf{h}$

$$\sqrt{(\mathbf{h} + \delta\mathbf{h})^2} = \sqrt{h^2 + \delta h^2} = h + \frac{\delta h^2}{2h} \quad + \quad \ldots .$$

Thus, the magnitude of the field is changed by only $\mathcal{O}(\delta h^2)$; for a small δh, this is equivalent to a rotation of the field through the angle $\frac{\delta h}{h}$ (Fig. 7.16).

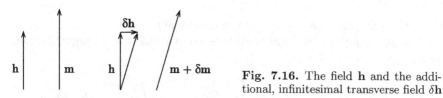

Fig. 7.16. The field $\mathbf{h}$ and the additional, infinitesimal transverse field $\delta\mathbf{h}$

---

[17] According to BCS theory, $\xi_0 \sim 0.18\frac{\hbar v_F}{kT_c}$. In pure metals, $m = m_e, v_F = 10^8 \frac{\text{cm}}{\text{s}}, T_c$ is low, $\xi_0 = 1000 - 16.000$ Å. The A-15 compounds Nb_3Sn and V_3Ga have flat bands, so that m is large, $v_F \doteq 10^6 \frac{\text{cm}}{\text{s}}, T_c$ is higher, and $\xi_0 = 50$ Å. The situation is different in high-T_c superconductors; there, $\xi_0 \sim$ Å.

The magnetization rotates through the same angle; this means that $\frac{\delta m}{m} = \frac{\delta h}{h}$, and we obtain for the transverse susceptibility,

$$\chi_\perp \equiv \frac{\delta m}{\delta h} = \frac{m}{h} . \tag{7.4.61}$$

The transverse correlation function in the Gaussian approximation (7.4.54a) is in agreement with this general result.

Remarks concerning the spatial dependence of the transverse correlation function $G_\perp(r)$:

(i)

$$G_\perp(r, h = 0) = \frac{1}{2\beta c} \int \frac{d^d k}{(2\pi)^d} \frac{e^{ikx}}{k^2} = A_d \left(\frac{\xi_\perp}{r}\right)^{d-2} , \quad \xi_\perp = (2\beta c)^{-\frac{1}{d-2}} \tag{7.4.62}$$

Employing the volume element $d^d k = dk\, k^{d-1}(\sin\theta)^{d-2} d\theta\, d\Omega_{d-1}$, the integral in (7.4.62) becomes

$$\frac{\Omega_{d-1}}{(2\pi)^d} \int_0^\infty dk k^{d-1} \frac{1}{2\beta c k^2} \int_0^\pi e^{ikr\cos\theta}(\sin\theta)^{d-2} d\theta$$

$$= \frac{K_{d-1}}{2\beta c\, 2\pi} \int_0^\infty dk\, k^{d-3} \Gamma\left(\frac{d}{2} - \frac{1}{2}\right) \Gamma\left(\frac{1}{2}\right) 2^{\frac{d}{2}-1} \frac{J_{\frac{d}{2}-1}(kr)}{(kr)^{\frac{d}{2}-1}}$$

$$\sim r^{-(d-2)} . \text{ [18]}$$

For dimensional reasons, $G_\perp(r)$ must be of the form

$$G_\perp(r) \sim M^2 \left(\frac{\xi}{r}\right)^{d-2} ,$$

i.e. the transverse correlation length from Eq. (7.4.62) is

$$\xi_\perp = \xi M^{\frac{2}{d-2}} \propto \tau^{-\nu} \tau^{\frac{2\beta}{d-2}} = \tau^{\eta\nu/(d-2)} , \tag{7.4.63}$$

where the exponent was rearranged using the scaling relations.
(ii) We also compute the local transverse fluctuations of the magnetization from

$$G_\perp(r = 0) \sim \int_0^\Lambda \frac{dk\, k^{d-1}}{\frac{h}{2m_1} + ck^2} \sim \left[\sqrt{\frac{2m_1}{h}} c\right]^{-d+2} \int_0^{\sqrt{\frac{2m_1}{h}}c\Lambda} dq \frac{q^{d-1}}{1 + q^2}$$

---

[18] I.S. Gradshteyn and I.M. Ryzhik, *Table of Integrals, Series, and Products* (Academic Press, New York 1980), Eq. 8.411.7

and consider the limit $h \to 0$: the result is

finite   for $d > 2$

$\log h$ for $d = 2$

$\left(\frac{m_1}{h}\right)^{\frac{2-d}{2}} \xrightarrow{h \to 0} \infty$ if $m_1 \neq 0$ for $d < 2$.

For $d \leq 2$, the transverse fluctuations diverge in the limit $h \to 0$. As a result, for $d \leq 2$, we must have $m_1 = 0$.

B) Longitudinal Correlation Function

In the Gaussian approximation, we found for $T < T_c$ in Eq. (7.4.54a) that

$$\lim_{k \to 0} \lim_{h \to 0} G_\|(k) = \frac{1}{-4\beta a}$$

as for $n = 1$. In fact, one would expect that the strong transverse fluctuations would modify the behavior of $G_\|(k)$. Going beyond the Gaussian approximation, we now calculate the contribution of orientation fluctuations to the longitudinal fluctuations. We consider a rotation of the magnetization at the point $\mathbf{x}$ and decompose the change $\delta\mathbf{m}$ into a component δm_1 parallel and a vector $\delta\mathbf{m}_\perp$ perpendicular to $\mathbf{m}_0$ (Fig. 7.17). The invariance of the length yields the condition

$$m_0^2 = m_0^2 + 2m_0\delta m_1 + \delta m_1^2 + (\delta\mathbf{m}_\perp)^2 \; ;$$

and it follows from this owing to $|\delta m_1| \ll m_0$ that

$$\delta m_1 = -\frac{1}{2m_0}(\delta\mathbf{m}_\perp)^2 \; . \tag{7.4.64}$$

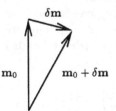

Fig. 7.17. The rotation of the spontaneous magnetization in isotropic systems

For the correlation of the longitudinal fluctuations, one obtains from this the following relation to the transverse fluctuations:

$$\langle \delta m_1(\mathbf{x})\delta m_1(0) \rangle = \frac{1}{4m_0^2}\langle \delta\mathbf{m}_\perp{}^2(\mathbf{x})\delta\mathbf{m}_\perp{}^2(0) \rangle \; . \tag{7.4.65}$$

We now factor this correlation function into the product of two transverse correlation functions, Eq. (7.4.54a), and obtain from it the Fourier-transformed longitudinal correlation function

$$G_\parallel(k=0) = \int d^d x \frac{e^{-2r/\sqrt{\frac{m_1}{h}}}}{r^{(d-2)2}} \sim \left(\sqrt{\frac{h}{m_1}}\right)^{d-4} \sim h^{\frac{d}{2}-2} \quad . \tag{7.4.66}$$

In three dimensions, we find from this for the longitudinal susceptibility

$$kT \frac{\partial m_1}{\partial h} = G_\parallel(k=0) \sim h^{-\frac{1}{2}} \quad . \tag{7.4.66'}$$

In the vicinity of the critical point T_c, we found $m \sim h^{\frac{1}{\delta}}$ (see just after Eq. (7.2.4b)); in contrast to (7.4.66), this yields

$$\frac{\partial m_1}{\partial h} \sim h^{-\frac{\delta-1}{\delta}} \quad .$$

In isotropic systems, the longitudinal susceptibility is not just singular only in the critical region, but instead in the whole coexistence region for $h \to 0$ (cf. Fig. 7.18). This is a result of rotational invariance.

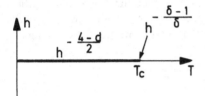

Fig. 7.18. Singularities in the longitudinal susceptibility in systems with internal rotational symmetry, $n \geq 2$

C) Coexistence Singularities

The term coexistence region denotes the region of the phase diagram with a finite magnetization in the limiting case of $h \to 0$. The coexistence singularities found in (7.4.54a), (7.4.62), and (7.4.66) for isotropic systems are exactly valid. This can be shown as follows: for $T < T_c^0$, the Ginzburg–Landau functional can be written in the form

$$\begin{aligned}
\mathcal{F}[\mathbf{m}] &= \int d^d x \left(\frac{1}{2} b \left(\mathbf{m}^2 - \frac{|a|}{b} \right)^2 + (\nabla \mathbf{m})^2 - h\mathbf{m} - \frac{|a|^2}{2b} \right) \\
&= \int d^d x \left(\frac{1}{2} b \left(m_1^2 + 2m_1 m_1'(\mathbf{x}) + m_1'(\mathbf{x})^2 + \mathbf{m}_\perp'(\mathbf{x})^2 - \frac{|a|}{b} \right)^2 \right. \tag{7.4.67} \\
&\quad \left. + c\left(\nabla m_1'(\mathbf{x})\right)^2 + c\left(\nabla \mathbf{m}_\perp'(\mathbf{x})\right)^2 - h\left(m_1 + m_1'(\mathbf{x})\right) - \frac{|a|^2}{2b} \right) .
\end{aligned}$$

In this expression, we have inserted (7.4.38) and have combined the components $m_i'(\mathbf{x}), i \geq 2$ into a vector of the transverse fluctuations $\mathbf{m}_\perp'(\mathbf{x}) = (0, m_2'(\mathbf{x}), \ldots, m_n'(\mathbf{x}))$. Using (7.4.18) and $m_1'(\mathbf{x}) \ll m_1$, one obtains

$$\mathcal{F}[\mathbf{m}] = \int d^d x \left(\frac{1}{2} b \left(2m_1 m_1' + \mathbf{m}_\perp'^2 + \frac{h}{2bm_1} \right)^2 \right.$$
$$\left. + c \left(\nabla m_1' \right)^2 + c \left(\nabla \mathbf{m}_\perp' \right)^2 - h \left(m_1 + m_1' \right) - \frac{|a|^2}{2b} \right) . \tag{7.4.68}$$

The terms which are nonlinear in the transverse fluctuations are absorbed into the longitudinal terms by making the substitution

$$m_1' = m_1'' - \frac{\mathbf{m}_\perp'^2}{2m_1} : \tag{7.4.69}$$

$$\mathcal{F}[\mathbf{m}] = \int d^d x \left(2bm_1^2 m_1''^2 + c \left(\nabla m_1'' \right)^2 \right.$$
$$\left. + \frac{h}{2m_1} \mathbf{m}_\perp'^2 + c \left(\nabla \mathbf{m}_\perp' \right)^2 - hm_1 + \frac{h^2}{8bm_1^2} - \frac{|a|^2}{2b} \right) . \tag{7.4.70}$$

The final result for the free energy is harmonic in the variables m_1'' and $\mathbf{m}_\perp'$. As a result, the transverse propagator in the coexistence region is given exactly by (7.4.54a). The longitudinal correlation function is

$$\langle m_1'(\mathbf{x}) m_1'(0) \rangle_C = \langle m_1''(\mathbf{x}) m_1''(0) \rangle + \frac{1}{4m_1^2} \langle \mathbf{m}'_\perp(\mathbf{x})^2 \mathbf{m}'_\perp(0)^2 \rangle_C . \tag{7.4.71}$$

In equation (7.4.70), terms of the form $(\nabla \frac{\mathbf{m}_\perp'^2}{m_1})^2$ and $\nabla m_1'' \nabla \frac{\mathbf{m}_\perp'^2}{m_1}$ have been neglected.

The second term in (7.4.69) leads to a reduction of the order parameter $-\frac{\langle \mathbf{m}_\perp'^2 \rangle}{2m_1}$. Eq. (7.4.71) gives the cumulant, i.e. the correlation function of the deviations from the mean value. Since (7.4.70) now contains only harmonic terms, the factorization of the second term in the sum in (7.4.71) is exact, as used in Eq. (7.4.65). One could still raise the objection to the derivation of (7.4.71) that a number of terms were neglected. However, using renormalization group theory[19], it can be shown that the anomalies of the coexistence region are described by a low-temperature fixed point at which $m_0 = \infty$. This means that the result is asymptotically exact.

7.4.4.2 First-Order Phase Transitions

There are systems in which not only the transition from one orientation of the order parameter to the opposite direction is of first order, but also the transition at T_c. This means that the order parameter jumps at T_c from zero to a finite value (an example is the ferroelectric transition in $BaTiO_3$). This situation can be described in the Ginzburg–Landau theory, if $b < 0$,

---

[19] I. D. Lawrie, J. Phys. A**14**, 2489 (1981); *ibid.*, A**18**, 1141 (1985); U. C. Täuber and F. Schwabl, Phys. Rev. B**46**, 3337 (1992).

and if a term of the form $\frac{1}{2}vm^6$ with $v > 0$ is added for stability. Then the Ginzburg–Landau functional takes the form

$$\mathcal{F} = \int d^d x \left\{ am^2 + c(\nabla m)^2 + \frac{1}{2}bm^4 + \frac{1}{2}vm^6 \right\} , \qquad (7.4.72)$$

where $a = a'(T - T_c^0)$. The free energy density is shown in Fig. 7.19 for a uniform order parameter.

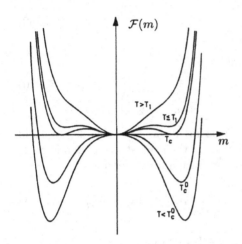

Fig. 7.19. The free energy density in the vicinity of a first-order phase transition at temperatures $T < T_c^0$, $T \approx T_c^0$, $T = T_c$, $T < T_1$, $T > T_1$

For $T > T_1$, there is only the minimum at $m = 0$, that is the non-ordered state. At T_1, a second relative minimum appears, which for $T \leq T_c$ finally becomes deeper than that at $m = 0$. For $T < T_c^0$, the $m = 0$ state is unstable. The stationarity condition is

$$\left(a + bm^2 + 3\frac{v}{2}m^4 \right) m = 0 , \qquad (7.4.73)$$

and the condition that a minimum is present is

$$\frac{1}{2}\frac{\partial^2 f}{\partial m^2} = a + 3bm^2 + 15\frac{v}{2}m^4 > 0 . \qquad (7.4.74)$$

The solutions of the stationarity condition are

$$m_0 = 0 \qquad (7.4.75a)$$

and

$$m_0^2 = -\frac{b}{3v} \underset{(-)}{+} \left(\frac{b^2}{9v^2} - \frac{2a}{3v} \right)^{1/2} . \qquad (7.4.75b)$$

We recall that $b < 0$. The nonzero solution with the minus sign corresponds to a maximum in the free energy and will be left out of further consideration.

The minimum (7.4.75b) exists for all temperatures for which the discriminant is positive, i.e. below the temperature T_1

$$T_1 = T_c^0 + \frac{b^2}{6va'} \, . \tag{7.4.76}$$

T_1 is the superheating temperature (see Fig. 7.19 and below). The transition temperature T_c is found from the condition that the free energy for (7.4.75b) is zero. At this temperature (see Fig. 7.19), the free energy has a double zero at $m^2 = m_0^2$ and thus has the form

$$\frac{v}{2}(m^2 - m_0^2)^2 m^2 = \left(a + \frac{b}{2}m^2 + \frac{v}{2}m^4\right)m^2$$
$$= \left(\frac{v}{2}\left(m^2 + \frac{b}{2v}\right)^2 - \frac{b^2}{8v} + a\right)m^2 = 0 \, .$$

It follows from this that $a = \frac{b^2}{8v}$ and $m^2 = -\frac{b}{2v}$, which both lead to

$$T_c = T_c^0 + \frac{b^2}{8va'} \, . \tag{7.4.77}$$

For $T < T_c^0$, there is a local maximum at $m = 0$. T_c^0 plays the role of a supercooling temperature. In the range $T_c^0 \leq T \leq T_1$, both phases can thus coexist, i.e. the supercooling or superheating of a phase is possible. Since for $T_c^0 \leq T < T_c$, the non-ordered phase ($m_0 = 0$) is metastable; for $T_1 \geq T > T_c$, in contrast, the ordered phase ($m_0 \neq 0$) is metastable. On slow cooling, so that the system attains the state of lowest free energy, m_0 jumps at T_c from 0 to

$$m_0^2(T_c) = -\frac{b}{3v} + \left(\frac{b^2}{9v^2} - \frac{b^2}{12v^2}\right)^{1/2} = -\frac{b}{2v} \, , \tag{7.4.78}$$

and, below T_c, it has the temperature dependence (Fig. 7.20)

$$m_0^2(T) = \frac{2}{3}m_0^2(T_c)\left[1 + \sqrt{1 - \frac{3}{4}\frac{(T - T_c^0)}{(T_c - T_c^0)}}\right] \, .$$

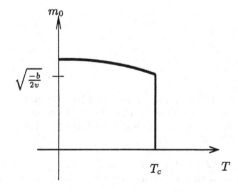

Fig. 7.20. The temperature dependence of the magnetization in a first-order phase transition

*7.4.5 The Momentum-Shell Renormalization Group

The RG theory can also be carried out in the framework of the G–L functional, with the following advantages compared to discrete spin models: the method is also practicable in higher dimensions, and various interactions and symmetries can be treated. One employs an expansion of the critical exponents in $\epsilon = 4 - d$. Here, we cannot go into the details of the necessary perturbation-theoretical techniques, but rather just show the essential structure of the renormalization group recursion relations and their consequences. For the detailed calculation, the reader is referred to more extensive descriptions[20,21] and to the literature at the end of this chapter.

7.4.5.1 Wilson's RG Scheme

We now turn to the renormalization group transformation for the Ginzburg–Landau functional (7.4.10). In order to introduce the notation which is usual in this context, we carry out the substitutions

$$m = \frac{1}{\sqrt{2c}}\phi, \quad a = rc, \quad b = uc^2 \quad \text{and} \quad \mathbf{h} \to \sqrt{2c}\,\mathbf{h}, \tag{7.4.79}$$

and obtain the so called *Landau–Ginzburg–Wilson functional*:

$$\mathcal{F}[\phi] = \int d^d x \left[\frac{r}{2}\phi^2 + \frac{u}{4}(\phi^2)^2 + \frac{1}{2}(\nabla\phi)^2 - \mathbf{h}\phi\right]. \tag{7.4.80}$$

An intuitively appealing method of proceeding was proposed by K. G. Wilson[20,21]. Essentially, the trace over the degrees of freedom with large k in momentum space is evaluated, and one thereby obtains recursion relations for the Ginzburg–Landau coefficients. Since it is to be expected that the detailed form of the short-wavelength fluctuations is not of great importance, the Brillouin zone can be approximated as simply a d-dimensional sphere of radius (cutoff) Λ, Fig. 7.21.

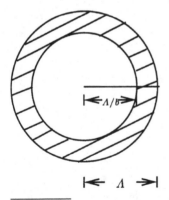

$\twoheadleftarrow\!\Lambda/b\!\twoheadrightarrow$

$\twoheadleftarrow\ \Lambda\ \twoheadrightarrow$

Fig. 7.21. The momentum-space RG: the partial trace is performed over the Fourier components ϕ_k with momenta within the shell $\Lambda/b < |k| < \Lambda$

[20] Wilson, K. G. and Kogut, J., Phys. Rep. **12 C**, 76 (1974).
[21] S. Ma, *Modern Theory of Critical Phenomena*, Benjamin, Reading, 1976.

The *momentum-shell RG transformation* then consists of the following steps:

(i) Evaluating the trace over all the Fourier components ϕ_k with $\Lambda/b < |k| < \Lambda$ (Fig. 7.21) eliminates these short-wavelength modes.

(ii) By means of a scale transformation[22]

$$k' = bk ,\tag{7.4.81}$$

$$\phi' = b^\zeta \phi ,\tag{7.4.82}$$

and therefore

$$\phi'_{k'} = b^{\zeta-d}\phi_k ,\tag{7.4.83}$$

the resulting effective Hamiltonian functional can be brought into a form resembling the original model, whereby effective scale-dependent coupling parameters are defined. Repeated application of this RG transformation (which represents a semigroup, since it has no inverse element) discloses the presumably universal properties of the long-wavelength regime. As in the real-space renormalization group transformation of Sect. 7.3.3, the fixed points of the transformation correspond to the various thermodynamic phases and the phase transitions between them. The eigenvalues of the linearized flow equations in the vicinity of the critical fixed point finally yield the critical exponents (see (7.3.41a,b,c)). Although a perturbational expansion (in terms of u) is in no way justifiable in the critical region, it is completely legitimate at some distance from the critical point, where the fluctuations are negligible. The important observation is now that the RG flow connects these quite different regions, so that the results of the perturbation expansion in the non-critical region can be transported to the vicinity of T_c, whereby the non-analytic singularities are consistently, controllably, and reliably taken into account by this mapping. Perturbation-theoretical methods can likewise be applied in the elimination of the short-wavelength degrees of freedom (step (i)).

7.4.5.2 Gaussian Model

We will now apply the concept described in the preceding section first to the Gaussian model, where $u = 0$ (see Sect. 7.4.3),

---

[22] If one considers (7.4.83) together with the field term in the Ginzburg–Landau functional (7.4.10), then it can be seen that the exponent ζ determines the transformation of the external field and is related to y_h from Sect. (7.3.4) via $\zeta = d - y_h$.

$$\mathcal{F}_0[\phi_k] = \int_{|k|<\Lambda} \frac{r+k^2}{2} |\phi_k|^2 \quad , \tag{7.4.84}$$

with $\int_k \equiv \int d^d k/(2\pi)^d$. Since (7.4.84) is diagonal in the Fourier modes, the elimination of the components with large k merely produces a constant contribution (independent of ϕ); the form of the effective Hamiltonian functional remains unchanged, provided that

$$\zeta = \frac{d-2}{2} \ , \ \text{i.e.} \ \eta = 0 \ , \tag{7.4.85}$$

and r transforms as

$$r' = b^2 r \ . \tag{7.4.86}$$

The fixed points of Eq. (7.4.86) are $r^* = \pm\infty$, corresponding to the high- and low-temperature phases, and the critical fixed point $r^* = 0$. The eigenvalue for the relevant temperature-direction at this critical fixed point is clearly $y_\tau = 2$, and therefore one obtains the same exponent $\nu = 1/2$ from (7.3.41c,d) as in the molecular field theory or the Gaussian approximation (Sect. 7.4.3).

7.4.5.3 Perturbation Theory and the ϵ-Expansion

The nonlinear interaction term in Eq. (7.4.80)

$$\mathcal{F}_{\text{int}}[\phi_k] = \frac{u}{4} \int_{|k_i|<\Lambda} \phi_{k_1} \phi_{k_2} \phi_{k_3} \phi_{-k_1-k_2-k_3} \tag{7.4.87}$$

can now be treated using perturbation theory, by expanding the exponential function in Eq. (7.4.6) in terms of u. If one separates the field variables into their parts in the inner and the outer momentum shell,

$$\phi_k = \phi_{k<} + \phi_{k>} \ ,$$

with $|k_<| < \Lambda/b$ and $\Lambda/b < |k_>| < \Lambda$, the result to first order in u includes terms of the following (symbolically written) form (from now on, we set kT equal to 1):

(i) $u \int \phi_<^4 e^{-\mathcal{F}_0}$ must merely be re-exponentiated, since these degrees of freedom are not eliminated;

(ii) all terms with an uneven number of $\phi_<$ or $\phi_>$, such as for example $u \int \phi_<^3 \phi_> e^{-\mathcal{F}_0}$, vanish;

(iii) $u \int \phi_>^4 e^{-\mathcal{F}_0}$ makes a constant contribution to the free energy and finally to $u \int \phi_<^2 \phi_>^2 e^{-\mathcal{F}_0}$, for which the Gaussian integral over the $\phi_>$ can be carried out with the aid of Eq. (7.4.47) for the *propagator* $\langle \phi_{k_>}^\alpha \phi_{-k_>'}^\beta \rangle_0 = \frac{\delta_{kk'}}{2(r+k^2)}$, an average value which is calculated with the statistical weight $e^{-\mathcal{F}_0}$.

Quite generally, *Wick's theorem*[20,21] states that expressions of the form

$$\left\langle \prod_{i}^{m} \phi_{k_i>} \right\rangle_0 \equiv \langle \phi_{k_1>} \phi_{k_2>} \cdots \phi_{k_m>} \rangle_0$$

factorize into a sum of products of all possible pairs $\langle \phi_{k_>} \phi_{-k_>} \rangle_0$ if m is even, and otherwise they yield zero. Especially in the treatment of higher orders of perturbation theory, the *Feynman diagrams* offer a very helpful representation of the large number of contributions which have to be summed in the perturbation expansion. In these diagrams, lines symbolize the propagators and *interaction vertices* stand for the nonlinear coupling u. With these means at our disposal, we can compute the two-point function $\langle \phi_{k_<} \phi_{-k_<} \rangle$ and the similarly defined four-point function. Using Eq. (7.4.47), one then obtains in the first non-trivial order (*"1-loop"*, a notation which derives from the graphical representation) the following recursion relation between the initial coefficients r, u and the transformed coefficients r', u' of the Ginzburg–Landau–Wilson functional[20,21]:

$$r' = b^2 \left(r + (n+2) A(r) u \right) , \tag{7.4.88}$$

$$u' = b^{4-d} u \left(1 - (n+8) C(r) u \right) , \tag{7.4.89}$$

where $A(r)$ and $C(r)$ refer to the integrals

$$A(r) = K_d \int_{\Lambda/b}^{\Lambda} (k^{d-1}/r + k^2) dk$$

$$= K_d \left[\Lambda^{d-2} (1 - b^{2-d})/(d-2) - r\Lambda^{d-4}(1 - b^{4-d})/(d-4) \right] + \mathcal{O}(r^2)$$

$$C(r) = K_d \int_{\Lambda/b}^{\Lambda} \left[k^{d-1}/(r + k^2)^2 \right] dk$$

$$= K_d \Lambda^{d-4} (1 - b^{4-d})/(d-4) + \mathcal{O}(r) ,$$

with $K_d = 1/2^{d-1} \pi^{d/2} \Gamma(d/2)$, and the factors depending on the number n of components of the order parameter field result from the combinatorial analysis in counting the equivalent possibilities for "contracting" the fields $\phi_{k_>}$, i.e. for evaluating the integrals over the large momenta. We note that here again, Eq. (7.4.85) applies.

Linearizing equations (7.4.88) and (7.4.89) at the Gaussian fixed point $r^* = 0$, $u^* = 0$, one immediately finds the eigenvalues $y_r = 2$ and $y_u = 4 - d$. Then for $d > d_c = 4$, the nonlinearity $\propto u$ is seen to be irrelevant, and the mean field exponents are valid, as already surmised in Sect. 7.4.4. For $d < 4$ ($d_c = 4$ is the upper critical dimension), the fluctuations however become relevant and each initial value $u \neq 0$ increases under the renormalization group transformation. In order to obtain the scaling behavior in this case, we must therefore search for a finite, non-trivial fixed point. This can be most easily done by introducing a differential flow, with $b^\ell = e^{\delta \ell}$ and $\delta \to 0$, making the number of RG steps effectively into a continuous variable, and studying the resulting differential recursion relations:

$$\frac{dr(\ell)}{d\ell} = 2r(\ell) + (n+2)u(\ell)K_d\Lambda^{d-2} - (n+2)r(\ell)u(\ell)K_d\Lambda^{d-4} , \quad (7.4.90)$$

$$\frac{du(\ell)}{d\ell} = (4-d)u(\ell) - (n+8)u(\ell)^2 K_d\Lambda^{d-4} . \quad (7.4.91)$$

Now, a fixed point is defined by the condition $dr/d\ell = 0 = du/d\ell$.

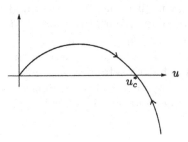

Fig. 7.22. Flow of the effective coupling $u(\ell)$, determined by the right-hand side of Eq. (7.4.91), which is plotted here as the ordinate. Both for initial values $u_0 > u_c^*$ and $0 < u_0 < u_c^*$, one finds $u(\ell) \to u_c^*$ for $\ell \to \infty$

Figure 7.22 shows the flow of $u(\ell)$ corresponding to Eq. (7.4.91); for any initial value $u_0 \neq 0$, one finds that asymptotically, i.e. for $\ell \to \infty$, the non-trivial fixed point

$$u_c^* K_d = \frac{\epsilon}{n+8}\Lambda^\epsilon , \quad \epsilon = 4-d \quad (7.4.92)$$

is approached; this should determine the universal critical properties of the model. As in the real-space renormalization in Sect. 7.3, the RG transformation via momentum-shell elimination generates new interactions; for example, terms $\propto \phi^6$ and $\nabla^2\phi^4$, etc., which again influence the recursion relations for r and u in the succeeding steps. It turns out, however, that up to order ϵ^3, these terms do not have to be taken into account.[20,21]

The original assumption that u should be small, which justified the perturbation expansion, now means in light of Eq. (7.4.92) that the effective expansion parameter here is the deviation from the upper critical dimension, ϵ. If one inserts (7.4.92) into Eq. (7.4.90) and includes terms up to $\mathcal{O}(\epsilon)$, the result is

$$r_c^* = -\frac{n+2}{2} u_c^* K_d\Lambda^{d-2} = -\frac{(n+2)\epsilon}{2(n+8)}\Lambda^2 . \quad (7.4.93)$$

The physical interpretation of this result is that fluctuations lead to a lowering of the transition temperature. With $\tau = r - r_c^*$, the differential form of the flow equation

$$\frac{d\tau(\ell)}{d\ell} = \tau(\ell)\Big(2 - (n+2)\, u\, K_d\,\Lambda^{d-4}\Big) \quad (7.4.94)$$

finally yields the eigenvalue $y_\tau = 2 - (n+2)\epsilon/(n+8)$ in the vicinity of the critical point (7.4.92). In the one-loop order which we have described here, $\mathcal{O}(\epsilon)$, one therefore finds for the critical exponent ν from Eq. (7.3.41e)

$$\nu = \frac{1}{2} + \frac{n+2}{4(n+8)}\, \epsilon + \mathcal{O}(\epsilon^2) \,. \tag{7.4.95}$$

Using the result $\eta = \mathcal{O}(\epsilon^2)$ and the scaling relations (7.3.41a–d), one obtains the following expressions (the difference from the result (7.4.35) of the Gaussian approximation is remarkable)

$$\alpha = \frac{4-n}{2(n+8)}\, \epsilon + \mathcal{O}(\epsilon^2) \,, \tag{7.4.96}$$

$$\beta = \frac{1}{2} - \frac{3}{2(n+8)}\, \epsilon + \mathcal{O}(\epsilon^2) \,, \tag{7.4.97}$$

$$\gamma = 1 + \frac{n+2}{2(n+8)}\, \epsilon + \mathcal{O}(\epsilon^2) \,, \tag{7.4.98}$$

$$\delta = 3 + \epsilon + \mathcal{O}(\epsilon^2) \tag{7.4.99}$$

to first order in the expansion parameter $\epsilon = 4 - d$. The first non-trivial contribution to the exponent η appears in the two-loop order,

$$\eta = \frac{n+2}{2(n+8)^2}\, \epsilon^2 + \mathcal{O}(\epsilon^3) \,. \tag{7.4.100}$$

The universality of these results manifests itself in the fact that they depend only on the spatial dimension d and the number of components n of the order parameter, and not on the original "microscopic" Ginzburg–Landau parameters.

Remarks:

(i) At the upper critical dimension, $d_c = 4$, an inverse power law is obtained as the solution of Eq. (7.4.91) instead of an exponential behavior, leading to logarithmic corrections to the mean-field exponents.

(ii) We also mention that for long-range interactions which exhibit power-law behavior $\propto |x|^{-(d+\sigma)}$, the critical exponents contain an additional dependence on the parameter σ.

(iii) In addition to the ϵ-expansion, an expansion in terms of powers of $1/n$ is also possible. Here, the limit $n \to \infty$ corresponds to the exactly solvable spherical model.[23] This $1/n$-expansion indeed helps to clarify some general aspects but its numerical accuracy is not very great, since precisely the small values of n are of practical interest.

---

[23] Shang-Keng Ma, *Modern Theory of Critical Phenomena*, Benjamin, Reading, 1976.

The differential recursion relations of the form (7.4.90) and (7.4.91) also serve as a basis for the treatment of more subtle issues such as the calculation of the scaling functions or the treatment of *crossover phenomena* within the framework of the RG theory. Thus, for example, an anisotropic perturbation in the n-component Heisenberg model favoring m directions leads to a crossover from the $O(n)$-Heisenberg fixed point[24] to the $O(m)$ fixed point.[25] The instability of the former is described by the crossover exponent. For small anisotropic disturbances, to be sure, the flow of the RG trajectory passes very close to the unstable fixed point. This means that one finds the behavior of an n-component system far from the transition temperature T_c, before the system is finally dominated by the anisotropic critical behavior.

The crossover from one RG fixed point to another can be represented (and measured) by the introduction of *effective exponents*. These are defined as logarithmic derivatives of suitable physical quantities. Other important perturbations which were treated within the RG theory are, on the one hand, cubic terms. They reflect the underlying crystal structure and contribute terms of fourth order in the cartesian components of ϕ to the Ginzburg–Landau–Wilson functional. On the other hand, dipolar interactions lead to a perturbation which alters the harmonic part of the theory.

7.4.5.4 More Advanced Field-Theoretical Methods

If one wishes to discuss perturbation theory in orders higher than the first or second, Wilson's momentum-shell renormalization scheme is not the best choice for practical calculations, in spite of its intuitively appealing properties. The technical reason for this is that the integrals in Fourier space involve nested momenta, which owing to the finite cutoff wavelength Λ are difficult to evaluate. It is then preferable to use a field-theoretical renormalization scheme with $\Lambda \to \infty$. However, this leads to additional ultraviolet (UV) divergences of the integrals for $d \geq d_c$. At the critical dimension d_c, both ultraviolet and infrared (IR) singularities occur in combination in logarithmic form, $[\propto \log(\Lambda^2/r)]$. The idea is now to treat the UV divergences with the methods originally developed in quantum field theory and thus to arrive at the correct scaling behavior for the IR limit. In the formal implementation, one takes advantage of the fact that the original unrenormalized theory does not depend on the arbitrarily chosen renormalization point; as a consequence, one obtains the *Callan–Symanzik-* or *RG equations*. These are partial differential equations which correspond to the differential flow equations in the Wilson scheme.

---

[24] $O(n)$ indicates invariance with respect to rotations in n-dimensional space, i.e. with respect to the group $O(n)$.

[25] See D. J. Amit, *Field Theory, the Renormalization Group and Critical Phenomena*, 2nd ed., World Scientific, Singapore, 1984, Chap. 5–3.

ϵ-expansions have been carried out up to the seventh order;[26] the series obtained is however only asymptotically convergent (the convergence radius of the perturbation expansion in u clearly must be zero, since $u < 0$ corresponds to an unstable theory). The combination of the results from expansions to such a high order with the divergent asymptotic behavior and Borel resummation techniques yields critical exponents with an impressive precision; cf. Table 7.4.

Table 7.4. The best estimates for the static critical exponents ν, β, and δ, for the $O(n)$-symmetric ϕ^4 model in $d = 2$ and $d = 3$ dimensions, from ϵ-expansions up to high order in connection with Borel summation techniques.[26] For comparison, the exact Onsager results for the 2d-Ising model are also shown. The limiting case $n = 0$ describes the statistical mechanics of polymers.

		γ	ν	β	η
$d = 2$	$n = 0$	1.39 ± 0.04	0.76 ± 0.03	0.065 ± 0.015	0.21 ± 0.05
	$n = 1$	1.73 ± 0.06	0.99 ± 0.04	0.120 ± 0.015	0.26 ± 0.05
2d Ising (exact)		1.75	1	0.125	0.25
$d = 3$	$n = 0$	1.160 ± 0.004	0.5885 ± 0.0025	0.3025 ± 0.0025	0.031 ± 0.003
	$n = 1$	1.239 ± 0.004	0.6305 ± 0.0025	0.3265 ± 0.0025	0.037 ± 0.003
	$n = 2$	1.315 ± 0.007	0.671 ± 0.005	0.3485 ± 0.0035	0.040 ± 0.003
	$n = 3$	1.390 ± 0.010	0.710 ± 0.007	0.368 ± 0.004	0.040 ± 0.003

*7.5 Percolation

Scaling theories and renormalization group theories also play an important role in other branches of physics, whenever the characteristic length tends to infinity and structures occur on every length scale. Examples are percolation in the vicinity of the percolation threshold, polymers in the limit of a large number of monomers, the self-avoiding random walk, growth processes, and driven dissipative systems in the limit of slow growth rates (self-organized criticality). As an example of such a system which can be described in the language of critical phenomena, we will consider percolation.

7.5.1 The Phenomenon of Percolation

The phenomenon of *percolation* refers to problems of the following type:

(i) Consider a landscape with hills and valleys, which gradually fills up with water. When the water level is low, lakes are formed; as the level rises, some of

---

[26] J. C. Le Guillou and J. C. Zinn-Justin, J. Phys. Lett. **46** L, 137 (1985)

the lakes join together until finally at a certain critical level (or critical area) of the water, a sea is formed which stretches from one end of the landscape to the other, with islands.

(ii) Consider a surface made of an electrical conductor in which circular holes are punched in a completely random arrangement (Fig. 7.23a). Denoting the fraction of remaining conductor area by p, we find for $p > p_c$ that there is still an electrical connection from one end of the surface to the other, while for $p < p_c$, the pieces of conducting area are reduced to islands and no longer form continuous bridges, so that the conductivity of this disordered medium is zero. One refers to p_c as the *percolation threshold*. Above p_c, there is an infinite "cluster"; below this limit, there are only finite clusters, whose average radius however diverges on approaching p_c. Examples (i) and (ii) represent continuum percolation. Theoretically, one can model such systems on a discrete d-dimensional lattice. In fact, such discrete models also occur in Nature, e.g. in alloys.

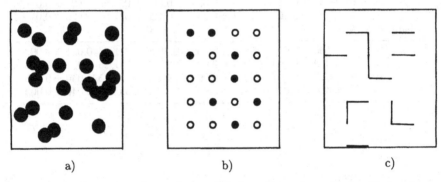

a)   b)   c)

Fig. 7.23. Examples of percolation **(a)** A perforated conductor (Swiss cheese model): continuum percolation; **(b)** site percolation; **(c)** bond percolation

(iii) Let us imagine a square lattice in which each site is occupied with a probability p and is unoccupied with the probability $(1 - p)$. 'Occupied' can mean in this case that an electrical conductor is placed there and 'unoccupied' implies an insulator, or that a magnetic ion or a nonmagnetic ion is present, cf. Fig. 7.23b. Staying with the first interpretation, we find the following situation: for small p, the conductors form only small islands (electric current can flow only between neighboring sites) and the overall system is an insulator. As p increases, the islands (clusters) of conducting sites get larger. Two lattice sites belong to the same cluster when there is a connection between them via occupied nearest neighbors. For large p ($p \lesssim 1$) there are many conducting paths between the opposite edges and the system is a good conductor. At an intermediate concentration p_c, the percolation threshold or critical concentration, a connection is just formed, i.e. current can percolate from one edge of the lattice to the other. The critical concentration separates the insulating phase below p_c from the conducting phase above p_c.

In the case of the magnetic example, at p_c a ferromagnet is formed from a paramagnet, presuming that the temperature is sufficiently low. A further example is the occupation of the lattice sites by superconductors or normal conductors, in which case a transition from the normal conducting to the superconducting state takes place.

We have considered here some examples of *site percolation*, in which the lattice sites are stochastically occupied, Fig. 7.23b. Another possibility is that bonds between the lattice sites are stochastically present or are broken. One then refers to *bond percolation* (cf. Fig. 7.23c). Here, clusters made up of existing bonds occur; two bonds belong to the same cluster if there is a connection between them via existing bonds. Two examples of bond percolation are: (i) a macroscopic system with percolation properties can be produced from a stochastic network of resistors and connecting wires; (ii) a lattice of branched monomers can form bonds between individual monomers with a probability p. For $p < p_c$, finite macromolecules are formed, and for $p > p_c$, a network of chemical bonds extends over the entire lattice. This gelation process from a solution to a gel state is called the sol-gel transition (example: cooking or "denaturing" of an egg or a pudding); see Fig. 7.23.

Remarks:

(i) Questions related to percolation are also of importance outside physics, e.g. in biology. An example is the spread of an epidemic or a forest fire. An affected individual can infect a still-healthy neighbor within a given time step, with a probability p. The individual dies after one time step, but the infected neighbors could transmit the disease to other still living, healthy neighbors. Below the critical probability p_c, the epidemic dies out after a certain number of time steps; above this probability, it spreads further and further. In the case of a forest fire, one can think of a lattice which is occupied by trees with a probability p. When a tree burns, it ignites the neighboring trees within one time step and is itself reduced to ashes. For small values of p, the fire dies out after several time steps. For $p > p_c$, the fire spreads over the entire forest region, assuming that all the trees along one boundary were ignited. The remains consist of burned-out trees, empty lattice sites, and trees which were separated from their surroundings by a ring of empty sites so that they were never ignited. For $p > p_c$, the burned-out trees form an infinite cluster.

(ii) In Nature, disordered systems often occur. Percolation is a simple example of this, in which the occupation of the individual lattice sites is uncorrelated among the sites.

As emphasized above, these models for percolation can also be introduced on a d-dimensional lattice. The higher the spatial dimension, the more possible connected paths there are between sites; therefore, the percolation threshold p_c decreases with increasing spatial dimension. The percolation threshold is also smaller for bond percolation than for site percolation, since a bond has more neighboring bonds than a lattice site has neighboring lattice sites (in a square lattice, 6 instead of 4). See Table 7.5.

Table 7.5. Percolation thresholds and critical exponents for some lattices

Lattice	p_c site	bond	β	ν	γ
one-dimensional	1	1	–	1	1
square	0.592	1/2	$\frac{5}{36}$	$\frac{4}{3}$	$\frac{43}{18}$
simple cubic	0.311	0.248	0.417	0.875	1.795
Bethe lattice	$\frac{1}{z-1}$	$\frac{1}{z-1}$	1	1	1
$d = 6$ hypercubic	0.107	0.0942	1	$\frac{1}{2}$	1
$d = 7$ hypercubic	0.089	0.0787	1	$\frac{1}{2}$	1

The *percolation transition*, in contrast to thermal phase transitions, has a geometric nature. When p increases towards p_c, the clusters become larger and larger; at p_c, an infinite cluster is formed. Although this cluster already extends over the entire area, the fraction of sites which it contains is still zero at p_c. For $p > p_c$, more and more sites join the infinite cluster at the expense of the finite clusters, whose average radii decrease. For $p = 1$, all sites naturally belong to the infinite cluster. The behavior in the vicinity of p_c exhibits many similarities to critical behavior in second-order phase transitions in the neighborhood of the critical temperature T_c. As discussed in Sect. 7.1, the magnetization increases below T_c as $M \sim (T_c - T)^{\beta}$. In the case of percolation, the quantity corresponding to the *order parameter* is the *probability* P_∞ that an occupied site (or an existing bond) belongs to the infinite cluster, Fig. (7.24). Accordingly,

$$P_\infty \propto \begin{cases} 0 & \text{for} \quad p < p_c \\ (p - p_c)^{\beta} & \text{for} \quad p > p_c . \end{cases} \tag{7.5.1}$$

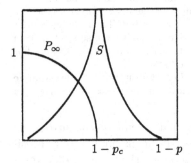

Fig. 7.24. P_∞: order parameter (the strength of the infinite clusters); S: average number of sites in a finite cluster

The *correlation length* ξ characterizes the linear dimension of the finite clusters (above and below p_c). More precisely, it is defined as the average distance between two occupied lattice sites in the same finite cluster. In the vicinity of p_c, ξ behaves as

$$\xi \sim |p - p_c|^{-\nu} . \tag{7.5.2}$$

A further variable is the average number of sites (bonds) in a finite cluster. It diverges as

$$S \sim |p - p_c|^{-\gamma} \tag{7.5.3}$$

and corresponds to the magnetic susceptibility χ; cf. Fig. (7.24).

Just as in a thermal phase transition, one expects that the critical properties (e.g. the values of β, ν, γ) are *universal*, i.e. that they do not depend on the lattice structure or the kind of percolation (site, bond, continuum percolation). These critical properties do, however, depend on the spatial dimension of the system. The values of the exponents are collected in Table 7.5 for several different lattices. One can map the percolation problem onto an s-state-Potts model, whereby the limit $s \to 1$ is to be taken.[27,28] From this relation, it is understandable that the upper critical dimension for percolation is $d_c = 6$. The Potts model in its field-theoretical Ginzburg–Landau formulation contains a term of the form ϕ^3; from it, following considerations analogous to the ϕ^4 theory, the characteristic dimension $d_c = 6$ is derived. The critical exponents β, ν, γ describe the geometric properties of the percolation transition. Furthermore, there are also dynamic exponents, which describe the transport properties such as the electrical conductivity of the perforated circuit board or of the disordered resistance network. Also the magnetic thermodynamic transitions in the vicinity of the percolation threshold can be investigated.

7.5.2 Theoretical Description of Percolation

We consider clusters of size s, i.e. clusters containing s sites. We denote the number of such s-clusters divided by the number of all lattice sites by n_s, and call this the (normalized) *cluster number*. Then $s\,n_s$ is the probability that an arbitrarily chosen site will belong to a cluster of size s. Below the percolation threshold $(p < p_c)$, we have

$$\sum_{s=1}^{\infty} s\,n_s = \frac{\text{number of all the occupied sites}}{\text{total number of lattice sites}} = p . \tag{7.5.4}$$

The number of clusters per lattice site, irrespective of their size, is

$$N_c = \sum_s n_s . \tag{7.5.5}$$

---

[27] C. M. Fortuin and P. W. Kasteleyn, Physica **57**, 536 (1972).

[28] The s-state-Potts model is defined as a generalization of the Ising model, which corresponds to the 2-state-Potts model: at each lattice site there are s states Z. The energy contribution of a pair is $-J\delta_{Z,Z'}$, i.e. $-J$ if both lattice sites are in the same state, and otherwise zero.

The average size (and also the average mass) of all finite clusters is

$$S = \sum_{s=1}^{\infty} s \, \frac{s \, n_s}{\sum_{s=1}^{\infty} s \, n_s} = \frac{1}{p} \sum_{s=1}^{\infty} s^2 \, n_s \, . \tag{7.5.6}$$

The following relation holds between the quantity P_∞ defined before (7.5.1) and n_s: we consider an arbitrary lattice site. It is either empty or occupied and belongs to a cluster of finite size, or it is occupied and belongs to the infinite cluster, that is $1 = 1 - p + \sum_{s=1}^{\infty} s \, n_s + p \, P_\infty$, and therefore

$$P_\infty = 1 - \frac{1}{p} \sum_s s \, n_s \, . \tag{7.5.7}$$

7.5.3 Percolation in One Dimension

We consider a one-dimensional chain in which every lattice site is occupied with the probability p. Since a single unoccupied site will interrupt the connection to the other end, i.e. an infinite cluster can be present only when all sites are occupied, we have $p_c = 1$. In this model we can thus study only the phase $p < p_c$.

We can immediately compute the normalized number of clusters n_s for this model. The probability that an arbitrarily chosen site belongs to a cluster of size s has the value $s \, p^s \, (1 - p)^2$, since a series of s sites must be occupied (factor p^s) and the sites at the left and right boundaries must be unoccupied (factor $(1 - p)^2$). Since the chosen site could be at any of the s locations within the clusters, the factor s occurs. From this and from the general considerations at the beginning of Sect. 7.5.2, it follows that:

$$n_s = p^s \, (1 - p)^2 \, . \tag{7.5.8}$$

With this expression and starting from (7.5.6), we can calculate the average cluster size:

$$S = \frac{1}{p} \sum s^2 \, n_s = \frac{1}{p} \sum_{s=1}^{\infty} s^2 p^s (1 - p)^2 = \frac{(1 - p)^2}{p} \left(p \frac{d}{dp} \right)^2 \sum_{s=1}^{\infty} p^s$$

$$= \frac{(1 - p)^2}{p} \left(p \frac{d}{dp} \right)^2 \frac{p}{1 - p} = \frac{1 + p}{1 - p} \quad \text{for} \quad p < p_c \, . \tag{7.5.9}$$

The average cluster size diverges on approaching the percolation threshold $p_c = 1$ as $1/(1 - p)$, i.e. in one dimension, the exponent introduced in (7.5.3) is $\gamma = 1$.

We now define the radial *correlation function* $g(r)$. Let the zero point be an occupied site; then $g(r)$ gives the average number of occupied sites at a distance r which belong to the same cluster as the zero point. This is also equal to the probability that a particular site at the distance r is occupied and belongs to the same cluster, multiplied by the number of sites at a distance r. Clearly, $g(0) = 1$. For a point to belong to the cluster requires that this point itself and all points lying between 0 and r be occupied, that is, the probability that the point r is occupied and belongs to the same cluster as 0 is p^r, and therefore we find

$$g(r) = 2\,p^r \quad \text{for} \quad r \geq 1 . \tag{7.5.10}$$

The factor of 2 is required because in a one-dimensional lattice there are two points at a distance r.

The correlation length is defined by

$$\xi^2 = \frac{\sum_{r=1}^{\infty} r^2\, g(r)}{\sum_{r=1}^{\infty} g(r)} = \frac{\sum_{r=1}^{\infty} r^2\, p^r}{\sum_{r=1}^{\infty} p^r} . \tag{7.5.11}$$

Analogously to the calculation in Eq. (7.5.9), one obtains

$$\xi^2 = \frac{1+p}{(1-p)^2} = \frac{1+p}{(p-p_c)^2} , \tag{7.5.11'}$$

i.e. here, the critical exponent of the correlation length is $\nu = 1$. We can also write $g(r)$ in the form

$$g(r) = 2\, e^{r \log p} = 2\, e^{-\frac{\sqrt{2}\,r}{\xi}} , \tag{7.5.10'}$$

where after the last equals sign, we have taken $p \approx p_c$, so that $\log p = \log(1 - (1 - p)) \approx -(1 - p)$. The correlation length characterizes the (exponential) decay of the correlation function.

The average cluster size previously introduced can also be represented in terms of the radial correlation function

$$S = 1 + \sum_{r=1}^{\infty} g(r) . \tag{7.5.12}$$

We recall the analogous relation between the static susceptibility and the correlation function, which was derived in the chapter on ferromagnetism, Eq. (6.5.42). One can readily convince oneself that (7.5.12) together with (7.5.10) again leads to (7.5.9).

7.5.4 The Bethe Lattice (Cayley Tree)

A further exactly solvable model, which has the advantage over the one-dimensional model that it is defined also in the phase region $p > p_c$, is percolation on a Bethe lattice. The Bethe lattice is constructed as follows: from the lattice site at the origin, z (coordination number) branches spread out, at whose ends again lattice sites are located, from each of which again $z - 1$ new branches emerge, etc. (see Fig. 7.25 for $z = 3$).

Fig. 7.25. A Bethe lattice with the coordination number $z = 3$

The first shell of lattice sites contains z sites, the second shell contains $z(z-1)$ sites, and the lth shell contains $z(z-1)^{l-1}$ sites. The number of lattice sites increases exponentially with the distance from the center point $\sim e^{l \log(z-1)}$, while in a d-dimensional Euclidean lattice, this number increases as l^{d-1}. This suggests that the critical exponents of the Bethe lattice would be the same as those of a usual Euclidean lattice for $d \to \infty$. Another particular difference between the Bethe lattice and Euclidean lattices is the property that it contains only branches but no closed loops. This is the reason for its exact solvability.

To start with, we calculate the radial correlation function $g(l)$, which as before is defined as the average number of occupied lattice sites within the same cluster at a distance l from an arbitrary occupied lattice site. The probability that a particular lattice site at the distance l is occupied as well as all those between it and the origin has the value p^l. The number of all the sites in the shell l is $z(z-1)^{l-1}$; from this it follows that:

$$g(l) = z(z-1)^{l-1} p^l = \frac{z}{z-1}(p(z-1))^l = \frac{z}{z-1}e^{l\log(p(z-1))} . \qquad (7.5.13)$$

From the behavior of the correlation function for large l, one can read off the percolation threshold for the Bethe lattice. For $p(z-1) < 1$, there is an exponential decrease, and for $p(z-1) > 1$, $g(l)$ diverges for $l \to \infty$ and there is an infinite cluster, which must not be included in calculating the correlation function of the finite clusters. It follows from (7.5.13) for p_c that

$$p_c = \frac{1}{z-1} . \qquad (7.5.14)$$

For $z = 2$, the Bethe lattice becomes a one-dimensional chain, and thus $p_c = 1$. From (7.5.13) it is evident that the correlation length is

$$\xi \propto \frac{-1}{\log[p(z-1)]} = \frac{-1}{\log\frac{p}{p_c}} \sim \frac{1}{p_c - p} \qquad (7.5.15)$$

for p in the vicinity of p_c, i.e. $\nu = 1$, as in one dimension[29]. The same result is found if one defines ξ by means of (7.5.11). For the average cluster size one finds for $p < p_c$

$$S = 1 + \sum_{l=1}^{\infty} g(l) = \frac{p_c(1+p)}{p_c - p} \quad \text{for} \quad p < p_c; \tag{7.5.16}$$

i.e. $\gamma = 1$.

The strength of the infinite cluster P_∞, i.e. the probability that an arbitrary occupied lattice site belongs to the infinite cluster, can be calculated in the following manner: the product pP_∞ is the probability that the origin or some other point is occupied and that a connection between occupied sites up to infinity exists. We first compute the probability Q that an arbitrary site is not connected to infinity via a particular branch originating from it. This is equal to the probability that the site at the end of the branch is not occupied, that is $(1 - p)$ plus the probability that this site is occupied but that none of the $z - 1$ branches which lead out from it connects to ∞, i.e.

$$Q = 1 - p + pQ^{z-1} .$$

This is a determining equation for Q, which we shall solve for simplicity for a coordination number $z = 3$. The two solutions of the quadratic equation are $Q = 1$ and $Q = \frac{1-p}{p}$.

The probability that the origin is occupied, that however no path leads to infinity, is on the one hand $p(1 - P_\infty)$ and on the other pQ^z, i.e. for $z = 3$:

$$P_\infty = 1 - Q^3 .$$

For the first solution, $Q = 1$, we obtain $P_\infty = 0$, obviously relevant for $p < p_c$; and for the second solution

$$P_\infty = 1 - \left(\frac{1-p}{p}\right)^3 , \tag{7.5.17}$$

for $p > p_c$. In the vicinity of $p_c = \frac{1}{2}$, the strength of the infinite clusters varies as

$$P_\infty \propto (p - p_c) , \tag{7.5.18}$$

that is $\beta = 1$. We will also obtain this result with Eq. (7.5.30) in a different manner.

---

[29] Earlier, it was speculated that hypercubic lattices of high spatial dimension have the same critical exponents as the Bethe lattice. The visible difference in ν seen in Table 7.5 is due to the fact that in the Bethe lattice, the topological (chemical) and in the hypercubic lattice the Euclidean distance was used. If one uses the chemical distance for the hypercubic lattice also, above $d = 6$, $\nu = 1$ is likewise obtained. See *Literature*: A. Bunde and S. Havlin, p. 71.

Now we will investigate the normalized cluster number n_s, which is also equal to the probability that a particular site belongs to a cluster of size s, divided by s. In one dimension, n_s could readily be determined. In general, the probability for a cluster with s sites and t (empty) boundary points is $p^s(1-p)^t$. The perimeter t includes external and internal boundary points of the cluster. For general lattices, such as e.g. the square lattice, there are various values of t belonging to one and the same value of s, depending on the shape of the cluster; the more stretched out the cluster, the larger is t, and the more nearly spherical the cluster, the smaller is t. In a square lattice, there are two clusters having the size 3, a linear and a bent cluster. The associated values of t are 8 and 7, and the number of orientations on the lattice are 2 and 4. For general lattices, the quantity g_{st} must therefore be introduced; it gives the number of clusters of size s and boundary t. Then the general expression for n_s is

$$n_s = \sum_t g_{st}\, p^s(1-p)^t \ . \tag{7.5.19}$$

For arbitrary lattices, a determination of g_{st} is in general not possible. For the Bethe lattice, there is however a unique connection between the size s of the cluster and the number of its boundary points t. A cluster of size 1 has $t = z$, and a cluster of $s = 2$ has $t = 2z - 2$. In general, a cluster of size s has $z - 2$ more boundary points than a cluster of size $s - 1$, i.e.

$$t(s) = z + (s-1)(z-2) = 2 + s(z-2) \ .$$

Thus, for the Bethe lattice,

$$n_s = g_s\, p^s (1-p)^{2+(z-2)s} \ , \tag{7.5.20}$$

where g_s is the number of configurations of clusters of the size s. In order to avoid the calculation of g_s, we will refer $n_s(p)$ to the distribution $n_s(p_c)$ at p_c.

We now wish to investigate the behavior of n_s in the vicinity of $p_c = (z-1)^{-1}$ as a function of the cluster size, and separate off the distribution at p_c,

$$n_s(p) = n_s(p_c)\left[\frac{1-p}{1-p_c}\right]^2 \left[\frac{p}{p_c}\left(\frac{(1-p)}{(1-p_c)}\right)^{z-2}\right]^s \ ; \tag{7.5.21}$$

we then expand around $p = p_c$

$$n_s(p) = n_s(p_c)\left[\frac{1-p}{1-p_c}\right]^2 \left[1 - \frac{(p-p_c)^2}{2\,p_c^2(1-p_c)} + \mathcal{O}((p-p_c)^3)\right]^s \tag{7.5.22}$$

$$= n_s(p_c)\,e^{-cs} \ ,$$

with $c = -\log\left(1 - \frac{(p-p_c)^2}{2p_c(1-p_c)}\right) \propto (p-p_c)^2$.

This means that the number of clusters of size s decreases exponentially. The second factor in (7.5.22) depends only on the combination $(p-p_c)^{\frac{1}{\sigma}}s$, with $\sigma = 1/2$. The exponent σ determines how rapidly the number of clusters decreases with increasing size s. At p_c, the s-dependence of n_s arises only from the prefactor $n_s(p_c)$. In analogy to critical points, we assume that $n_s(p_c)$ is a pure power law; in the case that ξ gives the only length scale, which is infinite at p_c; then at p_c there can be no characteristic lengths, cluster sizes, etc. That is, $n_s(p_c)$ can have only the form

$$n_s(p_c) \sim s^{-\tau} . \tag{7.5.23}$$

The complete function (7.5.22) is then of the form

$$n_s(p) = s^{-\tau} f\left((p - p_c)^{\frac{1}{\sigma}} s\right) , \tag{7.5.24}$$

and it is a homogeneous function of s and $(p-p_c)$. We can relate the exponent τ to already known exponents: the average cluster size is, from Eq. (7.5.6),

$$S = \frac{1}{p} \sum_s s^2 n_s(p) \propto \sum s^{2-\tau} e^{-cs}$$

$$\propto \int_1^\infty ds\, s^{2-\tau} e^{-cs} = c^{\tau-3} \int_c^\infty z^{2-\tau} e^{-z} dz . \tag{7.5.25}$$

For $\tau < 3$, the integral exists, even when its lower limit goes to zero: it is then

$$S \sim c^{\tau-3} = (p - p_c)^{\frac{\tau-3}{\sigma}} , \tag{7.5.26}$$

from which, according to (7.5.3), it follows that

$$\gamma = \frac{3 - \tau}{\sigma} . \tag{7.5.27}$$

Since for the Bethe lattice, $\gamma = 1$ and $\sigma = \frac{1}{2}$, we find $\tau = \frac{5}{2}$.

From (7.5.24) using the general relation (7.5.7) one can also determine P_∞. While the factor s^2 in (7.5.25) was sufficient to make the integral converge at its lower limit, this is not the case in (7.5.7). Therefore, we first write (7.5.7) in the form

$$P_\infty = 1 - \frac{1}{p} \sum_s s\left(n_s(p) - n_s(p_c)\right) - \frac{1}{p} \sum_s s\, n_s(p_c)$$

$$= \frac{1}{p} \sum_s s\left(n_s(p_c) - n_s(p)\right) + 1 - \frac{p_c}{p} , \tag{7.5.28}$$

where

$$P_\infty(p_c) = 0 = 1 - \frac{1}{p_c} \sum_s s\, n_s(p_c)$$

has been used. Now the first term in (7.5.28) can be replaced by an integral

$$
P_\infty = \text{const. } c^{\tau-2} \int_c^\infty z^{1-\tau} \left[1 - e^{-z}\right] dz + \frac{p - p_c}{p}
$$

$$
= \dots c^{\tau-2} + \frac{p - p_c}{p} .
$$

(7.5.29)

From this, we find for the exponent defined in Eq. (7.5.1)

$$
\beta = \frac{\tau - 2}{\sigma} .
$$

(7.5.30)

For the Bethe lattice, one finds once again $\beta = 1$, in agreement with (7.5.18). In the Bethe lattice, the first term in (7.5.29)) also has the form $p - p_c$, while in other lattices the first term, $(p - p_c)^\beta$, predominates relative to the second due to $\beta < 1$.

In (7.5.5), we also introduced the average number of clusters per lattice site, whose critical percolation behavior is characterized by an exponent α via

$$
N_c \equiv \sum_s n_s \sim |p - p_c|^{2-\alpha} .
$$

(7.5.31)

That is, this quantity plays an analogous role to that of the free energy in thermal phase transitions. We note that in the case of percolation there are no interactions, and the free energy is determined merely by the entropy. Again inserting (7.5.24) for the cluster number into (7.5.31), we find

$$
2 - \alpha = \frac{\tau - 1}{\sigma} ,
$$

(7.5.32)

which leads to $\alpha = -1$ for the Bethe lattice. In summary, the critical exponents for the Bethe lattice are

$$
\beta = 1 , \ \gamma = 1 , \ \alpha = -1 , \ \nu = 1 , \ \tau = 5/2 , \ \sigma = 1/2 .
$$

(7.5.33)

7.5.5 General Scaling Theory

In the preceding section, the exponents for the Bethe lattice (Cayley tree) were calculated. In the process, we made some use of a scaling assumption (7.5.24). We will now generalize that assumption and derive the consequences which follow from it.

We start with the general scaling hypothesis

$$
n_s(p) = s^{-\tau} f_\pm \left(|p - p_c|^{\frac{1}{\sigma}} s\right) ,
$$

(7.5.34)

where $\pm$ refers to $p \gtrless p_c$.[30] The relations (7.5.27), (7.5.30), and (7.5.32), which contain only the exponents $\alpha, \beta, \gamma, \sigma, \tau$, also hold for the general scaling hypothesis. The scaling relation for the correlation length and other characteristics of the extension of the finite clusters must be derived once more. The correlation length is the root mean square distance between all the occupied sites within the same finite cluster. For a cluster with s occupied sites, the root mean square distance between all pairs is

$$R_s^2 = \frac{1}{s^2} \sum_{i=1}^s \sum_{j=1}^i (\mathbf{x}_i - \mathbf{x}_j)^2 .$$

The correlation length ξ is obtained by averaging over all clusters

$$\xi^2 = \frac{\sum_{s=1}^\infty R_s^2 s^2 n_s}{\sum_{s=1}^\infty s^2 n_s} . \tag{7.5.35}$$

The quantity $\frac{1}{2} s^2 n_s$ is equal to the number of pairs in clusters n_s of size s, i.e. proportional to the probability that a pair (in the same cluster) belongs to a cluster of the size s.

The mean square cluster radius is given by

$$\overline{R^2} = \frac{\sum_{s=1}^\infty R_s^2 s n_s}{\sum_{s=1}^\infty s n_s} , \tag{7.5.36}$$

since $s n_s$ = the probability that an occupied site belongs to an s-cluster. The mean square distance increases with cluster size according to

$$R_s \sim s^{1/d_f} , \tag{7.5.37}$$

where d_f is the fractal dimension. Then it follows from (7.5.35) that

$$\xi^2 \sim \sum_{s=1}^\infty s^{\frac{2}{d_f} + 2 - \tau} f_\pm \left(|p - p_c|^{\frac{1}{\sigma}} s \right) \Big/ \sum_{s=1}^\infty s^{2-\tau} f_\pm \left(|p - p_c|^{\frac{1}{\sigma}} s \right)$$

$$\sim |p - p_c|^{-\frac{2}{d_f \sigma}} , \quad 2 < \tau < 2.5$$

$$\nu = \frac{1}{d_f \sigma} = \frac{\tau - 1}{d \sigma} ,$$

and from (7.5.36),

$$\overline{R^2} \sim \sum_{s=1}^\infty s^{\frac{2}{d_f} + 1 - \tau} f_\pm(|p - p_c|^{\frac{1}{\sigma}} s) \sim |p - p_c|^{-2\nu + \beta} .$$

---

[30] At the percolation threshold $p = p_c$, the distribution of clusters is a power law $n_s(p_c) = s^{-\tau} f_\pm(0)$. The cutoff function $f_\pm(x)$ goes to zero for $x \gtrsim 1$, for example as in (7.5.22) exponentially. The quantity $s_{max} = |p - p_c|^{-1/\sigma}$ refers to the largest cluster. Clusters of size $s \ll s_{max}$ are also distributed according to $s^{-\tau}$ for $p \neq p_c$, and for $s \gtrsim s_{max}$, $n_s(p)$ vanishes.

7.5.5.1 The Duality Transformation and the Percolation Threshold

The computation of p_c for bond percolation on a square lattice can be carried out by making use of a *duality transformation*. The definition of the dual lattice is illustrated in Fig. 7.26. The lattice points of the *dual lattice* are defined by the centers of the unit cells of the lattice. A bond in the dual lattice is placed wherever it does not cross a bond of the lattice; i.e. the probability for a bond in the dual lattice is

$$q = 1 - p .$$

In the dual lattice, there is likewise a bond percolation problem. For $p < p_c$, there is no infinite cluster on the lattice, however there is an infinite cluster on the dual lattice. There is a path from one end of the dual lattice to the other which cuts no bonds on the lattice; thus $q > p_c$. For $p \to p_c^-$ from below, $q \to p_c^+$ arrives at the percolation threshold from above, i.e.

$$p_c = 1 - p_c .$$

Thus, $p_c = \frac{1}{2}$. This result is exact for bond percolation.

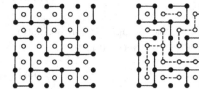

• lattice
∘ dual lattice
── bonds in the lattice
---- bonds in the dual lattice

Fig. 7.26. A lattice and its dual lattice. *Left side:* A lattice with bonds and the dual lattice. *Right side:* Showing also the bonds in the dual lattice

Remarks:

(i) By means of similar considerations, one finds also that the percolation threshold for site percolation on a triangular lattice is given by $p_c = \frac{1}{2}$.

(ii) For the two-dimensional Ising model, also, the transition temperatures for a series of lattice structures were already known from duality transformations before its exact solution had been achieved.

7.5.6 Real-Space Renormalization Group Theory

We now discuss a real-space renormalization-group transformation, which allows the approximate determination of p_c and the critical exponents.

In the decimation transformation shown in Fig. 7.27 for a square lattice, every other lattice site is eliminated; this leads again to a square lattice. In

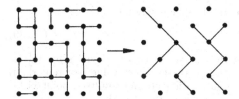

Fig. 7.27. A lattice and a decimated lattice

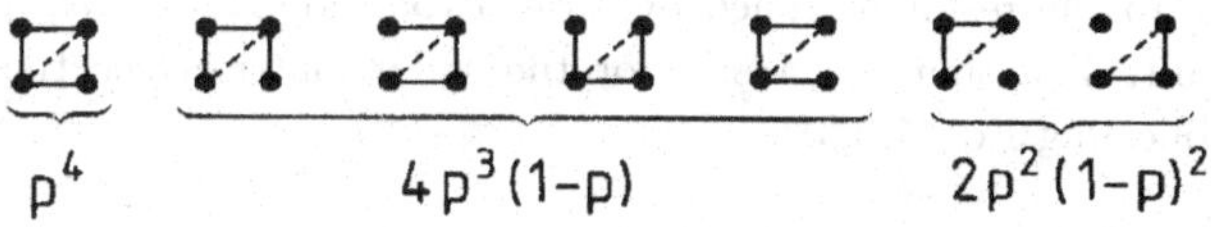

$$ \underbrace{\boxed{}}_{p^4} \quad \underbrace{\boxed{}\ \boxed{}\ \boxed{}\ \boxed{}}_{4p^3(1-p)} \quad \underbrace{\boxed{}\ \boxed{}}_{2p^2(1-p)^2} $$

Fig. 7.28. Bond configurations which lead to a bond (dashed) on the decimated lattice

the new lattice, a bond is placed between two remaining sites if at least one connection via two bonds existed on the original lattice (see Fig. 7.27). The bond configurations which lead to formation of a bond (shown as dashed lines) in the decimated lattice are indicated in Fig. 7.28. Below, the probability for these configurations is given. From the rules shown in Fig. 7.28, we find for the probability for the existence of a bond on the decimated lattice

$$ p' = p^4 + 4p^3(1-p) + 2p^2(1-p)^2 = 2p^2 - p^4 \ . \tag{7.5.38} $$

From this transformation law[31], one obtains the fixed-point equation $p^* = 2p^{*2} - p^{*4}$. It has the solutions $p^* = 0$, $p^* = 1$, which correspond to the high- and low-temperature fixed points for phase transitions; and in addition, the two fixed points $p^* = \dfrac{-1 \overset{+}{(-)} \sqrt{5}}{2}$, of which only $p^* = \dfrac{\sqrt{5}-1}{2} = 0.618\dots$ is acceptable. This value of the percolation threshold differs from the exact value found in the preceding section, $\frac{1}{2}$. The reasons for this are: (i) sites which were connected on the original lattice may not be connected on the decimated lattice; (ii) different bonds on the decimated lattice are no longer uncorrelated, since the existence of a bond on the original lattice can be responsible for the occurrence of several bonds on the decimated lattice.

The linearization of the recursion relation around the fixed point yields $\nu = 0.817$ for the exponent of the correlation length.

The treatment of site percolation on a triangular lattice in two dimensions is most simple. The lattice points of a triangle are combined into a cell. This cell is counted as occupied if all three sites are occupied, or if two sites are occupied and one is empty, since in both cases there is a path through the cell. For all other configurations (only one site occupied or none occupied), the cell is unoccupied. For the

---

[31] A. P. Young and R. B. Stinchcombe, J. Phys. C: Solid State Phys. **8**, L 535 (1975).

triangular lattice[32], one thus obtains as the recursion relation

$$p' = p^3 + 3p^2(1 - p) , \qquad (7.5.39)$$

with the fixed points $p^* = 0, 1, \frac{1}{2}$. This RG transformation thus yields $p_c = \frac{1}{2}$ for the percolation threshold, which is identical with the exact value (see remark (i) above). The linearization of the RG transformation around the fixed point yields the following result for the exponent ν of the correlation length:

$$\nu = \frac{\log \sqrt{3}}{\log \frac{3}{2}} = 1.3547 .$$

This is nearer to the result obtained by series expansion, $\nu = 1.34$, as well as to the exact result, $4/3$, than the result for the square lattice (see the remark on universality following Eq. (7.5.3)).

7.5.6.1 Definition of the Fractal Dimension

In a fractal object, the mass behaves as a function of the length L of a d-dimensional Euclidean section as

$$M(L) \sim L^{d_f} ,$$

and thus the density is

$$\rho(L) = \frac{M(L)}{L^d} \sim L^{d_f - d} .$$

An alternative definition of d_f is obtained from the number of hypercubes $N(L_m, \delta)$ which one requires to cover the fractal structure. We take the side length of the hypercubes to be δ, and the hypercube which contains the whole cluster to have the side length L_m:

$$N(L_m, \delta) = \left(\frac{L_m}{\delta} \right)^{d_f} ,$$

i.e.

$$d_f = - \lim_{\delta \to 0} \frac{\log N(L_m, \delta)}{\log \delta} .$$

Literature

D. J. Amit, *Field Theory, the Renormalization Group, and Critical Phenomena*, 2nd ed., World Scientific, Singapore 1984

P. Bak, C. Tang, and K. Wiesenfeld, Phys. Rev. Lett. **59**, 381 (1987)

K. Binder, Rep. Progr. Phys. **60**, 487 (1997)

---

[32] P. J. Reynolds, W. Klein, and H. E. Stanley, J. Phys. C: Solid State Phys. **10** L167 (1977).

J. J. Binney, N. J. Dowrick, A. J. Fisher, and M. E. J. Newman, *The Theory of Critical Phenomena*, 2nd ed., Oxford University Press, New York 1993

M. J. Buckingham and W. M. Fairbank, in: C. J. Gorter (Ed.), *Progress in Low Temperature Physics*, Vol. III, 80–112, North Holland Publishing Company, Amsterdam 1961

A. Bunde and S. Havlin, in: A. Bunde, S. Havlin (Eds.), *Fractals and Disordered Systems*, 51, Springer, Berlin 1991

Critical Phenomena, Lecture Notes in Physics **54**, Ed. J. Brey and R. B. Jones, Springer, Sitges, Barcelona 1976

M. C. Cross and P. C. Hohenberg (1994), Rev. Mod. Phys. **65**, 851–1112

P. G. De Gennes, *Scaling Concepts in Polymer Physics*, Cornell University Press, Ithaca, NY 1979

C. Domb and M. S. Green, *Phase Transitions and Critical Phenomena*, Academic Press, London 1972-1976

C. Domb and J. L. Lebowitz (Eds.), *Phase Transitions and Critical Phenomena*, Vols. 7–15, Academic Press, London 1983–1992

B. Drossel and F. Schwabl, Phys. Rev. Lett. **69**, 1629 (1992)

J. W. Essam, Rep. Prog. Phys. **43**, 843 (1980)

R. A. Ferrell, N. Menyhárd, H. Schmidt, F. Schwabl, and P. Szépfalusy, Ann. Phys. (New York) **47**, 565 (1968)

M. E. Fisher, Rep. Prog. Phys. **30**, 615–730 (1967)

M. E. Fisher, Rev. Mod. Phys. **46**, 597 (1974)

E. Frey and F. Schwabl, Adv. Phys. **43**, 577-683 (1994)

B. I. Halperin and P. C. Hohenberg, Phys. Rev. **177**, 952 (1969)

H. J. Jensen, *Self-Organized Criticality*, Cambridge University Press, Cambridge 1998

Shang-Keng Ma, *Modern Theory of Critical Phenomena*, Benjamin, Reading, Mass. 1976

S. Ma, in: C. Domb and M. S. Green (Eds.), *Phase Transitions and Critical Phenomena*, Vol. 6, 249–292, Academic Press, London 1976

T. Niemeijer and J. M. J. van Leeuwen, in: C. Domb and M. S. Green (Eds.), *Phase Transitions and Critical Phenomena*, Vol. 6, 425–505, Academic Press, London 1976

G. Parisi, *Statistical Field Theory*, Addison–Wesley, Redwood 1988

A. Z. Patashinskii and V. L. Prokovskii, *Fluctuation theory of Phase Transitions*, Pergamon Press, Oxford 1979

P. Pfeuty and G. Toulouse, *Introduction to the Renormalization Group and to Critical Phenomena*, John Wiley, London 1977

C. N. R. Rao and K. J. Rao, *Phase Transitions in Solids*, McGraw Hill, New York 1978

F. Schwabl and U. C. Täuber, *Phase Transitions: Renormalization and Scaling*, in *Encyclopedia of Applied Physics*, Vol. 13, 343, VCH (1995)

H. E. Stanley, *Introduction to Phase Transitions and Critical Phenomena*, Clarendon Press, Oxford 1971

D. Stauffer and A. Aharony, *Introduction to Percolation Theory*, Taylor and Francis, London and Philadelphia 1985

J. M. J. van Leeuwen in *Fundamental Problems in Statistical Mechanics III*, Ed. E. G. D. Cohen, North Holland Publishing Company, Amsterdam 1975

K. G. Wilson and J. Kogut, Phys. Rept. **12C**, 76 (1974)

K. G. Wilson, Rev. Mod. Phys. **47**, 773 (1975)

J. Zinn-Justin, *Quantum Field Theory and Critical Phenomena*, 3rd edition, Clarendon Press, Oxford 1996

Problems for Chapter 7

7.1 A generalized homogeneous function fulfills the relation

$$f(\lambda^{a_1}x_1, \lambda^{a_2}x_2) = \lambda^{a_f} f(x_1, x_2) .$$

Show that **(a)** the partial derivatives $\frac{\partial^j}{\partial x_1^j} \frac{\partial^k}{\partial x_2^k} f(x_1, x_2)$ and **(b)** the Fourier transform $g(k_1, x_2) = \int d^d x_1 e^{i k_1 x_1} f(x_1, x_2)$ of a generalized homogeneous function are likewise homogeneous functions.

7.2 Derive the relations (7.3.13') for A', K', L', and M'. Include in the starting model in addition an interaction between the second-nearest neighbors L. Compute the recursion relation to leading order in K and L, i.e. up to K^2 and L. Show that (7.3.15a,b) results.

7.3 What is the value of δ for the two-dimensional decimation transformation from Sect. 7.3.3?

7.4 Show, by Fourier transformation of the susceptibility $\chi(\mathbf{q}) = \frac{1}{q^{2-\eta}} \hat{\chi}(q\xi)$, that the correlation function assumes the form

$$G(\mathbf{x}) = \frac{1}{|\mathbf{x}|^{d-2+\eta}} \hat{G}(|\mathbf{x}|/\xi) .$$

7.5 Confirm Eq.(7.4.35).

7.6 Show that

$$m(x) = m_0 \tanh \frac{x - x_0}{2\xi_-}$$

is a solution of the Ginzburg–Landau equation (7.4.11). Calculate the free energy of the domain walls which it describes.

7.7 Tricritical phase transition point.
A *tricritical phase transition point* is described by the following Ginzburg–Landau functional:

$$\mathcal{F}[\phi] = \int d^d x \{ c(\nabla \phi)^2 + a\phi^2 + v\phi^6 - h\phi \}$$

$$\text{with } a = a'\tau , \quad \tau = \frac{T - T_c}{T_c} , \quad v \geq 0 .$$

Determine the uniform stationary solution ϕ_{st} with the aid of the variational derivative $(\frac{\delta \mathcal{F}}{\delta \phi} = 0)$ for $h = 0$ and the associated tricritical exponents $\alpha_t, \beta_t, \gamma_t$ and δ_t.

7.8 Consider the extended Ginzburg–Landau functional

$$\mathcal{F}[\phi] = \int d^d x \{ c(\nabla\phi)^2 + a\phi^2 + u\phi^4 + v\phi^6 - h\phi \} \ .$$

(a) Determine the critical exponents α, β, γ and δ for $u > 0$ in analogy to problem 7.7. They take on the same values as in the ϕ^4 model (see Sect. 4.6); the term $\sim \phi^6$ is *irrelevant*, i.e. it yields only corrections to the scaling behavior of the ϕ^4 model. Investigate the "crossover" of the tricritical behavior for $h = 0$ at small u. Consider the crossover function $\tilde{m}(x)$, which is defined as follows:

$$\phi_{eq}(u, \tau) = \phi_t(\tau) \cdot \tilde{m}(x) \quad \text{with} \quad \phi_t(\tau) = \phi_{eq}(u = 0, \tau) \sim \tau^{\beta t} \ , \quad x = \frac{u}{\sqrt{3|a|v}} \ .$$

(b) Now investigate the case $u < 0, h = 0$. Here, a *first-order phase transition* occurs; at T_c, the absolute minimum of $\mathcal{F}$ changes from $\phi = 0$ to $\phi = \phi^0$. Calculate the shift of the transition temperature $T_c - T_0$ and the height of the jump in the order parameter ϕ^0. Critical exponents can also be defined for the approach to the tricritical point by variation of u

$$\phi^0 \sim |u|^{\beta u} \ , \quad T_c - T_0 \sim |u|^{\frac{1}{\psi}} \ .$$

Give expressions for β_u and the "shift exponent" ψ.
(c) Calculate the second-order phase transition lines for $u < 0$ and $h \neq 0$ by deriving a parameter representation from the conditions

$$\frac{\partial^2 \mathcal{F}}{\partial \phi^2} = 0 = \frac{\partial^3 \mathcal{F}}{\partial \phi^3} \ .$$

(d) Show that the free energy in the vicinity of the tricritical point obeys a generalized scaling law

$$\mathcal{F}[\phi_{eq}] = |\tau|^{2-\alpha t} \hat{f}\left(\frac{u}{|\tau|^{\phi t}}, \frac{h}{|\tau|^{\delta t}} \right)$$

by inserting the crossover function found in (a) into $\mathcal{F}$ (ϕ_t is called the "crossover exponent"). Show that the *scaling relations*

$$\delta = 1 + \frac{\gamma}{\beta} \ , \quad \alpha + 2\beta + \gamma = 2$$

are obeyed in (a) and at the tricritical point (problem 7.7).
(e) Discuss the *hysteresis behavior* for a first-order phase transition ($u < 0$).

7.9 In the Ginzburg–Landau approximation, the spin-spin correlation function is given by

$$\langle m(\mathbf{x}) m(\mathbf{x}') \rangle = \frac{1}{L^d} \sum_{|\mathbf{k}| \leq \Lambda} e^{i\mathbf{k}(\mathbf{x} - \mathbf{x}')} \frac{1}{2\beta c(\xi^{-2} + k^2)} \ ; \quad \xi \propto (T - T_c)^{-\frac{1}{2}} \ .$$

(a) Replace the sum by an integral.
(b) show that in the limit $\xi \to \infty$, the following relation holds:

$$\langle m(\mathbf{x}) m(\mathbf{x}') \rangle \propto \frac{1}{|\mathbf{x} - \mathbf{x}'|^{d-2}} \ .$$

(c) Show that for $d = 3$ and large ξ,

$$\langle m(\mathbf{x}) m(\mathbf{x}') \rangle = \frac{1}{8\pi c\beta} \frac{e^{-|\mathbf{x} - \mathbf{x}'|/\xi}}{|\mathbf{x} - \mathbf{x}'|}$$

holds.

7.10 Investigate the behavior of the following integral in the limit $\xi \to \infty$:

$$I = \int_0^{\Lambda\xi} \frac{d^d q}{(2\pi)^d} \frac{\xi^{4-d}}{(1+q^2)^2} \,,$$

by demonstrating that:
(a) $I \propto \xi^{4-d}$, $d < 4$;
(b) $I \propto \ln \xi$, $d = 4$;
(c) $I \propto A - B\xi^{4-d}$, $d > 4$.

7.11 The phase transition of a molecular zipper *from C. Kittel, American Journal of Physics* **37**, *917, (1969).*
A greatly simplified model of the helix-coil transition in polypeptides or DNA, which describes the transition between hydrogen-bond stabilized helices and a molecular coil, is the "molecular zipper".

A molecular zipper consists of N bonds which can be broken from only one direction. It requires an energy ϵ to break bond $p+1$ if all the bonds $1, \dots, p$ are broken, but an infinite energy if the preceding bonds are not all broken. A broken bond is taken to have G orientations, i.e. its state is G–fold degenerate. The zipper is open when all $N-1$ bonds are broken.

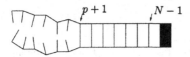

(a) Determine the partition function

$$Z = \frac{1 - x^N}{1 - x} \,; \quad x \equiv G \exp(-\epsilon\beta) \,.$$

(b) Determine the average number $\langle s \rangle$ of broken bonds. Investigate $\langle s \rangle$ in the vicinity of $x_c = 1$. Which value does $\langle s \rangle$ assume at x_c, and what is the slope there? How does $\langle s \rangle$ behave at $x \gg 1$ and $x \ll 1$?
(c) What would be the partition function if the zipper could be opened from both ends?

7.12 Fluctuations in the Gaussian approximation below T_c.
Expand the Ginzburg–Landau functional

$$\mathcal{F}[\mathbf{m}] = \int d^d x \left[a\mathbf{m}(x)^2 + \frac{b}{2}\mathbf{m}(x)^4 + c(\nabla\mathbf{m}(x))^2 - h\mathbf{m}(x) \right] \,,$$

which is $O(n)$-symmetrical for $h = 0$, up to second order in terms of the fluctuations of the order parameter $\mathbf{m}'(x)$. Below T_c,

$$\mathbf{m}(x) = m_1 \mathbf{e}_1 + \mathbf{m}'(x) \,, \quad h = 2(a + bm_1^2)m_1$$

holds.
(a) Show that for $h \to 0$, the long-wavelength ($k \to 0$) transverse fluctuations m_i' ($i = 2, \dots, n$) require no "excitation energy" (Goldstone modes), and determine the Gibbs free energy. In which cases do singularities occur?

(b) What is the expression for the specific heat $c_{h=0}$ below T_c in the harmonic approximation? Compare it with the result for the disordered phase.

(c) Calculate the longitudinal and transverse correlation functions relative to the spontaneous magnetization m_1

$$G_\|(x - x') = \langle m_1'(x)m_1'(x')\rangle \quad \text{and}$$
$$G_{\perp ij}(x - x') = \langle m_i'(x)m_j'(x')\rangle , \quad i, j = 2 \ldots, n$$

for $d = 3$ from its Fourier transform in the harmonic approximation. Discuss in particular the limiting case $h \to 0$.

7.13 The longitudinal correlation function below T_c.

The results from problem 7.12 lead us to expect that taking into account the transverse fluctuations just in a harmonic approximation will in general be insufficient. Anharmonic contributions can be incorporated if we fix the length of the vector $\mathbf{m}(x)$ ($h = 0$), as in the underlying Heisenberg model:

$$m_1(x)^2 + \sum_{i=2}^{n} m_i(x)^2 = m_0^2 = \text{const.}$$

Compute the Fourier transform $G_\|(k)$, by factorizing the four-spin correlation function in a suitable manner into two-spin correlation functions

$$G_\|(x - x') = \frac{1}{4m_0^2} \sum_{i,j=2}^{n} \langle m_i(x)^2 m_j(x')^2\rangle$$

and inserting

$$G_\perp(x - x') = \int \frac{d^d k}{(2\pi)^d} \frac{e^{ik(x-x')}}{2\beta ck^2} .$$

Remark: for $n \geq 2$ and $2 < d \leq 4$, the relations $G_\perp(k) \propto \frac{1}{k^2}$ and $G_\| \propto \frac{1}{k^{4-d}}$ are fulfilled exactly in the limit $k \to 0$.

7.14 Verify the second line in Eq. (7.5.22) .

7.15 The Hubbard–Stratonovich transformation: using the identity

$$\exp\left\{-\sum_{i,j} J_{ij} S_i S_j\right\} = \text{const.} \int_{-\infty}^{\infty} \left(\prod_i dm_i\right) \exp\left\{-\frac{1}{4}\sum_{i,j} m_i J_{ij}^{-1} m_j\right\},$$

show that the partition function of the Ising Hamiltonian $\mathcal{H} = \sum_{i,j} J_{ij} S_i S_j$ can be written in the form

$$Z = \text{const.} \int_{-\infty}^{\infty} \left(\prod_i dm_i\right) \exp\{\mathcal{H}'(\{m_i\})\} .$$

Give the expansion of $\mathcal{H}'$ in terms of m_i up to the order $\mathcal{O}(m^4)$. *Caveat:* the Ising Hamiltonian must be extended by terms with J_{ii} so that the matrix J_{ij} is positive definite.

7.16 Lattice-gas model. The partition function of a classical gas is to be mapped onto that of an Ising magnet.

Method: the d-dimensional configuration space is divided up into N cells. In each cell, there is at most one atom (hard core volume). One can imagine a lattice in which a cell is represented by a lattice site which is either empty or occupied ($n_i = 0$ or 1). The attractive interaction $U(x_i - x_j)$ between two atoms is to be taken into account in the energy by the term $\frac{1}{2}U_2(i,j)n_i n_j$.

(a) The grand partition function for this problem, after integrating out the kinetic energy, is given by

$$Z_G = \left(\prod_{i=1}^{N} \sum_{n_i=0,1}\right) \exp\left[-\beta\left(-\bar{\mu}\sum_i n_i + \frac{1}{2}\sum_{ij} U_2(i,j)n_i n_j\right)\right].$$

$$\bar{\mu} = kT \log z v_0 = \mu - kT \log\left(\frac{\lambda^d}{v_0}\right), \quad z = \frac{e^{\beta\mu}}{\lambda^d}, \quad \lambda = \frac{2\pi\hbar}{\sqrt{2\pi mkT}},$$

where v_0 is the volume of a cell.

(b) By introducing spin variables S_i ($n_i = \frac{1}{2}(1 + S_i)$, $S_i = \pm 1$), bring the grand partition function into the form

$$Z_G = \left(\prod_{i=1}^{N} \sum_{S_i=-1,1}\right) \exp\left[-\beta\left(E_0 - \sum_i hS_i - \sum_{ij} J_{ij}S_iS_j\right)\right].$$

Calculate the relations between E_0, h, J and μ, U_2, v_0.

(c) Determine the vapor-pressure curve of the gas from the phase-boundary curve $h = 0$ of the ferromagnet.

(d) Compute the particle-density correlation function for a lattice gas.

7.17 Demonstrate Eq. (7.4.63) using scaling relations.

7.18 Show that from (7.4.68) in the limit of small k and for $h = 0$, the longitudinal correlation function

$$G_\parallel(k) \propto \frac{1}{k^{d-2}}$$

follows.

7.19 Shift of T_c in the Ginzburg–Landau Theory. Start from Eq. (7.4.1) and use the so called quasi harmonic approximation in the paramagnetic phase. There the third (nonlinear) term in (7.4.1) is replaced by $6b < m(\mathbf{x})^2 > m(\mathbf{x})$.

(a) Justify this approximation.

(b) Compute the transition temperature T_c and show that $T_c < T_c^0$.

7.20 Determine the fixed points of the transformation equation (7.5.38).

8. Brownian Motion, Equations of Motion, and the Fokker–Planck Equations

The chapters which follow deal with nonequilibrium processes. First, in chapter 8, we treat the topic of the Langevin equations and the related Fokker–Planck equations. In the next chapter, the Boltzmann equation is discussed; it is fundamental for dealing with the dynamics of dilute gases and also for transport phenemona in condensed matter. In the final chapter, we take up general problems of irreversibility and the transition to equilibrium.

8.1 Langevin Equations

8.1.1 The Free Langevin Equation

8.1.1.1 Brownian Motion

A variety of situations occur in Nature in which one is not interested in the complete dynamics of a many-body system, but instead only in a subset of particular variables. The remaining variables lead through their equations of motion to relatively rapidly varying stochastic forces and to damping effects. Examples are the Brownian motion of a massive particle in a liquid, the equations of motion of conserved densities, and the dynamics of the order parameter in the vicinity of a critical point.

We begin by discussing the *Brownian motion* as a basic example of a stochastisic process. A heavy particle of mass m and velocity v is supposed to be moving in a liquid consisting of light particles. This "Brownian particle" is subject to random collisions with the molecules of the liquid (Fig. 8.1). The collisions with the molecules of the liquid give rise to an average frictional force on the massive particle, a *stochastic force* $f(t)$, which fluctuates around its average value as shown in Fig. 8.2. The first contribution $-m\zeta v$ to this force will be characterized by a *coefficient of friction* ζ. Under these physical conditions, the Newtonian equation of motion thus becomes the so called *Langevin* equation:

$$m\dot{v} = -m\zeta v + f(t). \tag{8.1.1}$$

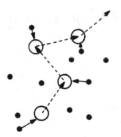

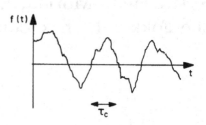

Fig. 8.1. The Brownian motion

Fig. 8.2. Stochastic forces in Brownian motion

Such equations are referred to as stochastic equations of motion and the processes they describe as stochastic processes.[1]

The *correlation time* τ_c denotes the time during which the fluctuations of the stochastic force remain correlated[2]. From this, we assume that the average force and its autocorrelation function have the following form at differing times[3]

$$\langle f(t) \rangle = 0$$
$$\langle f(t) f(t') \rangle = \phi(t - t') . \tag{8.1.2}$$

Here, $\phi(\tau)$ differs noticeably from zero only for $\tau < \tau_c$ (Fig. 8.3). Since we are interested in the motion of our Brownian particle over times of order t which are considerably longer than τ_c, we can approximate $\phi(\tau)$ by a delta function

$$\phi(\tau) = \lambda \delta(\tau) . \tag{8.1.3}$$

The coefficient λ is a measure of the strength of the mean square deviation of the stochastic force. Since friction also increases proportionally to the strength of the collisions, there must be a connection between λ and the coefficient of friction ζ. In order to find this connection, we first solve the Langevin equation (8.1.1).

---

[1] Due to the stochastic force in Eq. (8.1.1), the velocity is also a stochastic quantity, i.e. a random variable.

[2] Under the precondition that the collisions of the liquid molecules with the Brownian particle are completely uncorrelated, the correlation time is roughly equal to the duration of a collision. For this time, we obtain $\tau_c \approx \frac{a}{\bar{v}} = \frac{10^{-6} \text{ cm}}{10^5 \text{ cm/sec}} = 10^{-11}$ sec, where a is the radius of the massive particle and $\bar{v}$ the average velocity of the molecules of the medium.

[3] The mean value $\langle \; \rangle$ can be understood either as an average over independent Brownian particles or as an average over time for a single Brownian particle. In order to fix the higher moments of $f(t)$, we will later assume that $f(t)$ follows a Gaussian distribution, Eq. (8.1.26).

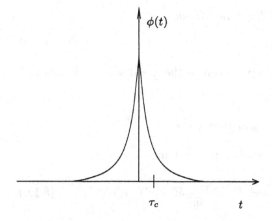

Fig. 8.3. The correlation of the stochastic forces

8.1.1.2 The Einstein Relation

The equation of motion (8.1.1) can be solved with the help of the *retarded Green's function* $G(t)$, which is defined by

$$\dot{G} + \zeta G = \delta(t) , \quad G(t) = \Theta(t)e^{-\zeta t} . \tag{8.1.4}$$

Letting v_0 be the initial value of the velocity, one obtains for $v(t)$

$$v(t) = v_0 e^{-\zeta t} + \int_0^\infty d\tau\, G(t-\tau)f(\tau)/m$$

$$= v_0 e^{-\zeta t} + e^{-\zeta t} \int_0^t d\tau\, e^{\zeta\tau} f(\tau)/m . \tag{8.1.5}$$

Since the dependence of $f(\tau)$ is known only statistically, we do not consider the average value of $v(t)$, but instead that of its square, $\langle v(t)^2 \rangle$

$$\langle v(t)^2 \rangle = e^{-2\zeta t} \int_0^t d\tau \int_0^t d\tau'\, e^{\zeta(\tau+\tau')} \phi(\tau-\tau')\frac{1}{m^2} + v_0^2 e^{-2\zeta t} ;$$

here, the cross term vanishes. With Eq. (8.1.3), we obtain

$$\langle v(t)^2 \rangle = \frac{\lambda}{2\zeta m^2}(1 - e^{-2\zeta t}) + v_0^2 e^{-2\zeta t} \xrightarrow{t \gg \zeta^{-1}} \frac{\lambda}{2\zeta m^2} . \tag{8.1.6}$$

For $t \gg \zeta^{-1}$, the contribution of v_0 becomes negligible and the memory of the initial value is lost. Hence ζ^{-1} plays the role of a *relaxation time*.

We *require* that our particle attain *thermal equilibrium* after long times, $t \gg \zeta^{-1}$, i.e. that the average value of the kinetic energy obey the equipartition theorem

$$\frac{1}{2}m\langle v(t)^2 \rangle = \frac{1}{2}kT . \tag{8.1.7}$$

From this, we find the so called *Einstein relation*

$$\lambda = 2\zeta mkT . \tag{8.1.8}$$

The coefficient of friction ζ is proportional to the mean square deviation λ of the stochastic force.

8.1.1.3 The Velocity Correlation Function

Next, we compute the velocity correlation function:

$$\langle v(t)v(t')\rangle = e^{-\zeta(t+t')}\int_0^t d\tau \int_0^{t'} d\tau'\, e^{\zeta(\tau+\tau')}\frac{\lambda}{m^2}\delta(\tau-\tau')+v_0^2 e^{-\zeta(t+t')} . \tag{8.1.9}$$

Since the roles of t and t' are arbitrarily interchangeable, we can assume without loss of generality that $t < t'$ and immediately evaluate the two integrals in the order given in this equation, with the result $\left(e^{2\zeta\min(t,t')}-1\right)\frac{\lambda}{2\zeta m^2}$, thus obtaining finally

$$\langle v(t)v(t')\rangle = \frac{\lambda}{2\zeta m^2}e^{-\zeta|t-t'|} + \left(v_0^2 - \frac{\lambda}{2\zeta m^2}\right)e^{-\zeta(t+t')} . \tag{8.1.10}$$

For $t, t' \gg \zeta^{-1}$, the second term in (8.1.10) can be neglected.

8.1.1.4 The Mean Square Deviation

In order to obtain the mean square displacement for $t \gg \zeta^{-1}$, we need only integrate (8.1.10) twice,

$$\langle x(t)^2\rangle = \int_0^t d\tau \int_0^t d\tau'\,\frac{\lambda}{2\zeta m^2}e^{-\zeta|\tau-\tau'|} . \tag{8.1.11}$$

Intermediate calculation for integrals of the type

$$I = \int_0^t d\tau \int_0^t d\tau'\, f(\tau - \tau') .$$

We denote the parent function of $f(\tau)$ by $F(\tau)$ and evaluate the integral over τ, $I = \int_0^t d\tau'\,(F(t - \tau') - F(-\tau'))$. Now we substitute $u = t - \tau'$ into the first term and obtain after integrating by parts

$$I = \int_0^t du\,(F(u) - F(-u)) = t(F(t) - F(-t)) - \int_0^t du\,u(f(u) + f(-u))$$

and from this the final result

$$\int_0^t d\tau \int_0^t d\tau'\, f(\tau - \tau') = \int_0^t du\,(t - u)(f(u) + f(-u)) . \tag{8.1.12}$$

With Eq. (8.1.12), it follows for (8.1.11) that

$$\langle x^2(t) \rangle = \frac{\lambda}{2\zeta m^2} 2 \int_0^t du\, (t-u)e^{-\zeta u} \approx \frac{\lambda}{\zeta^2 m^2} t$$

or

$$\langle x^2(t) \rangle = 2Dt \qquad\qquad (8.1.13)$$

with the diffusion constant

$$D = \frac{\lambda}{2\zeta^2 m^2} = \frac{kT}{\zeta m} . \qquad\qquad (8.1.14)$$

It can be seen that D plays the role of a diffusion constant by starting from the equation of continuity for the particle density

$$\dot{n}(x) + \boldsymbol{\nabla}\mathbf{j}(x) = 0 \qquad\qquad (8.1.15a)$$

and the current density

$$\mathbf{j}(x) = -D\boldsymbol{\nabla}n(x) . \qquad\qquad (8.1.15b)$$

The resulting diffusion equation

$$\dot{n}(x) = D\nabla^2\, n(x) \qquad\qquad (8.1.16)$$

has the one-dimensional solution

$$n(x,t) = \frac{N}{\sqrt{4\pi Dt}} e^{-\frac{x^2}{4Dt}} . \qquad\qquad (8.1.17)$$

The particle number density $n(x,t)$ from Eq. (8.1.17) describes the spreading out of N particles which were concentrated at $x = 0$ at the time $t = 0$ ($n(x,0) = N\delta(x)$). That is, the mean square displacement increases with time as $2Dt$. (More general solutions of (8.1.16) can be found from (8.1.17) by superposition.)

We can cast the Einstein relation in a more familiar form by introducing the *mobility* μ into (8.1.1) in place of the coefficient of friction. The Langevin equation then reads

$$m\ddot{x} = -\mu^{-1}\dot{x} + f \quad \text{with} \quad \mu = \frac{1}{\zeta m} , \qquad\qquad (8.1.18)$$

and the Einstein relation takes on the form

$$D = \mu kT . \qquad\qquad (8.1.19)$$

The diffusion constant is thus proportional to the mobility of the particle and to the temperature.

Remarks:

(i) In a simplified version of Einstein's[4] historical derivation of (8.1.19), we treat (instead of the osmotic pressure in a force field) the dynamic origin of the barometric pressure formula. The essential consideration is that in a gravitational field there are two currents which must compensate each other in equilibrium. They are the diffusion current $-D\frac{\partial}{\partial z}n(z)$ and the current of particles falling in the gravitational field, $\bar{v}n(z)$. Here, $n(z)$ is the particle number density and $\bar{v}$ is the mean velocity of falling, which, due to friction, is found from $\mu^{-1}\bar{v} = -mg$. Since the sum of these two currents must vanish, we find the condition

$$-D\frac{\partial}{\partial z}n(z) - mg\mu n(z) = 0 \,. \tag{8.1.20}$$

From this, the barometric pressure formula $n(z) \propto \mathrm{e}^{-\frac{mgz}{kT}}$ is obtained if the Einstein relation (8.1.19) is fulfilled.

(ii) In the Brownian motion of a sphere in a liquid with the viscosity constant η, the frictional force is given by Stokes' law, $F_{\mathrm{fr}} = 6\pi a\eta\dot{x}$, where a is the radius and $\dot{x}$ the velocity of the sphere. Then the diffusion constant is $D = kT/6\pi a\eta$ and the mean square displacement of the sphere is given by

$$\langle x^2(t)\rangle = \frac{kTt}{3\pi a\eta} \,. \tag{8.1.21}$$

Using this relation, an observation of $\langle x^2(t)\rangle$ allows the experimental determination of the *Boltzmann constant k*.

8.1.2 The Langevin Equation in a Force Field

As a generalization of the preceding treatment, we now consider the *Brownian motion* in an external *force field*

$$F(x) = -\frac{\partial V}{\partial x} \,. \tag{8.1.22a}$$

Then the Langevin equation is given by

$$m\ddot{x} = -m\zeta\dot{x} + F(x) + f(t) \,, \tag{8.1.22b}$$

where we assume that the collisions and frictional effects of the molecules are not modified by the external force and therefore the stochastic force $f(t)$ again obeys (8.1.2), (8.1.3), and (8.1.8).[5]

An important *special case* of (8.1.22b) is the limiting case of *strong damping* ζ. When the inequality $m\zeta\dot{x} \gg m\ddot{x}$ is fulfilled (as is the case e.g for periodic motion at low frequencies), it follows from (8.1.22b) that

---

[4] See the reference at the end of this chapter.

[5] We will later see that the Einstein relation (8.1.8) ensures that the function $\exp(-(\frac{p^2}{2m}+V(x))/kT)$ be an equilibrium distribution for this stochastic process.

$$\dot{x} = -\Gamma \frac{\partial V}{\partial x} + r(t) ,$$
(8.1.23)

where the damping constant Γ and the fluctuating force $r(t)$ are given by

$$\Gamma \equiv \frac{1}{m\zeta} \quad \text{and} \quad r(t) \equiv \frac{1}{m\zeta} f(t) .$$
(8.1.24)

The stochastic force $r(t)$, according to Eqns. (8.1.2) and (8.1.3), obeys the relation

$$\langle r(t) \rangle = 0$$
$$\langle r(t)r(t') \rangle = 2\Gamma kT \delta(t - t') .$$
(8.1.25)

For the characterization of the higher moments (correlation functions) of $r(t)$, we will further assume in the following that $r(t)$ follows a Gaussian distribution

$$\mathcal{P}[r(t)] = e^{-\int_{t_0}^{t_f} dt \, \frac{r^2(t)}{4\Gamma kT}} .$$
(8.1.26)

$\mathcal{P}[r(t)]$ gives the probability density for the values of $r(t)$ in the interval $[t_0, t_f]$, where t_0 and t_f are the initial and final times. To define the functional integration, we subdivide the interval into

$$N = \frac{t_f - t_0}{\Delta}$$

small subintervals of width Δ and introduce the discrete times

$$t_i = t_0 + i\Delta , \qquad i = 0, \dots, N - 1 .$$

The element of the functional integration $\mathcal{D}[r]$ is defined by

$$\mathcal{D}[r] \equiv \lim_{\Delta \to 0} \prod_{i=0}^{N-1} \left(dr(t_i) \sqrt{\frac{\Delta}{4\Gamma kT\pi}} \right) .$$
(8.1.27)

The normalization of the probability density is

$$\int \mathcal{D}[r] \, \mathcal{P}[r(t)] \equiv \lim_{\Delta \to 0} \prod_{i=0}^{N-1} \int \left(dr(t_i) \sqrt{\frac{\Delta}{4\Gamma kT\pi}} \right) e^{-\sum_i \Delta \frac{r^2(t_i)}{4\Gamma kT}} = 1 .$$
(8.1.28)

As a check, we calculate

$$\langle r(t_i)r(t_j) \rangle = \frac{4\Gamma kT}{2\Delta} \delta_{ij} = 2\Gamma kT \frac{\delta_{ij}}{\Delta} \to 2\Gamma kT \delta(t_i - t_j) ,$$

which is in agreement with Eqns. (8.1.2), (8.1.3) and (8.1.8).

Since Langevin equations of the type (8.1.23) occur in a variety of different physical situations, we want to add some elementary explanations. We first consider (8.1.23) without the stochastic force, i.e. $\dot{x} = -\Gamma \frac{\partial V}{\partial x}$. In regions of positive (negative) slope of $V(x)$, x will be shifted in the negative (positive) x direction. The coordinate x moves in the direction of one of the minima of $V(x)$ (see Fig. 8.4). At the extrema of $V(x)$, $\dot{x}$ vanishes. The effect of the stochastic force $r(t)$ is that the motion towards the minima becomes fluctuating, and even at its extreme positions the particle is not at rest, but instead is continually pushed away, so that the possibility exists of a transition from one minimum into another. The calculation of such transition rates is of interest for, among other applications, thermally activated hopping of impurities in solids and for chemical reactions (see Sect. 8.3.2).

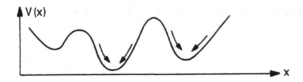

Fig. 8.4. The motion resulting from the equation of motion $\dot{x} = -\Gamma \partial V / \partial x$.

8.2 The Derivation of the Fokker–Planck Equation from the Langevin Equation

Next, we wish to derive equations of motion for the probability densities in the Langevin equations (8.1.1), (8.1.22b), and (8.1.23).

8.2.1 The Fokker–Planck Equation for the Langevin Equation (8.1.1)

We define

$$P(\xi, t) = \langle \delta(\xi - v(t)) \rangle , \tag{8.2.1}$$

the *probability density* for the event that the Brownian particle has the velocity ξ at the time t. This means that $P(\xi, t)d\xi$ is the probability that the velocity lies within the interval $[\xi, \xi + d\xi]$.

We now derive an equation of motion for $P(\xi, t)$:

$$\frac{\partial}{\partial t} P(\xi, t) = -\frac{\partial}{\partial \xi} \langle \delta(\xi - v(t)) \dot{v}(t) \rangle$$

$$= -\frac{\partial}{\partial \xi} \left\langle \delta(\xi - v(t)) \left(-\zeta v(t) + \frac{1}{m} f(t) \right) \right\rangle$$

$$= -\frac{\partial}{\partial \xi} \left\langle \delta(\xi - v(t)) \left(-\zeta \xi + \frac{1}{m} f(t) \right) \right\rangle$$

$$= \frac{\partial}{\partial \xi} (\zeta P(\xi, t) \xi) - \frac{1}{m} \frac{\partial}{\partial \xi} \langle \delta(\xi - v(t)) f(t) \rangle, \qquad (8.2.2)$$

where the Langevin equation (8.1.1) has been inserted in the second line. To compute the last term, we require the probability density for the stochastic force, assumed to follow a Gaussian distribution:

$$\mathcal{P}[f(t)] = e^{-\int_{t_0}^{t_f} dt\, \frac{f^2(t)}{4\zeta m k T}}. \qquad (8.2.3)$$

The averages $\langle \dots \rangle$ are given by the functional integral with the weight (8.2.3) (see Eq. (8.1.26)). In particular, for the last term in (8.2.2), we obtain

$$\langle \delta(\xi - v(t)) f(t) \rangle = \int \mathcal{D}[f(t')] \, \delta(\xi - v(t)) f(t) e^{-\int \frac{f(t')^2 dt'}{4\zeta m k T}}$$

$$= -2\zeta m k T \int \mathcal{D}[f(t')] \, \delta(\xi - v(t)) \frac{\delta}{\delta f(t)} e^{-\int \frac{f(t')^2 dt'}{4\zeta m k T}}$$

$$= 2\zeta m k T \int \mathcal{D}[f(t')] e^{-\int \frac{f(t')^2 dt'}{4\zeta m k T}} \frac{\delta}{\delta f(t)} \delta(\xi - v(t))$$

$$= 2\zeta m k T \left\langle \frac{\delta}{\delta f(t)} \delta(\xi - v(t)) \right\rangle = -2\zeta m k T \frac{\partial}{\partial \xi} \left\langle \delta(\xi - v(t)) \frac{\delta v(t)}{\delta f(t)} \right\rangle.$$
$$\qquad (8.2.4)$$

Here, we have to use the solution (8.1.5)

$$v(t) = v_0 e^{-\zeta t} + \int_0^\infty d\tau\, G(t - \tau) \frac{f(\tau)}{m} \qquad (8.1.5)$$

and take the derivative with respect to $f(t)$. With $\frac{\delta f(\tau)}{\delta f(t)} = \delta(\tau - t)$ and (8.1.4), we obtain

$$\frac{\delta v(t)}{\delta f(t)} = \int_0^t d\tau\, e^{-\zeta(t-\tau)} \frac{1}{m} \delta(t - \tau) = \frac{1}{2m}. \qquad (8.2.5)$$

The factor $\frac{1}{2}$ results from the fact that the integration interval includes only half of the δ-function. Inserting (8.2.5) into (8.2.4) and (8.2.4) into (8.2.2), we obtain the equation of motion for the probability density, the *Fokker–Planck equation*:

$$\frac{\partial}{\partial t} P(v, t) = \zeta \frac{\partial}{\partial v} v P(v, t) + \zeta \frac{kT}{m} \frac{\partial^2}{\partial v^2} P(v, t). \qquad (8.2.6)$$

Here, we have replaced the velocity ξ by v; it is not to be confused with the stochastic variable $v(t)$. This relation can also be written in the form of an equation of continuity

$$\frac{\partial}{\partial t}P(v,t) = -\zeta\frac{\partial}{\partial v}\left(-vP(v,t) - \frac{kT}{m}\frac{\partial}{\partial v}P(v,t)\right) . \qquad (8.2.7)$$

Remarks:

(i) The current density, the expression in large parentheses, is composed of a drift term and a diffusion current.

(ii) The current density vanishes if the probability density has the form $P(v,t) \propto \mathrm{e}^{-\frac{mv^2}{2kT}}$. The *Maxwell distribution* is thus (at least one) equilibrium distribution. Here, the Einstein relation (8.1.8) plays a decisive role. Conversely, we could have obtained the Einstein relation by requiring that the Maxwell distribution be a solution of the Fokker–Planck equation.

(iii) We shall see in Sect. 8.3.1 that $P(v,t)$ becomes the Maxwell distribution in the course of time, and that the latter is therefore the only equilibrium distribution of the Fokker–Planck equation (8.2.6).

8.2.2 Derivation of the Smoluchowski Equation for the Overdamped Langevin Equation, (8.1.23)

For the stochastic equation of motion (8.1.23),

$$\dot{x} = -\Gamma\frac{\partial V}{\partial x} + r(t), \qquad (8.1.23)$$

we can also define a probability density

$$P(\xi,t) = \langle\delta(\xi - x(t))\rangle , \qquad (8.2.8)$$

where $P(\xi,t)d\xi$ is the probability of finding the particle at time t at the position ξ in the interval $d\xi$. We now derive an equation of motion for $P(\xi,t)$, performing the operation $(F(x) \equiv -\frac{\partial V}{\partial x})$

$$\frac{\partial}{\partial t}P(\xi,t) = -\frac{\partial}{\partial\xi}\langle\delta(\xi - x(t))\dot{x}(t)\rangle$$

$$= -\frac{\partial}{\partial\xi}\langle\delta(\xi - x(t))(\Gamma K(x) + r(t))\rangle$$

$$= -\frac{\partial}{\partial\xi}(\Gamma P(\xi,t)K(\xi)) - \frac{\partial}{\partial\xi}\langle\delta(\xi - x(t))r(t)\rangle . \qquad (8.2.9)$$

The overdamped Langevin equation was inserted in the second line. For the last term, we find in analogy to Eq. (8.2.4)

$$\langle \delta(\xi - x(t))r(t)\rangle = 2\Gamma kT \left\langle \frac{\delta}{\delta r(t)}\delta(\xi - x(t))\right\rangle$$

$$= -2\Gamma kT \frac{\partial}{\partial \xi}\left\langle \delta(\xi - x(t))\frac{\delta x(t)}{\delta r(t)}\right\rangle = -\Gamma kT \frac{\partial}{\partial \xi}P(\xi, t) . \quad (8.2.10)$$

Here, we have integrated (8.1.23) between 0 and t,

$$x(t) = x(0) + \int_0^t d\tau \left(\Gamma K(x(\tau)) + r(\tau)\right) , \quad (8.2.11)$$

from which it follows that

$$\frac{\delta x(t)}{\delta r(t')} = \int_0^t \left(\frac{\partial \Gamma F(x(\tau))}{\partial x(\tau)}\frac{\delta x(\tau)}{\delta r(t')} + \delta(t' - \tau)\right) d\tau . \quad (8.2.12)$$

The derivative is $\frac{\delta x(\tau)}{\delta r(t')} = 0$ for $\tau < t'$ due to causality and is nonzero only for $\tau \geq t'$, with a finite value at $\tau = t'$. We thus obtain

$$\frac{\delta x(t)}{\delta r(t')} = \int_0^t \frac{\partial \Gamma F(x(\tau))}{\partial x(\tau)}\frac{\delta x(\tau)}{\delta r(t')}d\tau + 1 \quad \text{for } t' < t \quad (8.2.13a)$$

and

$$\frac{\delta x(t)}{\delta r(t')} = \underbrace{\int_0^t \frac{\partial \Gamma F(x(\tau))}{\partial x(\tau)}\frac{\delta x(\tau)}{\delta r(t')}}_{0 \text{ for } t'=t} + \frac{1}{2} = \frac{1}{2} \quad \text{for } t' = t . \quad (8.2.13b)$$

This demonstrates the last step in (8.2.10). From (8.2.10) and (8.2.9), we obtain the equation of motion for $P(\xi, t)$, the so called *Smoluchowski equation*

$$\frac{\partial}{\partial t}P(\xi, t) = -\frac{\partial}{\partial \xi}(\Gamma P(\xi, t)F(\xi)) + \Gamma kT \frac{\partial^2}{\partial \xi^2}P(\xi, t) . \quad (8.2.14)$$

Remarks:

(i) One can cast the Smoluchowski equation (8.2.14) in the form of an equation of continuity

$$\frac{\partial}{\partial t}P(x, t) = -\frac{\partial}{\partial x}j(x, t) , \quad (8.2.15a)$$

with the current density

$$j(x, t) = -\Gamma \left(kT \frac{\partial}{\partial x} - K(x)\right) P(x, t) . \quad (8.2.15b)$$

The current density $j(x, t)$ is composed of a diffusion term and a drift term, in that order.

(ii) Clearly,

$$P(x, t) \propto e^{-V(x)/kT} \quad (8.2.16)$$

is a stationary solution of the Smoluchowski equation. For this solution, $j(x, t)$ is zero.

8.2.3 The Fokker–Planck Equation for the Langevin Equation (8.1.22b)

For the general Langevin equation, (8.1.22b), we define the probability density

$$P(x, v, t) = \langle \delta(x - x(t))\delta(v - v(t)) \rangle . \tag{8.2.17}$$

Here, we must distinguish carefully between the quantities x and v and the stochastic variables $x(t)$ and $v(t)$. The meaning of the probability density $P(x, v, t)$ can be characterized as follows: $P(x, v, t)dxdv$ is the probability of finding the particle in the interval $[x, x + dx]$ with a velocity in $[v, v + dv]$. The equation of motion of $P(x, v, t)$, the generalized Fokker–Planck equation

$$\frac{\partial}{\partial t}P + v\frac{\partial P}{\partial x} + \frac{F(x)}{m}\frac{\partial P}{\partial v} = \zeta\left[\frac{\partial}{\partial v}vP + \frac{kT}{m}\frac{\partial^2 P}{\partial v^2}\right] \tag{8.2.18}$$

follows from a series of steps similar to those in Sect. 8.2.2; see problem 8.1.

8.3 Examples and Applications

In this section, the Fokker–Planck equation for free Brownian motion will be solved exactly. In addition, we will show in general for the Smoluchowski equation that the distribution function relaxes towards the equilibrium situation. In this connection, a relation to supersymmetric quantum mechanics will also be pointed out. Furthermore, two important applications of the Langevin equations or the Fokker–Planck equations will be given: the transition rates in chemical reactions and the dynamics of critical phenomena.

8.3.1 Integration of the Fokker–Planck Equation (8.2.6)

We now want to solve the Fokker–Planck equation for the free Brownian motion, (8.2.6):

$$\dot{P}(v) = \zeta\frac{\partial}{\partial v}\left\{Pv + \frac{kT}{m}\frac{\partial P}{\partial v}\right\} . \tag{8.3.1}$$

We expect that $P(v)$ will relax towards the Maxwell distribution, $e^{-\frac{mv^2}{2kT}}$, following the relaxation law $e^{-\zeta t}$. This makes it reasonable to introduce the variable $\rho = ve^{\zeta t}$ in place of v. Then we have

$$P(v, t) = P(\rho e^{-\zeta t}, t) \equiv Y(\rho, t) , \tag{8.3.2a}$$

$$\frac{\partial P}{\partial v} = \frac{\partial Y}{\partial \rho}e^{\zeta t}, \quad \frac{\partial^2 P}{\partial v^2} = \frac{\partial^2 Y}{\partial \rho^2}e^{2\zeta t}, \tag{8.3.2b}$$

$$\frac{\partial P}{\partial t} = \frac{\partial Y}{\partial \rho}\frac{\partial \rho}{\partial t} + \frac{\partial Y}{\partial t} = \frac{\partial Y}{\partial \rho}\zeta\rho + \frac{\partial Y}{\partial t} . \tag{8.3.2c}$$

Inserting (8.3.2a–c) into (8.2.6) or (8.3.1), we obtain

$$\frac{\partial Y}{\partial t} = \zeta Y + \zeta \frac{kT}{m} \frac{\partial^2 Y}{\partial \rho^2} e^{2\zeta t} . \tag{8.3.3}$$

This suggests the substitution $Y = \chi e^{\zeta t}$. Due to $\frac{\partial Y}{\partial t} = \frac{\partial \chi}{\partial t} e^{\zeta t} + \zeta Y$, it follows that

$$\frac{\partial \chi}{\partial t} = \zeta \frac{kT}{m} \frac{\partial^2 \chi}{\partial \rho^2} e^{2\zeta t} . \tag{8.3.4}$$

Now we introduce a new time variable by means of $d\vartheta = e^{2\zeta t} dt$

$$\vartheta = \frac{1}{2\zeta} \left(e^{2\zeta t} - 1 \right) , \tag{8.3.5}$$

where $\vartheta(t = 0) = 0$. We then find from (8.3.4) the diffusion equation

$$\frac{\partial \chi}{\partial \vartheta} = \zeta \frac{kT}{m} \frac{\partial^2 \chi}{\partial \rho^2} \tag{8.3.6}$$

with its solution known from (8.1.17),

$$\chi(\rho, \vartheta) = \frac{1}{\sqrt{4\pi q\vartheta}} e^{-\frac{(\rho - \rho_0)^2}{4q\vartheta}} \; ; \quad q = \zeta \frac{kT}{m} . \tag{8.3.7}$$

By returning to the original variables v and t, we find the following solution

$$P(v,t) = \chi e^{\zeta t} = \left\{ \frac{m}{2\pi kT(1 - e^{-2\zeta t})} \right\}^{\frac{1}{2}} e^{-\frac{m(v - v_0 e^{-\zeta t})^2}{2kT(1 - e^{-2\zeta t})}} \tag{8.3.8}$$

of the Fokker–Planck equation (8.2.6), which describes Brownian motion in the absence of external forces. The solution of the Smoluchowski equation (8.2.14) for a harmonic potential is also contained in (8.3.8).

We now discuss the most important properties and consequences of the solution (8.3.8):

In the limiting case $t \to 0$, we have

$$\lim_{t \to 0} P(v,t) = \delta(v - v_0) . \tag{8.3.9a}$$

In the limit of long times, $t \to \infty$, the result is

$$\lim_{t \to \infty} P(v,t) = e^{-mv^2/2kT} \left(\frac{m}{2\pi kT} \right)^{\frac{1}{2}} . \tag{8.3.9b}$$

Remark: Since $P(v,t)$ has the property (8.3.9a), we also have found the conditional probability density in (8.3.8)[6]

---

[6] The conditional probability $P(v,t|v_0,t_0)$ gives the probability that at time t the value v occurs, under the condition that it was v_0 at the time t_0.

$$P(v, t|v_0, t_0) = P(v, t - t_0) . \tag{8.3.10}$$

This is not surprising. Since, as a result of (8.1.1), (8.1.2) and (8.1.3), a Markov process[7] is specified, $P(v, t|v_0, t_0)$ likewise obeys the Fokker–Planck equation (8.2.6).

For an arbitrary integrable and normalized initial probability density $\rho(v_0)$ at time t_0

$$\int dv_0 \rho(v_0) = 1 \tag{8.3.11}$$

we find with (8.3.8) the time dependence

$$\rho(v, t) = \int dv_0 P(v, t - t_0) \rho(v_0) . \tag{8.3.12}$$

Clearly, $\rho(v, t)$ fulfills the initial condition

$$\lim_{t \to t_0} \rho(v, t) = \rho(v_0) , \tag{8.3.13a}$$

while for long times

$$\lim_{t \to \infty} \rho(v, t) = e^{-\frac{mv^2}{2kT}} \left(\frac{m}{2\pi kT} \right)^{\frac{1}{2}} \int dv_0 \rho(v_0) = e^{-\frac{mv^2}{2kT}} \left(\frac{m}{2\pi kT} \right)^{\frac{1}{2}} \tag{8.3.13b}$$

the Maxwell distribution is obtained. Therefore, for the Fokker–Planck equation (8.2.6), and for the Smoluchowski equation with an harmonic potential, (8.2.14), we have proved that an arbitrary initial distribution relaxes towards the Maxwell distribution, (8.3.13b).

The function (8.3.8) is also used, by the way, in Wilson's exact renormalization group transformation for the continuous partial elimination of short-wavelength critical fluctuations.[8]

8.3.2 Chemical Reactions

We now wish to calculate the thermally activated transition over a barrier (Fig. 8.5). An obvious physical application is the motion of an impurity atom in a solid from one local minimum of the lattice potential into another. Certain chemical reactions can also be described on this basis. Here, x refers to the reaction coordinate, which characterizes the state of the molecule. The vicinity of the point A can, for example, refer to an excited state of a molecule, while B signifies the dissociated molecule. The transition from A to B takes place via configurations which have higher energies and is made possible by the thermal energy supplied by the surrounding medium. We formulate the following calculation in the language of chemical reactions.

---

[7] A Markov process denotes a stochastic process in which all the conditional probabilities depend only on the last time which occurs in the conditions; e.g.

$$P(t_3, v_3|t_2, v_2; t_1, v_1) = P(t_3, v_3|t_2, v_2) ,$$

where $t_1 \le t_2 \le t_3$.

[8] K. G. Wilson and J. Kogut, Phys. Rep. **12C**, 75 (1974).

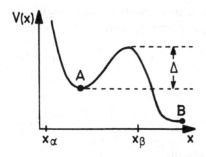

Fig. 8.5. A thermally activated transition over a barrier from the minimum A into the minimum B

We require the reaction rate (also called the transition rate), i.e. the transition probability per unit time for the conversion of type A into type B. We assume that friction is so strong that we can employ the Smoluchowski equation (8.2.15a,b),

$$\dot{P} = -\frac{\partial}{\partial x} j(x) \ . \tag{8.2.15a}$$

Integration of this equation between the points α and β yields

$$\frac{d}{dt} \int_{x_\alpha}^{x_\beta} dx P = -j(x_\beta) + j(x_\alpha) \ , \tag{8.3.14}$$

where x_β lies between the points A and B. It then follows that $j(x_\beta)$ is the *transition rate* between the states (the chemical species) A and B.

To calculate $j(x_\beta)$, we assume that the barrier is sufficiently high so that the transition rate is small. Then in fact all the molecules will be in the region of the minimum A and will occupy states there according to the thermal distribution. The few molecules which have reached state B can be imagined to be filtered out. The strategy of our calculation is to find a stationary solution $P(x)$ which has the properties

$$P(x) = \frac{1}{Z} e^{-V(x)/kT} \quad \text{in the vicinity of } A \tag{8.3.15a}$$

$$P(x) = 0 \qquad\qquad \text{in the vicinity of } B \ . \tag{8.3.15b}$$

From the requirement of stationarity, it follows that

$$0 = \Gamma \frac{\partial}{\partial x} \left(kT \frac{\partial}{\partial x} + \frac{\partial V}{\partial x} \right) P \ , \tag{8.3.16}$$

from which we find by integrating once

$$\Gamma \left(kT \frac{\partial}{\partial x} + \frac{\partial V}{\partial x} \right) P = -j_0 \ . \tag{8.3.17}$$

The integration constant j_0 plays the role of the current density which, owing to the fact that (8.2.14) is source-free between A and B, is independent of x.

This integration constant can be determined from the boundary conditions given above. We make use of the following Ansatz for $P(x)$:

$$P(x) = e^{-V/kT} \hat{P} \tag{8.3.18}$$

in equation (8.3.17)

$$\frac{\partial}{\partial x} \hat{P} = -\frac{j_0}{kT\Gamma} e^{V(x)/kT} . \tag{8.3.17'a}$$

Integrating this equation from A to x, we obtain

$$\hat{P}(x) = \text{const.} - \frac{j_0}{kT\Gamma} \int_A^x dx\, e^{V(x)/kT} . \tag{8.3.17'b}$$

The boundary condition at A, that there P follows the thermal equilibrium distribution, requires that

$$\text{const.} = \frac{1}{\int_A dx\, e^{-V/kT}} . \tag{8.3.19a}$$

Here, $\int_A$ means that the integral is evaluated in the vicinity of A. If the barrier is sufficiently high, contributions from regions more distant from the minimum are negligible[9]. The boundary condition at B requires

$$0 = e^{-V_B/kT} \left(\text{const.} - \frac{j_0}{kT\Gamma} \int_A^B dx\, e^{V/kT} \right) , \tag{8.3.19b}$$

so that

$$j_0 = \frac{kT\, \Gamma \left(\int_A dx\, e^{-V(x)/kT} \right)^{-1}}{\int_A^B dx\, e^{V(x)/kT}} . \tag{8.3.20}$$

For $V(x)$ in the vicinity of A, we set $V_A(x) \approx \frac{1}{2}(2\pi\nu)^2 x^2$, and, without loss of generality, take the zero point of the energy scale at the point A. We then find

$$\int_A dx\, e^{-V_A/kT} = \int_{-\infty}^{\infty} dx\, e^{-\frac{1}{2}(2\pi\nu)^2 x^2/kT} = \frac{\sqrt{kT}}{\sqrt{2\pi}\nu} .$$

Here, the integration was extended beyond the neighborhood of A out to $[-\infty, \infty]$, which is permissible owing to the rapid decrease of the integrand. The main contribution to the integral in the denominator of (8.3.20) comes

---

[9] Inserting (8.3.17'b) with (8.3.20) into (8.3.18), one obtains from the first term in the vicinity of point A just the equilibrium distribution, while the second term is negligible due to $\int_A^x dx\, e^{V/kT} / \int_A^B dx\, e^{V/kT} \ll 1$.

from the vicinity of the barrier, where we set $V(x) \approx \Delta - (2\pi\nu')^2 x^2/2$. Here, Δ is the height of the barrier and ν'^2 characterizes the barrier's curvature

$$\int_A^B dx\, e^{V/kT} \approx e^{\Delta/kT} \int_{-\infty}^{\infty} dx\, e^{-\frac{(2\pi\nu')^2 x^2}{2kT}} = e^{\frac{\Delta}{kT}} \frac{\sqrt{kT}}{\sqrt{2\pi}\nu'}.$$

This yields all together for the current density or the *transition rate*[10]

$$j_0 = 2\pi\nu\nu'\Gamma e^{-\Delta/kT}. \tag{8.3.21}$$

We point out some important aspects of the *thermally activated transition rate*: the decisive factor in this result is the *Arrhenius* dependence $e^{-\Delta/kT}$, where Δ denotes the barrier height, i.e. the activation energy. We can rewrite the prefactor by making the replacements $(2\pi\nu)^2 = m\omega^2, (2\pi\nu')^2 = m\omega'^2$ and $\Gamma = \frac{1}{m\zeta}$ (Eq. (8.1.24)):

$$j_0 = \frac{\omega\omega'}{2\pi\zeta} e^{-\Delta/kT}. \tag{8.3.21'}$$

If we assume that $\omega' \approx \omega$, then the prefactor is proportional to the square of the vibration frequency characterized by the potential well.[11]

8.3.3 Critical Dynamics

We have already pointed out in the introduction to Brownian motion that the theory developed to describe it has a considerably wider significance. Instead of the motion of a massive particle in a fluid of stochastically colliding molecules, one can consider quite different situations in which a small number of relatively slowly varying collective variables are interacting with many strongly varying, rapid degrees of freedom. The latter lead to a damping of the collective degrees of freedom.

This situation occurs in the *hydrodynamic* region. Here, the collective degrees of freedom represent the densities of the conserved quantities. The typical time scales for these hydrodynamic degrees of freedom increase with decreasing q proportionally to $1/q$ or $1/q^2$, where q is the wavenumber. In comparison, in the range of small wavenumbers all the remaining degrees of freedom are very rapid and can be regarded as stochastic noise in the equations of motion of the conserved densities. This then leads to the typical form of the hydrodynamic equations with damping terms proportional to q^2 or, in real space, $\sim \nabla^2$. We emphasize that "hydrodynamics" is by no means limited to the domain of liquids or gases, but instead, in an extension of its

---

[10] H. A. Kramers, Physica **7**, 284 (1940)

[11] ω is the frequency (attempt frequency) with which the particle arrives at the right side of the potential well, from where it has the possibility (with however a small probability $\sim e^{-\Delta/kT}$) of overcoming the barrier.

original meaning, it includes the general dynamics of conserved quantities depending on the particular physical situation (dielectrics, ferromagnets, liquid crystals, etc.).

A further important field in which this type of separation of time scales occurs is the dynamics in the neighborhood of critical points. As we know from the sections on static critical phenomena, the correlations of the local order parameter become long-ranged. There is thus a fluctuating order within regions whose size is of the order of the correlation length. As these correlated regions grow, the characteristic time scale also increases. Therefore, the remaining degrees of freedom of the system can be regarded as rapidly varying. In ferromagnets, the order parameter is the magnetization. In its motions, the other degrees of freedom such as those of the electrons and lattice vibrations act as rapidly varying stochastic forces.

In *ferromagnets*, the magnetic susceptibility behaves in the vicinity of the Curie point as

$$\chi \sim \frac{1}{T - T_c} \tag{8.3.22a}$$

and the correlation function of the magnetization as

$$G_{MM}(\mathbf{x}) \sim \frac{e^{-|\mathbf{x}|/\xi}}{|\mathbf{x}|} . \tag{8.3.22b}$$

In the neighborhood of the critical point of the *liquid-gas* transition, the isothermal compressibility diverges as

$$\kappa_T \sim \frac{1}{T - T_c} \tag{8.3.22c}$$

and the density-density correlation function has the dependence

$$g_{\rho\rho}(\mathbf{x}) \sim \frac{e^{-|\mathbf{x}|/\xi}}{|\mathbf{x}|} . \tag{8.3.22d}$$

In Eqns. (8.3.22 b,d), ξ denotes the correlation length, which behaves as $\xi \sim (T - T_c)^{-\frac{1}{2}}$ in the molecular field approximation, cf. Sects. 5.4 and 6.5.

A general model-independent approach to the theory of critical phenomena begins with a continuum description of the free energy, the Ginzburg–Landau expansion (see Sect. 7.4.1):

$$\mathcal{F}[M] = \int d^d x \left\{ \frac{a'}{2}(T - T_c)M^2 + \frac{b}{4}M^4 + \frac{c}{2}(\nabla M)^2 - Mh \right\} , \tag{8.3.23}$$

where $e^{-\mathcal{F}/kT}$ denotes the statistical weight of a configuration $M(\mathbf{x})$. The most probable configuration is given by

$$\frac{\delta \mathcal{F}}{\delta M} = 0 = a'(T - T_c)M - c\nabla^2 M + bM^3 - h . \tag{8.3.24}$$

It follows from this that the magnetization and the susceptibility in the limit $h \to 0$ are

$$M \sim (T_c - T)^{1/2}\Theta(T_c - T) \quad \text{and} \quad \chi \sim \frac{1}{T - T_c} \;.$$

Since the correlation length diverges on approaching the critical point, $\xi \to \infty$, the fluctuations also become slow. This suggests the following stochastic equation of motion for the magnetization[12]

$$\dot{M}(\mathbf{x}, t) = -\lambda \frac{\delta \mathcal{F}}{\delta M(\mathbf{x}, t)} + r(\mathbf{x}, t) \;. \tag{8.3.25}$$

The first term in the equation of motion causes relaxation towards the minimum of the free-energy functional. This thermodynamic force becomes stronger as the gradient $\delta \mathcal{F}/\delta M(\mathbf{x})$ increases. The coefficient λ characterizes the relaxation rate analogously to Γ in the Smoluchowski equation. Finally, $r(\mathbf{x}, t)$ is a stochastic force which is caused by the remaining degrees of freedom. Instead of a finite number of stochastic variables, we have here stochastic variables $M(\mathbf{x}, t)$ and $r(\mathbf{x}, t)$ which depend on a continuous index, the position $\mathbf{x}$.

Instead of $M(\mathbf{x})$, we can also introduce its Fourier transform

$$M_{\mathbf{k}} = \int d^d x \, e^{-i\mathbf{k}\mathbf{x}} M(\mathbf{x}) \tag{8.3.26}$$

and likewise for $r(\mathbf{x}, t)$. Then the equation of motion (8.3.25) becomes

$$\dot{M}_{\mathbf{k}} = -\lambda \frac{\partial \mathcal{F}}{\partial M_{-\mathbf{k}}} + r_{\mathbf{k}}(t) \;. \tag{8.3.25'}$$

Finally, we still have to specify the properties of the stochastic forces. Their average value is zero

$$\langle r(\mathbf{x}, t) \rangle = \langle r_{\mathbf{k}}(t) \rangle = 0$$

and furthermore they are correlated spatially and temporally only over short distances, which we can represent in idealized form by

$$\langle r_{\mathbf{k}}(t) r_{\mathbf{k}'}(t') \rangle = 2\lambda k T \delta_{\mathbf{k}, -\mathbf{k}'} \delta(t - t') \tag{8.3.27}$$

or

$$\langle r(\mathbf{x}, t) r(\mathbf{x}', t') \rangle = 2\lambda k T \delta(\mathbf{x} - \mathbf{x}') \delta(t - t') \;. \tag{8.3.27'}$$

For the mean square deviations of the force, we have postulated the Einstein relation, which guarantees that an equilibrium distribution is given by

---

[12] Also called the TDGL = time-dependent Ginzburg–Landau model.

$e^{-\beta \mathcal{F}[M]}$. We also assume that the probability density for the stochastic forces $r(\mathbf{x},t)$ is a Gaussian distribution (cf. (8.1.26)). This has the result that the odd correlation functions for $r(\mathbf{x},t)$ vanish and the even ones factor into products of (8.3.27′) (sum over all the pairwise contractions). We will now investigate the equation of motion (8.3.25′) for $T > T_c$. In what follows, we use the *Gaussian approximation*, i.e. we neglect the anharmonic terms; then the equation of motion simplifies to

$$\dot{M}_{\mathbf{k}} = -\lambda \big(a'(T - T_c) + ck^2 \big) M_{\mathbf{k}} + r_{\mathbf{k}} \ . \tag{8.3.28}$$

Its solution is already familiar from the elementary theory of Brownian motion:

$$M_{\mathbf{k}}(t) = e^{-\gamma_{\mathbf{k}} t} M_{\mathbf{k}}(0) + e^{-\gamma_{\mathbf{k}} t} \int_0^t dt'\, r_{\mathbf{k}}(t') e^{\gamma_{\mathbf{k}} t'} \ , \tag{8.3.29}$$

as is the resulting correlation function

$$\langle M_{\mathbf{k}}(t) M_{\mathbf{k}'}(t') \rangle = e^{-\gamma_{\mathbf{k}}|t-t'|} \frac{\lambda k T}{\gamma_k} \delta_{\mathbf{k},-\mathbf{k}'} + \mathcal{O}(e^{-\gamma_{\mathbf{k}}(t+t')}) \tag{8.3.30}$$

or, for times $t, t' > \gamma_{\mathbf{k}}^{-1}$,

$$\langle M_{\mathbf{k}}(t) M_{\mathbf{k}'}(t') \rangle = \delta_{\mathbf{k},-\mathbf{k}'} \frac{kT}{a'(T - T_c) + ck^2} e^{-\gamma_{\mathbf{k}}|t-t'|} \ . \tag{8.3.31}$$

Here, we have introduced the relaxation rate

$$\gamma_{\mathbf{k}} = \lambda \big(a'(T - T_c) + ck^2 \big) \ . \tag{8.3.32a}$$

In particular, for $k = 0$ we find

$$\gamma_0 \sim (T - T_c) \sim \xi^{-2} \ . \tag{8.3.32b}$$

As we suspected at the beginning, the relaxation rate decreases drastically on approaching the critical point. One denotes this situation as "critical slowing down".

As we already know from Chap. 7, the interaction bM^4 between the critical fluctuations leads to a modification of the critical exponents, e.g. $\xi \to (T - T_c)^{-\nu}$. Likewise, in the framework of dynamic renormalization group theory it is seen that these interactions lead in the dynamics to

$$\gamma_0 \to \xi^{-z} \tag{8.3.33}$$

with a *dynamic critical exponent* z [13] which differs from 2.

---

[13] See e.g. F. Schwabl and U. C. Täuber, Encyclopedia of Applied Physics, Vol. 13, 343 (1995), VCH.

Remark:

According to Eq. (8.3.25′), the dynamics of the order parameter are relaxational. For *isotropic ferromagnets*, the magnetization is conserved and the coupled precessional motion of the magnetic moments leads to spin waves. In this case, the equations of motion are given by[14]

$$\dot{\mathbf{M}}(\mathbf{x},t) = -\lambda M(\mathbf{x},t) \times \frac{\delta \mathcal{F}}{\delta \mathbf{M}}(\mathbf{x},t) + \Gamma \nabla^2 \frac{\delta \mathcal{F}}{\delta \mathbf{M}}(\mathbf{x},t) + \mathbf{r}(\mathbf{x},t) , \qquad (8.3.34)$$

with

$$\langle \mathbf{r}(\mathbf{x},t) \rangle = 0 , \qquad (8.3.35)$$

$$\langle r_i(\mathbf{x},t) r_j(\mathbf{x},t) \rangle = -2\Gamma k T \nabla^2 \delta^{(3)}(\mathbf{x} - \mathbf{x}') \delta(t - t') \delta_{ij} , \qquad (8.3.36)$$

which leads to spin diffusion above the Curie temperature and to spin waves below it (cf. problem 8.9). The first term on the right-hand side of the equation of motion produces the precessional motion of the local magnetization $\mathbf{M}(\mathbf{x},t)$ around the local field $\delta \mathcal{F}/\delta \mathbf{M}(\mathbf{x},t)$ at the point $\mathbf{x}$. The second term gives rise to the damping. Since the magnetization is conserved, it is taken to be proportional to ∇^2, i.e. in Fourier space it is proportional to k^2. These equations of motion are known as the Bloch equations or Landau–Lifshitz equations and, without the stochastic term, have been applied in solid-state physics since long before the advent of interest in critical dynamic phenomena. The stochastic force $\mathbf{r}(\mathbf{x},t)$ is due to the remaining, rapidly fluctuating degrees of freedom. The functional of the free energy is

$$\mathcal{F}[\mathbf{M}(\mathbf{x},t)] = \frac{1}{2} \int d^3x \left[a'(T - T_c)\mathbf{M}^2(\mathbf{x},t) + \frac{b}{2}\mathbf{M}^4(\mathbf{x},t) \right.$$

$$\left. + c(\nabla \mathbf{M}(\mathbf{x},t))^2 - \mathbf{h}\mathbf{M}(\mathbf{x},t) \right] . \qquad (8.3.37)$$

*8.3.4 The Smoluchowski Equation and Supersymmetric Quantum Mechanics

8.3.4.1 The Eigenvalue Equation

In order to bring the Smoluchowski equation (8.2.14) ($V' \equiv \partial V/\partial x) \equiv -F$

$$\frac{\partial P}{\partial t} = \Gamma \frac{\partial}{\partial x} \left(kT \frac{\partial}{\partial x} + V' \right) P \qquad (8.3.38)$$

into a form which contains only the second derivative with respect to x, we apply the Ansatz

---

[14] S. Ma and G. F. Mazenko, Phys. Rev. B**11**, 4077 (1975).

$$P(x,t) = e^{-V(x)/2kT} \rho(x,t) \,, \tag{8.3.39}$$

obtaining

$$\frac{\partial \rho}{\partial t} = kT\Gamma \left(\frac{\partial^2}{\partial x^2} + \frac{V''}{2kT} - \frac{V'^2}{4(kT)^2} \right) \rho \,. \tag{8.3.40}$$

This is a Schrödinger equation with an imaginary time

$$i\hbar \frac{\partial \rho}{\partial(-i\hbar 2kT\Gamma t)} = \left(-\frac{1}{2}\frac{\partial^2}{\partial x^2} + V^0(x) \right) \rho \,. \tag{8.3.41}$$

with the potential

$$V^0(x) = \frac{1}{2} \left\{ \frac{V'^2}{4(kT)^2} - \frac{V''}{2kT} \right\} \,. \tag{8.3.42}$$

Following the separation of the variables

$$\rho(x,t) = e^{-2kT\Gamma E_n t} \varphi_n(x) \,, \tag{8.3.43}$$

we obtain from Eq. (8.3.40) the eigenvalue equation

$$\frac{1}{2}\varphi_n'' = \left(-E_n + V^0(x) \right) \varphi_n(x) \,. \tag{8.3.44}$$

Formally, equation (8.3.44) is identical with a time-independent Schrödinger equation.[15] In (8.3.43) and (8.3.44), we have numbered the eigenfunctions and eigenvalues which follow from (8.3.44) with the index n.

The ground state of (8.3.44) is given by

$$\varphi_0 = \mathcal{N}e^{-\frac{V}{2kT}}, \quad E_0 = 0 \,, \tag{8.3.45}$$

where $\mathcal{N}$ is a normalization factor. Inserting in (8.3.39), we find for $P(x,t)$ the equilibrium distribution

$$P(x,t) = \mathcal{N}e^{-V(x)/kT} \,. \tag{8.3.45'}$$

From (8.3.42), we can immediately see the connection with *supersymmetric quantum mechanics*. The supersymmetric partner[16] to V^0 has the potential

$$V^1 = \frac{1}{2} \left[\frac{V'^2}{4(kT)^2} + \frac{V''}{2kT} \right] \,. \tag{8.3.46}$$

---

[15] N. G. van Kampen, J. Stat. Phys. **17**, 71 (1977).

[16] M. Bernstein and L. S. Brown, Phys. Rev. Lett. **52**, 1933 (1984); F. Schwabl, QM I, Chap. 19, Springer 2005. The quantity Φ introduced there is connected to the ground state wavefunctions φ_0 and the potential V as follows: $\Phi = -\varphi_0'/\varphi_0 = V'/2kT$.

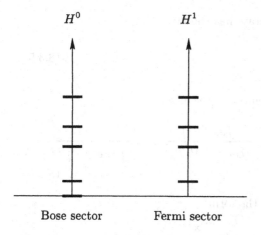

H^0 H^1

Bose sector   Fermi sector

Fig. 8.6. The excitation spectra of the two Hamiltonians H^0 and H^1, from QM I, pp. 353 and 361

The excitation spectra of the two Hamiltonians

$$H^{0,1} = -\frac{1}{2}\frac{d^2}{dx^2} + V^{0,1}(x) \tag{8.3.47}$$

are related in the manner shown in Fig. 8.6. One can advantageously make use of this connection if the problem with H^1 is simpler to solve than that with H^0.

8.3.4.2 Relaxation towards Equilibrium

We can now solve the initial value problem for the Smoluchowski equation in general. Starting with an arbitrarily normalized initial distribution $P(x)$, we can calculate $\rho(x)$ and expand in the eigenfunctions of (8.3.44)

$$\rho(x) = e^{V(x)/2kT}P(x) = \sum_n c_n\varphi_n(x) , \tag{8.3.48}$$

with the expansion coefficients

$$c_n = \int dx\, \varphi_n^*(x)e^{V(x)/2kT}P(x) . \tag{8.3.49}$$

From (8.3.43), we find the time dependence

$$\rho(x,t) = \sum_n e^{-2kT\Gamma E_n t}c_n\varphi_n(x) , \tag{8.3.50}$$

from which, with (8.3.39),

$$P(x,t) = e^{-V(x)/2kT}\sum_{n=0}^{\infty} c_n e^{-2kT\Gamma E_n t}\varphi_n(x) \tag{8.3.51}$$

follows. The normalized ground state has the form

$$\varphi_0 = \frac{e^{-V(x)/2kT}}{\left(\int dx\, e^{-V(x)/kT}\right)^{1/2}} \, . \tag{8.3.52}$$

Therefore, the expansion coefficient c_0 is given by

$$c_0 = \int dx\, \varphi_0^* e^{V(x)/2kT} P(x) = \frac{\int dx\, P(x)}{\left(\int dx\, e^{-V(x)/kT}\right)^{1/2}} = \frac{1}{\left(\int dx\, e^{-V(x)/kT}\right)^{1/2}} \, . \tag{8.3.53}$$

This allows us to cast (8.3.51) in the form

$$P(x,t) = \frac{e^{-V(x)/kT}}{\int dx\, e^{-V(x)/kT}} + e^{-V(x)/2kT} \sum_{n=1}^{\infty} c_n e^{-2kT\Gamma E_n t} \varphi_n(x) \, . \tag{8.3.54}$$

With this, the initial-value problem for the Smoluchowski equation is solved in general. Since $E_n > 0$ for $n \geq 1$, it follows from this expansion that

$$\lim_{t\to\infty} P(x,t) = \frac{e^{-V(x)/kT}}{\int dx\, e^{-V(x)/kT}} \, , \tag{8.3.55}$$

which means that, starting from an arbitrary initial distribution, $P(x,t)$ develops at long times towards the equilibrium distribution (8.3.45') or (8.3.55).

Literature

A. Einstein, Ann. d. Physik **17**, 182 (1905); reprinted in Annalen der Physik **14**, Supplementary Issue (2005).
R. Becker, *Theorie der Wärme*, 3. Aufl., Springer Verlag, Heidelberg 1985, Chap. 7;
R. Becker, *Theory of Heat*, 2nd Ed., Springer, Berlin, Heidelberg, New York 1967
H. Risken, *The Fokker–Planck Equation*, Springer Verlag, Heidelberg, 1984
N. G. van Kampen, *Stochastic Processes in Physics and Chemistry*, North Holland, Amsterdam, 1981

Problems for Chapter 8

8.1 Derive the generalized Fokker–Planck equation, (8.2.18).

8.2 A particle is moving with the step length l along the x-axis. Within each time step it hops to the right with the probability p_+ and to the left with the probability p_- $(p_+ + p_- = 1)$. How far is it from the starting point on the average after t time steps if $p_+ = p_- = 1/2$, or if $p_+ = 3/4$ and $p_- = 1/4$?

8.3 Diffusion and Heat Conductivity
(a) Solve the diffusion equation

$$\dot{n} = D\Delta n$$

for $d = 1$, 2 and 3 dimensions with the initial condition

$$n(\mathbf{x}, t = 0) = N\delta^d(\mathbf{x}) \,.$$

Here, n is the particle density, N the particle number, and D is the diffusion constant.
(b) Another form of the diffusion equation is the heat conduction equation

$$\Delta T = \frac{c\rho}{\kappa} \frac{\partial T}{\partial t}$$

where T is the temperature, κ the coefficient of thermal conductivity, c the specific heat, and ρ the density.

Solve the following problem as an application: potatoes are stored at $+5°C$ in a broad trench which is covered with a loose layer of earth of thickness d. Right after they are covered, a cold period suddenly begins, with a steady temperature of $-10°C$, and it lasts for two months. How thick does the earth layer have to be so that the potatoes will have cooled just to $0°C$ at the end of the two months? Assume as an approximation that the same values hold for the earth and for the potatoes: $\kappa = 0.4 \frac{W}{m \cdot K}$, $c = 2000 \frac{J}{kg \cdot K}$, $\rho = 1000 \frac{kg}{m^3}$.

8.4 Consider the Langevin equation of an overdamped harmonic oscillator

$$\dot{x}(t) = -\Gamma x(t) + h(t) + r(t),$$

where $h(t)$ is an external force and $r(t)$ a stochastic force with the properties (8.1.25). Compute the correlation function

$$C(t, t') = \langle x(t) x(t') \rangle_{h=0} \,,$$

the response function

$$\chi(t, t') = \frac{\delta \langle x(t) \rangle}{\delta h(t')} \,,$$

and the Fourier transform of the response function.

8.5 Damped Oscillator
(a) Consider the damped harmonic oscillator

$$m\ddot{x} + m\zeta\dot{x} + m\omega_0^2 x = f(t)$$

with the stochastic force $f(t)$ from Eq. (8.1.25). Calculate the correlation function and the dynamic susceptibility. Discuss in particular the position of the poles and the line shape. What changes relative to the limiting cases of the non-damped oscillator or the overdamped oscillator?
(b) Express the stationary solution $\langle x(t) \rangle$ under the action of a periodic external force $f_e(t) = f_0 \cos \frac{2\pi}{T} t$ in terms of the dynamic susceptibility. Use it to compute the power dissipated, $\frac{1}{T} \int_0^T dt \, f_e(t) \langle \dot{x}(t) \rangle$.

8.6 Diverse physical systems can be described as a subsystem capable of oscillations that is coupled to a relaxing degree of freedom, whereby both systems are in contact with a heat bath (e.g. the propagation of sound waves in a medium in which chemical reactions are taking place, or the dynamics of phonons taking energy/heat diffusion into account). As a simple model, consider the following system of coupled equations:

$$\dot{x} = \frac{1}{m}p$$

$$\dot{p} = -m\omega_0^2 x - \Gamma p + by + R(t)$$

$$\dot{y} = -\gamma y - \frac{b}{m}p + r(t) \, .$$

Here, x and p describe the vibrational degrees of freedom (with the eigenfrequency ω_0), and y is the relaxational degree of freedom. The subsystems are mutually linearly coupled with their coupling strength determined by the parameter b. The coupling to the heat bath is accomplished by the stochastic forces R and r for each subsystem, with the usual properties (vanishing of the average values and the Einstein relations), and the associated damping coefficients Γ and γ.
(a) Calculate the dynamic susceptibility $\chi_x(\omega)$ for the vibrational degree of freedom.
(b) Discuss the expression obtained in the limiting case of $\gamma \to 0$, i.e. when the relaxation time of the relaxing system is very long.

8.7 An example of an application of the overdamped Langevin equation is an electrical circuit consisting of a capacitor of capacity C and a resistor of resistance R which is at the temperature T. The voltage drop U_R over the resistor depends on the current I via $U_R = RI$, and the voltage U_C over the capacitor is related to the capacitor charge Q via $U_C = \frac{Q}{C}$. On the average, the sum of the two voltages is zero, $U_R + U_C = 0$. In fact, the current results from the motion of many electrons, and collisions with the lattice ions and with phonons cause fluctuations which are modeled by a noise term V_{th} in the voltage balance $(J = \dot{Q})$

$$R\dot{Q} + \frac{1}{C}Q = V_{\text{th}}$$

or

$$\dot{U}_c + \frac{1}{RC}U_c = \frac{1}{RC}V_{\text{th}} \, .$$

(a) Assume the Einstein relation for the stochastic force and calculate the spectral distribution of the voltage fluctuations

$$\phi(\omega) = \int_{-\infty}^{\infty} dt \, e^{i\omega t} \langle U_c(t)U_c(0)\rangle \, .$$

(b) Compute

$$\langle U_c^2 \rangle \equiv \langle U_c(t)U_c(t)\rangle \equiv \int_{-\infty}^{\infty} d\omega \, \phi(\omega)$$

and interpret the result, $\frac{1}{2}C\langle U_c^2 \rangle = \frac{1}{2}kT$.

8.8 In a generalization of problem 8.7, let the circuit now contain also a coil or inductor of self-inductance L with a voltage drop $U_L = L\dot{I}$. The equation of motion for the charge on the capacitor is

$$\ddot{Q} + R\dot{Q} + \frac{1}{C}Q = V_{\text{th}} \,.$$

By again assuming the Einstein relation for the noise voltage V_{th}, calculate the spectral distribution for the current $\int_{-\infty}^{\infty} dt\, e^{i\omega t} \langle I(t)I(0) \rangle$.

8.9 Starting from the equations of motion for an isotropic ferromagnet (Eq. 8.3.34), investigate the ferromagnetic phase, in which

$$\mathbf{M}(\mathbf{x},t) = \hat{e}_z M_0 + \delta\mathbf{M}(\mathbf{x},t)$$

holds.
(a) Linearize the equations of motion in $\delta\mathbf{M}(\mathbf{x},t)$, and determine the transverse and longitudinal excitations relative to the z-direction.
(b) Calculate the dynamic susceptibility

$$\chi_{ij}(\mathbf{k},\omega) = \int d^3x\, dt\, e^{-i(\mathbf{kx}-\omega t)} \frac{\partial M_i(\mathbf{x},t)}{\partial h_j(0,0)}$$

and the correlation function

$$G_{ij}(\mathbf{k},\omega) = \int d^3x\, dt\, e^{-i(\mathbf{kx}-\omega t)} \langle \delta M_i(\mathbf{x},t)\delta M_j(0,0)\rangle \,.$$

8.10 Solve the Smoluchowski equation

$$\frac{\partial P(x,t)}{\partial t} = \Gamma \frac{\partial}{\partial x}\left(kT\frac{\partial}{\partial x} + \frac{\partial V(x)}{\partial x}\right) P(x,t)$$

for an harmonic potential and an inverted harmonic potential $V(x) = \pm\frac{m\omega^2}{2}x^2$, by solving the corresponding eigenvalue problem.

8.11 Justify the Ansatz of Eq. (8.3.39) and carry out the rearrangement to give Eq. (8.3.40).

8.12 Solve the Smoluchowski equation for the model potential

$$V(x) = 2kT\log(\cosh x)$$

using supersymmetric quantum mechanics, by transforming as in Chapter 8.3.4 to a Schrödinger equation. (*Literature: F. Schwabl, Quantum Mechanics, 3*rd *ed., Chap. 19 (Springer Verlag, Heidelberg, New York, corrected printing 2005.)*

8.13 Stock-market prices as a stochastic process.
    Assume that the logarithm $l(t) = \log S(t)$ of the price $S(t)$ of a stock obeys the Langevin equation (on a sufficiently rough time scale)

$$\frac{d}{dt}l(t) = r + \Gamma(t)$$

where r is a constant and Γ is a Gaussian "random force" with $\langle \Gamma(t)\Gamma(t')\rangle = \sigma^2\delta(t - t')$.

(a) Explain this approach. *Hints:* What does the assumption that prices in the future cannot be predicted from the price trends in the past imply? Think first of a process which is discrete in time (e.g. the time dependence of the daily closing rates). Should the transition probability more correctly be a function of the price difference or of the price ratio?

(b) Write the Fokker–Planck equation for l, and based on it, the equation for S.

(c) What is the expectation value for the market price at the time t, when the stock is being traded at the price S_0 at time $t_0 = 0$? *Hint:* Solve the Fokker–Planck equation for $l = \log S$.

9. The Boltzmann Equation

9.1 Introduction

In the Langevin equation (Chap. 8), irreversibility was introduced phenomenologically through a damping term. *Kinetic theories* have the goal of explaining and quantitatively calculating transport processes and dissipative effects due to scattering of the atoms (or in a solid, of the quasiparticles). The object of these theories is the single-particle distribution function, whose time development is determined by the kinetic equation.

In this chapter, we will deal with a monatomic classical gas consisting of particles of mass m; we thus presume that the thermal wavelength $\lambda_T = 2\pi\hbar/\sqrt{2\pi mkT}$ and the volume per particle $v = n^{-1}$ obey the inequality

$$\lambda_T \ll n^{-1/3} \,,$$

i.e. the wavepackets are so strongly localized that the atoms can be treated classically.

Further characteristic quantities which enter include the *duration of a collision* τ_c and the *collision time* τ (this is the mean time between two collisions of an atom; see (9.2.12)). We have $\tau_c \approx r_c/\bar{v}$ and $\tau \approx 1/nr_c^2\bar{v}$, where r_c is the range of the potentials and $\bar{v}$ is the average velocity of the particles. In order to be able to consider independent two-particle collisions, we need the additional condition

$$\tau_c \ll \tau \,,$$

i.e. the duration of a collision is short in comparison to the collision time. This condition is fulfilled in the low-density limit, $r_c \ll n^{-1/3}$. Then collisions of more than two particles can be neglected.

The kinetic equation which describes the case of a dilute gas considered here is the *Boltzmann equation*[1]. The Boltzmann equation is one of the most fundamental equations of non-equilibrium statistical mechanics and is applied in areas far beyond the case of the dilute gas[2].

---

[1] Ludwig Boltzmann, Wien. Ber. **66**, 275 (1872); *Vorlesungen über Gastheorie*, Leipzig, 1896; *Lectures on Gas Theory*, translated by S. Brush, University of California Press, Berkeley, 1964

[2] See e.g. J. M. Ziman, *Principles of the Theory of solids*, 2nd ed, Cambridge Univ. Press, Cambridge, 1972.

In this chapter we will introduce the Boltzmann equation using the classical derivation of Boltzmann[1]. Next, we discuss some fundamental questions relating to irreversibility based on the H theorem. As an application of the Boltzmann equation we then determine the hydrodynamic equations and their eigenmodes (sound, heat diffusion). The transport coefficients are derived systematically from the linearized Boltzmann equation using its eigenmodes and eigenfrequencies.

9.2 Derivation of the Boltzmann Equation

We presume that only one species of atoms is present. For these atoms, we seek the equation of motion of the single-particle distribution function.

Definition: The single-particle distribution function $f(\mathbf{x}, \mathbf{v}, t)$ is defined by
$f(\mathbf{x}, \mathbf{v}, t) \, d^3x \, d^3v =$ the number of particles which are found at time t in the volume element d^3x around the point $\mathbf{x}$ and d^3v around the velocity $\mathbf{v}$.

$$\int d^3x \, d^3v \, f(\mathbf{x}, \mathbf{v}, t) = N \ . \tag{9.2.1}$$

The single-particle distribution function $f(\mathbf{x}, \mathbf{v}, t)$ is related to the N-particle distribution function $\rho(\mathbf{x}_1, \mathbf{v}_1, \dots, \mathbf{x}_N, \mathbf{v}_N, t)$ (Eq. (2.3.1)) through
$f(\mathbf{x}_1, \mathbf{v}_1, t) = N \int d^3x_2 \, d^3v_2 \dots \int d^3x_N \, d^3v_N \, \rho(\mathbf{x}_1, \mathbf{v}_1, \dots, \mathbf{x}_N, \mathbf{v}_N, t)$.

Remarks:

1. In the kinetic theory, one usually takes the velocity as variable instead of the momentum, $\mathbf{v} = \mathbf{p}/m$.
2. The 6-dimensional space generated by $\mathbf{x}$ and $\mathbf{v}$ is called μ space.
3. The volume elements d^3x and d^3v are supposed to to be of small linear dimensions compared to the macroscopic scale or to the mean velocity $\bar{v} = \sqrt{kT/m}$, but large compared to the microscopic scale, so that many particles are to be found within each element. In a gas under standard conditions ($T = 1°C$, $P = 1\,\text{atm}$), the number of molecules per cm^3 is $n = 3 \times 10^{19}$. In a cube of edge length $10^{-3}\,\text{cm}$, i.e. a volume element of the size $d^3x = 10^{-9}\,\text{cm}^3$, which for all experimental purposes can be considered to be pointlike, there are still 3×10^{10} molecules. If we choose $d^3v \approx 10^{-6} \times \bar{v}^3$, then from the Maxwell distribution

$$f^0(\mathbf{v}) = n \left(\frac{m}{2\pi kT} \right)^{3/2} e^{-\frac{mv^2}{2kT}} \ ,$$

in this element of μ space, there are $f^0 \, d^3x \, d^3v \approx 10^4$ molecules.

To derive the Boltzmann equation, we follow the motion of a volume element in μ space during the time interval $[t, t + dt]$; cf. Fig. 9.1. Since those

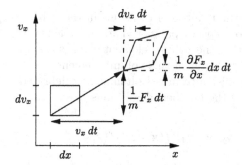

Fig. 9.1. Deformation of a volume element in μ space during the time interval dt.

particles with a higher velocity move more rapidly, the volume element is deformed in the course of time. However, the consideration of the sizes of the two parallelepipeds[3] yields

$$d^3x' \, d^3v' = d^3x \, d^3v \,. \tag{9.2.2}$$

The number of particles at the time t in $d^3x \, d^3v$ is $f(\mathbf{x}, \mathbf{v}, t) \, d^3x \, d^3v$, and the number of particles in the volume element which develops after the time interval dt is $f(\mathbf{x} + \mathbf{v}dt, \mathbf{v} + \frac{1}{m}\mathbf{F}dt, t + dt) \, d^3x' \, d^3v'$. If the gas atoms were collision-free, these two numbers would be the same. A change in these particle numbers can only occur through collisions. We thus obtain

$$\left[f(\mathbf{x} + \mathbf{v}\, dt, \mathbf{v} + \frac{1}{m}\mathbf{F}\, dt, t + dt) - f(\mathbf{x}, \mathbf{v}, t) \right] d^3x \, d^3v =$$

$$= \frac{\partial f}{\partial t} \bigg)_{\text{coll}} dt \, d^3x \, d^3v \,, \tag{9.2.3}$$

i.e. the change in the particle number is equal to its change due to collisions. The expansion of this balance equation yields

$$\left[\frac{\partial}{\partial t} + \mathbf{v}\boldsymbol{\nabla}_x + \frac{1}{m}\mathbf{F}(\mathbf{x})\boldsymbol{\nabla}_v \right] f(\mathbf{x}, \mathbf{v}, t) = \frac{\partial f}{\partial t} \bigg)_{\text{coll}} \,. \tag{9.2.4}$$

The left side of this equation is termed the *flow term*[4]. The *collision term* $\frac{\partial f}{\partial t}\big)_{\text{coll}}$ can be represented as the difference of *gain* and *loss* processes:

$$\frac{\partial f}{\partial t} \bigg)_{\text{coll}} = \mathsf{g} - \mathsf{l} \,. \tag{9.2.5}$$

Here, $\mathsf{g} \, d^3x \, d^3v \, dt$ is the number of particles which are scattered during the time interval dt into the volume $d^3x \, d^3v$ by collisions, and $\mathsf{l} \, d^3x \, d^3v \, dt$ is the

---

[3] The result obtained here from geometric considerations can also be derived by using Liouville's theorem (L. D. Landau and E. M. Lifshitz, *Course of Theoretical Physics*, Vol. I: Mechanics, Pergamon Press, Oxford 1960, Eq. (4.6.5)).

[4] In Remark (i), p. 441, the flow term is derived in a different way.

number which are scattered out, i.e. the number of collisions in the volume element d^3x in which one of the two collision partners had the velocity $\mathbf{v}$ before the collision. We assume here that the volume element d^3v is so small in velocity space that every collision leads out of this volume element.

The following expression for the collision term is Boltzmann's celebrated *Stosszahlansatz* (assumption regarding the number of collisions):

$$\left.\frac{\partial f}{\partial t}\right)_{\text{coll}} = \int d^3v_2\, d^3v_3\, d^3v_4\, W(\mathbf{v}, \mathbf{v}_2; \mathbf{v}_3, \mathbf{v}_4)[f(\mathbf{x}, \mathbf{v}_3, t)f(\mathbf{x}, \mathbf{v}_4, t)$$
$$- f(\mathbf{x}, \mathbf{v}, t)f(\mathbf{x}, \mathbf{v}_2, t)] . \quad (9.2.6)$$

Here, $W(\mathbf{v}, \mathbf{v}_2; \mathbf{v}_3, \mathbf{v}_4)$ refers to the transition probability $\mathbf{v}, \mathbf{v}_2 \rightarrow \mathbf{v}_3, \mathbf{v}_4$,

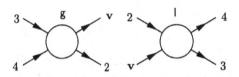

Fig. 9.2. Gain and loss processes, g and l

i.e. the probability that in a collision two particles with the velocities $\mathbf{v}$ and $\mathbf{v}_2$ will have the velocities $\mathbf{v}_3$ and $\mathbf{v}_4$ afterwards. The number of collisions which lead out of the volume element considered is proportional to the number of particles with the velocity $\mathbf{v}$ and the number of particles with velocity $\mathbf{v}_2$, and proportional to $W(\mathbf{v}, \mathbf{v}_2; \mathbf{v}_3, \mathbf{v}_4)$; a sum is carried out over all values of $\mathbf{v}_2$ and of the final velocities $\mathbf{v}_3$ and $\mathbf{v}_4$. The number of collisions in which an additional particle is in the volume element d^3v after the collision is given by the number of particles with the velocities $\mathbf{v}_3$ and $\mathbf{v}_4$ whose collision yields a particle with the velocity $\mathbf{v}$. Here, the transition probability $W(\mathbf{v}_3, \mathbf{v}_4; \mathbf{v}, \mathbf{v}_2)$ has been expressed with the help of (9.2.8e).

The *Stosszahlansatz* (9.2.6), together with the balance equation (9.2.4), yields the *Boltzmann equation*

$$\left[\frac{\partial}{\partial t} + \mathbf{v}\boldsymbol{\nabla}_x + \frac{1}{m}\mathbf{F}(\mathbf{x})\boldsymbol{\nabla}_v\right] f(\mathbf{x}, \mathbf{v}, t) =$$
$$\int d^3v_2 \int d^3v_3 \int d^3v_4\, W(\mathbf{v}, \mathbf{v}_2; \mathbf{v}_3, \mathbf{v}_4)\big(f(\mathbf{x}, \mathbf{v}_3, t)f(\mathbf{x}, \mathbf{v}_4, t)$$
$$- f(\mathbf{x}, \mathbf{v}, t)f(\mathbf{x}, \mathbf{v}_2, t)\big) . \quad (9.2.7)$$

It is a nonlinear integro-differential equation.

The transition probability W has the following symmetry properties:

- Invariance under particle exchange:

$$W(\mathbf{v}, \mathbf{v}_2; \mathbf{v}_3, \mathbf{v}_4) = W(\mathbf{v}_2, \mathbf{v}; \mathbf{v}_4, \mathbf{v}_3) . \quad (9.2.8a)$$

- Rotational and reflection invariance: with an orthogonal matrix D we have

$$W(D\mathbf{v}, D\mathbf{v}_2; D\mathbf{v}_3, D\mathbf{v}_4) = W(\mathbf{v}, \mathbf{v}_2; \mathbf{v}_3, \mathbf{v}_4) \ . \tag{9.2.8b}$$

This relation contains also inversion symmetry:

$$W(-\mathbf{v}, -\mathbf{v}_2; -\mathbf{v}_3, -\mathbf{v}_4) = W(\mathbf{v}, \mathbf{v}_2; \mathbf{v}_3, \mathbf{v}_4) \ . \tag{9.2.8c}$$

- Time-reversal invariance:

$$W(\mathbf{v}, \mathbf{v}_2; \mathbf{v}_3, \mathbf{v}_4) = W(-\mathbf{v}_3, -\mathbf{v}_4; -\mathbf{v}, -\mathbf{v}_2) \ . \tag{9.2.8d}$$

The combination of inversion and time reversal yields the relation which we have already used in (9.2.6) for $\left(\frac{\partial f}{\partial t}\right)_{\text{coll}}$:

$$W(\mathbf{v}_3, \mathbf{v}_4; \mathbf{v}, \mathbf{v}_2) = W(\mathbf{v}, \mathbf{v}_2; \mathbf{v}_3, \mathbf{v}_4) \ . \tag{9.2.8e}$$

From the conservation of momentum and energy, it follows that

$$W(\mathbf{v}_1, \mathbf{v}_2; \mathbf{v}_3, \mathbf{v}_4) = \sigma(\mathbf{v}_1, \mathbf{v}_2; \mathbf{v}_3, \mathbf{v}_4)\delta^{(3)}(\mathbf{p}_1 + \mathbf{p}_2 - \mathbf{p}_3 - \mathbf{p}_4)$$
$$\times \delta\left(\frac{\mathbf{p}_1^2}{2m} + \frac{\mathbf{p}_2^2}{2m} - \frac{\mathbf{p}_3^2}{2m} - \frac{\mathbf{p}_3^2}{2m}\right) \ , \tag{9.2.8f}$$

as one can see explicitly from the microscopic calculation of the two-particle collision in Eq. (9.5.21). The form of the scattering cross-section σ depends on the interaction potential between the particles. For all the general, fundamental results of the Boltzmann equation, the exact form of σ is not important. As an explicit example, we calculate σ for the interaction potential of hard spheres (Eq. (9.5.15)) and for a potential which falls off algebraically (problem 9.15, Eq. (9.5.29)).

To simplify the notation, in the following we shall frequently use the abbreviations

$$\begin{aligned} &f_1 \equiv f(\mathbf{x}, \mathbf{v}_1, t) \text{ with } \mathbf{v}_1 = \mathbf{v}, \\ &f_2 \equiv f(\mathbf{x}, \mathbf{v}_2, t), \qquad f_3 \equiv f(\mathbf{x}, \mathbf{v}_3, t), \qquad \text{and} \qquad f_4 \equiv f(\mathbf{x}, \mathbf{v}_4, t) \ . \end{aligned} \tag{9.2.9}$$

Remarks:

(i) The flow term in the Boltzmann equation can also be derived by setting up an equation of continuity for the fictitious case of collision-free, non-interacting gas atoms. To do this, we introduce the six-dimensional velocity vector

$$\mathbf{w} = \left(\mathbf{v} = \dot{\mathbf{x}}, \dot{\mathbf{v}} = \frac{\mathbf{F}}{m}\right) \tag{9.2.10}$$

and the current density $\mathbf{w}f(\mathbf{x}, \mathbf{v}, t)$. For a collision-free gas, f fulfills the equation of continuity

$$\frac{\partial f}{\partial t} + \text{div } \mathbf{w}f = 0 \ . \tag{9.2.11}$$

Using Hamilton's equations of motion, Eq. (9.2.11) takes on the form

$$\left(\frac{\partial}{\partial t} + \mathbf{v}\boldsymbol{\nabla}_x + \frac{1}{m}\mathbf{F}(\mathbf{x})\boldsymbol{\nabla}_v\right) f(\mathbf{x}, \mathbf{v}, t) = 0 \qquad (9.2.11')$$

of the flow term in Eqns. (9.2.4) and (9.2.7).

(ii) With a collision term of the form (9.2.6), the presence of correlations between two particles has been neglected. It is assumed that at each instant the number of particles with velocities $\mathbf{v}_3$ and $\mathbf{v}_4$, or $\mathbf{v}$ and $\mathbf{v}_2$, is uncorrelated, an assumption which is also referred to as *molecular chaos*. A statistical element is introduced here. As a justification, one can say that in a gas of low density, a binary collision between two molecules which had already interacted either directly or indirectly through a common set of molecules is extremely improbable. In fact, molecules which collide come from quite different places within the gas and previously underwent collisions with completely different molecules, and are thus quite uncorrelated. The assumption of molecular chaos is required only for the particles before a collision. After a collision, the two particles are correlated (they move apart in such a manner that if all motions were reversed, they would again collide); however, this does not enter into the equation. It is possible to derive the Boltzmann equation approximately from the Liouville equation. To this end, one derives from the latter the equations of motion for the single-, two-, etc. -particle distribution functions. The structure of these equations, which is also called the BBGKY (Bogoliubov, Born, Green, Kirkwood, Yvon) hierarchy, is such that the equation of motion for the r-particle distribution function ($r = 1, 2, \ldots$) contains in addition also the $(r+1)$-particle distribution function[5]. In particular, the equation of motion for the single-particle distribution function $f(\mathbf{x}, \mathbf{v}, t)$ has the form of the left side of the Boltzmann equation. The right side however contains f_2, the two-particle distribution function, and thus includes correlations between the particles. Only by an approximate treatment, i.e. by truncating the equation of motion for f_2 itself, does one obtain an expression which is identical with the collision term of the Boltzmann equation[6].

It should be mentioned that terms beyond those in the Boltzmann equation lead to phenomena which do not exhibit the usual exponential decay in their relaxation behavior, but instead show a much slower, algebraic behavior; these time dependences are called "long time tails". Considered microscopically, they result from so called ring collisions; see the reference by J. A. McLennan at the end of this chapter. Quantitatively, these effects are in reality immeasurably small; up to now, they have been observed only in computer experiments. In this sense, they have a similar fate to the deviations from exponential decay of excited quantum levels which occur in quantum mechanics.

(iii) To calculate the *collision time* τ, we imagine a cylinder whose length is equal to the distance which a particle with thermal velocity travels in unit time, and whose basal area is equal to the total scattering cross-section. An atom with a thermal velocity passes through this cylinder in a unit time and collides with all the other atoms within the cylinder. The number of atoms within the cylinder and thus the number of collisions of an atom per second is $\sigma_{\text{tot}}\bar{v}n$, and it follows that the mean collision time is

$$\tau = \frac{1}{\sigma_{\text{tot}}\bar{v}n} \, . \qquad (9.2.12)$$

---

[5] The r-particle distribution function is obtained from the N-particle distribution function by means of $f_r(\mathbf{x}_1, \mathbf{v}_1, \ldots \mathbf{x}_r, \mathbf{v}_r, t) \equiv \frac{N!}{(N-r)!} \int d^3x_{r+1}d^3v_{r+1}d^3x_N d^3v_N$ $\rho(\mathbf{x}_1, \mathbf{v}_1, \ldots \mathbf{x}_N, \mathbf{v}_N, t)$. The combinatorial factor results from the fact that it is not important which of the particles is at the μ-space positions $\mathbf{x}_1, \mathbf{v}_1, \ldots$.

[6] See references at the end of this chapter, e.g. K. Huang, S. Harris.

The *mean free path* l is defined as the distance which an atom typically travels between two successive collisions; it is given by

$$l \equiv \bar{v}\tau = \frac{1}{\sigma_{\text{tot}} n} . \tag{9.2.13}$$

(iv) Estimates of the lengths and times which play a role in setting up the Boltzmann equation: the range r_c of the potential must be so short that collisions occur between only those molecules which are within the same volume element d^3x: $r_c \ll dx$. This inequality is obeyed for the numerical example $r_c \approx 10^{-8}$ cm, $dx = 10^{-3}$ cm. With $\bar{v} \approx 10^5 \frac{\text{cm}}{\text{sec}}$, we obtain for the time during which the particle is within d^3x the value $\tau_{d^3x} \approx \frac{10^{-3}\,\text{cm}}{10^5\,\frac{\text{cm}}{\text{sec}}} \approx 10^{-8}$ sec. The duration of a collision is $\tau_c \approx \frac{10^{-8}\,\text{cm}}{10^5\,\frac{\text{cm}}{\text{sec}}} \approx 10^{-13}$ sec, the collision time $\tau \approx (r_c^2 n \bar{v})^{-1} \approx (10^{-16}\,\text{cm}^2 \times 3 \times 10^{19}\,\text{cm}^{-3} \times 10^5\,\text{cm}\,\text{sec}^{-1})^{-1} \approx 3 \times 10^{-9}$ sec.

9.3 Consequences of the Boltzmann Equation

9.3.1 The H-Theorem[7] and Irreversibility

The goal of this section is to show that the Boltzmann equation shows irreversible behavior, and the distribution function tends towards the Maxwell distribution. To do this, Boltzmann introduced the quantity H, which is related to the negative of the entropy: [7]

$$H(\mathbf{x}, t) = \int d^3v\, f(\mathbf{x}, \mathbf{v}, t) \log f(\mathbf{x}, \mathbf{v}, t) . \tag{9.3.1}$$

For its time derivative, one obtains from the Boltzmann equation (9.2.7)

$$\begin{aligned}
\dot{H}(\mathbf{x}, t) &= \int d^3v\, (1 + \log f) \dot{f} \\
&= -\int d^3v\, (1 + \log f) \left(\mathbf{v}\boldsymbol{\nabla}_x + \frac{1}{m}\mathbf{F}\boldsymbol{\nabla}_v \right) f - I \\
&= -\boldsymbol{\nabla}_x \int d^3v\, (f \log f)\, \mathbf{v} - I .
\end{aligned} \tag{9.3.2}$$

The second term in the large brackets in the second line is proportional to $\int d^3v\, \boldsymbol{\nabla}_v (f \log f)$ and vanishes, since there are no particles with infinite velocities, i.e. $f \to 0$ for $v \to \infty$.

---

[7] Occasionally, the rumor makes the rounds that according to Boltzmann, this should actually be called the Eta-Theorem. In fact, Boltzmann himself (1872) used E (for entropy), and only later (S. H. Burbury, 1890) was the Roman letter H adopted (D. Flamm, private communication, and S. G. Brush, *Kinetic Theory*, Vol. 2, p. 6, Pergamon Press, Oxford, 1966).

The contribution of the collision term

$$I = \int d^3v_1 \, d^3v_2 \, d^3v_3 \, d^3v_4 \, W(\mathbf{v}_1, \mathbf{v}_2; \mathbf{v}_3, \mathbf{v}_4)(f_1 f_2 - f_3 f_4)(1 + \log f_1) \quad (9.3.3)$$

is found by making use of the invariance of W with respect to the exchanges $1, 3 \leftrightarrow 2, 4$ and $1, 2 \leftrightarrow 3, 4$ to be

$$I = \frac{1}{4} \int d^3v_1 \, d^3v_2 \, d^3v_3 \, d^3v_4 \, W(\mathbf{v}_1, \mathbf{v}_2; \mathbf{v}_3, \mathbf{v}_4)(f_1 f_2 - f_3 f_4) \log \frac{f_1 f_2}{f_3 f_4} \,. \quad (9.3.4)$$

The rearrangement which leads from (9.3.3) to (9.3.4) is a special case of the general identity

$$\int d^3v_1 \, d^3v_2 \, d^3v_3 \, d^3v_4 \, W(\mathbf{v}_1, \mathbf{v}_2; \mathbf{v}_3, \mathbf{v}_4)(f_1 f_2 - f_3 f_4)\varphi_1$$

$$= \frac{1}{4} \int d^3v_1 \, d^3v_2 \, d^3v_3 \, d^3v_4 \, W(\mathbf{v}_1, \mathbf{v}_2; \mathbf{v}_3, \mathbf{v}_4) \times$$

$$\times (f_1 f_2 - f_3 f_4)(\varphi_1 + \varphi_2 - \varphi_3 - \varphi_4) \,, \quad (9.3.5)$$

which follows from the symmetry relations (9.2.8), and where $\varphi_i = \varphi(\mathbf{x}, \mathbf{v}_i, t)$ (problem 9.1).

From the inequality $(x - y) \log \frac{x}{y} \geq 0$, it follows that

$$I \geq 0 \,. \quad (9.3.6)$$

The time derivative of H, Eq. (9.3.2), can be written in the form

$$\dot{H}(\mathbf{x}, t) = -\boldsymbol{\nabla}_x \, \mathbf{j}_H(\mathbf{x}, t) - I \,, \quad (9.3.7)$$

where

$$\mathbf{j}_H = \int d^3v \, f \log f \, \mathbf{v} \quad (9.3.8)$$

is the associated current density. The first term on the right-hand side of (9.3.7) gives the change in H due to the entropy flow and the second gives its change due to entropy production.

Discussion:

a) If no external forces are present, $\mathbf{F}(\mathbf{x}) = 0$, then the simplified situation may occur that $f(\mathbf{x}, \mathbf{v}, t) = f(\mathbf{v}, t)$ is independent of $\mathbf{x}$. Since the Boltzmann equation then contains no $\mathbf{x}$-dependence, f remains independent of position for all times and it follows from (9.3.7), since $\boldsymbol{\nabla}_x \mathbf{j}_H(\mathbf{x}, t) = 0$, that

$$\dot{H} = -I \leq 0 \,. \quad (9.3.9)$$

The quantity H *decreases* and *tends towards a minimum*, which is finite, since the function $f \log f$ has a lower bound, and the integral over $\mathbf{v}$ exists.[8] At the minimum, the equals sign holds in (9.3.9). In Sect. 9.3.3, we show that at the minimum, f becomes the *Maxwell distribution*

$$f^0(\mathbf{v}) = n \left(\frac{m}{2\pi kT} \right)^{3/2} e^{-\frac{mv^2}{2kT}} . \tag{9.3.10}$$

b) When $\mathbf{F}(\mathbf{x}) \neq 0$, and we are dealing with a closed system of volume V, then

$$\int_V d^3x \, \boldsymbol{\nabla}_x \mathbf{j}_H(\mathbf{x}, t) = \int_{O(V)} d\mathbf{O} \, \mathbf{j}_H(\mathbf{x}, t) = 0$$

holds. The flux of H through the surface of this volume vanishes if the surface is an ideal reflector; then for each contribution $\mathbf{v} \, d\mathbf{O}$ there is a corresponding contribution $-\mathbf{v} \, d\mathbf{O}$, and it follows that

$$\frac{d}{dt} H_{\text{tot}} \equiv \frac{d}{dt} \int_V d^3x H(\mathbf{x}, t) = - \int_V d^3x I \leq 0 . \tag{9.3.11}$$

H_{tot} decreases, we have *irreversibility*. The fact that irreversibility follows from an equation derived from Newtonian mechanics, which itself is time-reversal invariant, was met at first with skepticism. However, the *Stosszahlansatz* contains a probabilistic element, as we will demonstrate in detail following Eq. (9.3.14).

As already mentioned, H is closely connected with the entropy. The calculation of H for the equilibrium distribution $f^0(\mathbf{v})$ for an ideal gas yields (see problem 9.3) $H = n \left[\log \left(n \left(\frac{m}{2\pi kT} \right)^{3/2} \right) - \frac{3}{2} \right]$. The total entropy S of the ideal gas (Eq. (2.7.27)) is thus

$$S = -VkH - kN \left(3 \log \frac{2\pi\hbar}{m} - 1 \right) . \tag{9.3.12a}$$

Here, $\hbar$ is Planck's quantum of action. Expressed locally, the relation between the entropy per unit volume, H, and the particle number density n is

$$S(\mathbf{x}, t) = -kH(\mathbf{x}, t) - k \left(3 \log \frac{2\pi\hbar}{m} - 1 \right) n(\mathbf{x}, t) . \tag{9.3.12b}$$

---

[8] One can readily convince oneself that $H(t)$ cannot decrease without limit. Due to $\int d^3v \, f(\mathbf{x}, \mathbf{v}, t) < \infty$, $f(\mathbf{x}, \mathbf{v}, t)$ is bounded everywhere and a divergence of $H(t)$ could come only from the range of integration $\mathbf{v} \to \infty$. For $v \to \infty$, $f \to 0$ must hold and as a result, $\log f \to -\infty$. Comparison of $H(t) = \int d^3v \, f \log f$ with $\int d^3v \, v^2 f(\mathbf{x}, \mathbf{v}, t) < \infty$ shows that a divergence requires $|\log f| > v^2$. Then, however, $f < e^{-v^2}$, and H remains finite.

The associated current densities are

$$\mathbf{j}_S(\mathbf{x},t) = -k\mathbf{j}_H(\mathbf{x},t) - k\left(3\log\frac{2\pi\hbar}{m} - 1\right)\mathbf{j}(\mathbf{x},t) \tag{9.3.12c}$$

and fulfill

$$\dot{S}(\mathbf{x},t) = -\boldsymbol{\nabla}\mathbf{j}_S(\mathbf{x},t) + kI \ . \tag{9.3.12d}$$

Therefore, kI has the meaning of the local entropy production.

*9.3.2 Behavior of the Boltzmann Equation under Time Reversal

In a classical time-reversal transformation $\mathcal{T}$ (also motion reversal), the momenta (velocities) of the particles are reversed $(\mathbf{v} \rightarrow -\mathbf{v})$[9]. Consider a system which, beginning with an initial state at the positions $\mathbf{x}_n(0)$ and the velocities $\mathbf{v}_n(0)$, evolves for a time t, to the state $\{\mathbf{x}_n(t), \mathbf{v}_n(t)\}$, then at time t_1 experiences a motion-reversal transformation $\{\mathbf{x}_n(t_1), \mathbf{v}_n(t_1)\} \rightarrow \{\mathbf{x}_n(t_1), -\mathbf{v}_n(t_1)\}$; then if the system is invariant with respect to time reversal, the further motion for time t_1 will lead back to the motion-reversed initial state $\{\mathbf{x}_n(0), -\mathbf{v}_n(0)\}$. The solution of the equations of motion in the second time period $(t > t_1)$ is

$$\mathbf{x}'_n(t) = \mathbf{x}(2t_1 - t) \tag{9.3.13}$$
$$\mathbf{v}'_n(t) = -\mathbf{v}(2t_1 - t) \ .$$

Here, we have assumed that no external magnetic field is present. Apart from a translation by $2t_1$, the replacement $t \rightarrow -t, \mathbf{v} \rightarrow -\mathbf{v}$ is thus made. Under this transformation, the Boltzmann equation (9.2.7) becomes

$$\left(\frac{\partial}{\partial t} + \mathbf{v}\nabla_{\mathbf{x}} + \frac{1}{m}\mathbf{F}(\mathbf{x})\nabla_{\mathbf{v}}\right) f(\mathbf{x}, -\mathbf{v}, -t) = -I\left[f(\mathbf{x}, -\mathbf{v}, -t)\right] \ . \tag{9.3.14}$$

The notation of the collision term should indicate that all distribution functions have the time-reversed arguments. The Boltzmann equation is therefore *not time-reversal invariant*; $f(\mathbf{x}, -\mathbf{v}, -t)$ is not a solution of the Boltzmann equation, but instead of an equation which has a negative sign on its right-hand side $(-I\left[f(\mathbf{x}, -\mathbf{v}, -t)\right])$.

The fact that an equation which was derived from Newtonian mechanics, which is time-reversal invariant, is itself not time-reversal invariant and exhibits irreversible behavior (Eq. (9.3.11)) may initially appear surprising. Historically, it was a source of controversy. In fact, the Stosszahlansatz contains a probabilistic element which goes beyond Newtonian mechanics. Even if one assumes uncorrelated particle numbers, the numbers of particles with

---

[9] See e.g. QM II, Sect. 11.4.1

the velocities $\mathbf{v}$ and $\mathbf{v}_2$ will fluctuate: they will sometimes be larger and sometimes smaller than would be expected from the single-particle distribution
functions f_1 and f_2. The most probable value of the collisions is $f_1 \cdot f_2$, and
the time-averaged value of this number will in fact be $f_1 \cdot f_2$. The Boltzmann
equation thus yields the typical evolution of typical configurations of the
particle distribution. Configurations with small statistical weights, in which
particles go from a (superficially) probable configuration to a less probable
one (with lower entropy) – which is possible in Newtonian mechanics – are
not described by the Boltzmann equation. We will consider these questions in
more detail in the next chapter (Sect. 10.7), independently of the Boltzmann
equation.

9.3.3 Collision Invariants and the Local Maxwell Distribution

9.3.3.1 Conserved Quantities

The following conserved densities can be calculated from the single-particle
distribution function: the *particle-number density* is given by

$$n(\mathbf{x}, t) \equiv \int d^3 v \, f \, . \tag{9.3.15a}$$

The *momentum density*, which is also equal to the product of the mass and
the current density, is given by

$$m \, \mathbf{j}(\mathbf{x}, t) \equiv m \, n(\mathbf{x}, t) \mathbf{u}(\mathbf{x}, t) \equiv m \int d^3 v \, \mathbf{v} f \, . \tag{9.3.15b}$$

Equation (9.3.15b) also defines the average local velocity $\mathbf{u}(\mathbf{x}, t)$. Finally, we
define the *energy density*, which is composed of the kinetic energy of the local
convective flow at the velocity $\mathbf{u}(\mathbf{x}, t)$, i.e. $n(\mathbf{x}, t) m \mathbf{u}(\mathbf{x}, t)^2 / 2$, together with
the average kinetic energy in the local rest system[10], $n(\mathbf{x}, t) e(\mathbf{x}, t)$:

$$n(\mathbf{x}, t) \left[\frac{m \mathbf{u}(\mathbf{x}, t)^2}{2} + e(\mathbf{x}, t) \right] \equiv \int d^3 v \, \frac{m v^2}{2} f = \int d^3 v \, \frac{m}{2} \left(\mathbf{u}^2 + \phi^2 \right) f \, . \tag{9.3.15c}$$

Here, the relative velocity $\phi = \mathbf{v} - \mathbf{u}$ has been introduced, and $\int d^3 v \, \phi f = 0$,
which follows from Eq. (9.3.15b), has been used. For $e(\mathbf{x}, t)$, the internal
energy per particle in the local rest system (which is moving at the velocity
$\mathbf{u}(\mathbf{x}, t)$), it follows from (9.3.15c) that

$$n(\mathbf{x}, t) \, e(\mathbf{x}, t) = \frac{m}{2} \int d^3 v (\mathbf{v} - \mathbf{u}(\mathbf{x}, t))^2 f \, . \tag{9.3.15c'}$$

---

[10] We note that for a dilute gas, the potential energy is negligible relative to the
kinetic energy, so that the internal energy per particle $e(\mathbf{x}, t) = \bar{e}(\mathbf{x}, t)$ is equal
to the average kinetic energy.

9.3.3.2 Collisional Invariants

The collision integral I of Eq. (9.3.3) and the collision term in the Boltzmann equation vanish if the distribution function f fulfills the relation

$$f_1 f_2 - f_3 f_4 = 0 \qquad (9.3.16)$$

for all possible collisions (restricted by the conservation laws contained in (9.2.8f), i.e. if

$$\log f_1 + \log f_2 = \log f_3 + \log f_4 \qquad (9.3.17)$$

holds. Note that all the distribution functions f_i have the same **x**-argument. Due to conservation of momentum, energy, and particle number, each of the five so called *collisional invariants*

$$\chi^i = mv_i , \qquad i = 1, 2, 3 \qquad (9.3.18a)$$

$$\chi^4 = \epsilon_{\mathbf{v}} \equiv \frac{mv^2}{2} \qquad (9.3.18b)$$

$$\chi^5 = 1 \qquad (9.3.18c)$$

obeys the relation (9.3.17). There are no other collisional invariants apart from these five[11]. Thus the logarithm of the most general distribution function for which the collision term vanishes is a linear combination of the collisional invariants with position-dependent prefactors:

$$\log f^\ell(\mathbf{x}, \mathbf{v}, t) = \alpha(\mathbf{x}, t) + \beta(\mathbf{x}, t) \left(\mathbf{u}(\mathbf{x}, t) \cdot m\mathbf{v} - \frac{m}{2} \mathbf{v}^2 \right) , \qquad (9.3.19)$$

or

$$f^\ell(\mathbf{x}, \mathbf{v}, t) = n(\mathbf{x}, t) \left(\frac{m}{2\pi k T(\mathbf{x}, t)} \right)^{\frac{3}{2}} \exp\left[-\frac{m}{2kT(\mathbf{x}, t)} (\mathbf{v} - \mathbf{u}(\mathbf{x}, t))^2 \right] .$$
$$(9.3.19')$$

Here, the quantities $T(\mathbf{x}, t) = (k\beta(\mathbf{x}, t))^{-1}$, $n(\mathbf{x}, t) = \left(\frac{2\pi}{m\beta(\mathbf{x},t)} \right)^{\frac{3}{2}} \exp\left[\alpha(\mathbf{x}, t)\right.$ $+ \beta(\mathbf{x}, t) mu^2(\mathbf{x}, t)/2\big]$ and $\mathbf{u}(\mathbf{x}, t)$ represent the *local temperature*, the *local particle-number density*, and the *local velocity*. One refers to $f^\ell(\mathbf{x}, \mathbf{v}, t)$ as the *local Maxwell distribution* or the local equilibrium distribution function, since it is identical locally to the Maxwell distribution, (9.3.10) or (2.6.13). If we insert (9.3.19') into the expressions (9.3.15a–c) for the conserved quantities, we can see that the quantities $n(\mathbf{x}, t)$, $\mathbf{u}(\mathbf{x}, t)$, and $T(\mathbf{x}, t)$ which occur on the right-hand side of (9.3.19') refer to the local density, velocity, and temperature, respectively, with the last quantity related to the mean kinetic energy via

---

[11] H. Grad, Comm. Pure Appl. Math. **2**, 331 (1949).

$$e(\mathbf{x}, t) = \frac{3}{2}kT(\mathbf{x}, t) ,$$

i.e. by the caloric equation of state of an ideal gas.

The local equilibrium distribution function $f^{\ell}(\mathbf{x}, \mathbf{v}, t)$ is in general not a solution of the Boltzmann equation, since for it, only the collision term but not the flow term vanishes[12]. The local Maxwell distribution is in general a solution of the Boltzmann equation only when the coefficients are constant, i.e. in global equilibrium. Together with the results from Sect. 9.3.1, it follows that a gas with an arbitrary inhomogeneous initial distribution $f(\mathbf{x}, \mathbf{v}, 0)$ will finally relax into a Maxwell distribution (9.3.10) with a constant temperature and density. Their values are determined by the initial conditions.

9.3.4 Conservation Laws

With the aid of the collisional invariants, we can derive equations of continuity for the conserved quantities from the Boltzmann equation. We first relate the conserved densities (9.3.15a–c) to the collisional invariants (9.3.18a–c). The *particle-number density*, the *momentum density*, and the *energy density* can be represented in the following form:

$$n(\mathbf{x}, t) \equiv \int d^3 v \, \chi^5 f , \tag{9.3.20}$$

$$m \, j_i(\mathbf{x}, t) \equiv m \, n(\mathbf{x}, t) u_i(\mathbf{x}, t) = \int d^3 v \, \chi^i f , \tag{9.3.21}$$

and

$$n(\mathbf{x}, t) \left[\frac{m\mathbf{u}(\mathbf{x}, t)^2}{2} + e(\mathbf{x}, t) \right] = \int d^3 v \, \chi^4 f . \tag{9.3.22}$$

Next, we want to derive the equations of motion for these quantities from the Boltzmann equation (9.2.7) by multiplying the latter by $\chi^{\alpha}(\mathbf{v})$ and integrating over $\mathbf{v}$. Using the general identity (9.3.7), we find

$$\int d^3 v \, \chi^{\alpha}(\mathbf{v}) \left[\frac{\partial}{\partial t} + \mathbf{v} \boldsymbol{\nabla}_x + \frac{1}{m} \mathbf{F}(\mathbf{x}) \boldsymbol{\nabla}_v \right] f(\mathbf{x}, \mathbf{v}, t) = 0 . \tag{9.3.23}$$

By inserting χ^5, $\chi^{1,2,3}$, and χ^4 in that order, we obtain from (9.3.23) the following three conservation laws:

---

[12] There are special local Maxwell distributions for which the flow term likewise vanishes, but they have no physical relevance. See G. E. Uhlenbeck and G. W. Ford, *Lectures in Statistical Mechanics*, American Mathematical Society, Providence, 1963, p. 86; S. Harris, *An Introduction to the Theory of the Boltzmann Equation*, Holt Rinehart and Winston, New York, 1971, p. 73; and problem 9.16.

Conservation of Particle Number:

$$\frac{\partial}{\partial t}n + \nabla \mathbf{j} = 0 . \qquad (9.3.24)$$

Conservation of Momentum:

$$m\frac{\partial}{\partial t}j_i + \nabla_{x_j}\int d^3v\, m\, v_j v_i f - F_i(\mathbf{x})n(\mathbf{x}) = 0 . \qquad (9.3.25)$$

For the third term, an integration by parts was used. If we again employ the substitution $\mathbf{v} = \mathbf{u} - \boldsymbol{\phi}$ in (9.3.25), we obtain

$$m\frac{\partial}{\partial t}j_i + \frac{\partial}{\partial x_j}(m\, n\, u_i u_j + P_{ji}) = nF_i , \qquad (9.3.25')$$

where we have introduced the *pressure tensor*

$$P_{ji} = P_{ij} = m\int d^3v\,\phi_i\phi_j f . \qquad (9.3.26)$$

Conservation of Energy:

Finally, setting $\chi^4 = \frac{mv^2}{2}$ in (9.3.23), we obtain

$$\frac{\partial}{\partial t}\int d^3v\,\frac{m}{2}v^2 f + \nabla_{x_i}\int d^3v\,(u_i+\phi_i)\frac{m}{2}(u^2+2u_j\phi_j+\phi^2)f - \mathbf{j}\cdot\mathbf{F} = 0 , \quad (9.3.27)$$

where an integration by parts was used for the last term. Applying (9.3.22) and (9.3.26), we obtain the equation of continuity for the energy density

$$\frac{\partial}{\partial t}\left[n\left(\frac{m}{2}u^2 + e\right)\right] + \nabla_i\left[nu_i\left(\frac{m}{2}u^2 + e\right) + u_j P_{ji} + q_i\right] = \mathbf{j}\cdot\mathbf{F} . \quad (9.3.28)$$

Here, along with the internal energy density e defined in (9.3.15c'), we have also introduced the *heat current density*

$$\mathbf{q} = \int d^3v\,\boldsymbol{\phi}\left(\frac{m}{2}\phi^2\right)f . \qquad (9.3.29)$$

Remarks:

(i) (9.3.25') and (9.3.28) in the absence of external forces ($\mathbf{F} = \mathbf{0}$) take on the usual form of equations of continuity, like (9.3.24).

(ii) In the momentum density, according to Eq. (9.3.25'), the tensorial current density is composed of a convective part and the pressure tensor P_{ij}, which gives the microscopic momentum current in relation to the coordinate system moving at the average velocity $\mathbf{u}$.

(iii) The energy current density in Eq. (9.3.28) contains a macroscopic convection current, the work which is performed by the pressure, and the heat current $\mathbf{q}$ (= mean energy flux in the system which is moving with the liquid).

(iv) The conservation laws do not form a complete system of equations as long as the current densities are unknown. In the hydrodynamic limit, it is possible to express the current densities in terms of the conserved quantities.

The conservation laws for momentum and energy can also be written as equations for $\mathbf{u}$ and e. To this end, we employ the rearrangement

$$\frac{\partial}{\partial t} j_i + \nabla_j (nu_j u_i) = n\frac{\partial}{\partial t} u_i + u_i \frac{\partial}{\partial t} n + u_i \nabla_j nu_j + nu_j \nabla_j u_i$$

$$= n\left(\frac{\partial}{\partial t} + u_j \nabla_j\right) u_i \tag{9.3.30}$$

using (9.3.21) and the conservation law for the particle-number density (9.3.21), which yields for (9.3.25′)

$$mn\left(\frac{\partial}{\partial t} + u_j \nabla_j\right) u_i = -\nabla_j P_{ji} + nF_i . \tag{9.3.31}$$

From this, taking the hydrodynamic limit, we obtain the Navier–Stokes equations. Likewise, starting from Eq. (9.3.28), we can show that

$$n\left(\frac{\partial}{\partial t} + u_j \nabla_j\right) e + \nabla \mathbf{q} = -P_{ij} \nabla_i u_j . \tag{9.3.32}$$

9.3.5 Conservation Laws and Hydrodynamic Equations for the Local Maxwell Distribution

9.3.5.1 Local Equilibrium and Hydrodynamics

In this section, we want to collect and explain some concepts which play a role in nonequilibrium theory.

The term *local equilibrium* describes the situation in which the thermodynamic quantities of the system such as density, temperature, pressure, etc. can vary spatially and with time, but in each volume element the thermodynamic relations between the values which apply locally there are obeyed. The resulting dynamics are quite generally termed *hydrodynamics* in condensed-matter physics, in analogy to the dynamic equations which are valid in this limit for the flow of gases and liquids. The conditions for local equilibrium are

$$\omega\tau \ll 1 \quad \text{and} \quad kl \ll 1 , \tag{9.3.33}$$

where ω is the frequency of the time-dependent variations and k their wavenumber, τ is the collision time and l the mean free path. The first condition guarantees that the variations with time are sufficiently slow that the system has time to reach equilibrium *locally* through collisions of its atoms. The second condition presumes that the particles move along a distance l without changing their momenta and energies. The local values of momentum and energy must therefore in fact be constant over a distance l.

Beginning with an arbitrary initial distribution function $f(\mathbf{x}, \mathbf{v}, 0)$, according to the Boltzmann equation, the following relaxation processes occur: the collision term causes the distribution function to *approach* a local Maxwell distribution within the characteristic time τ. The flow term causes an equalization in space, which requires a longer time. These two approaches towards equilibrium – in velocity space and in configuration space – come to an end only when global equilibrium has been reached. If the system is subject only to perturbations which vary slowly in space and time, it will be in local equilibrium after the time τ. This temporally and spatially slowly varying distribution function will differ from the local Maxwellian function (9.3.19′), which does not obey the Boltzmann equation.

9.3.5.2 Hydrodynamic Equations without Dissipation

In order to obtain explicit expressions for the current densities $\mathbf{q}$ and P_{ij}, these quantities must be calculated for a distribution function $f(\mathbf{x}, \mathbf{v}, t)$ which at least approximately obeys the Boltzmann equation. In this section, we will employ the local Maxwell distribution as an approximation. In Sect. 9.4, the Boltzmann equation will be solved systematically in a linear approximation.

Following the preceding considerations concerning the different relaxation behavior in configuration space and in velocity space, we can expect that in local equilibrium, the actual distribution function will not be very different from the local Maxwellian distribution. If we use the latter as an approximation, we will be neglecting dissipation.

Using the local Maxwell distribution, Eq. (9.3.19′),

$$f^\ell = n(\mathbf{x}, t) \left(\frac{m}{2\pi kT(\mathbf{x}, t)}\right)^{\frac{3}{2}} \exp\left[-\frac{m\left(\mathbf{v} - \mathbf{u}(\mathbf{x}, t)\right)^2}{2kT(\mathbf{x}, t)}\right] , \qquad (9.3.34)$$

with position- and time-dependent density n, temperature T, and flow velocity $\mathbf{u}$, we find from (9.3.15a), (9.3.15b), and (9.3.15c′)

$$\mathbf{j} = n\mathbf{u} \qquad (9.3.35)$$

$$ne = \frac{3}{2}nkT \qquad (9.3.36)$$

$$P_{ij} \equiv \int d^3v\, m\phi_i\phi_j f^\ell = \delta_{ij} nkT \equiv \delta_{ij} P , \qquad (9.3.37)$$

where the local pressure P was introduced; from (9.3.37), it is given by

$$P = nkT \ . \tag{9.3.38}$$

The equations (9.3.38) and (9.3.36) express the local thermal and caloric equations of state of the ideal gas. The pressure tensor P_{ij} contains no dissipative contribution which would correspond to the viscosity of the fluid, as seen from Eq. (9.3.37). The heat current density (9.3.29) vanishes ($\mathbf{q} = 0$) for the local Maxwell distribution.

With these results, we obtain for the equations of continuity (9.3.24), (9.3.25'), and (9.3.32)

$$\frac{\partial}{\partial t} n = -\boldsymbol{\nabla} n\mathbf{u} \tag{9.3.39}$$

$$m\,n \left(\frac{\partial}{\partial t} + \mathbf{u}\boldsymbol{\nabla} \right) \mathbf{u} = -\boldsymbol{\nabla} P + n\mathbf{F} \tag{9.3.40}$$

$$n \left(\frac{\partial}{\partial t} + \mathbf{u}\boldsymbol{\nabla} \right) e = -P\boldsymbol{\nabla}\mathbf{u} \ . \tag{9.3.41}$$

Here, (9.3.40) is *Euler's equation*, well-known in hydrodynamics[13]. The equations of motion (9.3.39)–(9.3.41) together with the local thermodynamic relations (9.3.36) and (9.3.38) represent a complete system of equations for n, $\mathbf{u}$, and e.

9.3.5.3 Propagation of Sound in Gases

As an application, we consider the *propagation of sound*. In this process, the gas undergoes small oscillations of its density n, its pressure P, its internal energy e, and its temperature T around their equilibrium values and around $\mathbf{u} = 0$. In the following, we shall follow the convention that thermodynamic quantities for which no position or time dependence is given are taken to have their equilibrium values, that is we insert into Eqns. (9.3.39)–(9.3.41)

$$n(\mathbf{x},t) = n + \delta n(\mathbf{x},t), \qquad P(\mathbf{x},t) = P + \delta P(\mathbf{x},t),$$
$$e(\mathbf{x},t) = e + \delta e(\mathbf{x},t), \qquad T(\mathbf{x},t) = T + \delta T(\mathbf{x},t) \tag{9.3.42}$$

and expand with respect to the small deviations indicated by δ:

$$\frac{\partial}{\partial t} \delta n = -n\boldsymbol{\nabla}\mathbf{u} \tag{9.3.43a}$$

$$m\,n\frac{\partial}{\partial t}\mathbf{u} = -\boldsymbol{\nabla}\delta P \tag{9.3.43b}$$

$$n\frac{\partial}{\partial t}\delta e = -P\boldsymbol{\nabla}\mathbf{u} \ . \tag{9.3.43c}$$

---

[13] Euler's equation describes nondissipative fluid flow; see L. D. Landau and E. M. Lifshitz, *Course of Theoretical Physics*, Vol. IV: Hydrodynamics, Pergamon Press, Oxford 1960, p. 4.

The flow velocity $\mathbf{u}(\mathbf{x}, t) \equiv \delta\mathbf{u}(\mathbf{x}, t)$ is small. Insertion of Eq. (9.3.36) and (9.3.38) into (9.3.43c) leads us to

$$\frac{3}{2}\frac{\partial}{\partial t}\delta T = -T\nabla\mathbf{u} \, ,$$

which, together with (9.3.43a), yields

$$\frac{\partial}{\partial t}\left[\frac{\delta n}{n} - \frac{3}{2}\frac{\delta T}{T}\right] = 0 \, . \tag{9.3.44}$$

Comparison with the entropy of an ideal gas,

$$S = kN\left(\frac{5}{2} + \log\frac{(2\pi mkT)^{3/2}}{nh^3}\right) \, , \tag{9.3.45}$$

shows that the time independence of S/N (i.e. of the entropy per particle or per unit mass) follows from (9.3.44). By applying $\partial/\partial t$ to (9.3.43a) and ∇ to (9.3.43b) and eliminating the term containing $\mathbf{u}$, we obtain

$$\frac{\partial^2 \delta n}{\partial t^2} = m^{-1}\nabla^2\delta P \, . \tag{9.3.46}$$

It follows from Eq. (9.3.38) that

$$\delta P = nk\delta T + \delta nkT \, ,$$

and, together with (9.3.44), we obtain $\frac{\partial}{\partial t}\delta P = \frac{5}{3}kT\frac{\partial}{\partial t}\delta n$. With this, the equation of motion (9.3.46) can be brought into the form

$$\frac{\partial^2 \delta P}{\partial t^2} = \frac{5kT}{3m}\nabla^2\delta P \, . \tag{9.3.47}$$

The sound waves (pressure waves) which are described by the wave equation (9.3.47) have the form

$$\delta P \propto e^{i(\mathbf{k}\mathbf{x}\pm c_s|\mathbf{k}|t)} \tag{9.3.48}$$

with the *adiabatic sound velocity*

$$c_s = \sqrt{\frac{1}{mn\kappa_S}} = \sqrt{\frac{5kT}{3m}} \, . \tag{9.3.49}$$

Here, κ_S is the adiabatic compressibility (Eq. (3.2.3b)), which according to Eq. (3.2.28) is given by

$$\kappa_S = \frac{3}{5P} = \frac{3V}{5NkT} \tag{9.3.50}$$

for an ideal gas.

Notes:

The result that the entropy per particle S/N or the entropy per unit mass s for a sound wave is time-independent remains valid not only for an ideal gas but in general. If one takes the second derivative with respect to time of the following thermodynamic relation which is valid for local equilibrium[14]

$$\delta n = \left(\frac{\partial n}{\partial P}\right)_{S/N} \delta P + \left(\frac{\partial n}{\partial S/N}\right)_P \delta\left(\frac{S}{N}\right) , \qquad (9.3.51)$$

obtaining $\frac{\partial^2 \delta n}{\partial t^2} = \left(\frac{\partial n}{\partial P}\right)_{S/N} \frac{\partial^2 P}{\partial t^2} + \left(\frac{\partial n}{\partial S/N}\right)_P \underbrace{\frac{\partial^2 S/N}{\partial t^2}}_{=0}$, then one obtains togther

with (9.3.43a) and (9.3.43b) the result

$$\frac{\partial^2 P(\mathbf{x},t)}{\partial t^2} = m^{-1}\left(\frac{\partial P}{\partial n}\right)_{S/N} \boldsymbol{\nabla}^2 P(\mathbf{x},t) , \qquad (9.3.52)$$

which again contains the adiabatic sound velocity

$$c_s^2 = m^{-1}\left(\frac{\partial P}{\partial n}\right)_{S/N} = m^{-1}\left(\frac{\partial P}{\partial N/V}\right)_S$$
$$= m^{-1}N^{-1}(-V^2)\left(\frac{\partial P}{\partial V}\right)_S = \frac{1}{m\,n\kappa_s} . \qquad (9.3.53)$$

Following the third equals sign, the particle number N was taken to be fixed.

For local Maxwell distributions, the collision term vanishes; there is no damping. Between the regions of different local equilibria, reversible oscillation processes take place. Deviations of the actual local equilibrium distribution functions $f(\mathbf{x},\mathbf{v},t)$ from the local Maxwell distribution $f^l(\mathbf{x},\mathbf{v},t)$ lead as a result of the collision term to local, irreversible relaxation effects and, together with the flow term, to diffusion-like equalization processes which finally result in global equilibrium.

*9.4 The Linearized Boltzmann Equation

9.4.1 Linearization

In this section, we want to investigate systematically the solutions of the Boltzmann equation in the limit of small deviations from equilibrium. The Boltzmann equation can be linearized and from its linearized form, the hydrodynamic equations can be derived. These are equations of motion for the conserved quantities, whose region of validity is at long wavelengths and

---

[14] Within time and space derivatives, $\delta n(\mathbf{x},t)$, etc. can be replaced by $n(\mathbf{x},t)$ etc.

low frequencies. It will occasionally be expedient to use the variables $(\mathbf{k}, \omega)$ (wavenumber and frequency) instead of $(\mathbf{x}, t)$. We will also take an external potential, which vanishes for early times, into account:

$$\lim_{t \to -\infty} V(\mathbf{x}, t) = 0 . \tag{9.4.1}$$

Then the distribution function is presumed to have the property

$$\lim_{t \to -\infty} f(\mathbf{x}, \mathbf{v}, t) = f^0(\mathbf{v}) \equiv n \left(\frac{m}{2\pi kT} \right)^{\frac{3}{2}} e^{-\frac{m\mathbf{v}^2}{2kT}} , \tag{9.4.2}$$

where f^0 is the global spatially uniform Maxwellian equilibrium distribution[15].

For small deviations from global equilibrium, we can write $f(\mathbf{x}, \mathbf{v}, t)$ in the form

$$f(\mathbf{x}, \mathbf{v}, t) = f^0(\mathbf{v}) \left(1 + \frac{1}{kT} \nu(\mathbf{x}, \mathbf{v}, t) \right) \equiv f^0 + \delta f \tag{9.4.3}$$

and linearize the Boltzmann equation in δf or ν. The linearization of the collision term (9.2.6) yields

$$\left. \frac{\partial f}{\partial t} \right)_{\text{coll}} = -\int d^3 v_2 \, d^3 v_3 \, d^3 v_4 \, W(f_1^0 f_2^0 - f_3^0 f_4^0 + f_1^0 \, \delta f_2 + f_2^0 \, \delta f_1 - f_3^0 \, \delta f_4 - f_4^0 \, \delta f_3)$$

$$= -\frac{1}{kT} \int d^3 v_2 \, d^3 v_3 \, d^3 v_4 \, W(\mathbf{v} \, \mathbf{v}_1; \mathbf{v}_3 \mathbf{v}_4) f^0(\mathbf{v}_1) f^0(\mathbf{v}_2)(\nu_1 + \nu_2 - \nu_3 - \nu_4) , \tag{9.4.4}$$

since $f_3^0 f_4^0 = f_1^0 f_2^0$ owing to energy conservation, which is contained in $W(\mathbf{v} \, \mathbf{v}_1; \mathbf{v}_3 \mathbf{v}_4)$. We also use the notation $\mathbf{v}_1 \equiv \mathbf{v}$, $f_1^0 = f^0(\mathbf{v})$ etc. The flow term has the form

$$\left[\frac{\partial}{\partial t} + \mathbf{v} \boldsymbol{\nabla}_x + \frac{1}{m} \mathbf{F}(\mathbf{x}, t) \boldsymbol{\nabla}_v \right] \left(f^0 + \frac{f^0}{kT} \nu \right)$$

$$= \frac{f^0(\mathbf{v})}{kT} \left[\frac{\partial}{\partial t} + \mathbf{v} \boldsymbol{\nabla}_x \right] \nu(\mathbf{x}, \mathbf{v}, t) + \mathbf{v} \cdot \left(\boldsymbol{\nabla} V(\mathbf{x}, t) \right) f^0(\mathbf{v}) / kT . \tag{9.4.5}$$

All together, the linearized Boltzmann equation is given by:

$$\left[\frac{\partial}{\partial t} + \mathbf{v} \boldsymbol{\nabla}_x \right] \nu(\mathbf{x}, \mathbf{v}, t) + \mathbf{v}(\boldsymbol{\nabla} V(\mathbf{x}, t)) = -\mathcal{L}\nu \tag{9.4.6}$$

---

[15] We write here the index which denotes an equilibrium distribution as an upper index, since later the notation $f_i^0 \equiv f^0(\mathbf{v}_i)$ will also be employed.

with the *linear collision operator* $\mathcal{L}$:

$$\mathcal{L}\nu = \frac{kT}{f^0(\mathbf{v})} \int d^3v_2\, d^3v_3\, d^3v_4\, W'(\mathbf{v}, \mathbf{v}_2; \mathbf{v}_3, \mathbf{v}_4)(\nu + \nu_2 - \nu_3 - \nu_4) \quad (9.4.7)$$

and

$$W'(\mathbf{v}\,\mathbf{v}_2; \mathbf{v}_3\,\mathbf{v}_4) = \frac{1}{kT}\left(f^0(\mathbf{v})f^0(\mathbf{v}_2)f^0(\mathbf{v}_3)f^0(\mathbf{v}_4)\right)^{\frac{1}{2}} W(\mathbf{v}\,\mathbf{v}_2; \mathbf{v}_3\,\mathbf{v}_4)\ , \quad (9.4.8)$$

where conservation of energy, contained in W, has been utilized.

9.4.2 The Scalar Product

For our subsequent investigations, we introduce the *scalar product* of two functions $\psi(\mathbf{v})$ and $\chi(\mathbf{v})$,

$$\langle \psi | \chi \rangle = \int d^3v\, \psi(\mathbf{v}) \frac{f^0(\mathbf{v})}{kT} \chi(\mathbf{v})\ ; \quad (9.4.9)$$

it possesses the usual properties. The collisional invariants are special cases:

$$\langle \chi^5 | \chi^5 \rangle \equiv \langle 1 | 1 \rangle = \int d^3v\, \frac{f^0(\mathbf{v})}{kT} = \frac{n}{kT}\ , \quad (9.4.10\text{a})$$

$$\langle \chi^4 | \chi^5 \rangle \equiv \langle \epsilon | 1 \rangle = \int d^3v\, \frac{mv^2}{2} \frac{f^0(\mathbf{v})}{kT} = \frac{ne}{kT} = \frac{3}{2}n \quad (9.4.10\text{b})$$

with $\epsilon \equiv \frac{mv^2}{2}$ and

$$\langle \chi^4 | \chi^4 \rangle \equiv \langle \epsilon | \epsilon \rangle = \int d^3v\, \left(\frac{mv^2}{2}\right)^2 \frac{f^0(\mathbf{v})}{kT} = \frac{15}{4} nkT\ . \quad (9.4.10\text{c})$$

The collision operator $\mathcal{L}$ introduced in (9.4.7) is a linear operator, and obeys the relation

$$\langle \chi | \mathcal{L}\nu \rangle = \frac{1}{4} \int d^3v_1\, d^3v_2\, d^3v_3\, d^3v_4\, W'(\mathbf{v}_1\,\mathbf{v}_2; \mathbf{v}_3\,\mathbf{v}_4)$$

$$\times (\nu_1 + \nu_2 - \nu_3 - \nu_4)(\chi^1 + \chi^2 - \chi^3 - \chi^4)\ . \quad (9.4.11)$$

It follows from this that $\mathcal{L}$ is self-adjoint and positive semidefinite,

$$\langle \chi | \mathcal{L}\nu \rangle = \langle \mathcal{L}\chi | \nu \rangle\ , \quad (9.4.12)$$

$$\langle \nu | \mathcal{L}\nu \rangle \geq 0\ . \quad (9.4.13)$$

9.4.3 Eigenfunctions of $\mathcal{L}$ and the Expansion of the Solutions of the Boltzmann Equation

The eigenfunctions of $\mathcal{L}$ are denoted as χ^λ

$$\mathcal{L}\chi^\lambda = \omega_\lambda \chi^\lambda , \qquad \omega_\lambda \geq 0 . \tag{9.4.14}$$

The collisional invariants $\chi^1, \chi^2, \chi^3, \chi^4, \chi^5$ are eigenfunctions belonging to the eigenvalue 0.

It will prove expedient to use orthonormalized eigenfunctions:

$$\left\langle \hat{\chi}^\lambda | \hat{\chi}^{\lambda'} \right\rangle = \delta^{\lambda\lambda'} . \tag{9.4.15}$$

For the collisional invariants, this means the introduction of

$$\hat{\chi}^i \equiv \hat{\chi}^{u_i} = \frac{v_i}{\sqrt{\langle v_i | v_i \rangle}} = \frac{v_i}{\sqrt{n/m}} , \qquad i = 1, 2, 3 ; \tag{9.4.16a}$$

$$\langle v_i | v_i \rangle = \frac{1}{3} \int d^3v\, \mathbf{v}^2\, f^0(\mathbf{v})/kT \qquad \text{(here not summed over } i) ;$$

$$\hat{\chi}^5 \equiv \hat{\chi}^n = \frac{1}{\sqrt{\langle 1|1 \rangle}} = \frac{1}{\sqrt{n/kT}} ; \qquad \text{and} \tag{9.4.16b}$$

$$\hat{\chi}^4 \equiv \hat{\chi}^T = \frac{\epsilon \langle 1|1 \rangle - 1 \langle 1|\epsilon \rangle}{\sqrt{\langle 1|1 \rangle \left(\langle 1|1 \rangle \langle \epsilon|\epsilon \rangle - \langle 1|\epsilon \rangle^2 \right)}} = \frac{\epsilon - \frac{3}{2}kT}{\sqrt{\frac{3}{2}nkT}} . \tag{9.4.16c}$$

The eigenfunctions χ^λ with $\omega_\lambda > 0$ are orthogonal to the functions (9.4.16a–c) and in the case of degeneracy are orthonormalized among themselves. An arbitrary solution of the linearized Boltzmann equation can be represented as a superposition of the eigenfunctions of $\mathcal{L}$ with position- and time-dependent prefactors[16]

$$\nu(\mathbf{x}, \mathbf{v}, t) = a^5(\mathbf{x}, t)\hat{\chi}^n + a^4(\mathbf{x}, t)\hat{\chi}^T + a^i(\mathbf{x}, t)\hat{\chi}^{u_i} + \sum_{\lambda=6}^{\infty} a^\lambda(\mathbf{x}, t)\hat{\chi}^\lambda . \tag{9.4.17}$$

Here, the notation indicates the particle-number density $n(\mathbf{x}, t)$, the temperature $T(\mathbf{x}, t)$, and the flow velocity $u_i(\mathbf{x}, t)$:

$$\hat{T}(\mathbf{x}, t) \equiv a^4(\mathbf{x}, t) = \left\langle \hat{\chi}^T | \nu \right\rangle = \int d^3v \left(\frac{f^0}{kT}\nu \right) \hat{\chi}^T \equiv \int d^3v\, \delta f(\mathbf{x}, \mathbf{v}, t)\hat{\chi}^T$$

$$= \frac{\delta e - \frac{3}{2}kT\delta n}{\sqrt{\frac{3}{2}nkT}} = \sqrt{\frac{3n}{2kT}}\delta T(\mathbf{x}, t) . \tag{9.4.18a}$$

---

[16] Here we assume that the eigenfunctions χ^λ form a complete basis. For the explicitly known eigenfunctions of the Maxwell potential (repulsive r^{-4} potential), this can be shown directly. For repulsive r^{-n} potentials, completeness was proved by Y. Pao, Comm. Pure Appl. Math. **27**, 407 (1974).

The identification of $\delta T(\mathbf{x}, t)$ with local fluctuations of the temperature, apart from the normalization factor, can be justified by considering the local internal energy

$$e + \delta e = \frac{3}{2}(n + \delta n)k(T + \delta T) \,,$$

from which, neglecting second-order quantities, it follows that

$$\delta e = \frac{3}{2}nk\delta T + \frac{3}{2}kT\delta n \quad \Rightarrow \quad \delta T = \frac{\delta e - \frac{3}{2}\delta nkT}{\frac{3}{2}nk} \,. \tag{9.4.19}$$

Similarly, we obtain for

$$\hat{n}(\mathbf{x}, t) \equiv a^5(\mathbf{x}, t) = \langle \hat{\chi}^n | \nu \rangle = \int d^3v \, \delta f(\mathbf{x}, \mathbf{v}, t) \frac{1}{\sqrt{n/kT}} = \frac{\delta n}{\sqrt{n/kT}} \,, \tag{9.4.18b}$$

and

$$\hat{u}_i(\mathbf{x}, t) \equiv a^i(\mathbf{x}, t) = \langle \hat{\chi}^{u_i} | \nu \rangle = \int d^3v \, \frac{v_i}{\sqrt{n/m}} \delta f(\mathbf{x}, \mathbf{v}, t)$$

$$= \int d^3v \, \frac{v_i}{\sqrt{n/m}}(f^0 + \delta f) = \frac{nu_i(\mathbf{x}, t)}{\sqrt{n/m}} \,, \quad i = 1, 2, 3 \,. \tag{9.4.18c}$$

These expressions show the relations to the density and momentum fluctuations. We now insert the expansion (9.4.17) into the linearized Boltzmann equation (9.4.6)

$$\left(\frac{\partial}{\partial t} + \mathbf{v}\boldsymbol{\nabla}\right)\nu(\mathbf{x}, \mathbf{v}, t) = -\sum_{\lambda'=6}^{\infty} a^{\lambda'}(\mathbf{x}, t)\omega_{\lambda'}\hat{\chi}^{\lambda'}(\mathbf{v}) - \mathbf{v}\boldsymbol{\nabla}V(\mathbf{x}, t) \,. \tag{9.4.20}$$

Only terms with $\lambda' \geq 6$ contribute to the sum, since the collisional invariants have the eigenvalue 0. Multiplying this equation by $\hat{\chi}^\lambda f^0(\mathbf{v})/kT$ and integrating over $\mathbf{v}$, we obtain, using the orthonormalization of $\hat{\chi}^\lambda$ from Eq. (9.4.15),

$$\frac{\partial}{\partial t}a^\lambda(\mathbf{x}, t) + \boldsymbol{\nabla}\sum_{\lambda'=1}^{\infty}\left\langle \hat{\chi}^\lambda | \mathbf{v}\hat{\chi}^{\lambda'} \right\rangle a^{\lambda'}(\mathbf{x}, t)$$

$$= -\omega_\lambda a^\lambda(\mathbf{x}, t) - \left\langle \hat{\chi}^\lambda | \mathbf{v} \right\rangle \boldsymbol{\nabla}V(\mathbf{x}, t) \,. \tag{9.4.21}$$

Fourier transformation

$$a^\lambda(\mathbf{x}, t) = \int \frac{d^3k}{(2\pi)^3}\frac{d\omega}{2\pi} e^{i(\mathbf{k}\cdot\mathbf{x}-\omega t)}a^\lambda(\mathbf{k}, \omega) \tag{9.4.22}$$

yields

$$(\omega + i\omega_\lambda)a^\lambda(\mathbf{k}, \omega) - \mathbf{k}\sum_{\lambda'=1}^{\infty}\left\langle\hat{\chi}^\lambda|\mathbf{v}\hat{\chi}^{\lambda'}\right\rangle a^{\lambda'}(\mathbf{k}, \omega) - \mathbf{k}\left\langle\hat{\chi}^\lambda|\mathbf{v}\right\rangle V(\mathbf{k}, \omega) = 0\;.$$

$$(9.4.23)$$

Which quantities couple to each other depends on the scalar products $\left\langle\hat{\chi}^\lambda|\mathbf{v}\hat{\chi}^{\lambda'}\right\rangle$, whereby the symmetry of the $\hat{\chi}^\lambda$ clearly plays a role.

Since $\omega_\lambda = 0$ for the modes $\lambda = 1$ to 5, i.e. momentum, energy, and particle-number density, the structure of the conservation laws for these quantities in (9.4.23) can already be recognized at this stage. The term containing the external force obviously couples only to $\hat{\chi}^i \equiv \hat{\chi}^{u_i}$ for reasons of symmetry

$$\left\langle\hat{\chi}^i|v^j\right\rangle = \frac{\left\langle v^i|v^j\right\rangle}{\sqrt{n/m}} = \sqrt{n/m}\,\delta^{ij}\;.\tag{9.4.24}$$

For the modes with $\lambda \leq 5$,

$$\omega a^\lambda(\mathbf{k}, \omega) - \mathbf{k}\sum_{\lambda'=1}^{\infty}\left\langle\hat{\chi}^\lambda|\mathbf{v}\hat{\chi}^{\lambda'}\right\rangle a^{\lambda'}(\mathbf{k}, \omega) - \mathbf{k}\left\langle\hat{\chi}^\lambda|\mathbf{v}\right\rangle V(\mathbf{k}, \omega) = 0 \quad (9.4.25)$$

holds, and for the non-conserved degrees of freedom[17] $\lambda \geq 6$, we have

$$a^\lambda(\mathbf{k}, \omega) = \frac{k_i}{\omega + i\omega_\lambda}\Bigg(\sum_{\lambda'=1}^{5}\left\langle\hat{\chi}^\lambda|v_i\hat{\chi}^{\lambda'}\right\rangle a^{\lambda'}(\mathbf{k}, \omega)$$

$$+ \sum_{\lambda'=6}^{\infty}\left\langle\hat{\chi}^\lambda|v_i\hat{\chi}^{\lambda'}\right\rangle a^{\lambda'}(\mathbf{k}, \omega) + \left\langle\hat{\chi}^\lambda|v_i\right\rangle V(\mathbf{k}, \omega)\Bigg)\;. \quad (9.4.26)$$

This difference, which results from the different time scales, forms the basis for the elimination of the non-conserved degrees of freedom.

9.4.4 The Hydrodynamic Limit

For low frequencies ($\omega \ll \omega^\lambda$) and ($\mathbf{vk} \ll \omega^\lambda$), $a^\lambda(\mathbf{k}, \omega)$ with $\lambda \geq 6$ is of higher order in these quantities than are the conserved quantities $\lambda = 1, \ldots, 5$. Therefore, in leading order we can write for (9.4.26)

$$a^\lambda(\mathbf{k}, \omega) = -\frac{ik_i}{\omega_\lambda}\Bigg(\sum_{\lambda'=1}^{5}\left\langle\hat{\chi}^\lambda|v_i\hat{\chi}^{\lambda'}\right\rangle a^{\lambda'}(\mathbf{k}, \omega) + \left\langle\hat{\chi}^\lambda|v_i\right\rangle V(\mathbf{k}, \omega)\Bigg)\;. \quad (9.4.27)$$

Inserting this into (9.4.25) for the conserved (also called the *hydrodynamic*) variables, we find

---

[17] Here, the Einstein summation convention is employed: repeated indices i, j, l, r are to be summed over from 1 to 3.

$$\omega a^\lambda(\mathbf{k}, \omega) - k_i \sum_{\lambda'=1}^{5} \left\langle \hat{\chi}^\lambda | v_i \hat{\chi}^{\lambda'} \right\rangle a^{\lambda'}(\mathbf{k}, \omega)$$

$$+ i k_i k_j \sum_{\lambda'=1}^{5} \sum_{\mu=6}^{\infty} \left\langle \hat{\chi}^\lambda | v_i \hat{\chi}^\mu \right\rangle \frac{1}{\omega_\mu} \left\langle \hat{\chi}^\mu | v_j \hat{\chi}^{\lambda'} \right\rangle a^{\lambda'}(\mathbf{k}, \omega) - k_i \left\langle \hat{\chi}^\lambda | v_i \right\rangle V(\mathbf{k}, \omega)$$

$$- k_i \sum_{\lambda'=6}^{\infty} \left\langle \hat{\chi}^\lambda | v_i \hat{\chi}^{\lambda'} \right\rangle \left(\frac{-ik_j}{\omega_{\lambda'}} \right) \left\langle \hat{\chi}^{\lambda'} | v_j \right\rangle V(\mathbf{k}, \omega) = 0 \; ; \quad (9.4.28)$$

this is a closed system of hydrodynamic equations of motion. The second term in these equations leads to motions which propagate like sound waves, the third term to damping of these oscillations. The latter results formally from the elimination of the infinite number of non-conserved variables which was possible due to the separation of the time scales of the hydrodynamic variables (typical frequency ck, Dk^2) from the that of the non-conserved variables (typical frequency $\omega_\mu \propto \tau^{-1}$).

The structure which is visible in Eq. (9.4.28) is of a very general nature and can be derived from the Boltzmann equations for other physical systems, such as phonons and electrons or magnons in solids.

Now we want to further evaluate Eq. (9.4.28) for a dilute gas without the effect of an external potential. We first compute the scalar products in the second term (see Eqns. (9.4.16a–c))

$$\left\langle \hat{\chi}^n | v_i \hat{\chi}^j \right\rangle = \int d^3v \frac{f^0(\mathbf{v})}{kT} \frac{v_i v_j}{\sqrt{n^2/kTm}} = \delta_{ij} \sqrt{\frac{kT}{m}} \qquad (9.4.29\text{a})$$

$$\left\langle \hat{\chi}^T | v_i \hat{\chi}^j \right\rangle = \int d^3v \frac{f^0(\mathbf{v})}{kT} v_i v_j \frac{\left(\frac{mv^2}{2} - \frac{3}{2}kT \right)}{\sqrt{\frac{n}{m}\frac{3}{2}nkT}} = \delta_{ij} \sqrt{\frac{2kT}{3m}} \; . \qquad (9.4.29\text{b})$$

These scalar products and $\left\langle \hat{\chi}^j | v_i \hat{\chi}^{n,T} \right\rangle = \left\langle \hat{\chi}^{n,T} | v_i \hat{\chi}^j \right\rangle$ are the only finite scalar products which result from the flow term in the equation of motion.

We now proceed to analyze the equations of motion for the particle-number density, the energy density, and the velocity. In the equation of motion for the *particle-number density*, $\lambda \equiv 5$ (9.4.28), there is a coupling to $a^i(\mathbf{k}, \omega)$ due to the second term. As noted above, all the other scalar products vanish. The third term vanishes completely, since $\left\langle \hat{\chi}^n | v_i \hat{\chi}^\mu \right\rangle \propto \left\langle v_i | \hat{\chi}^\mu \right\rangle = 0$ for $\mu \geq 6$ owing to the orthonormalization. We thus find

$$\omega \hat{n}(\mathbf{k}, \omega) - k_i \sqrt{\frac{kT}{m}} \hat{u}^i(\mathbf{k}, \omega) = 0 \; , \qquad (9.4.30)$$

or, due to (9.4.18),

$$\omega \delta n(\mathbf{k}, \omega) - k_i n u^i(\mathbf{k}, \omega) = 0 \; , \qquad (9.4.30')$$

or in real space

$$\frac{\partial}{\partial t}n(\mathbf{x},t) + \boldsymbol{\nabla}n\mathbf{u}(\mathbf{x},t) = 0 \ . \tag{9.4.30''}$$

This equation of motion is identical with the equation of continuity for the density, (9.3.24), except that here, $n(\mathbf{x},t)$ in the gradient term is replaced by n because of the linearization.

The equation of motion for the *local temperature*, making use of (9.4.28), (9.4.18a), and (9.4.29b), can be cast in the form

$$\omega\sqrt{\frac{3n}{2kT}}k\delta T(\mathbf{k},\omega) - k_i\sqrt{\frac{2kT}{3m}}\frac{nu_i(\mathbf{k},\omega)}{\sqrt{n/m}}$$

$$+ ik_ik_j\sum_{\lambda'=1}^{5}\sum_{\mu=6}^{\infty}\langle\hat\chi^4|v_i\hat\chi^\mu\rangle\frac{1}{\omega_\mu}\langle\hat\chi^\mu|v_j\hat\chi^{\lambda'}\rangle a^{\lambda'}(\mathbf{k},\omega) = 0 \ . \tag{9.4.31}$$

In the sum over λ', the term $\lambda' = 5$ makes no contribution, since $\langle\hat\chi^\mu|v_j\hat\chi^5\rangle \propto \langle\hat\chi^\mu|v_j\rangle = 0$. Due to the fact that $\hat\chi^4$ transforms as a scalar, $\hat\chi^\mu$ must transform like v_i, so that due to the second factor, $\hat\chi^{\lambda'} = \hat\chi^i$ also makes no contribution, leaving only $\hat\chi^{\lambda'} = \hat\chi^4$. Finally, only the following expression remains from the third term of Eq. (9.4.31):

$$ik_ik_j\sum_{\mu=6}^{\infty}\langle\hat\chi^4|v_i\hat\chi^\mu\rangle\frac{1}{\omega_\mu}\langle\hat\chi^\mu|v_j\hat\chi^4\rangle a^4(\mathbf{k},\omega)$$

$$\approx ik_ik_j\tau\sum_{\mu=6}^{\infty}\langle\hat\chi^4|v_i\hat\chi^\mu\rangle\langle\hat\chi^\mu|v_j\hat\chi^4\rangle a^4(\mathbf{k},\omega)$$

$$= ik_ik_j\tau\Big(\langle\hat\chi^4|v_iv_j\hat\chi^4\rangle - \sum_{\lambda=1}^{5}\langle\hat\chi^4|v_i\hat\chi^\lambda\rangle\langle\hat\chi^\lambda|v_j\hat\chi^4\rangle\Big)a^4(\mathbf{k},\omega)$$

$$= ik_ik_j\tau(\langle\hat\chi^4|v_iv_j\hat\chi^4\rangle - \langle\hat\chi^4|v_i\hat\chi^i\rangle\langle\hat\chi^i|v_j\hat\chi^4\rangle)a^4(\mathbf{k},\omega) \ . \tag{9.4.32}$$

In this expression, all the ω_μ^{-1} were replaced by the collision time, $\omega_\mu^{-1} = \tau$, and we have employed the completeness relation for the eigenfunctions of $\mathcal{L}$ as well as the symmetry properties. We now have

$$\langle\hat\chi^4|v_i\hat\chi^i\rangle = \sqrt{\frac{2kT}{3m}} \ , \tag{9.4.33a}$$

where here, we do not sum over i, and

$$\langle\hat\chi^4|v_iv_j\hat\chi^4\rangle = \delta_{ij}\frac{1}{3}\int d^3v\, f^0(\mathbf{v})\,\mathbf{v}^2\frac{\left(\frac{m\mathbf{v}^2}{2}\right)^2 - \frac{m\mathbf{v}^2}{2}3kT + \left(\frac{3}{2}kT\right)^2}{\frac{3}{2}n(kT)^2}$$

$$= \delta_{ij}\frac{7kT}{3m} \ . \tag{9.4.33b}$$

Thus the third term in Eq. (9.4.31) becomes $ik^2 D\sqrt{3n/2kT}\,k\delta T$, with the coefficient

$$D \equiv \frac{5}{3}\frac{kT\tau}{m} = \frac{\kappa}{mc_v} , \tag{9.4.34}$$

where

$$c_v = \frac{3}{2}nk \tag{9.4.35}$$

is the specific heat at constant volume, and

$$\kappa = \frac{5}{2}nk^2 T\tau \tag{9.4.36}$$

refers to the *heat conductivity*. All together, using (9.4.32)–(9.4.34), we obtain for the equation of motion (9.4.31) of the local temperature

$$\omega a^4(\mathbf{k},\omega) - k_i\sqrt{\frac{2kT}{3m}}a^i(\mathbf{k},\omega) + ik^2 Da^4(\mathbf{k},\omega) = 0 , \tag{9.4.37}$$

or

$$\omega\delta T - \frac{2}{3}\frac{T}{n}\mathbf{k}\cdot n\mathbf{u} + ik^2 D\delta T = 0 , \tag{9.4.37'}$$

or in real space,

$$\frac{\partial}{\partial t}T(\mathbf{x},t) + \frac{2T}{3n}\boldsymbol{\nabla}n\mathbf{u}(\mathbf{x},t) - D\boldsymbol{\nabla}^2 T(\mathbf{x},t) = 0 . \tag{9.4.37''}$$

Connection with phenomenological considerations:
The time variation of the quantity of heat δQ is

$$\delta\dot{Q} = -\boldsymbol{\nabla}\mathbf{j}_Q \tag{9.4.38a}$$

with the heat current density $\mathbf{j}_Q$. In local equilibrium, the thermodynamic relation

$$\delta Q = c_P\delta T \tag{9.4.38b}$$

holds. Here, the specific heat at constant pressure appears, because heat diffusion is isobaric owing to $c_s k \gg D_s k^2$ in the limit of small wavenumbers with the velocity of sound c_s and the thermal diffusion constant D_s. The heat current flows in the direction of decreasing temperature, which implies

$$\mathbf{j}_Q = -\frac{\kappa}{m}\boldsymbol{\nabla}T \tag{9.4.38c}$$

with the thermal conductivity κ. Overall, we thus obtain

$$\frac{d}{dt}T = \frac{\kappa}{mc_P}\boldsymbol{\nabla}^2 T , \tag{9.4.38d}$$

a diffusion equation for the temperature.

Finally, we determine the equation of motion of the *momentum density*, i.e. for a^j, $j = 1, 2, 3$. For the reversible terms (the first and second terms in Eq. (9.4.28)), we find by employing (9.4.18b–c) and $\left\langle \hat{\chi}^j | v_i \hat{\chi}^{j'} \right\rangle = 0$ the result

$$
\begin{aligned}
\omega a^j(\mathbf{k}, \omega) &- k_i \left(\left\langle \hat{\chi}^j | v_i \hat{\chi}^5 \right\rangle a^5(\mathbf{k}, \omega) + \left\langle \hat{\chi}^j | v_i \hat{\chi}^4 \right\rangle a^4(\mathbf{k}, \omega) \right) \\
&= \sqrt{\frac{m}{n}} \left(\omega n u^j(\mathbf{k}, \omega) - k_j \frac{kT}{m} \delta n(\mathbf{k}, \omega) - k_j \frac{n}{m} k \delta T(\mathbf{k}, \omega) \right) \\
&= \sqrt{\frac{m}{n}} \left(\omega n u^j(\mathbf{k}, \omega) - \frac{1}{m} k_j \delta P(\mathbf{k}, \omega) \right) ,
\end{aligned}
\tag{9.4.39}
$$

where, from $P(\mathbf{x}, t) = n(\mathbf{x}, t) kT(\mathbf{x}, t) = (n + \delta n(\mathbf{x}, t)) k(T + \delta T(\mathbf{x}, t))$, it follows that

$$
\delta P = nk\delta T + kT\delta n ,
$$

which was used above. For the damping term in the equation of motion of the momentum density, we obtain from (9.4.28) using the approximation $\omega_\mu = 1/\tau$ the result:

$$
\begin{aligned}
ik_i k_l &\sum_{\lambda'=1}^{5} \sum_{\mu=6}^{\infty} \left\langle \hat{\chi}^j | v_i \hat{\chi}^\mu \right\rangle \frac{1}{\omega_\mu} \left\langle \hat{\chi}^\mu | v_l \hat{\chi}^{\lambda'} \right\rangle a^{\lambda'}(\mathbf{k}, \omega) \\
&= ik_i k_l \sum_{\mu=6}^{\infty} \left\langle \frac{v_j}{\sqrt{n/m}} \Big| v_i \hat{\chi}^\mu \right\rangle \frac{1}{\omega_\mu} \left\langle \hat{\chi}^\mu \Big| v_l \frac{v_r}{\sqrt{n/m}} \right\rangle a^r(\mathbf{k}, \omega) \\
&\approx ik_i k_l \tau \left(\left\langle \frac{v_j}{\sqrt{n/m}} \Big| v_i v_l \frac{v_r}{\sqrt{n/m}} \right\rangle - \right. \\
&\qquad\qquad \left. \sum_{\lambda=1}^{5} \left\langle \frac{v_j}{\sqrt{n/m}} \Big| v_i \hat{\chi}^\lambda \right\rangle \left\langle \hat{\chi}^\lambda \Big| v_l \frac{v_r}{\sqrt{n/m}} \right\rangle \right) a^r(\mathbf{k}, \omega) .
\end{aligned}
\tag{9.4.40}
$$

In the second line, we have used the fact that the sum over λ' reduces to $r = 1, 2, 3$. For the first term in the curved brackets we obtain:

$$
\begin{aligned}
\left\langle \frac{v_j}{\sqrt{n/m}} v_i \Big| v_l \frac{v_r}{\sqrt{n/m}} \right\rangle &= \frac{m}{nkT} \int d^3v \, f^0(\mathbf{v}) v_j v_i v_l v_r \\
&= \frac{kT}{m} (\delta_{ji}\delta_{lr} + \delta_{jl}\delta_{ir} + \delta_{jr}\delta_{il}) .
\end{aligned}
$$

For the second term in the curved brackets in (9.4.40), we need the results of problem 9.12, leading to $\delta_{ij}\delta_{lr} \frac{5kT}{3m}$. As a result, the overall damping term (9.4.40) is given by

$$ik_ik_l \sum_{\lambda'=1}^{5} \sum_{\mu=6}^{\infty} \left\langle \hat{\chi}^j | v_i \hat{\chi}^\mu \right\rangle \frac{1}{\omega_\mu} \left\langle \hat{\chi}^\mu | v_l \hat{\chi}^{\lambda'} \right\rangle a^{\lambda'}(\mathbf{k}, \omega)$$

$$= ik_ik_l \tau \frac{kT}{m} \left(\delta_{ji}\delta_{lr} + \delta_{jl}\delta_{ir} + \delta_{jr}\delta_{il} - \frac{5}{3}\delta_{ij}\delta_{lr} \right) a^r(\mathbf{k}, \omega)$$

$$= i \left(k_j k_l u_l(\mathbf{k}, \omega) \left(-\frac{2}{3} \right) + k_i k_j u_i(\mathbf{k}, \omega) + k_i k_i u_j(\mathbf{k}, \omega) \right) \tau kT \sqrt{\frac{n}{m}}$$

$$= i \left(\frac{1}{3}k_j \left(\mathbf{k} \cdot \mathbf{u}(\mathbf{k}, \omega) \right) + \mathbf{k}^2 u_j(\mathbf{k}, \omega) \right) \tau kT \sqrt{\frac{n}{m}} .$$

$$(9.4.40')$$

Defining the *shear viscosity* as

$$\eta \equiv n\tau kT , \tag{9.4.41}$$

we find with (9.4.39) and (9.4.40') the following equivalent forms of the equation of motion for the momentum density:

$$\omega n u_j(\mathbf{k}, \omega) - \frac{1}{m}k_j \delta P(\mathbf{k}, \omega) + i\frac{\eta}{m} \left(\frac{1}{3}k_j(\mathbf{k}\mathbf{u}(\mathbf{k}, \omega)) + \mathbf{k}^2 u_j(\mathbf{k}, \omega) \right) = 0 ,$$

$$(9.4.42)$$

or, in terms of space and time,

$$\frac{\partial}{\partial t}mnu_j(\mathbf{x}, t) + \nabla_j P(\mathbf{x}, t) - \eta \left(\frac{1}{3}\nabla_j (\nabla \cdot \mathbf{u}(\mathbf{x}, t)) + \nabla^2 u_j(\mathbf{x}, t) \right) = 0 \quad (9.4.42')$$

or

$$\frac{\partial}{\partial t}mnu_j(\mathbf{x}, t) + P_{jk,k}(\mathbf{x}, t) = 0 \tag{9.4.42''}$$

with the pressure tensor ($P_{jk,k} \equiv \nabla_k P_{jk}$, etc.)

$$P_{jk}(\mathbf{x}, t) = \delta_{jk}P(\mathbf{x}, t) - \eta \left(u_{j,k}(\mathbf{x}, t) + u_{k,j}(\mathbf{x}, t) - \frac{2}{3}\delta_{jk}u_{l,l}(\mathbf{x}, t) \right) . \quad (9.4.43)$$

We can compare this result with the general pressure tensor of hydrodynamics:

$$P_{jk}(\mathbf{x}, t) = \delta_{jk}P(\mathbf{x}, t) - \eta \left(u_{j,k}(\mathbf{x}, t) + u_{k,j}(\mathbf{x}, t) - \frac{2}{3}\delta_{jk}u_{l,l}(\mathbf{x}, t) \right) -$$

$$- \zeta\delta_{jk}u_{l,l}(\mathbf{x}, t) . \quad (9.4.44)$$

Here, ζ is the bulk viscosity, also called the *compressional viscosity*. As a result of Eq. (9.4.44), the bulk viscosity vanishes according to the Boltzmann equation for simple monatomic gases. The expression (9.4.41) for the viscosity can also be written in the following form (see Eqns. (9.2.12) and (9.2.13)):

$$\eta = \tau n k T = \tau n \frac{m v_{th}^2}{3} = \frac{1}{3} n m v_{th} l = \frac{m v_{th}}{3 \sigma_{tot}} , \tag{9.4.45}$$

where $v_{th} = \sqrt{3kT/m}$ is the thermal velocity from the Maxwell distribution; i.e. the viscosity is independent of the density.

It is instructive to write the hydrodynamic equations in terms of the normalized functions $\hat{n} = \frac{n}{\sqrt{\langle n^2 \rangle / kT}}$, etc. instead of the usual quantities $n(\mathbf{x}, t)$, $T(\mathbf{x}, t)$, $u_i(\mathbf{x}, t)$. From Eqns. (9.4.30), (9.4.37), and (9.4.42') it follows that

$$\dot{\hat{n}}(\mathbf{x}, t) = -c_n \nabla_i \hat{u}^i(\mathbf{x}, t) \tag{9.4.46a}$$

$$\dot{\hat{T}}(\mathbf{x}, t) = -c_T \nabla_i \hat{u}^i(\mathbf{x}, t) + D \nabla^2 \hat{T}(\mathbf{x}, t) \tag{9.4.46b}$$

$$\dot{\hat{u}}^i(\mathbf{x}, t) = -c_n \nabla_i \hat{n} - c_T \nabla_i \hat{T} + \frac{\eta}{mn} \nabla^2 \hat{u}^i + \frac{\eta}{3mn} \nabla_i (\nabla \cdot \hat{\mathbf{u}}) \tag{9.4.46c}$$

with the coefficients $c_n = \sqrt{kT/m}$, $c_T = \sqrt{2kT/3m}$, D and η from Eqns. (9.4.34) and (9.4.41). Note that with the orthonormalized quantities, the coupling of the degrees of freedom in the equations of motion is symmetric.

9.4.5 Solutions of the Hydrodynamic Equations

The periodic solutions of (9.4.46a–c), which can be found using the ansatz $\hat{n}(\mathbf{x}, t) \propto \hat{u}^i(\mathbf{x}, t) \propto \hat{T}(\mathbf{x}, t) \propto e^{i(\mathbf{k}\mathbf{x} - \omega t)}$, are particularly interesting. The acoustic resonances which follow from the resulting secular determinant and the thermal diffusion modes have the frequencies

$$\omega = \pm c_s k - \frac{i}{2} D_s k^2 \tag{9.4.47a}$$

$$\omega = -i D_T k^2 \tag{9.4.47b}$$

with the sound velocity c_s, the acoustic attenuation constant D_s, and the heat diffusion constant (thermal diffusivity) D_T

$$c_s = \sqrt{c_n^2 + c_T^2} = \sqrt{\frac{5}{3} \frac{kT}{m}} \equiv \frac{1}{\sqrt{mn\kappa_s}} \tag{9.4.48a}$$

$$D_s = \frac{4\eta}{3mn} + \frac{\kappa}{mn} \left(\frac{1}{c_v} - \frac{1}{c_P} \right) \tag{9.4.48b}$$

$$D_T = D \frac{c_v}{c_P} = \frac{\kappa}{mc_P} . \tag{9.4.48c}$$

In this case, the specific heat at constant pressure enters; for an ideal gas, it is given by

$$c_P = \frac{5}{2} nk . \tag{9.4.49}$$

The two transverse components of the momentum density undergo a purely diffusive shearing motion:

$$D_\eta = \frac{\eta k^2}{mn} .$$

(9.4.50)

The resonances (9.4.47a,b) express themselves for example in the density-density correlation function, $S_{nn}(\mathbf{k}, \omega)$. The calculation of dynamic susceptibilities and correlation functions (problem 9.11) starting from equations of motion with damping terms is described in QM II, Sect. 4.7. The coupled system of hydrodynamic equations of motion for the density, the temperature, and the longitudinal momentum density yields the density-density correlation function:

$$S_{nn}(\mathbf{k}, \omega) = 2kTn \left(\frac{\partial n}{\partial P}\right)_T$$

$$\times \left\{ \frac{\frac{c_v}{c_P}(c_s k)^2 D_s k^2 + \left(1 - \frac{c_v}{c_P}\right)(\omega^2 - c_s^2 k^2) D_T k^2}{(\omega^2 - c_s^2 k^2)^2 + (\omega D_s k^2)^2} + \frac{\left(1 - \frac{c_v}{c_P}\right) D_T k^2}{\omega^2 + (D_T k^2)^2} \right\}.$$

(9.4.51)

The density-density correlation function for fixed $\mathbf{k}$ is shown schematically as a function of ω in Fig. 9.3.

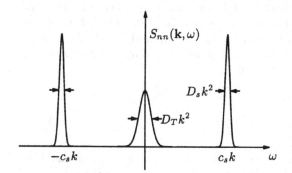

Fig. 9.3. The density-density correlation function for fixed $\mathbf{k}$ as a function of ω

The positions of the resonances are determined by the real parts and their widths by the imaginary parts of the frequencies (9.4.47a, b). In addition to the two resonances representing longitudinal acoustic phonons at $\pm c_s k$, one finds a resonance at $\omega = 0$ related to heat diffusion. The area below the curve shown in Fig. 9.3, which determines the overall intensity in inelastic scattering experiments, is proportional to the isothermal compressibility $\left(\frac{\partial n}{\partial P}\right)_T$. The relative strength of the diffusion compared to the two acoustic resonances is given by the ratio of the specific heats, $\frac{c_P - c_V}{c_V}$. This ratio is also called the Landau–Placzek ratio, and the diffusive resonance in $S_{nn}(\mathbf{k}, \omega)$ is the Landau–Placzek peak.

Since the specific heat at constant pressure diverges as $(T - T_c)^{-\gamma}$, while that at constant volume diverges only as $(T - T_c)^{-\alpha}$ (p. 256, p. 255), this ratio becomes increasingly large on approaching T_c. The expression (9.4.51), valid in the limit of small $\mathbf{k}$ (scattering in the forward direction), exhibits the phenomenon of critical opalescence, as a result of $(\partial n / \partial P)_T \propto (T - T_c)^{-\gamma}$.

*9.5 Supplementary Remarks

9.5.1 Relaxation-Time Approximation

The general evaluation of the eigenvalues and eigenfunctions of the linear collision operator is complicated. On the other hand, since not all the eigenfunctions contribute to a particular diffusion process and certainly the ones with the largest weight are those whose eigenvalues ω_λ are especially small, we can as an approximation attempt to characterize the collision term through only one characteristic frequency,

$$\left(\frac{\partial}{\partial t} + \mathbf{v} \boldsymbol{\nabla} \right) f(\mathbf{x}, \mathbf{v}, t) = -\frac{1}{\tau}(f(\mathbf{x}, \mathbf{v}, t) - f^\ell(\mathbf{x}, \mathbf{v}, t)) \,. \qquad (9.5.1)$$

This approximation is called the *conserved relaxation time approximation*, since the right-hand side represents the difference between the distribution function and a local Maxwell distribution. This takes into account the fact that the collision term vanishes when the distribution function is equal to the local Maxwell distribution. The local quantities $n(\mathbf{x}, t)$, $u^i(\mathbf{x}, t)$ and $e(\mathbf{x}, t)$ which occur in $f^\ell(\mathbf{x}, \mathbf{v}, t)$ can be calculated from $f(\mathbf{x}, \mathbf{v}, t)$ using Eqns. (9.3.15a), (9.3.15b), and (9.3.15c′).

Our goal is now to calculate f or $f - f^\ell$. We write

$$\left(\frac{\partial}{\partial t} + \mathbf{v} \boldsymbol{\nabla} \right) (f - f^\ell) + \left(\frac{\partial}{\partial t} + \mathbf{v} \boldsymbol{\nabla} \right) f^\ell = -\frac{1}{\tau}(f - f^\ell) \,. \qquad (9.5.2)$$

In the hydrodynamic region, $\omega \tau \ll 1$, $\mathbf{v} k \tau \ll 1$, we can neglect the first term on the left-hand side of (9.5.2) compared to the term on the right side, obtaining $f - f^\ell = \tau \left(\frac{\partial}{\partial t} + \mathbf{v} \boldsymbol{\nabla} \right) f^\ell$. Therefore, the distribution function has the form

$$f = f^\ell + \tau \left(\frac{\partial}{\partial t} + \mathbf{v} \boldsymbol{\nabla} \right) f^\ell \,, \qquad (9.5.3)$$

and, using this result, one can again calculate the current densities, in an extension of Sect. 9.3.5.2. In zeroth order, we obtain the expressions found in (9.3.35) and (9.3.36) for the reversible parts of the pressure tensor and the remaining current densities. The second term gives additional contributions to the pressure tensor, and also yields a finite heat current. Since f^ℓ depends on $\mathbf{x}$ and t only through the three functions $n(\mathbf{x}, t)$, $T(\mathbf{x}, t)$, and $\mathbf{u}(\mathbf{x}, t)$, the second term depends on these and their derivatives. The time derivatives of f^ℓ or n, T, and $\mathbf{u}$ can be replaced by the zero-oder equations of motion. The corrections therefore are of the form $\boldsymbol{\nabla} n(\mathbf{x}, t)$, $\boldsymbol{\nabla} T(\mathbf{x}, t)$, and $\boldsymbol{\nabla} u_i(\mathbf{x}, t)$. Along with the derivatives of P_{ij} and $\mathbf{q}$ which already occur in the equations of motion, the additional terms in the equations are of the type $\tau \boldsymbol{\nabla}^2 T(\mathbf{x}, t)$ etc. (See problem 9.13).

9.5.2 Calculation of $W(\mathbf{v}_1, \mathbf{v}_2; \mathbf{v}'_1, \mathbf{v}'_2)$

The general results of the Boltzmann equation did not depend on the precise form of the collision probability, but instead only the general relations (9.2.8a–f) were required. For completeness, we give the relation between $W(\mathbf{v}_1, \mathbf{v}_2; \mathbf{v}'_1, \mathbf{v}'_2)$ and the scattering cross-section for two particles[18]. It is assumed that the two colliding particles interact via a central potential $w(\mathbf{x}_1 - \mathbf{x}_2)$. We treat the scattering process

$$\mathbf{v}_1, \mathbf{v}_2 \;\Rightarrow\; \mathbf{v}'_1, \mathbf{v}'_2 \,,$$

in which particles 1 and 2, with velocities $\mathbf{v}_1$ and $\mathbf{v}_2$ before the collision, are left with the velocities $\mathbf{v}'_1$ and $\mathbf{v}'_2$ following the collision (see Fig. 9.4). The conservation laws for momentum and energy apply; owing to the equality of the two masses, they are given by

$$\mathbf{v}_1 + \mathbf{v}_2 = \mathbf{v}'_1 + \mathbf{v}'_2 \tag{9.5.4a}$$

$$\mathbf{v}_1^2 + \mathbf{v}_2^2 = {\mathbf{v}'_1}^2 + {\mathbf{v}'_2}^2 \,. \tag{9.5.4b}$$

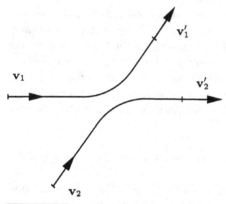

Fig. 9.4. The collision of two particles

---

[18] The theory of scattering in classical mechanics is given for example in L. D. Landau and E. M. Lifshitz, *Course of Theoretical Physics*, Vol. I: Mechanics, 3rd Ed. (Butterworth–Heinemann, London 1976), or H. Goldstein, *Classical Mechanics*, 2nd Ed. (Addison–Wesley, New York 1980).

It is expedient to introduce the center-of-mass and relative velocities; before the collision, they are

$$\mathbf{V} = \frac{1}{2}(\mathbf{v}_1 + \mathbf{v}_2), \quad \mathbf{u} = \mathbf{v}_1 - \mathbf{v}_2, \tag{9.5.5a}$$

and after the collision,

$$\mathbf{V}' = \frac{1}{2}(\mathbf{v}_1' + \mathbf{v}_2'), \quad \mathbf{u}' = \mathbf{v}_1' - \mathbf{v}_2'. \tag{9.5.5b}$$

Expressed in terms of these velocities, the two conservation laws have the form

$$\mathbf{V} = \mathbf{V}' \tag{9.5.6a}$$

and

$$|\mathbf{u}| = |\mathbf{u}'|. \tag{9.5.6b}$$

In order to recognize the validity of (9.5.6b), one need only subtract the square of (9.5.4a) from two times Eq. (9.5.4b). The center-of-mass velocity does not change as a result of the collision, and the (asymptotic) relative velocity does not change its magnitude, but it is rotated in space. For the velocity transformations to the center-of-mass frame before and after the collision given in (9.5.5a) and (9.5.5b), the volume elements in velocity space obey the relations

$$d^3v_1 d^3v_2 = d^3V d^3u = d^3V' d^3u' = d^3v_1' d^3v_2' \tag{9.5.7}$$

due to the fact that the Jacobians have unit value.

The scattering cross-section can be most simply computed in the center-of-mass frame. As is known from classical mechanics,[18] the relative coordinate $\mathbf{x}$ obeys an equation of motion in which the mass takes the form of a reduced mass μ (here $\mu = \frac{1}{2}m$) and the potential enters as a central potential $w(\mathbf{x})$. Hence, one obtains the scattering cross-section in the center-of-mass frame from the scattering of a fictitious particle of mass μ by the potential $w(\mathbf{x})$. We first write down the velocities of the two particles in the center-of-mass frame before and after the collision

$$\mathbf{v}_{1s} = \mathbf{v}_1 - \mathbf{V} = \frac{1}{2}\mathbf{u}, \quad \mathbf{v}_{2s} = -\frac{1}{2}\mathbf{u}, \quad \mathbf{v}_{1s}' = \frac{1}{2}\mathbf{u}', \quad \mathbf{v}_{2s}' = -\frac{1}{2}\mathbf{u}'. \tag{9.5.8}$$

We now recall some concepts from scattering theory. The equivalent potential scattering problem is represented in Fig. 9.5, and we can use it to define the scattering cross-section. The orbital plane of the particle is determined by the asymptotic incoming velocity $\mathbf{u}$ and position of the scattering center $\mathcal{O}$. This follows from the conservation of angular momentum in the central potential. The z-axis of the coordinate system drawn in Fig. 9.5 passes

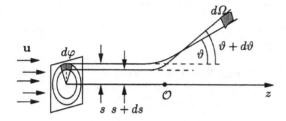

Fig. 9.5. Scattering by a fixed potential, with collision parameter s and scattering center $\mathcal{O}$. The particles which impinge on the surface element $s\,ds\,d\varphi$ are deflected into the solid angle element $d\Omega$

through the scattering center $\mathcal{O}$ and is taken to be parallel to **u**. The orbit of the incoming particle is determined by the collision parameter s and the angle φ. In Fig. 9.5, the orbital plane which is defined by the angle φ lies in the plane of the page. We consider a uniform beam of particles arriving at various distances s from the axis with the asymptotic incoming velocity **u**. The intensity I of this beam is defined as the number of particles which impinge per second on one cm^2 of the perpendicular surface shown. Letting n be the number of particles per cm^3, then $I = n|\mathbf{u}|$. The particles which impinge upon the surface element defined by the collision parameters s and $s + ds$ and the differential element of angle $d\varphi$ are deflected into the solid-angle element $d\Omega$. The number of particles arriving in $d\Omega$ per unit time is denoted by $dN(\Omega)$. The differential scattering cross-section $\sigma(\Omega, u)$, which of course also depends upon u, is defined by $dN(\Omega) = I\sigma(\Omega, u)d\Omega$, or

$$\sigma(\Omega, u) = I^{-1}\frac{dN(\Omega)}{d\Omega} . \tag{9.5.9}$$

Owing to the cylinder symmetry of the beam around the z-axis, $\sigma(\Omega, u) = \sigma(\vartheta, u)$ is independent of φ. The scattering cross-section in the center-of-mass system is obtained by making the replacement $u = |\mathbf{v}_1 - \mathbf{v}_2|$.

The collision parameter s uniquely determines the orbital curve, and therefore the scattering angle:

$$dN(\Omega) = Isd\varphi(-ds) . \tag{9.5.10}$$

From this it follows using $d\Omega = \sin\vartheta d\vartheta d\varphi$ that

$$\sigma(\Omega, u) = -\frac{1}{\sin\vartheta}s\frac{ds}{d\vartheta} = -\frac{1}{\sin\vartheta}\frac{1}{2}\frac{ds^2}{d\vartheta} . \tag{9.5.11}$$

From $\vartheta(s)$ or $s(\vartheta)$, we obtain the scattering cross-section. The scattering angle ϑ and the asymptotic angle φ_a are related by

$$\vartheta = \pi - 2\varphi_a \quad \text{or} \quad \varphi_a = \frac{1}{2}(\pi - \vartheta) \tag{9.5.12}$$

(cf. Fig. 9.6).

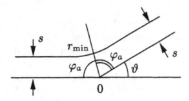

Fig. 9.6. The scattering angle (deflection angle) ϑ and the asymptotic angle φ_a

In classical mechanics, the conservation laws for energy and angular momentum give

$$\varphi_a = \int_{r_{min}}^{\infty} dr \frac{l}{r^2 \sqrt{2\mu(E - w(r)) - \frac{l^2}{r^2}}} = \int_{r_{min}}^{\infty} dr \frac{s}{r^2 \sqrt{1 - \frac{s^2}{r^2} - \frac{2w(r)}{\mu u^2}}} \; ;$$

(9.5.13)

here, we use

$$l = \mu s u \qquad\qquad (9.5.14\text{a})$$

to denote the angular momentum and

$$E = \frac{\mu}{2} u^2 \qquad\qquad (9.5.14\text{b})$$

for the energy, expressed in terms of the asymptotic velocity. The distance r_{min} of closest approach to the scattering center is determined from the condition ($\dot{r} = 0$):

$$w(r_{min}) + \frac{l^2}{2\mu r_{min}^2} = E \; . \qquad\qquad (9.5.14\text{c})$$

As an example, we consider the scattering of two *hard spheres* of radius R. In this case, we have

$$s = 2R \sin \varphi_a = 2R \sin \left(\frac{\pi}{2} - \frac{\vartheta}{2} \right) = 2R \cos \frac{\vartheta}{2} \; ,$$

from which, using (9.5.11), we find

$$\sigma(\vartheta, u) = R^2 \; . \qquad\qquad (9.5.15)$$

In this case, the scattering cross-section is independent of the deflection angle and of u, which is otherwise not the case, as is known for example from Rutherford scattering[18].

After this excursion into classical mechanics, we are in a position to calculate the transition probability $W(\mathbf{v}, \mathbf{v}_2; \mathbf{v}_3, \mathbf{v}_4)$ for the loss and gain processes in Eqns. (9.2.5) and (9.2.6). To calculate the loss rate, we recall the following assumptions:

(i) The forces are assumed to be short-ranged, so that only particles within the same volume element d^3x_1 will scatter each other.

(ii) When particle 1 is scattered, it leaves the velocity element d^3v_1.

To calculate the *loss rate* I, we pick out a molecule in d^3x which has the velocity $\mathbf{v}_1$ and take it to be the scattering center on which molecule 2 with velocity $\mathbf{v}_2$ in the velocity element d^3v_2 impinges. The flux of such particles is $f(\mathbf{x}, \mathbf{v}_2, t)|\mathbf{v}_2 - \mathbf{v}_1|d^3v_2$. The number of particles which impinge on the surface element $(-s\,ds)d\varphi$ per unit time is

$$f(\mathbf{x}, \mathbf{v}_2, t)|\mathbf{v}_2 - \mathbf{v}_1|d^3v_2(-s\,ds)d\varphi =$$
$$= f(\mathbf{x}, \mathbf{v}_2, t)|\mathbf{v}_2 - \mathbf{v}_1|d^3v_2\sigma(\Omega, |\mathbf{v}_1 - \mathbf{v}_2|)d\Omega \ .$$

In order to obtain the number of collisions which the particles within $d^3xd^3v_1$ experience in the time interval dt, we have to multiply this result by $f(\mathbf{x}, \mathbf{v}_1, t)d^3xd^3v_1dt$ and then integrate over $\mathbf{v}_2$ and all deflection angles $d\Omega$:

$$\mathsf{I}d^3xd^3v_1dt = \int d^3v_2 \int d\Omega f(\mathbf{x}, \mathbf{v}_1, t)f(\mathbf{x}, \mathbf{v}_2, t)|\mathbf{v}_2 - \mathbf{v}_1| \times$$
$$\times \sigma(\Omega, |\mathbf{v}_1 - \mathbf{v}_2|)d^3xd^3v_1dt \ . \quad (9.5.16)$$

To calculate the *gain rate* g, we consider scattering processes in which a molecule of given velocity $\mathbf{v}_1'$ is scattered into a state with velocity $\mathbf{v}_1$ by a collision with some other molecule:

$$\mathsf{g}d^3xd^3v_1dt = \int d\Omega \int d^3v_1'd^3v_2'|\mathbf{v}_1' - \mathbf{v}_2'|\,\sigma(\Omega, |\mathbf{v}_1' - \mathbf{v}_2'|) \times$$
$$\times f(\mathbf{x}, \mathbf{v}_1', t)f(\mathbf{x}, \mathbf{v}_2', t)d^3xdt \ . \quad (9.5.17)$$

The limits of the velocity integrals are chosen so that the velocity $\mathbf{v}_1$ lies within the element d^3v_1. Using (9.5.7), we obtain for the right side of (9.5.17)

$$d^3v_1 \int d^3v_2 \int d\Omega |\mathbf{v}_1 - \mathbf{v}_2|\,\sigma(\Omega, |\mathbf{v}_1 - \mathbf{v}_2|)f(\mathbf{x}, \mathbf{v}_1', t)f(\mathbf{x}, \mathbf{v}_2', t)d^3xdt \ ,$$

i.e.

$$\mathsf{g} = \int d^3v_2 \int d\Omega |\mathbf{v}_1 - \mathbf{v}_2|\,\sigma(\Omega, |\mathbf{v}_1 - \mathbf{v}_2|)f(\mathbf{x}, \mathbf{v}_1', t)f(\mathbf{x}, \mathbf{v}_2', t) \ . \quad (9.5.18)$$

Here, we have also taken account of the fact that the scattering cross-section for the scattering of $\mathbf{v}_1', \mathbf{v}_2' \rightarrow \mathbf{v}_1, \mathbf{v}_2$ is equal to that for $\mathbf{v}_1, \mathbf{v}_2 \rightarrow \mathbf{v}_1', \mathbf{v}_2'$, since the two events can be transformed into one another by a reflection in space and time.

As a result, we find for the total collision term:

$$\left(\frac{\partial f}{\partial t}\right)_{\text{coll}} = \mathsf{g}-\mathsf{I} = \int d^3v_2\,d\Omega\,|\mathbf{v}_2 - \mathbf{v}_1|\,\sigma(\Omega, |\mathbf{v}_2 - \mathbf{v}_1|)\left(f_1'f_2'-f_1f_2\right) \ . \quad (9.5.19)$$

The deflection angle ϑ can be expressed as follows in terms of the asymptotic relative velocities[18]:

$$\vartheta = \arccos \frac{(\mathbf{v}_1 - \mathbf{v}_2)(\mathbf{v}_1' - \mathbf{v}_2')}{|\mathbf{v}_1 - \mathbf{v}_2||\mathbf{v}_1' - \mathbf{v}_2'|} \ .$$

The integral $\int d\Omega$ refers to an integration over the direction of $\mathbf{u}'$. With the rearrangements

$$\mathbf{u}'^2 - \mathbf{u}^2 = \mathbf{v}_1'^{\,2} - 2\mathbf{v}_1'\mathbf{v}_2' + \mathbf{v}_2'^{\,2} - \mathbf{v}_1^2 + 2\mathbf{v}_1\mathbf{v}_2 - \mathbf{v}_2^2$$
$$= -4\mathbf{V}'^2 + 2\mathbf{v}_1'^{\,2} + 2\mathbf{v}_2'^{\,2} + 4\mathbf{V}^2 - 2\mathbf{v}_1^2 - 2\mathbf{v}_2^2 = 2(\mathbf{v}_1'^{\,2} + \mathbf{v}_2'^{\,2} - \mathbf{v}_1^2 - \mathbf{v}_2^2)$$

and

$$\int d\Omega\, |\mathbf{v}_2 - \mathbf{v}_1| = \int d\Omega\, u = \int du'\, d\Omega\, \delta(u' - u)u'$$

$$= \int du'\, u'^2\, d\Omega\, \delta\left(\frac{u'^2}{2} - \frac{u^2}{2}\right)$$

$$= \int d^3u'\, \delta\left(\frac{u'^2}{2} - \frac{u^2}{2}\right) \int d^3V'\, \delta^{(3)}\left(\mathbf{V}' - \mathbf{V}\right)$$

$$= 4 \int d^3v_1'\, d^3v_2'\, \delta\left(\frac{\mathbf{v}_1'^{\,2} + \mathbf{v}_2'^{\,2}}{2} - \frac{\mathbf{v}_1^2 + \mathbf{v}_2^2}{2}\right) \delta^{(3)}\left(\mathbf{v}_1' + \mathbf{v}_2' - \mathbf{v}_1 - \mathbf{v}_2\right) \ ,$$

which also imply the conservation laws, we obtain

$$\mathbf{g} - 1 = \int d^3v_2\, d^3v_1'\, d^3v_2'\, W(\mathbf{v}_1, \mathbf{v}_2; \mathbf{v}_1', \mathbf{v}_2')(f_1' f_2' - f_1 f_2) \ . \tag{9.5.20}$$

In this expression, we use

$$W(\mathbf{v}_1, \mathbf{v}_2; \mathbf{v}_1', \mathbf{v}_2') = 4\sigma(\Omega, |\mathbf{v}_2 - \mathbf{v}_1|)\delta\left(\frac{\mathbf{v}_1'^{\,2} + \mathbf{v}_2'^{\,2}}{2} - \frac{\mathbf{v}_1^2 + \mathbf{v}_2^2}{2}\right) \times$$

$$\times \delta^{(3)}\left(\mathbf{v}_1' + \mathbf{v}_2' - \mathbf{v}_1 - \mathbf{v}_2\right) \ . \tag{9.5.21}$$

Comparison with Eq. (9.2.8f) yields

$$\sigma(\mathbf{v}_1, \mathbf{v}_2; \mathbf{v}_1', \mathbf{v}_2') = 4m^4 \sigma(\Omega, |\mathbf{v}_2 - \mathbf{v}_1|) \ . \tag{9.5.22}$$

From the loss term in (9.5.19), we can read off the total scattering rate for particles of velocity $\mathbf{v}_1$:

$$\frac{1}{\tau(\mathbf{x}, \mathbf{v}, t)} = \int d^3v_2 \int d\Omega\, |\mathbf{v}_2 - \mathbf{v}_1|\, \sigma(\Omega, |\mathbf{v}_2 - \mathbf{v}_1|) f(\mathbf{x}, \mathbf{v}_2, t) \ . \tag{9.5.23}$$

The expression for τ^{-1} corresponds to the estimate in Eq. (9.2.12), which was derived by elementary considerations: $\tau^{-1} = n v_{th} \sigma_{tot}$, with

$$\sigma_{tot} = \int d\Omega \sigma(\Omega, |\mathbf{v}_2 - \mathbf{v}_1|) = 2\pi \int_0^{r_{max}} ds\, s \ . \tag{9.5.24}$$

r_{max} is the distance from the scattering center for which the scattering angle goes to zero, i.e. for which no more scattering occurs. In the case of hard spheres, from Eq. (9.5.15) we have

$$\sigma_{tot} = 4\pi R^2 \ . \tag{9.5.25}$$

For potentials with infinite range, r_{max} diverges. In this case, the collision term has the form

$$\left.\frac{\partial f}{\partial t}\right)_{coll} = \int d^3 v_2 \int_0^\infty ds\, s \int_0^{2\pi} d\varphi (f_1' f_2' - f_1 f_2)|\mathbf{v}_1 - \mathbf{v}_2| \ . \tag{9.5.26}$$

Although the individual contributions to the collision term diverge, the overall term remains finite:

$$\lim_{r_{max}\to\infty} \int_0^{r_{max}} ds\, s\, (f_1' f_2' - f_1 f_2) = \text{finite} \ ,$$

since for $s \to \infty$, the deflection angle tends to 0, and $\mathbf{v}_1' - \mathbf{v}_1 \to 0$ and $\mathbf{v}_2' - \mathbf{v}_2 \to 0$, so that

$$(f_1' f_2' - f_1 f_2) \to 0 \ .$$

Literature

P. Résibois and M. De Leener, *Classical Kinetic Theory of Fluids* (John Wiley, New York, 1977).

K. Huang, *Statistical Mechanics*, 2nd Ed. (John Wiley, New York, 1987).

L. Boltzmann, *Vorlesungen über Gastheorie, Vol. 1: Theorie der Gase mit einatomigen Molekülen, deren Dimensionen gegen die mittlere Weglänge verschwinden* (Barth, Leipzig, 1896); or *Lectures on Gas Theory*, transl. by S. Brush, University of California Press, Berkeley 1964.

R. L. Liboff, *Introduction to the Theory of Kinetic Equations*, Robert E. Krieger publishing Co., Huntington, New York 1975.

S. Harris, *An Introduction to the Theory of the Boltzmann Equation*, Holt, Rinehart and Winston, New York 1971.

J. A. McLennan, *Introduction to Non-Equilibrium Statistical Mechanics*, Prentice-Hall, Inc., London 1988.

K. H. Michel and F. Schwabl, *Hydrodynamic Modes in a Gas of Magnons*, Phys. Kondens. Materie **11**, 144 (1970).

Problems for Chapter 9

9.1 Symmetry Relations. Demonstrate the validity of the identity (9.3.5) used to prove the H theorem:

$$\int d^3v_1 \int d^3v_2 \int d^3v_3 \int d^3v_4 \, W(\mathbf{v}_1, \mathbf{v}_2; \mathbf{v}_3, \mathbf{v}_4)(f_1 f_2 - f_3 f_4)\varphi_1$$

$$= \frac{1}{4} \int d^3v_1 \int d^3v_2 \int d^3v_3 \int d^3v_4 \, W(\mathbf{v}_1, \mathbf{v}_2; \mathbf{v}_3, \mathbf{v}_4)$$

$$\times (f_1 f_2 - f_3 f_4)(\varphi_1 + \varphi_2 - \varphi_3 - \varphi_4) . \quad (9.5.27)$$

9.2 The Flow Term in the Boltzmann Equation. Carry out the intermediate steps which lead from the equation of continuity (9.2.11) for the single-particle distribution function in μ–space to Eq. (9.2.11′).

9.3 The Relation between H and S. Calculate the quantity

$$H(\mathbf{x}, t) = \int d^3v \, f(\mathbf{x}, \mathbf{v}, t) \log f(\mathbf{x}, \mathbf{v}, t)$$

for the case that $f(\mathbf{x}, \mathbf{v}, t)$ is the Maxwell distribution.

9.4 Show that in the absence of an external force, the equation of continuity (9.3.28) can be brought into the form (9.3.32)

$$n(\partial_t + u_j \partial_j)e + \partial_j q_j = -P_{ij}\partial_i u_j .$$

9.5 The Local Maxwell Distribution. Confirm the statements made following Eq. (9.3.19′) by inserting the local Maxwell distribution (9.3.19′) into (9.3.15a)–(9.3.15c).

9.6 The Distribution of Collision Times. Consider a spherical particle of radius r, which is passing with velocity v through a cloud of similar particles with a particle density n. The particles deflect each other only when they come into direct contact. Determine the probability distribution for the event in which the particle experiences its first collision after a time t. How long is the mean time between two collisions?

9.7 Equilibrium Expectation Values. Confirm the results (G.1c) and (G.1g) for

$$\int d^3v \left(\frac{mv^2}{2}\right)^s f^0(v) \quad \text{and} \quad \int d^3v \, v_k v_i v_j v_l \, f^0(v) .$$

9.8 Calculate the scalar products used in Sect. 9.4.2: $\langle 1|1\rangle$, $\langle \epsilon|1\rangle$, $\langle \epsilon|\epsilon\rangle$, $\langle v_i|v_j\rangle$, $\langle \hat{\chi}^5|\hat{\chi}^4\rangle$, $\langle \hat{\chi}^4|v_i\hat{\chi}^j\rangle$, $\langle \hat{\chi}^5|v_i\hat{\chi}^j\rangle$, $\langle \hat{\chi}^4|v_i^2\hat{\chi}^4\rangle$, and $\langle v_j|v_i\hat{\chi}^4\rangle$.

9.9 Sound Damping. In (9.4.30″), (9.4.37″) and (9.4.42″), the linearized hydrodynamic equations for an ideal gas were derived. For real gases and liquids with general equations of state $P = P(n, T)$, analogous equations hold:

$$\frac{\partial}{\partial t}n(\mathbf{x}, t) + n\nabla \cdot \mathbf{u}(\mathbf{x}, t) = 0$$

$$mn\frac{\partial}{\partial t}u_j(\mathbf{x}, t) + \partial_i P_{ji}(\mathbf{x}, t) = 0$$

$$\frac{\partial}{\partial t}T(\mathbf{x}, t) + n\left(\frac{\partial T}{\partial n}\right)_S \nabla \cdot \mathbf{u}(\mathbf{x}, t) - D\nabla^2 T(\mathbf{x}, t) = 0 .$$

The pressure tensor P_{ij}, with components

$$P_{ij} = \delta_{ij}P - \eta\left(\nabla_j u_i + \nabla_i u_j\right) + \left(\frac{2}{3}\eta - \zeta\right)\delta_{ij}\nabla\cdot\mathbf{u}$$

now however contains an additional term on the diagonal, $-\zeta\nabla\cdot\mathbf{u}$. This term results from the fact that real gases have a nonvanishing bulk viscosity (or compressional viscosity) ζ in addition to their shear viscosity η. Determine and discuss the modes.

Hint: Keep in mind that the equations partially decouple if one separates the velocity field into transverse and longitudinal components: $\mathbf{u} = \mathbf{u}_t + \mathbf{u}_l$ with $\nabla\cdot\mathbf{u}_t = 0$ and $\nabla\times\mathbf{u}_l = 0$. (This can be carried out simply in Fourier space without loss of generality by taking the wavevector to lie along the z-direction.)

In order to evaluate the dispersion equations (eigenfrequencies $\omega(k)$) for the Fourier transforms of n, $\mathbf{u}_l$, and T, one can consider approximate solutions for $\omega(k)$ of successively increasing order in the magnitude of the wavevector k. A useful abbreviation is

$$mc_s^2 = \left(\frac{\partial P}{\partial n}\right)_S = \left(\frac{\partial P}{\partial n}\right)_T\left[1 - \left(\frac{\partial T}{\partial n}\right)_S\Big/\left(\frac{\partial T}{\partial n}\right)_P\right] = \left(\frac{\partial P}{\partial n}\right)_T\frac{c_P}{c_V}.$$

Here, c_s is the adiabatic velocity of sound.

9.10 Show that

$$\left\langle\frac{v_j}{\sqrt{n/m}}\Big|v_i\frac{1}{\sqrt{n/kT}}\right\rangle = \delta_{ji}\sqrt{kT/m}\,, \qquad \left\langle\frac{1}{\sqrt{n/kT}}\Big|v_l\frac{v_r}{\sqrt{n/m}}\right\rangle = \delta_{lr}\sqrt{kT/m}\,,$$

$$\left\langle\frac{v_j}{\sqrt{n/m}}\Big|v_i\hat{\chi}^4\right\rangle = \delta_{ij}\left\langle v_i|\hat{\chi}^i\hat{\chi}^4\right\rangle = \delta_{ij}\sqrt{\frac{2kT}{3m}}, \qquad \left\langle\hat{\chi}^4\Big|v_l\frac{v_r}{\sqrt{n/m}}\right\rangle = \delta_{lr}\sqrt{\frac{2kT}{3m}}$$

and verify (9.4.40′).

9.11 Calculate the density-density correlation function $S_{nn}(\mathbf{k},\omega) = \int d^3x\int dt\, e^{-i(\mathbf{kx}-\omega t)}\langle n(\mathbf{x},t)n(\mathbf{0},0)\rangle$ and confirm the result in (9.4.51) by transforming to Fourier space and expressing the fluctuations at a given time in terms of thermodynamic derivatives (see also QM II, Sect. 4.7).

9.12 The Viscosity of a Dilute Gas. In Sect. 9.4, the solution of the linearized Boltzmann equation was treated by using an expansion in terms of the eigenfunctions of the collision operator. Complete the calculation of the dissipative part of the momentum current, Eq. (9.4.40). Show that

$$\sum_{\lambda=1}^{5}\langle\frac{v_j}{\sqrt{n/m}}|v_i\hat{\chi}^\lambda\rangle\langle\hat{\chi}^\lambda|v_l\frac{v_r}{\sqrt{n/m}}\rangle = \delta_{ij}\delta_{lr}\frac{5kT}{3m}\,.$$

9.13 Heat Conductivity Using the Relaxation-Time Approach. A further possibility for the approximate determination of the dissipative contributions to the equations of motion for the conserved quantities particle number, momentum and energy is found in the relaxation-time approach introduced in Sect. 9.5.1:

$$\left.\frac{\partial f}{\partial t}\right)_{\text{collision}} = -\frac{f - f^\ell}{\tau}\,.$$

For $g = f - f^\ell$, one obtains in lowest order from the Boltzmann equation (9.5.1)

$$g(\mathbf{x}, \mathbf{v}, t) = -\tau \left(\partial_t + \mathbf{v} \cdot \nabla + \frac{1}{m} \mathbf{F} \cdot \nabla_\mathbf{v} \right) f^\ell(\mathbf{x}, \mathbf{v}, t) \,.$$

Eliminate the time derivative of f^ℓ by employing the non-dissipative equations of motion obtained from f^ℓ and determine the heat conductivity by inserting $f = f^\ell + g$ into the expression for the heat current $\mathbf{q}$ derived in (9.3.29).

9.14 The Relaxation-Time Approach for the Electrical Conductivity. Consider an infinite system of charged particles immersed in a positive background. The collision term describes collisions of the particles among themselves as well as with the (fixed) ions of the background. Therefore, the collision term no longer vanishes for general local Maxwellian distributions $f^\ell(\mathbf{x}, \mathbf{v}, t)$. Before the application of a weak homogeneous electric field E, take $f = f^0$, where f^0 is the position- and time-independent Maxwell distribution. Apply the relaxation-time approach $\partial f / \partial t|_{\text{coll}} = -(f - f^0)/\tau$ and determine the new equilibrium distribution f to first order in E after application of the field. What do you find for $\langle v \rangle$? Generalize to a time-dependent field $E(t) = E_0 \cos(\omega t)$. Discuss the effects of the relaxation-time approximation on the conservation laws (see e.g. John M. Ziman, *Principles of the Theory of Solids*, 2nd Ed. (Cambridge University Press, Cambridge 1972)).

9.15 An example which is theoretically easy to treat but is unrealistic for atoms is the purely repulsive potential[19]

$$w(r) = \frac{\kappa}{\nu - 1} \frac{1}{r^{\nu - 1}} \,, \quad \nu \geq 2, \kappa > 0 \,. \tag{9.5.28}$$

Show that the corresponding scattering cross-section has the form

$$\sigma(\vartheta, |\mathbf{v}_1 - \mathbf{v}_2|) = \left(\frac{2\kappa}{m} \right)^{\frac{2}{\nu - 1}} |\mathbf{v}_1 - \mathbf{v}_2|^{-\frac{4}{\nu - 1}} F_\nu(\vartheta) \,, \tag{9.5.29}$$

with functions $F_\nu(\vartheta)$ which depend on ϑ and the power ν. For the special case of the so called Maxwell potential ($\nu = 5$), $|\mathbf{v}_1 - \mathbf{v}_2| \sigma(\vartheta, |\mathbf{v}_1 - \mathbf{v}_2|)$ is independent of $|\mathbf{v}_1 - \mathbf{v}_2|$.

9.16 Find the special local Maxwell distributions

$$f^0(\mathbf{v}, \mathbf{x}, t) = \exp \left(A + \mathbf{B} \cdot \mathbf{v} + C \frac{v^2}{2m} \right)$$

which are solutions of the Boltzmann equation, by comparing the coefficients of the powers of $\mathbf{v}$. The result is $A = A_1 + \mathbf{A}_2 \cdot \mathbf{x} + C_3 \mathbf{x}^2$, $\mathbf{B} = \mathbf{B}_1 - \mathbf{A}_2 t - (2C_3 t + C_2)\mathbf{x} + \boldsymbol{\Omega} \times \mathbf{x}$, $C = C_1 + C_2 t + C_3 t^2$.

9.17 Let an external force $\mathbf{F}(\mathbf{x}) = -\boldsymbol{\nabla} V(\mathbf{x})$ act in the Boltzmann equation. Show that the collision term and the flow term vanish for the case of the Maxwell distribution function

$$f(\mathbf{v}, \mathbf{x}) \propto n \left(\frac{m}{2\pi kT} \right)^{3/2} \exp \left[-\frac{1}{kT} \left(\frac{m(\mathbf{v} - \mathbf{u})^2}{2} + V(\mathbf{x}) \right) \right] \,.$$

9.18 Verify Eq. (9.4.33b).

---

[19] Landau/Lifshitz, *Mechanics*, p. 51, *op. cit.* in footnote 18.

10. Irreversibility and the Approach to Equilibrium

10.1 Preliminary Remarks

In this chapter, we will consider some basic aspects related to irreversible processes and their mathematical description, and to the derivation of macroscopic equations of motion from microscopic dynamics: classically from the Newtonian equations, and quantum-mechanically from the Schrödinger equation. These microscopic equations of motion are time-reversal invariant, and the question arises as to how it is possible that such equations can lead to expressions which do not exhibit time-reversal symmetry, such as the Boltzmann equation or the heat diffusion equation. This apparent incompatibility, which historically was raised in particular by Loschmidt as an objection to the Boltzmann equation, is called the *Loschmidt paradox*. Since during his lifetime the reality of atoms was not experimentally verifiable, the apparent contradiction between the time-reversal invariant (time-reversal symmetric) mechanics of atoms and the irreversibility of non-equilibrium thermodynamics was used by the opponents of Boltzmann's ideas as an argument against the very existence of atoms[1]. A second objection to the Boltzmann equation and to a purely mechanical foundation for thermodynamics came from the fact – which was proved with mathematical stringence by Poincaré – that every finite system, no matter how large, must regain its initial state periodically after a so called recurrence time. This objection was named the *Zermelo paradox*, after its most vehement protagonist. Boltzmann was able to refute both of these objections. In his considerations, which were carried further by his student P. Ehrenfest[2], probability arguments play an important role, as they do in all areas of statistical mechanics – a way of thinking that was however foreign to the mechanistic worldview of physics at that time. We mention at this point that the entropy which is defined in Eq. (2.3.1) in terms of the density matrix does not change within a closed system. In this chapter, we will denote the entropy defined in this way as the *Gibbs' entropy*. Boltzmann's

---

[1] See also the preface by H. Thirring in E. Broda, *Ludwig Boltzmann*, Deuticke, Wien, 1986.

[2] See P. Ehrenfest and T. Ehrenfest, *Begriffliche Grundlagen der statistischen Auffassung in der Mechanik*, Encykl. Math. Wiss. 4 (32) (1911); English translation by M. J. Moravcsik: *The Conceptual Foundations of the Statistical Approach in Mechanics*, Cornell University Press, Ithaca, NY 1959.

concept of entropy, which dates from an earlier time, associates a particular value of the entropy not only to an ensemble but also to each microstate, as we shall show in more detail in Sect. 10.6.2. In equilibrium, Gibbs' entropy is equal to *Boltzmann's entropy*. To eliminate the recurrence-time objection, we will estimate the recurrence time on the basis of a simple model. Using a second simple model of the Brownian motion, we will investigate how its time behavior depends on the particle number and the different time scales of the constituents. This will lead us to a general derivation of macroscopic hydrodynamic equations with dissipation from time-reversal invariant microscopic equations of motion. Finally, we will consider the tendency of a dilute gas to approach equilibrium, and its behavior under time reversal. In this connection, the influence of external perturbations will also be taken into account. In addition, this chapter contains an estimate of the size of statistical fluctuations and a derivation of Pauli's master equations.

In this chapter, we treat a few significant aspects of this extensive area of study. On the one hand, we will examine some simple models, and on the other, we will present qualitative considerations which will shed light on the subject from various sides.

In order to illuminate the problem arising from the Loschmidt paradox, we show the time development of a gas in Fig. 10.1. The reader may conjecture that the time sequence is a,b,c, in which the gas expands to fill the total available volume. If on the other hand a motion reversal is carried out at configuration c, then the atoms will move back via stage b into configuration a, which has a lower entropy. Two questions arise from this situation: (i) Why is the latter sequence (c,b,a) in fact never observed? (ii) How are we to understand the derivation of the H theorem, according to which the entropy always increases?

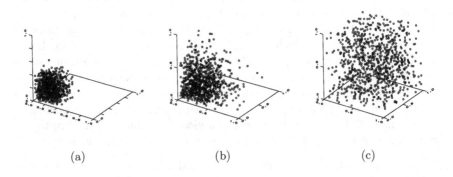

(a)   (b)   (c)

Fig. 10.1. Expansion or contraction of a gas: total volume V, subvolume V_1 (cube in lower-left corner)

10.2 Recurrence Time

Zermelo (1896)[3] based his criticism of the Boltzmann equation on Poincaré's recurrence-time theorem[4]. It states that a closed, finite, conservative system will return arbitrarily closely to its initial configuration within a finite time, the Poincaré recurrence time τ_P. According to Zermelo's paradox, $H(t)$ could not decrease monotonically, but instead must finally again increase and regain the value $H(0)$.

To adjudge this objection, we will estimate the recurrence time with the aid of a model[5]. We consider a system of classical harmonic oscillators (linear chain) with displacements q_n, momenta p_n and the Hamiltonian (see QM II, Sect. 12.1):

$$\mathcal{H} = \sum_{n=1}^{N} \left\{ \frac{1}{2m} p_n^2 + \frac{m\Omega^2}{2} \left(q_n - q_{n-1} \right)^2 \right\} . \tag{10.2.1}$$

From this, the equations of motion are obtained:

$$\dot{p}_n = m\ddot{q}_n = m\Omega^2 \left(q_{n+1} + q_{n-1} - 2q_n \right) . \tag{10.2.2}$$

Assuming periodic boundary conditions, $q_0 = q_N$, we are dealing with a translationally invariant problem, which is diagonalized by the Fourier transformation

$$q_n = \frac{1}{(mN)^{1/2}} \sum_s e^{isn} Q_s , \quad p_n = \left(\frac{m}{N} \right)^{1/2} \sum_s e^{-isn} P_s . \tag{10.2.3}$$

Q_s and (P_s) are called the normal coordinates (and momenta). The periodic boundary conditions require that $1 = e^{isN}$, i.e. $s = \frac{2\pi l}{N}$ with integral l. The values of s for which l differs by N are equivalent. A possible choice of values of l, e.g. for odd N, would be: $l = 0, \pm1, \ldots, \pm(N-1)/2$. Since q_n and p_n are real, it follows that

$$Q_s^* = Q_{-s} \text{ and } P_s^* = P_{-s} .$$

The Fourier coefficients obey the orthogonality relations

$$\frac{1}{N} \sum_{n=1}^{N} e^{isn} e^{-is'n} = \Delta(s - s') = \begin{cases} 1 & \text{for } s - s' = 2\pi h \text{ with } h \text{ integral} \\ 0 & \text{otherwise} \end{cases}$$
$$\tag{10.2.4}$$

---

[3] E. Zermelo, Wied. Ann. **57**, 485 (1896); ibid. **59**, 793 (1896).
[4] H. Poincaré, Acta Math. **13**, 1 (1890)
[5] P. C. Hemmer, L. C. Maximon, and H. Wergeland, Phys. Rev. **111**, 689 (1958).

and the completeness relation

$$\frac{1}{N} \sum_s e^{-isn} e^{isn'} = \delta_{nn'} . \tag{10.2.5}$$

Insertion of the transformation to normal coordinates yields

$$\mathcal{H} = \frac{1}{2} \sum_s \left(P_s P_s^* + \omega_s^2 Q_s Q_s^* \right) \tag{10.2.6}$$

with the dispersion relation

$$\omega_s = 2\Omega \left| \sin \frac{s}{2} \right| . \tag{10.2.7}$$

We thus find N non-coupled oscillators with eigenfrequencies[6] ω_s. The motion of the normal coordinates can be represented most intuitively by introducing complex vectors

$$Z_s = P_s + i\omega_s Q_s , \tag{10.2.8}$$

which move on a unit circle according to

$$Z_s = a_s e^{i\omega_s t} \tag{10.2.9}$$

with a complex amplitude a_s (Fig. 10.2).

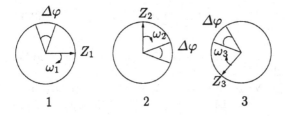

Fig. 10.2. The motion of the normal coordinates

We assume that the frequencies ω_s of $N-1$ such normal coordinates are incommensurate, i.e. their ratios are not rational numbers. Then the phase vectors Z_s rotate independently of one another, without coincidences. We now wish to calculate how much time passes until all N vectors again come into their initial positions, or more precisely, until all the vectors lie within an interval $\Delta\varphi$ around their initial positions. The probability that the vector Z_s lies within $\Delta\varphi$ during one rotation is given by $\Delta\varphi/2\pi$, and the probability that all the vectors lie within their respective prescribed intervals is $(\Delta\varphi/2\pi)^{N-1}$. The number of rotations required for this recurrence is therefore $(2\pi/\Delta\varphi)^{N-1}$. The recurrence time is found by multiplying by the typical

---

[6] The normal coordinate with $s = 0$, $\omega_s = 0$ corresponds to a translation and need not be considered in the following.

rotational period[7] $\frac{1}{\omega}$:

$$\tau_P \approx \left(\frac{2\pi}{\Delta\varphi}\right)^{N-1} \cdot \frac{1}{\omega} . \tag{10.2.10}$$

Taking $\Delta\varphi = \frac{2\pi}{100}$, $N = 10$ and $\omega = 10$ Hz, we obtain $\tau_P \approx 10^{12}$ years, i.e. more than the age of the Universe. These times of course become much longer if we consider a macroscopic system with $N \approx 10^{20}$. The recurrence thus exists theoretically, but in practice it plays no role. We have thereby eliminated Zermelo's paradox.

Remark: We consider further the time dependence of the solution for the coupled oscillators. From (10.2.3) and (10.2.9) we obtain

$$q_n(t) = \sum_s \frac{e^{isn}}{\sqrt{Nm}} \left(Q_s(0)\cos\omega_s t + \frac{\dot{Q}_s(0)}{\omega_s}\sin\omega_s t\right) , \tag{10.2.11}$$

from which the following solution of the general initial-value problem is found:

$$q_n(t) = \frac{1}{N}\sum_{s,n'}\left(q_{n'}(0)\cos\big(s(n-n')-\omega_s t\big) + \frac{\dot{q}_{n'}(0)}{\omega_s}\sin\big(s(n-n')-\omega_s t\big)\right) . \tag{10.2.12}$$

As an example, we consider the particular initial condition $q_{n'} = \delta_{n',0}$, $\dot{q}_{n'}(0) = 0$, for which only the oscillator at the site 0 is displaced initially, leading to

$$q_n(t) = \frac{1}{N}\sum_s \cos\left(sn - 2\Omega t \left|\sin\frac{s}{2}\right|\right) . \tag{10.2.13}$$

As long as N is finite, the solution is quasiperiodic. On the other hand, in the limit $N \to \infty$

$$q_n(t) = \frac{1}{2\pi}\int_{-\pi}^{\pi} ds \cos\left(sn - 2\Omega t \left|\sin\frac{s}{2}\right|\right) = \frac{1}{\pi}\int_0^{\pi} ds \cos\left(s2n - 2\Omega t \sin s\right)$$

$$= J_{2n}(2\Omega t) \sim \sqrt{\frac{1}{\pi\Omega t}}\cos\left(2\Omega t - \pi n - \frac{\pi}{4}\right) \text{ for long } t . \tag{10.2.14}$$

J_n are Bessel functions[8]. The excitation does not decay exponentially, but instead algebraically as $t^{-1/2}$.

We add a few more remarks concerning the properties of the solution (10.2.13) for finite N. If the zeroth atom in the chain is released at the time $t = 0$, it swings back and its neighbors begin to move upwards. The excitation propagates along the chain at the velocity of sound, $a\Omega$; the n-th atom, at a distance $d = na$ from the origin, reacts after a time of about $t \sim \frac{n}{\Omega}$. Here, a is the lattice constant.

---

[7] A more precise formula by P. C. Hemmer, L. C. Maximon, and H. Wergeland,

$$op. \; cit. \; 5, \text{ yields } \tau_P = \frac{\Pi_{s=1}^{N-1}\dfrac{2\pi}{\Delta\varphi_s}}{\sum_{s=1}^{N-1}\dfrac{\omega_s}{\Delta\varphi_s}} \propto \frac{1}{N}\Delta\varphi^{2-N}.$$

[8] I. S. Gradshteyn and I. M. Ryzhik, *Table of Integrals, Series and Products*, Academic Press, New York, 1980, 8.4.11 and 8.4.51

The displacement amplitude remains largest for the zeroth atom. In a finite chain, there would be echo effects. For periodic boundary conditions, the radiated oscillations come back again to the zeroth atom. The limit $N \to \infty$ prevents Poincaré recurrence. The displacement energy of the zeroth atom initially present is divided up among the infinitely many degrees of freedom. The decrease of the oscillation amplitude of the initially excited atom is due to energy transfer to its neighbors.

10.3 The Origin of Irreversible Macroscopic Equations of Motion

In this section, we investigate a microscopic model of Brownian motion. We will find the appearance of irreversibility in the limit of infinitely many degrees of freedom. The derivation of hydrodynamic equations of motion in analogy to the Brownian motion will be sketched at the end of this section and is given in more detail in Appendix H..

10.3.1 A Microscopic Model for Brownian Motion

As a microscopic model for Brownian motion, we consider a harmonic oscillator which is coupled to a harmonic lattice[9]. Since the overall system is harmonic, the Hamiltonian function or the Hamiltonian operator as well as the equations of motion and their solutions have the same form classically and quantum mechanically. We start with the quantum-mechanical formulation. In contrast to the Langevin equation of Sect. 8.1, where a stochastic force was assumed to act on the Brownian particle, we now take explicit account of the many colliding particles of the lattice in the Hamiltonian operator and in the equations of motion. The Hamiltonian of this system is given by

$$\mathcal{H} = \mathcal{H}_O + \mathcal{H}_F + \mathcal{H}_I \ ,$$

$$\mathcal{H}_O = \frac{1}{2M}P^2 + \frac{M\Omega^2}{2}Q^2 \ , \ \mathcal{H}_F = \frac{1}{2m}\sum_{\mathbf{n}} p_{\mathbf{n}}^2 + \frac{1}{2}\sum_{\mathbf{nn'}} \Phi_{\mathbf{nn'}} q_{\mathbf{n}} q_{\mathbf{n'}} \ , \quad (10.3.1)$$

$$\mathcal{H}_I = \sum_{\mathbf{n}} c_{\mathbf{n}} q_{\mathbf{n}} Q \ ,$$

where $\mathcal{H}_O$ is the Hamiltonian of the oscillator of mass M and frequency Ω. Furthermore, $\mathcal{H}_F$ is the Hamiltonian of the lattice[10] with masses m, momenta $p_{\mathbf{n}}$, and displacements $q_{\mathbf{n}}$ from the equilibrium positions, where we take $m \ll M$. The harmonic interaction coefficients of the lattice atoms are $\Phi_{\mathbf{nn'}}$. The interaction of the oscillator with the lattice atoms is given by $\mathcal{H}_I$;

---

[9] The coupling to a bath of oscillators as a mechanism for damping has been investigated frequently, e.g. by F. Schwabl and W. Thirring, Ergeb. exakt. Naturwiss. **36**, 219 (1964); A. Lopez, Z. Phys. **192**, 63 (1965); P. Ullersma, Physica **32**, 27 (1966).

[10] We use the index F, since in the limit $N \to \infty$ the lattice becomes a field.

the coefficients $c_{\mathbf{n}}$ characterize the strength and the range of the interactions of the oscillator which is located at the origin of the coordinate system. The vector $\mathbf{n}$ enumerates the atoms of the lattice. The equations of motion which follow from (10.3.1) are given by

$$M\ddot{Q} = -M\Omega^2 Q - \sum_{\mathbf{n}} c_{\mathbf{n}} q_{\mathbf{n}}$$

and

$$m\ddot{q}_{\mathbf{n}} = -\sum_{\mathbf{n}'} \Phi_{\mathbf{n}\mathbf{n}'} q_{\mathbf{n}'} - c_{\mathbf{n}} Q \ . \tag{10.3.2}$$

We take periodic boundary conditions, $q_{\mathbf{n}} = q_{\mathbf{n}+\mathbf{N}_i}$, with $\mathbf{N}_1 = (N_1, 0, 0)$, $\mathbf{N}_2 = (0, N_2, 0)$, and $\mathbf{N}_3 = (0, 0, N_3)$, where N_i is the number of atoms in the direction $\hat{\mathbf{e}}_i$. Due to the translational invariance of $\mathcal{H}_F$, we introduce the following transformations to normal coordinates and momenta:

$$q_{\mathbf{n}} = \frac{1}{\sqrt{mN}} \sum_{\mathbf{k}} e^{i\mathbf{k}\mathbf{a_n}} Q_{\mathbf{k}} \ , \quad p_{\mathbf{n}} = \sqrt{\frac{m}{N}} \sum_{\mathbf{k}} e^{-i\mathbf{k}\mathbf{a_n}} P_{\mathbf{k}} \ . \tag{10.3.3}$$

The inverse transformation is given by

$$Q_{\mathbf{k}} = \sqrt{\frac{m}{N}} \sum_{\mathbf{n}} e^{-i\mathbf{k}\mathbf{a_n}} q_{\mathbf{n}} \ , \quad P_{\mathbf{k}} = \frac{1}{\sqrt{mN}} \sum_{\mathbf{n}} e^{i\mathbf{k}\mathbf{a_n}} p_{\mathbf{n}} \ . \tag{10.3.4}$$

The Fourier coefficients obey orthogonality and completeness relations

$$\frac{1}{N} \sum_{\mathbf{n}} e^{i(\mathbf{k}-\mathbf{k}')\cdot\mathbf{a_n}} = \Delta(\mathbf{k} - \mathbf{k}') \ , \quad \frac{1}{N} \sum_{\mathbf{k}} e^{i\mathbf{k}\cdot(\mathbf{a_n}-\mathbf{a_{n'}})} = \delta_{\mathbf{n},\mathbf{n}'} \tag{10.3.5a,b}$$

with the generalized Kronecker delta $\Delta(\mathbf{k}) = \begin{cases} 1 & \text{for } \mathbf{k} = \mathbf{g} \\ 0 & \text{otherwise} \end{cases}$.

From the periodic boundary conditions we find the following values for the wavevector:

$$\mathbf{k} = \mathbf{g}_1 \frac{r_1}{N_1} + \mathbf{g}_2 \frac{r_2}{N_2} + \mathbf{g}_3 \frac{r_3}{N_3} \quad \text{with } r_i = 0, \pm 1, \pm 2, \ldots \ .$$

Here, we have introduced the reciprocal lattice vectors which are familiar from solid-state physics:

$$\mathbf{g}_1 = (\frac{2\pi}{a}, 0, 0) \ , \ \mathbf{g}_2 = (0, \frac{2\pi}{a}, 0) \ , \ \mathbf{g}_3 = (0, 0, \frac{2\pi}{a}) \ .$$

The transformation to normal coordinates (10.3.3) converts the Hamiltonian for the lattice into the Hamiltonian for N decoupled oscillators, viz.

$$\mathcal{H}_F = \frac{1}{2} \sum_{\mathbf{k}} (P_{\mathbf{k}}^\dagger P_{\mathbf{k}} + \omega_{\mathbf{k}}^2 Q_{\mathbf{k}}^\dagger Q_{\mathbf{k}}) \ , \tag{10.3.6}$$

with the frequencies[11] (see Fig. 10.3)

$$\omega_{\mathbf{k}}^2 = \frac{1}{m} \sum_{\mathbf{n}} \Phi(\mathbf{n}) \, e^{-i\mathbf{k}a_{\mathbf{n}}} \ . \tag{10.3.7}$$

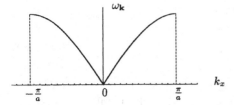

Fig. 10.3. The frequencies $\omega_{\mathbf{k}}$ along one of the coordinate axes, $\omega_{\max} = \omega_{\pi/a}$

From the invariance of the lattice with respect to infinitesimal translations, we obtain the condition $\sum_{\mathbf{n}'} \Phi(\mathbf{n}, \mathbf{n}') = 0$, and from translational invariance with respect to lattice vectors $\mathbf{t}$, it follows that $\Phi(\mathbf{n} + \mathbf{t}, \mathbf{n}' + \mathbf{t}) = \Phi(\mathbf{n}, \mathbf{n}') = \Phi(\mathbf{n} - \mathbf{n}')$. The latter relation was already used in (10.3.7). From the first of the two relations, we find $\lim_{\mathbf{k} \to 0} \omega_{\mathbf{k}}^2 = 0$, i.e. the oscillations of the lattice are acoustic phonons. Expressed in terms of the normal coordinates, the equations of motion are (10.3.2)

$$M\ddot{Q} = -M\Omega^2 Q - \frac{1}{\sqrt{mN}} \sum_{\mathbf{k}} c(\mathbf{k})^* Q_{\mathbf{k}} \tag{10.3.8a}$$

$$m\ddot{Q}_{\mathbf{k}} = -m\omega_{\mathbf{k}}^2 Q_{\mathbf{k}} - \sqrt{\frac{m}{N}} \, c(\mathbf{k}) \, Q \tag{10.3.8b}$$

with

$$c(\mathbf{k}) = \sum_{\mathbf{n}} c_{\mathbf{n}} e^{-i\mathbf{k} \, a_{\mathbf{n}}} \ . \tag{10.3.9}$$

For the further treatment of the equations of motion (10.3.8a,b) and the solution of the initial-value problem, we introduce the half-range Fourier transform (Laplace transform) of $Q(t)$:

$$\tilde{Q}(\omega) \equiv \int_0^\infty dt \, e^{i\omega t} Q(t) = \int_{-\infty}^\infty dt \, e^{i\omega t} \Theta(t) Q(t) \ . \tag{10.3.10a}$$

---

[11] We assume that the harmonic potential for the heavy oscillator is based on the same microscopic interaction as that for the lattice atoms, $\Phi(\mathbf{n}, \mathbf{n}')$. If we denote its strength by g, then we find $\Omega = \sqrt{\frac{g}{M}}$ and $\omega_{max} = \sqrt{\frac{g}{m}}$, and therefore $\Omega \ll \omega_{max}$. The order of magnitude of the velocity of sound is $c = a\omega_{max}$.

The inverse of this equation is given by

$$\Theta(t)Q(t) = \int_{-\infty}^{\infty} d\omega\, e^{-i\omega t}\tilde{Q}(\omega) .$$
(10.3.10b)

For free oscillatory motions, (10.3.10a) contains δ_+ distributions. For their convenient treatment, it is expedient to consider

$$\tilde{Q}(\omega + i\eta) = \int_0^{\infty} dt\, e^{i(\omega+i\eta)t}\, Q(t) ,$$
(10.3.11a)

with $\eta > 0$. If (10.3.10a) exists, then with certainty so does (10.3.11a) owing to the factor $e^{-\eta t}$. The inverse of (10.3.11a) is given by

$$e^{-\eta t}\, Q(t) = \int_{-\infty}^{\infty} d\omega\, e^{-i\omega t}\, \tilde{Q}(\omega + i\eta) , \text{ i.e.}$$

$$Q(t)\Theta(t) = \int_{-\infty}^{\infty} d\omega\, e^{-i(\omega+i\eta)t}\, \tilde{Q}(\omega + i\eta) .$$
(10.3.11b)

For the complex frequency appearing in (10.3.11a,b) we introduce $z \equiv \omega + i\eta$. The integral (10.3.11b) implies an integration path in the complex z-plane which lies $i\eta$ above the real axis

$$Q(t)\Theta(t) = \int_{-\infty+i\eta}^{\infty+i\eta} dz\, e^{-izt}\, \tilde{Q}(z) .$$
(10.3.11b')

The half-range Fourier transformation of the equation of motion (10.3.8a) yields for the first term

$$\int_0^{\infty} dt\, e^{izt}\frac{d^2}{dt^2}Q(t) = e^{izt}\dot{Q}(t)|_0^{\infty} - iz\int_0^{\infty} dt\, e^{izt}\dot{Q}(t)$$
$$= -\dot{Q}(0) + izQ(0) - z^2\tilde{Q}(z) .$$

All together, for the half-range Fourier transform of the equations of motion (10.3.8a,b) we obtain

$$M\left(-z^2 + \Omega^2\right)\tilde{Q}(z) = -\frac{1}{\sqrt{mN}}\sum_{\mathbf{k}} c(\mathbf{k})^* \tilde{Q}_{\mathbf{k}}(z) + M\left(\dot{Q}(0) - iz\, Q(0)\right)$$
(10.3.12)

$$m\left(-z^2 + \omega_{\mathbf{k}}^2\right)\tilde{Q}_{\mathbf{k}}(z) = -\sqrt{\frac{m}{N}}\, c(\mathbf{k})\,\tilde{Q}(z) + m\left(\dot{Q}_{\mathbf{k}}(0) - izQ_{\mathbf{k}}(0)\right) .$$
(10.3.13)

The elimination of $\tilde{Q}_{\mathbf{k}}(z)$ and replacement of the initial values $Q_{\mathbf{k}}(0), \dot{Q}_{\mathbf{k}}(0)$ by $q_{\mathbf{n}}(0), \dot{q}_{\mathbf{n}}(0)$ yields

$$D(z)\,\tilde{Q}(z) = M\left(\dot{Q}(0) - iz\,Q(0)\right)$$

$$- \frac{m}{N}\sum_{\mathbf{n}}\sum_{\mathbf{k}} c(\mathbf{k})^* \frac{e^{-ik\,a_{\mathbf{n}}}}{m(-z^2 + \omega_{\mathbf{k}}^2)}\left(\dot{q}_{\mathbf{n}}(0) - iz\,q_{\mathbf{n}}(0)\right) \quad (10.3.14)$$

with

$$D(z) \equiv \left(M\left(-z^2 + \Omega^2\right) + \frac{1}{N}\sum_{\mathbf{k}} \frac{|c(\mathbf{k})|^2}{m(z^2 - \omega_{\mathbf{k}}^2)}\right). \qquad (10.3.15)$$

Now we restrict ourselves to the classical case, and insert the particular initial values for the lattice atoms $q_{\mathbf{n}}(0) = 0$, $\dot{q}_{\mathbf{n}}(0) = 0$ for all the $\mathbf{n}$[12], then we find

$$\tilde{Q}(z) = \frac{M\left(\dot{Q}(0) - iz Q(0)\right)}{-Mz^2 + M\Omega^2 - \sum_{\mathbf{k}} \frac{|c(\mathbf{k})|^2}{m\,N}/(-z^2 + \omega_{\mathbf{k}}^2)}. \qquad (10.3.16)$$

From this, in the time representation, we obtain

$$\Theta(t)Q(t) = \int \frac{d\omega}{2\pi}\, e^{-izt}\tilde{Q}(z) = -i\sum_{\nu} g(\omega_{\nu})\, e^{-i\omega_{\nu}t}\,, \qquad (10.3.17)$$

where ω_{ν} are the poles of $\tilde{Q}(z)$ and $g(\omega_{\nu})$ are the residues[13]. The solution is thus *quasiperiodic*. One could use this to estimate the Poincaré time in analogy to the previous section.

In the limit of a large particle number N, the sums over $\mathbf{k}$ can be replaced by integrals and a different analytic behavior may result:[14]

$$D(z) = -Mz^2 + M\Omega^2 + \frac{a^3}{m}\int \frac{d^3k}{(2\pi)^3} \frac{|c(\mathbf{k})|^2}{z^2 - \omega_{\mathbf{k}}^2}\,. \qquad (10.3.18)$$

The integral over $\mathbf{k}$ spans the first Brillouin zone: $-\frac{\pi}{a} \leq k_i \leq \frac{\pi}{a}$. For a simple evaluation of the integral over $\mathbf{k}$, we replace the region of integration by a sphere of the same volume having a radius $\Lambda = \left(\frac{3}{4\pi}\right)^{1/3} \frac{2\pi}{a}$ and substitute

---

[12] In the quantum-mechanical treatment, we would have to use the expectation value of (10.3.14) instead and insert $\langle q_{\mathbf{n}}(0)\rangle = \langle \dot{q}_{\mathbf{n}}(0)\rangle = 0$. In problem 10.6, the force on the oscillator due to the lattice particles is investigated when the latter are in thermal equilibrium.

[13] The poles of $\tilde{Q}(z)$, $z \equiv \omega + i\eta$ are real, i.e. they lie in the complex ω-plane below the real axis. (10.3.17) follows with the residue theorem by closure of the integration in the lower half-plane.

[14] In order to determine what the ratio of t and N must be to permit the use of the limit $N \to \infty$ even for finite N, the N−dependence of the poles ω_{ν} must be found from $D(z) = 0$. The distance between the poles ω_{ν} is $\Delta\omega_{\nu} \sim \frac{1}{N}$, and the values of the residues are of $\mathcal{O}\left(\frac{1}{N}\right)$. The frequencies ω_{ν} obey $\omega_{\nu+1} - \omega_{\nu} \sim \frac{\omega_{max}}{N}$. For $t \ll \frac{N}{\omega_{max}}$, the phase factors $e^{i\omega_{\nu}t}$ vary only weakly as a function of ν, and the sum over ν in (10.3.17) can be replaced by an integral.

the dispersion relation by $\omega_{\mathbf{k}} = c|\mathbf{k}|$ where c is the velocity of sound. It then follows that

$$\frac{a^3}{m} \frac{1}{2\pi^2 c^3} \int_0^{\Lambda c} \frac{d\nu \, \nu^2}{z^2 - \nu^2} |c(\nu)|^2 = \frac{a^3}{m} \frac{1}{2\pi^2 c^3} \left[-\int_0^{\Lambda c} d\nu |c(\nu)|^2 + \right.$$
$$\left. + z^2 \int_0^\infty \frac{d\nu \, |c(\nu)|^2}{z^2 - \nu^2} - z^2 \int_{\Lambda c}^\infty \frac{d\nu \, |c(\nu)|^2}{z^2 - \nu^2} \right] \qquad (10.3.19)$$

with $\nu = c|\mathbf{k}|$. We now discuss the last equation term by term making use of the simplification $|c(\nu)|^2 = g^2$ corresponding to $c_{\mathbf{n}} = g\delta_{\mathbf{n},0}$.
1st term of (10.3.19):

$$-\frac{a^3}{m} \frac{1}{2\pi^2 c^3} \int_0^{\Lambda c} d\nu |c(\nu)|^2 = -g^2 \Lambda c . \qquad (10.3.20)$$

This yields a renormalization of the oscillator frequency

$$\bar{\omega} = \sqrt{\Omega^2 - g^2 \Lambda c \frac{a^3}{m2\pi^2 c^3} \frac{1}{M}} . \qquad (10.3.21)$$

2nd term of (10.3.19) and evaluation using the theorem of residues:

$$\frac{a^3}{m} \frac{1}{2\pi^2 c^3} g^2 z^2 \int_0^\infty \frac{d\nu}{z^2 - \nu^2} = -M \Gamma \mathrm{i} \, z \qquad (10.3.22)$$

$$\Gamma = \frac{g^2 a^3}{4\pi m c^3} \frac{1}{M} = c\Lambda \frac{m}{M} . \qquad (10.3.23)$$

The third term of (10.3.19) is due to the high frequencies and affects the behavior at very short times. This effect is treated in problem 10.5, where a continuous cutoff function is employed. If we neglect it, we obtain from (10.3.16)

$$\left(-z^2 + \bar{\omega}^2 - \mathrm{i}\Gamma z\right) \tilde{Q}(z) = M\left(\dot{Q}(0) - \mathrm{i}zQ(0)\right) , \qquad (10.3.24)$$

and, after transformation into the time domain for $t > 0$, we have the following equation of motion for $Q(t)$:

$$\left(\frac{d^2}{dt^2} + \bar{\omega}^2 + \Gamma \frac{d}{dt}\right) Q(t) = 0 . \qquad (10.3.25)$$

The coupling to the bath of oscillators leads to a *frictional term* and to *irreversible damped motion*. For example, let the initial values be $Q(0) = 0$, $\dot{Q}(t = 0) = \dot{Q}(0)$ (for the lattice oscillators, we have already set $q_{\mathbf{n}}(0) = \dot{q}_{\mathbf{n}}(0) = 0$); then from Eq. (10.3.24) it follows that

$$\Theta(t) Q(t) = \int_{-\infty}^\infty \frac{d\omega}{2\pi} \frac{e^{-\mathrm{i}zt} \dot{Q}(0)}{-z^2 + \bar{\omega}^2 - \mathrm{i}\Gamma z} \qquad (10.3.26)$$

and, using the theorem of residues,

$$Q(t) = e^{-\Gamma t/2} \frac{\sin \omega_0 t}{\omega_0} \dot{Q}(0) \,, \tag{10.3.27}$$

with $\omega_0 = \sqrt{\bar{\omega}^2 - \frac{\Gamma^2}{4}}$.

The *conditions for the derivation* of the irreversible equation of motion (10.3.25) were:

a) A limitation to times $t \ll \frac{N}{\omega_{max}}$[15]. This implies practically no limitation for large N, since the exponential decay is much more rapid.

b) The separation into macroscopic variables $\equiv$ massive oscillator (of mass M) and microscopic variables $\equiv$ lattice oscillators (of mass m) leads, owing to $\frac{m}{M} \ll 1$, to a separation of time scales

$$\Omega \ll \omega_{max} \,, \quad \Gamma \ll \omega_{max} \,.$$

The time scales of the macroscopic variables are $1/\Omega$, $1/\Gamma$.

The irreversibility (exponential damping) arises in going to the limit $N \to \infty$. In order to obtain irreversibility even at arbitrarily long times, the limit $N \to \infty$ must first be taken.

10.3.2 Microscopic Time-Reversible and Macroscopic Irreversible Equations of Motion, Hydrodynamics

The derivation of *hydrodynamic* equations of motion (Appendix H.) directly from the microscopic equations is based on the following elements:

(i) The point of departure is represented by the equations of motion for the conserved quantities and the equations of motion for the infinitely many nonconserved quantities.

(ii) An important precondition is the separation of time scales $ck \ll \omega_{n.c.}$, i.e. the characteristic frequencies of the conserved quantities ck are much slower than the typical frequencies of the nonconserved quantities $\omega_{n.c.}$, analogous to the ω_λ ($\lambda > 5$) in the Boltzmann equation, Sect. 9.4.4. This permits the elimination of the rapid variables.

In the analytic treatment in Appendix H., one starts from the equations of motion for the so called Kubo relaxation function ϕ and obtains equations of motion for the relaxation functions of the conserved quantities. From the one-to-one correspondence of equations of motion for ϕ and the time-dependent expectation values of operators under the influence of a perturbation, the hydrodynamic equations for the conserved quantities are obtained. The remaining variables express themselves in the form of damping terms, which can be expressed by Kubo formulas.

---

[15] These times, albeit long, are much shorter than the Poincaré recurrence time.

*10.4 The Master Equation and Irreversibility in Quantum Mechanics[16]

We consider an isolated system and its density matrix at the time t, with probabilities $w_i(t)$

$$\varrho(t) = \sum_i w_i(t) |i\rangle \langle i| . \qquad (10.4.1)$$

The states $|i\rangle$ are eigenstates of the Hamiltonian $\mathcal{H}_0$. We let the quantum numbers i represent the energy E_i and a series of additional quantum numbers ν_i. A perturbation V also acts on the system or within it and causes transitions between the states; thus the overall Hamiltonian is

$$\mathcal{H} = \mathcal{H}_0 + V . \qquad (10.4.2)$$

For example, in a nearly ideal gas, $\mathcal{H}_0$ could be the kinetic energy and V the interaction which results from collisions of the atoms. We next consider the time development of ϱ on the basis of (10.4.1) and denote the time-development operator by $U(\tau)$. After the time τ the density matrix has the form

$$
\begin{aligned}
\varrho(t+\tau) &= \sum_i w_i(t) U(\tau) |i\rangle \langle i| U^\dagger(\tau) \\
&= \sum_i \sum_{j,k} w_i(t) |j\rangle \langle j| U(\tau) |i\rangle \langle i| U^\dagger(\tau) |k\rangle \langle k| \\
&= \sum_i \sum_{j,k} w_i(t) |j\rangle \langle k| U_{ji}(\tau) U^*_{ki}(\tau) ,
\end{aligned}
\qquad (10.4.3)
$$

where the matrix elements

$$U_{ji}(\tau) \equiv \langle j| U(\tau) |i\rangle \qquad (10.4.4)$$

have been introduced. We assume that the system, even though it is practically isolated, is in fact subject to a phase averaging at each instant as a result of weak contacts to other macroscopic systems. This corresponds to taking the trace over other, unobserved degrees of freedom which are coupled to the system[17]. Then the density matrix (10.4.3) is transformed to

---

[16] W. Pauli, *Sommerfeld Festschrift*, S. Hirzel, Leipzig, 1928, p. 30.

[17] If for example every state $|j\rangle$ of the system is connected with a state $|2, j\rangle$ of these other macroscopic degrees of freedom, so that the contributions to the total density matrix are of the form $|2, j\rangle |j\rangle \langle k| \langle 2, k|$, then taking the trace over 2 leads to the diagonal form $|j\rangle \langle j|$. This stochastic nature, which is introduced through contact to the system's surroundings, is the decisive and subtle step in the derivation of the master equation. Cf. N. G. van Kampen, Physica **20**, 603 (1954), and Fortschritte der Physik **4**, 405 (1956).

$$\sum_i \sum_j w_i(t) \, |j\rangle \, \langle j| \, U_{ji}(\tau) U_{ji}^*(\tau) \, . \tag{10.4.5}$$

Comparison with (10.4.1) shows that the probability for the state $|j\rangle$ at the time $t + \tau$ is thus

$$w_j(t + \tau) = \sum_i w_i(t) |U_{ji}(\tau)|^2 \, ,$$

and the change in the probability is

$$w_j(t + \tau) - w_j(t) = \sum_i (w_i(t) - w_j(t)) |U_{ji}(\tau)|^2 \, , \tag{10.4.6}$$

where we have used $\sum_i |U_{ji}(\tau)|^2 = 1$. On the right-hand side, the term $i = j$ vanishes. We thus require only the nondiagonal elements of $U_{ij}(\tau)$, for which we can use the Golden Rule[18]:

$$|U_{ji}(\tau)|^2 = \frac{1}{\hbar^2} \left(\frac{\sin \omega_{ij} \, \tau/2}{\omega_{ij}/2} \right)^2 |\langle j| \, V \, |i\rangle \, |^2 = \tau \frac{2\pi}{\hbar} \delta(E_i - E_j) |\langle j| \, V \, |i\rangle \, |^2 \tag{10.4.7}$$

with $\omega_{ij} = (E_i - E_j)/\hbar$. The limit of validity of the Golden Rule is $\Delta E \gg \frac{2\pi\hbar}{\tau} \gg \delta\varepsilon$, where ΔE is the width of the energy distribution of the states and $\delta\varepsilon$ is the spacing of the energy levels. From (10.4.6) and (10.4.7), it follows that

$$\frac{dw_j(t)}{dt} = \sum_i (w_i(t) - w_j(t)) \frac{2\pi}{\hbar} \delta(E_i - E_j) |\langle j| \, V \, |i\rangle \, |^2 \, .$$

As already mentioned at the beginning of this section, the index $i \equiv (E_i, \nu_i)$ includes the quantum numbers of the energy and the ν_i, the large number of all remaining quantum numbers. The sum over the energy eigenvalues on the right-hand side can be replaced by an integral with the density of states $\varrho(E_i)$ according to

$$\sum_{E_i} \cdots = \int dE_i \, \varrho(E_i) \, \cdots$$

so that, making use of the δ-function, we obtain:

$$\frac{dw_{E_j \nu_j}(t)}{dt} = \sum_{\nu_i} (w_{E_j, \nu_i} - w_{E_j, \nu_j}) \frac{2\pi}{\hbar} \varrho(E_j) |\langle E_j, \nu_j| \, V \, |E_j, \nu_i\rangle \, |^2 \, . \tag{10.4.8}$$

---

[18] QM I, Eq. (16.36)

With the coefficients

$$\lambda_{E_j,\nu_j;\nu_i} = \frac{2\pi}{\hbar} \varrho(E_j) |\langle E_j, \nu_j| V |E_j, \nu_i\rangle|^2 , \tag{10.4.9}$$

Pauli's master equation follows:

$$\frac{dw_{E_j\nu_j}(t)}{dt} = \sum_{\nu_i} \lambda_{E_j,\nu_j;\nu_i} \left(w_{E_j,\nu_i}(t) - w_{E_j,\nu_j}(t)\right) . \tag{10.4.10}$$

This equation has the general structure

$$\dot{p}_n = \sum_{n'} (W_{n'\,n}\, p_{n'} - W_{n'\,n}\, p_n) , \tag{10.4.11}$$

where the transition rates $W_{n'\,n} = W_{n\,n'}$ obey the so called *detailed balance* condition[19]

$$W_{n'\,n}\, p_{n'}^{\text{eq}} = W_{n\,n'}\, p_n^{\text{eq}} \tag{10.4.12}$$

for the microcanonical ensemble, $p_{n'}^{\text{eq}} = p_n^{\text{eq}}$ for all n and n'. One can show in general that Eq. (10.4.11) is irreversible and that the entropy

$$S = -\sum_n p_n \log p_n \tag{10.4.13}$$

increases. With (10.4.11), we have

$$\dot{S} = -\sum_{n,\,n'} (p_n \log p_n)'\, W_{n'\,n}(p_{n'} - p_n)$$

$$= \sum_{n,n'} W_{n'\,n}\, p_{n'} \left((p_n \log p_n)' - (p_{n'} \log p_{n'})'\right) .$$

By permutation of the summation indices n and n' and using the symmetry relation $W_{n\,n'} = W_{n'\,n}$, we obtain

$$\dot{S} = \frac{1}{2} \sum_{n,n'} W_{n\,n'}(p_{n'} - p_n)\left((p_{n'} \log p_{n'})' - (p_n \log p_n)'\right) > 0 , \tag{10.4.14}$$

where the inequality follows from the convexity of $x\log x$ (Fig. 10.4). The entropy continues to increase until $p_n = p_{n'}$ for all n and n'. Here we assume that all the n and n' are connected via a chain of matrix elements. The *isolated* system described by the master equation (10.4.10) approaches the *microcanonical equilibrium*.

---

[19] See QM II, following Eq. (4.2.17).

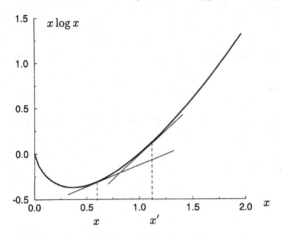

Fig. 10.4. The function $f(x) = x \log x$ is convex, $(x'-x)(f'(x')-f'(x)) > 0$

10.5 Probability and Phase-Space Volume

*10.5.1 Probabilities and the Time Interval of Large Fluctuations

In the framework of equilibrium statistical mechanics one can calculate the probability that the system spontaneously takes on a constraint. In the context of the Gay-Lussac experiment, we found the probability that a system with a fixed particle number N with a total volume V would be found only within the subvolume V_1 (Eq. (3.5.5)):

$$W(E, V_1) = e^{-(S(E,V)-S(E,V_1))/k} . \tag{10.5.1}$$

For an ideal gas[20], the entropy is $S(E,V) = kN(\log \frac{V}{N\lambda_T^3} + \frac{5}{2})$. Since $E = \frac{3}{2}NkT$, λ_T remains unchanged on expansion and it follows that $W(E, V_1) = e^{-N \log \frac{V}{V_1}}$. This gives $\log \frac{V}{V_1} = \log \frac{V}{V-(V-V_1)} \approx \frac{V-V_1}{V}$ for low compressions. At higher compressions, $V_1 \ll V$, the factor $\log \frac{V}{V_1}$ becomes larger. The dependence on $N \approx 10^{20}$ is dominant in Eq. (10.5.1) for macroscopic systems. For general constraints, from Eq. (3.5.5) we obtain

$$W(E, Z) = e^{-(S(E)-S(E,Z))/k} . \tag{10.5.2}$$

As an example, we consider in more detail the density fluctuations in an ideal gas. The N gas atoms are distributed between the subvolume V_1 and the subvolume $V - V_1$ with the probabilities $p = \frac{V_1}{V}$ and $1 - p$ corresponding to a binomial distribution[21]. The mean square deviation of the particle number $(\Delta n)^2$ within the subvolume V_1 is related to the average of the particle number in this volume, $\bar{n} = pN$:

---

[20] Chapter 2
[21] Sect. 1.5.1

$$\frac{(\Delta n)^2}{\bar{n}^2} = \frac{1}{\bar{n}} , \quad \text{i.e.} \quad \frac{(\Delta n)}{\bar{n}} = \frac{1}{\sqrt{\bar{n}}} . \tag{10.5.3}$$

For $\bar{n} = 10^{20}$, this ratio is $\frac{(\Delta n)}{\bar{n}} = 10^{-10}$. For large N, n and $N - n$, from the binomial distribution we obtain the probability density for the particle number n by expansion and using the Stirling approximation; it is a Gaussian distribution

$$w(n) = \frac{1}{\sqrt{2\pi\bar{n}}} e^{-\frac{(n-\bar{n})^2}{2\bar{n}}} . \tag{10.5.4}$$

This last formula is also a result of the central limit theorem[22]. The probability density is normalized to 1: $\int dn \, w(n) = 1$.

We are now interested in the case that n is greater than the value $\bar{n} + \delta n$, i.e. in the occurrence of fluctuations which are greater than a given δn. Its probability is obtained by integrating (10.5.4) over the interval $[\bar{n} + \delta n, \infty]$

$$w_\delta(\delta n) = \frac{1}{\sqrt{2\pi\bar{n}}} \int_{\delta n}^\infty d\nu \, e^{-\frac{\nu^2}{2\bar{n}}} = \frac{1}{\sqrt{\pi}} \int_{\delta n/\sqrt{2\bar{n}}}^\infty dx \, e^{-x^2}$$

$$= \frac{1}{2}\left(1 - \Phi\left(\frac{\delta n}{\sqrt{2\bar{n}}}\right)\right) , \tag{10.5.5}$$

with the error integral

$$\Phi(x) = \frac{2}{\sqrt{\pi}} \int_0^x dy \, e^{-y^2} , \tag{10.5.6}$$

where already for values $x \gtrsim 1$ the approximation formula $\frac{1}{2}(1 - \Phi(x)) \approx \frac{e^{-x^2}}{2\sqrt{\pi}x}$ may be employed.

In Fig. 10.5, we show a possible time dependence of the particle number $n(t)$. Knowing the probability for the occurrence of a fluctuation, we can also make predictions about the typical time interval between fluctuations. First of all, the probability $w_\delta(\delta n)$ has the following meaning in terms of the time: denoting by $t_{\delta n}$ the *overall time* during which $n(t)$ remains *above* $\bar{n} + \delta n$ during the time period t_0 of *observation*, then we find

$$w_\delta(\delta n) = \frac{t_{\delta n}}{t_0} . \tag{10.5.7}$$

We are interested in the average time required until a density increase which is greater than δn occurs, and term it the *waiting time* $\vartheta_{\delta n}$. Here, we must also introduce the time τ_0 which is required for a fluctuation, once it occurs, to again degrade. Typically, $\tau_0 = L/c$, where L is the linear dimension of the system and c is the velocity of sound, e.g. $L = 1\,\text{cm}$, $c = 10^5\,\text{cm/s}$, $\tau_0 = 10^{-5}\,\text{s}$. The time introduced above, $t_{\delta n}$, within which the deviation exceeds δn, is

---

[22] Sect. 1.2.2

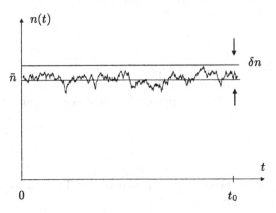

Fig. 10.5. A possible time dependence of the particle number within the subvolume V_1

$t_{\delta n} = \tau_0 \times$ (the number of fluctuations above δn). For relatively large δn, $t_0 - t_{\delta n} \approx t_0$ and thus the waiting time $\vartheta_{\delta n}$ is equal to the ratio of t_0 to the number of fluctuations larger than δn

$$\vartheta_{\delta n} = \frac{t_0}{t_{\delta n}}\tau_0 = \frac{\tau_0}{w_\delta(\delta n)} = \tau_0 \Big/ \frac{1}{2}\left(1 - \varPhi\Big(\frac{\delta n}{\bar{n}}\sqrt{\frac{\bar{n}}{2}}\Big)\right), \qquad (10.5.8)$$

where (10.5.5) was used.

As an illustration, the waiting times until a density fluctuation upwards of more than δn occurs are collected in Table 10.1 for several values of the relative deviation $\frac{\delta n}{\bar{n}\sqrt{2}}$ and $x \equiv \frac{\delta n}{\bar{n}}\sqrt{\frac{\bar{n}}{2}}$ for $\bar{n} = 10^{20}$. Deviations smaller than 3×10^{-10} regularly occur; those larger than 7×10^{-10} *never* occur! And even these in fact small deviations (7×10^{-10}) are macroscopically unobservable. This small interval encompasses the whole range from frequent to never.

Now we return to the fluctuation curve (Fig. 10.5). We consider a relatively large δn. To the right of the maximum, there is a monotonic decrease. There are just as many intersection points to the left of the maximum as to its right. For large δn, the intersection point must in fact lie on the maximum. But to the left of the maximum, $n(t)$ must have increased; this is so improbable that it practically never occurs. In practice, such an improbable state is never generated by a spontaneous fluctuation, but instead is determined as the initial state in an experiment by the lifting of constraints. Starting from

Table 10.1. Waiting Times

x	$\frac{1}{2}(1 - \varPhi(x))$	Relative deviation $\frac{\delta n}{\bar{n}\sqrt{2}}$	Waiting time $\vartheta_{\delta n} = \frac{10^{-5}}{\frac{1}{2}(1-\varPhi(x))}$
3	1×10^{-5}	3×10^{-10}	1 s
5	8×10^{-13}	5×10^{-10}	1.3×10^7 s $= 5$ months
7	2×10^{-23}	7×10^{-10}	5×10^{17} s $= 2 \times 10^{10}$ years

such an improbable configuration, the system always relaxes towards $\bar{n}$ and never assumes an even more improbable state, e.g. an even greater density on expansion of a gas. Small fluctuations are superposed onto this relaxation.

10.5.2 The Ergodic Theorem

The *ergodic theorem* played an important role in classical physics, providing the justification for the statistical description of matter. It was proposed by Boltzmann, and states that an arbitrary trajectory passes through every point in phase space and that therefore an average over time is identical with an average over phase space. In this form, the theorem is not tenable; rather, its must be modified as follows (it is then referred to as the *quasi-ergodic theorem*): every trajectory, apart from a set of null measure, approaches each point in phase space arbitrarily closely. This also yields the result that *a time average is equal to an ensemble average*. If a system is *ergodic*[23], i.e. the quasi-ergodic theorem applies to it, then the following expression holds for functions in phase space:

$$\lim_{T \to \infty} \frac{1}{T} \int_0^T dt \; f(q(t), p(t)) = \frac{1}{\Omega(E)} \int d\Gamma \; f(q, p) \; ; \tag{10.5.9}$$

"time average is equal to ensemble average".

Choosing $f(q, p) = \Theta(q, p \in G)$, where G is a region of the energy shell, we obtain from (10.5.9) in the limit of long T

$$\frac{\tau_G}{T} = \frac{|\Gamma_G|}{\Omega(E)} \; , \tag{10.5.10}$$

where τ_G is the time during which the trajectory remains within G and $|\Gamma_G|$ is the phase-space volume of the region G. It is thus the fraction of time which the system spends in G, equal to the ratio of the volumes in phase space. The time spent by a system within an improbable region (in an atypical configuration) is short.

---

[23] L. Boltzmann, as well as P. and T. Ehrenfest, associated something more like the idea of what we would now call "mixing" with the concept "ergodic", which they also circumscribed as "stirred apart" (Ger. 'zerrühren').

10.6 The Gibbs and the Boltzmann Entropies and their Time Dependences

10.6.1 The Time Derivative of Gibbs' Entropy

The microscopic entropy of an isolated system with the Hamilton $\mathcal{H}$, introduced in Sect. 2.3, which in the present connection we call Gibbs' entropy S_G,

$$S_G = -k \, \mathrm{Tr} \, (\varrho \log \varrho) \tag{10.6.1}$$

does not change in the course of time. The von Neumann equation for this system is given by

$$\dot{\varrho} = \frac{i}{\hbar} [\varrho, \mathcal{H}] \,, \tag{10.6.2}$$

so that

$$-\dot{S}_G/k = \mathrm{Tr} \, (\varrho \log \varrho)^{\cdot} = \mathrm{Tr} \, (\dot{\varrho} \log \varrho) + \mathrm{Tr} \, \dot{\varrho}$$
$$= \frac{i}{\hbar} \mathrm{Tr} \, ([\varrho, \mathcal{H}] \log \varrho) = \frac{i}{\hbar} \mathrm{Tr} \, ([\log \varrho, \varrho] \, \mathcal{H}) = 0 \,.$$

We have used $\mathrm{Tr} \, \dot{\varrho} = 0$, which follows from $\mathrm{Tr} \, \varrho = 1$. In this derivation, $\mathcal{H}$ may be time dependent, e.g. $\mathcal{H}$ could contain a time-dependent external parameter such as the available volume, fixed by the wall potentials. The density matrix varies with time, but

$$\dot{S}_G = 0 \tag{10.6.3}$$

still holds. The Gibbs entropy remains constant and yields no information about the irreversible motion; it has significance only for the equilibrium state of the system.

The derivation of $\dot{S}_G = 0$ from classical statistics with

$$S_G = -k \int d\Gamma \, \varrho \log \varrho \,,$$

making use of the Liouville equation, is the topic of problem 10.7. This result is based on the fact that a particular region in phase space does indeed change, but its volume remains the same according to Liouville's theorem.

10.6.2 Boltzmann's Entropy

To each macrostate[24] M and thus to each microstate X or each point in Γ-space which represents the macrostate, $M = M(X)$, a *Boltzmann entropy* S_B can be associated:

---
[24] See Chaps. 1 and 2.

$$S_B(M) = k \log |\Gamma_M| \, , \qquad (10.6.4)$$

where Γ_M is the phase-space region of M and $|\Gamma_M|$ is its volume.

The *Gibbs entropy* S_G is defined for an ensemble with the distribution function $\varrho(X)$ by

$$S_G[\varrho] = -k \int d\Gamma \, \varrho(X) \log \varrho(X) \, . \qquad (10.6.5)$$

For a microcanonical ensemble (Sect. 2.2), we have

$$\varrho_{MC} = \begin{cases} |\Gamma_M|^{-1} & X \in \Gamma_M \\ 0 & \text{otherwise} \end{cases} \, .$$

In this case,

$$S_G[\varrho_{MC}] = k \log |\Gamma_M| = S_B(M) \, . \qquad (10.6.6)$$

holds. In equilibrium, the Boltzmann and the Gibbs entropies are thus equal. More generally, the two entropies are equal when the particle density, the energy density, and the momentum density vary only slowly on a microscopic scale, and the system is in equilibrium in every small macroscopic region, i.e. it is in a state of local equilibrium. When however the system is not in complete equilibrium, in which case M and ϱ would no longer change, then the time developments of S_B and S_G, even starting from local equilibrium, are quite different. As we have shown, S_G remains constant, while $S_B(M)$ changes. Let us consider e.g. the expansion of a gas. Initially, $S_B = S_G$. Then, typically, S_B increases while S_G remains constant, and from S_G alone the tendency towards equilibrium is not at all apparent. This is due to the fact that the size of the volume of phase space remains the same as in the initial state throughout the entire time development.

*10.6.2.1 Boltzmann's Calculation of S_B and its Connection with the μ-Space of the Boltzmann Equation[25]

We consider a dilute gas with N particles and introduce a division of μ-space into cells $\omega_1, \omega_2, \ldots$ of size $|\omega|$, which we enumerate by the index i. Let the cell i be occupied by n_i particles. The volume in phase space is

$$|\Gamma| = \frac{N!}{n_1! n_2! \ldots} |\omega|^N \, , \qquad (10.6.7)$$

see below. From this, using Stirling's approximation, we obtain

$$\log |\Gamma| \approx N \log N - \sum_i n_i \log n_i + N \log |\omega| \, , \qquad (10.6.8)$$

---

[25] P. and T. Ehrenfest, *op. cit.* 2;
M. Kac, *Probability and Related Topics in Science*, Interscience Publishers, London, 1953

and for Boltzmann's entropy,

$$S_{\mathrm{B}} = k \log |\Gamma| \, .$$

The relation to the distribution function f and Boltzmann's H-function: $n_i = \int_{\omega_i} d^3x \, d^3v \, f$ is the number of particles in cell i. In the case that $f(x,v)$ varies slowly (it is a smooth function), then $n_i = |\omega| \times f(\text{in cell } i) \Rightarrow f(\text{in cell } i) = \frac{n_i}{|\omega|}$ [26]. The ω_i are taken to be small; on the other hand, the n_i have to be so large that Stirling's approximation may be used. For the H-function which we introduced in connection with the Boltzmann equation, we then obtain

$$H_{\mathrm{tot}} = \int d^3x \, d^3v \, f \log f \approx \sum_i n_i \log \frac{n_i}{|\omega|} = \sum_i n_i \log n_i - N \log |\omega| \, . \quad (10.6.9)$$

Comparison with Eq. (10.6.8) for a dilute gas (with negligible interactions) then yields

$$S_{\mathrm{B}} = -k \, H_{\mathrm{tot}} \, , \quad (10.6.10)$$

apart from the term $N \log N$, which is independent of the configuration.

Demonstration of the formula (10.6.7) for $|\Gamma|$: For each point in Γ-space, there are associated image points of the N molecules in μ-space. For example: the kth molecule has the image point $m^{(k)}$. To each point in Γ-space, there corresponds a distribution $n_1, n_2, \ldots$. On the other hand, to each distribution of states $n_1, n_2, \ldots$ there corresponds a continuum of Γ-points. (i) Each of the image points can be arbitrarily shifted within the cell ω_i in which it is located. This gives the volume $|\omega|^N$ in Γ. (ii) Let a point in Γ-space be given. Every permutation of the image points leads to a Γ-point with the same distribution $n_1, n_2, \ldots$ All together, there are $N!$ permutations. (iii) Permutations which only exchange the image points within a cell have already been taken into account in the shift permutations (i). For every Γ-point, there are $n_1! \, n_2! \, \ldots$ such permutations. The number of permutations which lead to new image points ("combinations") is therefore $\frac{N!}{n_1! n_2! \ldots}$. As we know from the discussion of Gibbs' paradox, the term $N \log N$ in (10.6.8) is no longer present after division by $N!$, the "correct Boltzmann counting", as follows from quantum statistics (Bose and Fermi statistics).

10.7 Irreversibility and Time Reversal

10.7.1 The Expansion of a Gas

We now have acquired the necessary fundamentals to be able to discuss the expansion of a gas and "Loschmidt's paradox", which results from time reversal [27]. We assume that initially, at time $t = 0$, the gas occupies only a sub-volume V_1 of the total volume V, and the velocities correspond to a Maxwell

---

[26] This is certainly fulfilled in local equilibrium.

[27] As an illustration, we refer to a computer experiment with $N = 864$ atoms, which interact via a Lennard–Jones potential (Eq. (5.3.16)); see Fig. 10.1. The time development is determined by methods of molecular dynamics, i.e. by numerical solution of the discretized Newtonian equations (B. Kaufmann, Master's Thesis, TU München, 1995). The times are given in units of a characteristic time $\sqrt{m\sigma^2/\epsilon} = 2.15 \times 10^{-12}$ sec for the potential and the mass (argon) considered.

distribution. One can imagine that this initial state is produced by removal of previously present separating partitions[28]. The time which the particles typically require in order to pass through the volume ballistically is $\tau_0 = \frac{L}{v}$. For $L = 1$ cm and $v = 10^5$ cm sec^{-1}, we find $\tau_0 = 10^{-5}$ sec. In comparison, the collision time is very short, $\tau \approx 10^{-9}$ sec. It thus typically requires several τ_0 for the gas to spread out ballistically (with reflections by the walls) within the total volume. After about 10 τ_0, the gas is uniformly distributed over the whole volume and can be considered to be in equilibrium in macroscopic terms.

In Fig. 10.1, the configurations in the computer simulation[27] are shown for the times $t = 4.72$, 14.16, and 236. This expansion is accompanied by a monotonic increase of the "coarse-grained" Boltzmann entropy, which is based only on the spatial distribution of the particle-number density, $n(\mathbf{x})$:

$$S = - \int d^3x \; n(\mathbf{x}, t) \log n(\mathbf{x}, t) \, , \tag{10.7.1}$$

which attains its equilibrium value after $t = 50$. This observed behavior is exactly as predicted by the Boltzmann equation. The entropy increases monotonically, cf. Fig. 10.6(1)[29]. However, if one reverses all the particle velocities in this gas at a particular time t, whatever time has elapsed, then all the particles will return along their original paths and after a further time interval t will again arrive at their initial configuration, and the entropy will decrease to its initial value; cf. Fig. 10.6(2). Although the gas represented in Fig. 10.1 c) appears completely disordered with respect to its spatial distribution, and no special features can be seen in its motion along the positive direction of time, nevertheless due to the special initial state from which it came, which occupies only a fraction of the phase space and is subject to a high degree of spatial constraints, it is in a state which contains subtle correlations of the particle velocities. Following a time reversal $\mathbf{v} \to -\mathbf{v}$, the particles move in a "conspiratorial" fashion, so that they finally all come back

---

[28] In the following discussion, we shall denote the initial state in the subvolume V_1 by X, the microstate after the time t by $T_t X$, and the time-reversed state by $\mathcal{T} T_t X$. We have $T_t \mathcal{T} T_t X = \mathcal{T} X$. Here, T_t denotes the time development operator for the time interval t, and $\mathcal{T}$ is the time reversal-operator, which reverses all velocities.

[29] As already emphasized at the end of Sect. 10.6, the Gibbs entropy S_G remains constant and gives no indication of the irreversible expansion. This is due to the fact that the size of the phase-space volume of the initial states remains constant with time. These microstates are however *no longer typical* of the macrostate $M(t)$ (or of the local equilibrium state) which is present for $t > 0$. The phase space of these states is the same size as the phase space at the time $t = 0$; it is thus considerably smaller than that of all the states which represent the macrostate at times $t > 0$. The state $T_t X$ contains complex correlations. The typical microstates of $M(t)$ lack these correlations. They become apparent upon time reversal. In the forward direction of time, in contrast, the future of such atypical microstates is just the same as that of the typical states.

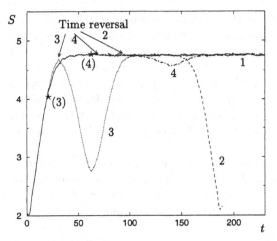

Fig. 10.6. The entropy as a function of time in the expansion of a computer gas consisting of 864 atoms. In the initial stages, all the curves lie on top of one another. (1) The unperturbed expansion of V_1 to V (solid curve). (2) Time reversal at $t = 94.4$ (dashed curve), the system returns to its initial state and the entropy to its initial value. (3) A perturbation $\star$ at $t = 18.88$ and time reversal at $t = 30.68$. The system approaches its initial state closely (dotted curve). (4) A perturbation $\star$ at $t = 59$ and time reversal at $t = 70.8$ (chain curve). Only for a short time after the time reversal does the entropy decrease; it then increases towards its equilibrium value.[32]

together within the original subvolume[30]. It is apparent that the initial state which we defined at the beginning leads in the course of time to a state which is not typical of a gas with the density shown in Fig. 10.1 c) and a Maxwell distribution. A typical microstate for such a gas would never compress itself into a subvolume after a time reversal. States which develop in such a correlated manner and which are not typical will be termed *quasi-equilibrium states*[31], also called local quasi-equilibrium states during the intermediate stages of the time development. Quasi-equilibrium states have the property that their macroscopic appearance is not invariant under time reversal. Although these quasi-equilibrium states of isolated systems doubtless exist and their time-reversed counterparts can be visualized in the computer experiment, the latter would seem to have no significance in reality. Thus, why was Boltzmann nevertheless correct in his statement that the entropy S_B always increases monotonically apart from small fluctuations?

---

[30] The associated "coarse-grained" Boltzmann entropy (10.7.1) decreases following the time reversal, curve (2) in Fig. 10.6. A time dependence of this type is not described by the Boltzmann equation and is also never observed in Nature.

[31] J. M. Blatt, *An Alternative Approach to the Ergodic Problem*, Prog. Theor. Phys. **22**, 745 (1959)

[32] We must point out an unrealistic feature of the computer experiments here. The sample is so small that within an equilibration time $10\,\tau_0$ only a few collisions

One must first realize that the number of quasi-equilibrium states X which belong to a particular macrostate is much smaller than the number of typical microstates which represent this macrostate. The phase-space volume of the macrostate M with volume V is $|\Gamma_{M(V)}|$. In contrast, the phase-space volume of the quasi-equilibrium states is equal to the phase-space volume $|\Gamma_{M(V_1)}|$ from which they derived by expansion and to which they return after a time reversal, and $|\Gamma_{M(V_1)}| \ll |\Gamma_{M(V)}|$. This means that if one prepares a system in a particular macrostate, the microstate which thereby appears will never of its own coincidentally or intentionally be one of the time-reversed quasi-equilibrium states such as $T\,T_t\,X$. The only possibility of generating such an atypical state is in fact to allow a gas to expand and then to reverse all the particle velocities, i.e. to prepare $T\,T_t\,X$. Thus one could refute Loschmidt's paradox by making the laconic remark that in practice (in a real experiment) it is not possible to reverse the velocities of 10^{20} particles. There is however an additional impossibility which prevents states with decreasing entropy from occurring. We have so far not taken into account the fact that in reality, it is impossible to produce a totally isolated system. There are always *external perturbations* present, such as radiation, sunspots or the variable gravitational influence of the surrounding matter. The latter effect is estimated in Sect. 10.7.3. If it were in fact possible to reverse all the velocities, the entropy would indeed decrease for a short time, but then owing to the external perturbations the system would within a very short time (ca. $10\,\tau$) be affected in such a way that its entropy would again increase. External perturbations transform quasi-equilibrium states into more typical representatives of the macrostate. Even though external perturbations may be so weak that they play no role in the energy balance, still quantum mechanically they lead to a randomization of the phases and classically to small deviations of the trajectories, so that the system loses its memory of the initial state after only a small number of collisions. This drastic effect of external perturbations is closely connected with the sensitive dependence on the initial conditions which is well known in classical mechanics and is responsible for the phenomenon of deterministic chaos.

In curves 3 and 4 of Fig. 10.6, the system was perturbed at the times $t = 23.6$ and $t = 59$ and thereafter the time-reversal transformation was carried out. The perturbation consisted of a small change in the directions of the particle velocities, such as could be produced by energetically negligible gravitational influences (Sect. 10.7.3). For the shorter time, the system still closely approaches its original initial state and the entropy decreases, but then it again increases. In the case of the longer time, a decrease of the entropy occurs for only a brief period.

It is intuitively clear that every perturbation leads away from the atypical region, since the phase space of the typical states is so much larger. This also implies that *the perturbations are all the more efficient*, the closer to quasi-equilibrium the state is. For, considered statistically, the imbalance of

---

occur. The decrease of the entropy following a perturbation and time reversal is due primarily to atoms which have undergone no collisions at all. This is the reason for the great difference between curves (3) and (4).

the number of typical and of atypical states is then all the greater and the probability of a perturbation leading to a much more typical state is higher. More significant, however, is the fact that the number of collisions undergone by the particles of the system increases enormously with time. Per τ_0, the number of collisions is $\tau_0/\tau \approx 10^5$. And all of these collisions would have to be run through in precisely the reverse direction after a time-reversal transformation in order for the initial state to again be reached.

Remarks:

(i) Stability of the irreversible macroscopic relaxation: the weak *perturbations* investigated have no effect on the macroscopic time development in the future direction. The state $T_t X$ is converted by the perturbation to a more typical state, which relaxes further towards equilibrium (quasi-equilibrium) just like $T_t X$. The time reversal of such a more typical state leads however to a state exhibiting an entropy decrease for at most a short time; thereafter, the entropy again increases. (If one conversely first performs the time reversal and then applies the perturbation, the effects are still similar.) In the time development $T_{t'}(T_t X)$, the particles spread apart spatially; in Γ-space, X moves into regions with all together a larger phase-space volume. This is not changed by external disturbances. For $T_{t'}(TT_t X)$, the particles move together into a more confined region. All velocities and positions must be correlated with one another in order for the improbable initial state to be produced once again. In the forward direction, the macroscopic *time development is stable with respect to perturbations* but in the time-reversed direction, it is very unstable.

(ii) Following Boltzmann's arguments, the explanation of irreversiblilty is probabilistic. The basic laws of physics are not irreversible, but the initial state of the system in an expansion experiment as described above is very peculiar. This initial state is quite improbable; this means that it corresponds to only a very small volume in phase space and to a correspondingly small entropy. Its time development then leads into regions with a large total volume (and also greater entropy), corresponding to a more probable macrostate of the system with a longer dwell time. In principle, the system would return to its improbable initial state after an unrealistically long time; however, we will never observe this. As soon as the limit of infinite particle number is introduced into the theory, this recurrence time in fact tends towards infinity. In this limit, there is no eternal return and one has complete irreversibility.

(iii) The significance of the external perturbations for the relaxation towards equilibrium instead of only towards quasi-equilibrium goes hand in hand with the justification in Chap. 2 of the necessity of describing real systems in terms of statistical ensembles. One could also ask what would happen in an idealized strictly isolated system. Its microstate would develop in the course of time into a quasi-equilibrium state which in terms of its macroscopic behavior would be indistinguishable from the typical microstates of the macrostate formed. In this situation, one could for convenience of computation still employ a density matrix instead of the single state.

We have completed the most important considerations of the irreversible transition towards equilibrium and the accompanying increase in the Boltzmann entropy. The next sections contain some additional observations and numerical estimates.

10.7.2 Description of the Expansion Experiment in μ-Space

It is instructive to describe the expansion experiment with an isolated gas also in Boltzmann's μ-space, and to compare the Gibbs and the Boltzmann entropies in detail; cf. Fig. 10.7. In the initial state, all the gas atoms are in the small subvolume V_1. The single-particle distribution function is uniform within this volume and vanishes outside it. The particles are to a large extent uncorrelated, i.e. the two-particle distribution function obeys $f_2(\mathbf{x}_1, \mathbf{v}_1, \mathbf{x}_2, \mathbf{v}_2) - f(\mathbf{x}_1, \mathbf{v}_1)f(\mathbf{x}_2, \mathbf{v}_2) = 0$, and higher correlation functions vanish as well. During the expansion, f spreads throughout the entire volume. As we have already mentioned in connection with the Stosszahlansatz for deriving the Boltzmann equation, two colliding particles are correlated after their collision (their velocities are such that they would once again collide if the gas were subjected to a time reversal = motion reversal). The information contained in the initial state: "all particles in one corner", i.e. the distribution which is concentrated within V_1 (and thus is spatially constrained) shifts to produce subtle correlations of the particles among themselves. The longer the time that passes, the more collisions will have occurred, and the higher the order of the correlation functions which take on nonzero values. All this information is contained in the time-dependent N-particle distribution function $\varrho(\mathbf{x}_1, \mathbf{v}_1, \ldots \mathbf{x}_N, \mathbf{v}_N, t)$ on which the Gibbs entropy is based. On the other hand, in the Boltzmann entropy only the macroscopic manifestation, in the simplest case the single-particle distribution function, is considered. The Boltzmann entropy increases.

The time which is typically required by the particles in order to pass through the volume ballistically is $\tau_0 = \frac{L}{v}$. For $L = 1$ cm and $v = 10^5$ cm sec^{-1}, $\tau_0 = 10^{-5}$ sec. In comparison, the collision time $\tau = 10^{-9}$ sec is much shorter. It thus typically requires several τ_0 for the gas to spread out

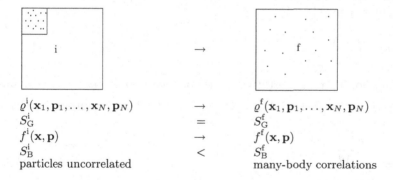

$$\varrho^{\mathrm{i}}(\mathbf{x}_1, \mathbf{p}_1, \ldots, \mathbf{x}_N, \mathbf{p}_N) \quad \rightarrow \quad \varrho^{\mathrm{f}}(\mathbf{x}_1, \mathbf{p}_1, \ldots, \mathbf{x}_N, \mathbf{p}_N)$$
$$S^{\mathrm{i}}_{\mathrm{G}} \quad = \quad S^{\mathrm{f}}_{\mathrm{G}}$$
$$f^{\mathrm{i}}(\mathbf{x}, \mathbf{p}) \quad \rightarrow \quad f^{\mathrm{f}}(\mathbf{x}, \mathbf{p})$$
$$S^{\mathrm{i}}_{\mathrm{B}} \quad < \quad S^{\mathrm{f}}_{\mathrm{B}}$$

particles uncorrelated   many-body correlations

Fig. 10.7. Expansion experiment: N-particle and single-particle distribution functions ρ and f, the Boltzmann and Gibbs entropies of the initial state i and the final state f

throughout the whole volume ballistically (with reflections from the walls). After around $10\,\tau_0$, about $\frac{10 \times 10^{-5}}{10^{-9}} = 10^5$ collisions have also taken place. Even after this short time, i.e. 10^{-4} sec, the initial configuration (all the particles uncorrelated within a corner of the total volume) has been transferred to correlation functions of the order of 10000 particles.

10.7.3 The Influence of External Perturbations on the Trajectories of the Particles

In the following section, the influence of an external perturbation on the relative motion of two particles is estimated, along with the change in the collisions which follow such a perturbation. We consider two particles which collide with each other and investigate the effect of an additional external force on their relative distance and its influence on their trajectories. The two atoms are presumed to be initially at a distance l (mean free path). Owing to the spatial variation of the force $\mathbf{F}$, it acts differently on the two atoms, $\Delta \mathbf{F} = \mathbf{F}_1 - \mathbf{F}_2$. The Newtonian equation of motion for the relative coordinate $\Delta \ddot{\mathbf{x}} = \frac{\Delta \mathbf{F}}{m}$ leads to $\Delta \dot{\mathbf{x}} \approx \frac{\Delta \mathbf{F} t}{m}$ and finally to

$$\Delta \mathbf{x} \approx \frac{\Delta \mathbf{F} t^2}{m} \approx \frac{\Delta \mathbf{F}}{m} \left(\frac{l}{v}\right)^2 . \tag{10.7.2}$$

This yields an angular change in the trajectory after a path of length l of

$$\Delta \vartheta \approx \frac{|\Delta \mathbf{x}|}{l} \approx \frac{|\Delta \mathbf{F}|}{m} \frac{l}{v^2} . \tag{10.7.3}$$

Even if this angular change is very small, it will be amplified by the collisions which follow. In the first collision, there is a change in the deflection angle of $\Delta \vartheta_1 = \frac{l}{r_c} \Delta \vartheta$ (Fig. 10.8), where it must be kept in mind that $l \gg r_c$. Here, r_c is the range of the potentials or the radius of a hard sphere. After k collisions, the angular change is

$$\Delta \vartheta_k = \left(\frac{l}{r_c}\right)^k \Delta \vartheta . \tag{10.7.4}$$

The condition that the perturbed trajectory have no connection to the unperturbed one is given by $\Delta \vartheta_k = 2\pi = \left(\frac{l}{r_c}\right)^k \Delta \vartheta$. It follows from this that

$$k = \frac{\log \dfrac{2\pi}{\Delta \vartheta}}{\log \dfrac{l}{r_c}} . \tag{10.7.5}$$

We consider for example the influence of an experimenter of mass $M = 80\,\mathrm{kg}$ at a distance of $d = 1\,\mathrm{m}$ on a sample of helium gas (1 mole) due

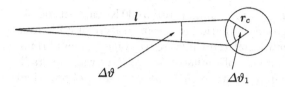

Fig. 10.8. $\Delta\vartheta_1 = \frac{l}{r_c}\,\Delta\vartheta$

to the gravitational force, $W = -\frac{GMNm}{d}$; $G = 6.67 \times 10^{-11}\mathrm{m}^3\mathrm{kg}^{-1}\mathrm{s}^{-2}$, $N = 6 \times 10^{23}$, $m = 6.7 \times 10^{-23}$ g. The additional energy $W \approx -2 \times 10^{-10}$ J is negligible compared to the total energy of the gas, $E \approx 3\,\mathrm{kJ}$. The difference in the force on the two particles spaced a distance l apart owing to the additional mass M is

$$|\Delta \mathbf{F}| = \frac{GMm}{d^2} - \frac{GMm}{(d+l)^2} \approx \frac{GMml}{d^3} \,,$$

and the resulting angular change is

$$\Delta\vartheta \approx \frac{GMml}{d^3}\frac{l}{mv^2} = \frac{GM}{d^3}\left(\frac{l}{v}\right)^2 .$$

For the numerical values given, one finds $\Delta\vartheta \approx 4 \times 10^{-28}$. To determine the number of collisions which lead to completely different trajectories, this value is inserted into equation (10.7.5) along with $l \approx 1400$ Å and $r_c \approx 1.5$ Å resulting in $k \approx 10$. In spite of the smallness of $\Delta\vartheta$, due to the logarithmic dependence a relatively small number of collisions is sufficient. Even much smaller masses at much greater distances lead to a similarly drastic effect. *Energetically completely negligible perturbations lead to a randomization of the trajectory.*

*10.8 Entropy Death or Ordered Structures?

We close with some qualitative remarks on the consequences of the global increase of the entropy. Boltzmann himself considered the evolution of the cosmos and feared that it would end in a state of thermal equilibrium (heat death). Our Earth and the surrounding cosmos show no signs of this: (i) How does the extreme thermal non-equilibrium within the galaxies come about? (ii) What allows the existence of ordered and highly organized structures on our planet? (iii) Where will further evolution lead?

In its early period at temperatures above $3\,000$ K, the Universe did not consist of galaxies and stars but rather of an ionized and undifferentiated soup of matter and radiation. By the time the temperature of the Universe had decreased to $3\,000$ K (around $300\,000$ years after the Big Bang), nucleons and electrons began to bind together to form atoms. Matter then became transparent to electromagnetic radiation. The radiation, which at that time obeyed a Planck distribution with the temperature 3 000 K, can still be observed today as cosmic background radiation at a temperature of 2.7 K (due to the red shift), and it indicates that the Universe was once in equilibrium. The decisive effect of the decoupling of radiation and matter was that radiation pressure was no longer important, so that the *gravitational force* had only to overcome the pressure of matter to form stars. Thus, the stars, stone and life forms of today's Universe could come into being.

Gas clouds contract as a result of gravitational attraction. In this process, their potential energy decreases and due to energy conservation, their kinetic energy increases. It follows from the equipartition theorem that they become hotter. The hot clouds emit radiation, thus lowering their energy, and contract more and more, becoming hotter and hotter. This heating up follows from the negativity of the specific heat in systems with gravitational interactions below the gravitational instability[33]. This means that with decreasing energy their temperature increases. This feature is fundamental for stellar evolution. The instability mentioned can also be observed in computer experiments[34]. The thermal instability due to gravitation destroys the thermal equilibrium and leads to hot clusters, the stars. The temperature differences which result permit the formation of ordered structures including life. As was already suspected by Boltzmann, on the Earth this is a result of the fact that solar radiation is rich in energy and poor in entropy. As was shown in Sect. 4.5.4, the entropy of a photon gas is roughly equal to the product of the Boltzmann constant and the photon number, and the energy per photon is kT. Photons with the thermal energy of the solar surface of $\approx$ 6 000 K can be split up into 20 photons through processes on the Earth's surface, each with an energy corresponding to 300 K ($\approx$ temperature of the Earth's surface). One need only recall that the energy of the Earth does not change; just as much energy reaches the Earth from solar radiation in the visible region as is again reradiated in the form of long-wavelength infrared photons. In this process, the entropy of the photons increases by a factor of 20. Even if as a result of these processes structures are formed which are ordered and have entropies lower than the equilibrium value, the entropy balance still remains positive, i.e. the total entropy continues to increase. It is in the end the thermodynamic instability of gravitational systems which makes life possible.

In the further long-term evolution (10^{10} years), stars will collapse after they have used up their nuclear fuel, forming neutron stars or, if their masses are sufficient, black holes. The phase space of black holes and thus their entropies are so large that the ratio of phase space volumes in the final state $|\Gamma_f|$ and in the initial state $|\Gamma_i|$ will attain a value (according to an estimate

---

[33] J. Messer, Lecture Notes in Physics **147** (1981); P. Hertel and W. Thirring, Ann. Phys. (N.Y.) **63**, 520 (1971)

[34] H. Posch, H. Narnhofer, and W. Thirring, J. Stat. Phys. **65**, 555 (1991); Phys. Rev. A **42**, 1880 (1990)

by Penrose[35]) of

$$\frac{|\Gamma_f|}{|\Gamma_i|} = 10^{10^{123}} .$$

The enormous increase in entropy associated with the gravitational instability can be accompanied by local entropy decreases, leaving room for a multiplicity of highly organized structures such as gorgons, mermaids, and Black Clouds[36]...

Literature

R. Balian, *From Microphysics to Macrophysics II*, Springer, Berlin, 1982

R. Becker, *Theory of Heat*, 2nd ed., Springer, Berlin, 1967

J. M. Blatt, *An Alternative Approach to the Ergodic Problem*, Prog. Theor. Phys. **22**, 745 (1959)

K. Huang, *Statistical Mechanics*, 2nd ed., John Wiley, New York, 1987

M. Kac, *Probability and Related Topics in Physical Science*, Interscience Publishers, London, 1953

H. J. Kreuzer, *Nonequilibrium Thermodynamics and its Statistical Foundations*, Clarendon Press, Oxford, 1981

J. L. Lebowitz, *Boltzmann's Entropy and Time's Arrow*, Physics Today, Sept. 1993, p. 32; and *Macroscopic Law and Microscopic Dynamics*, Physica A **194**, 1 (1993)

O. Penrose, *Foundations of Statistical Mechanics*, Pergamon, Oxford, 1970

D. Ruelle, *Chance and Chaos*, Princeton University Press, Princton, 1991

W. Thirring, *A Course in Mathematical Physics 4, Quantum Mechanics of Large Systems*, Springer, New York, Wien 1980

S. Weinberg, *The First Three Minutes*, Basic Books, New York 1977.

Problems for Chapter 10

10.1 Verify the equations (10.2.11) and (10.2.12).

10.2 Solve the equation of motion for a chain of atoms (10.2.2) by introducing the coordinates $x_{2n} = \sqrt{m}\,\frac{dq_n}{dt}$ and $x_{2n+1} = \sqrt{m}\,\Omega(q_n - q_{n+1})$. This leads to the equations of motion

$$\frac{dx_n}{dt} = -\,\Omega(x_{n+1} - x_{n-1}) ,$$

whose solution can be found by comparing with the recursion relations for the Bessel functions (see e.g. Abramowitz/Stegun, *Handbook of Mathematical Functions*). *Reference:* E. Schrödinger, Ann. der Physik **44**, 916 (1914).

---

[35] R. Penrose, *The Emperor's New Mind*, Oxford Univ. Press, Oxford, 1990, chapter 7; based on the Bekenstein–Hawking formula under the assumption that the final state consists of a single black hole.

[36] F. Hoyle, *The Black Cloud*, Harper, New York, 1957.

10.3 Recurrence time.

Complete the intermediate steps of the calculation given in 10.3.1 for the recurrence time in a chain of N harmonic oscillators. Use the following representation of the Bessel function for integral n:

$$J_n(x) = \frac{(-i)^n}{\pi} \int_0^\pi d\phi\, e^{ix \cos \phi} \cos n\phi$$

$$x \to \infty \atop \sim \sqrt{\frac{2}{\pi x}} \left[\cos(x - n\pi/2 - \pi/4) + \mathcal{O}(1/x) \right].$$

10.4 Calculate the integral which occurs in Eq. (10.3.19), $g^2 z^2 \int_0^\infty \frac{d\nu}{z^2 - \nu^2}$, with $z = \omega + i\eta$, $\eta > 0$, by applying the theorem of residues.

10.5 A microscopic model of Brownian motion.

(a) Calculate the inverse Green's function for the model treated in Sect. 10.3 using a continuous cutoff function $c(\mathbf{k})$:

$$D(z) = -M(z^2 + \Omega^2) + \frac{a^3}{m(2\pi)^3} \int d^3k \frac{|c(\mathbf{k})|^2}{z^2 - |c\mathbf{k}|^2}$$

with

$$|c(\mathbf{k})|^2 = g^2 \frac{\Lambda^2}{k^2 + \Lambda^2}.$$

(b) Determine the poles of $D(z)^{-1}$ for large values of Λ.
(c) Carry out the integration in the solution of the equation of motion

$$\Theta(t)Q(t) = \int_{-\infty}^\infty \frac{d\omega}{2\pi} e^{-izt} D(z)^{-1} M(\dot{Q}(0) - izQ(0)) \,,$$

by expanding the residues for large Λ up to the order $\mathcal{O}(\Lambda^{-2})$. Cf. also P. C. Aichelburg and R. Beig, Ann. Phys. **98**, 264 (1976).

10.6 Stochastic forces in a microscopic model of Brownian motion.

(a) Show that for the model discussed in Sect. 10.3 of a heavy particle coupled to a bath of light particles, the heavy particle is subjected to a force $F(t)$ which depends on the initial conditions of the light particles and whose half-range Fourier transform

$$\tilde{F}(z) = \int_0^\infty dt\, e^{izt} F(t)$$

(with $z = \omega + i\eta$, $\eta > 0$) has the form

$$\tilde{F}(z) = \frac{1}{\sqrt{mN}} \sum_k c(\mathbf{k}) \frac{\dot{Q}_\mathbf{k}(0) - izQ_\mathbf{k}(0)}{\omega_\mathbf{k}^2 - z^2} \,.$$

(b) Compute the correlation function $\langle F(t)F(t')\rangle$ under the assumption that the light particles are in thermal equilibrium at the time $t = 0$:

$$\langle \dot{Q}_{\mathbf{k}}(0)\dot{Q}_{\mathbf{k}'}(0)\rangle = \delta_{\mathbf{k},-\mathbf{k}'}\,kT\,, \quad \langle Q_{\mathbf{k}}(0)Q_{\mathbf{k}'}(0)\rangle = \delta_{\mathbf{k},-\mathbf{k}'}\frac{kT}{\omega_{\mathbf{k}}^2}\,.$$

Express the sum over $\mathbf{k}$ which occurs in terms of an integral with a cutoff Λ, as in (10.3.18), and assume that the heavy particle couples only to the light particle at the origin: $c(\mathbf{k}) = g$. Discuss the correlation function obtained and find a relation between its prefactor and the damping constant Γ.

10.7 The time-independence of the Gibbs entropy.

Let $\rho(p,q)$ with $(p,q) = (p_1,\dots,p_{3N},q_1,\dots q_{3N})$ be an arbitrary distribution function in phase space. Show using the microscopic equation of motion (Liouville equation):

$$\dot{\rho} = -\{\mathcal{H},\rho\} = -\frac{\partial\mathcal{H}}{\partial p_i}\frac{\partial\rho}{\partial q_i} + \frac{\partial\rho}{\partial p_i}\frac{\partial\mathcal{H}}{\partial q_i}\,,$$

that the Gibbs entropy $S_G = -k\int d\Gamma\,\rho\log\rho$ is stationary: $\dot{S}_G = 0$.

10.8 The urn model[37].

Consider the following stochastic process: N numbered balls $1,2,\dots N$ are divided between two urns. In each step, a number between 1 and N is drawn and the corresponding ball is taken out of the urn where it is found and put into the other urn. We consider the number n of balls in the first urn as a statistical variable.

Calculate the conditional probability (transition probability) $T_{n,n'}$ of finding n' balls in the first urn if it contained n balls in the preceding step.

10.9 For Ehrenfest's urn model defined in problem 10.8, consider the probability $P(n,t)$ of finding n balls in the first urn after t steps.
(a) Can you find an equilibrium distribution $P_{eq}(n,t)$? Does detailed balance apply?
(b) How does the conditional probability $P(0,n_0|t,n)$ behave for $t \to \infty$? Discuss this result. *Hint:* the matrix $T_{n,n'}$ of transition probabilities per time step has the eigenvalues $\lambda_k = 1 - 2k/N$, $k = 0,1,\dots,N$. Furthermore, $T_{n,n'}$ anticommutes with a suitably chosen diagonal matrix. (Definition: $P(t_0,n_0|t,n) =$ the probability that at the time t the first urn contains n balls, under the condition that at time t_0 there were n_0 balls in it.)

10.10 The urn model and the paramagnet.

The urn model with N balls (problem 10.8 and problem 10.9) can be considered as a model for the dynamics of the total magnetization N of non-interacting Ising spins. Explain this.

---

[37] A model for the Boltzmann equation and for irreversibility in which the typical behavior can be calculated in a simple manner is the urn model. Although in principle one of the urns could fill up at the expense of the other, this path is so improbable that the system tends with an overwhelming probability towards the state with equipartition of the balls and then exhibits small fluctuations around that state. The urn model is analyzed here in a series of problems.

10.11 The urn model and the H theorem.
Let X_t be the number of balls in urn 1 after t time steps, and

$$H_t = \frac{X_t}{N} \log \frac{X_t}{N} + \frac{N - X_t}{N} \log \frac{N - X_t}{N} .$$

Study the time dependence of H_t for a system with $X_0 = N$ by carrying out a computer simulation. Plot the time development of $\Delta_t \equiv X_t/N - 1/2$ for several runs of this stochastic process. What do you observe? Discuss the relation between your observation and the Second Law.

10.12 The urn model for large N.

(a) Calculate the average time dependence of Δ_t from the preceding problem for very large values of N. It is expedient to introduce a quasi-continuous time $\tau = t/N$ and to consider the quantity $f(\tau) = \Delta_{N\tau}$. Write a difference equation for $\langle f(\tau + 1/N) \rangle_{f(\tau)=f}$ based on the equation of motion of the probability $P_n(t)$ of finding n of the N balls in the first urn after the time step t. Taking the limit $N \to \infty$ and averaging over f, you will obtain a differential equation for $\langle f(\tau) \rangle$.
(b) Calculate also the mean square deviation $v(\tau) \equiv \langle f(\tau)^2 \rangle - \langle f(\tau) \rangle^2$. What do you conclude for the time dependence of the non-averaged quantity $f(\tau)$?
(c) Compare the result obtained with the result of the simulation from the preceding problem. Explain the connection.
    Reference: A. Martin-Löf, *Statistical Mechanics and the Foundation of Thermodynamics*, Springer Lecture Notes in Physics **101** (1979).

10.13 The Fokker–Planck equation and the Langevin equation for the urn model.

In Ehrenfest's urn model with N balls, let X_t be the number of balls which are in the left-hand urn after t steps. Consider the time development of $x(\tau) := \sqrt{N}f(\tau)$, where $f(\tau) = X_{N\tau}/N - \frac{1}{2}$ (see problem 10.12).
(a) Set up the Fokker–Planck equation for $P(x, \tau)$ by calculating the average and mean square jump length $\langle x(\tau + \frac{1}{N}) - x(\tau) \rangle_{x(\tau)=x}$ and $\langle [x(\tau + \frac{1}{N}) - x(\tau)]^2 \rangle_{x(\tau)=x}$. For this, you can use the intermediate results obtained in problem 10.12.
(b) Do you recognize the equation obtained? Give its solution for $P(x, \tau)$ by comparing with the case treated in Chap. 8, and read off the results for $\langle f(\tau) \rangle$ and $v(\tau) = \langle [f(\tau) - \langle f(\tau) \rangle]^2 \rangle$ obtained in a different way in problem 10.12 (each under the condition that $f(\tau = 0) = f_0$).
(c) What is the associated Langevin equation? Interpret the forces which appear. Compare the potential which corresponds to the non-stochastic part of the force with the Boltzmann entropy $S_B(x) = k \log |\Gamma_x|$, where Γ_x is the set of microstates characterized by x. (Use the fact that the binomial distribution can be approximated by a Gaussian distribution for large N.)

Appendix

A. Nernst's Theorem (Third Law)

A.1 Preliminary Remarks on the Historical Development of Nernst's Theorem

Based on experimental results[1], Nernst (1905) originally postulated that changes in the entropy ΔS in isothermal processes (chemical reactions, phase transitions, pressure changes or changes in external fields for $T = $ const.) have the property

$$\Delta S \to 0$$

in the limit $T \to 0$. This postulate was formulated in a more stringent way by Planck, who made the statement $S \to 0$, or, more precisely,

$$\lim_{T \to 0} \frac{S(T)}{N} = 0 , \tag{A.1}$$

where, depending on the physical situation, N is the number of particles or of lattice sites. One refers to (A.1) as Nernst's theorem or the Third Law of thermodynamics[2].

According to statistical mechanics, the value of the entropy at absolute zero, $T = 0$, depends on the degeneracy of the ground state. We assume that the ground state energy E_0 is g_0-fold degenerate. Let P_0 be the projection

---

[1] The determination of the entropy as a function of the temperature T is carried out by measuring the specific heat $C_X(T)$ in the interval $[T_0, T]$ and integrating according to the equation $S(T) = S_0 + \int_{T_0}^{T} dT \frac{C_X(T)}{T}$, where the value S_0 at the initial temperature T_0 is required. Nernst's Theorem in the form (A.1) states that this constant for all systems at $T = 0$ has the value zero.

[2] Nernst's theorem is understandable only in the framework of quantum mechanics. The entropy of classical gases and solids does not obey it. Classically, the energy levels would be continuous, e.g. for a harmonic oscillator, $E = \frac{1}{2}\left(\frac{p^2}{m} + m\omega^2 q^2\right)$ instead of $E = \hbar\omega\left(n + \frac{1}{2}\right)$. The entropy of a classical crystal, effectively a system of harmonic oscillators, would diverge at $T = 0$, since per vibrational degree of freedom, $S = k + k \log T$. In this sense, Nernst's theorem can certainly be regarded as visionary.

operator onto states with $E = E_0$. Then the density matrix of the canonical ensemble can be cast in the form

$$\rho = \frac{e^{-\beta H}}{\operatorname{Tr} e^{-\beta H}} = \frac{\sum_n e^{-\beta E_n} |n\rangle \langle n|}{\sum_n e^{-\beta E_n}} = \frac{P_0 + \sum_{E_n > E_0} e^{-\beta(E_n - E_0)} |n\rangle \langle n|}{g_0 + \sum_{E_n > E_0} e^{-\beta(E_n - E_0)}} .$$

$$(A.2)$$

For $T = 0$, this leads to $\rho(T = 0) = \frac{P_0}{g_0}$, and thus for the entropy to

$$S(T = 0) = -k\langle \log \rho \rangle = k \log g_0 . \tag{A.3}$$

The general opinion in mathematical physics is that the ground state of interacting systems should not be degenerate, or that the degree of degeneracy in any case should be considerably less than the number of particles. If $g_0 = \mathcal{O}(1)$ or even if $g_0 = \mathcal{O}(N)$, we find

$$\lim_{N \to \infty} \frac{S(T = 0)}{kN} = 0 , \tag{A.4}$$

i.e. for such degrees of degeneracy, Nernst's theorem follows from quantum statistics.

In Sect. A.2, the Third Law is formulated generally taking into account the possibility of a residual entropy. This is in practice necessary for the following reasons: (i) there are model systems with greater ground-state degeneracies (ice, non-interacting magnetic moments); (ii) a very weak lifting of the degeneracy might make itself felt only at extremely low temperatures; (iii) a disordered metastable state can be 'frozen in' by rapid cooling and retains a finite residual entropy. We will discuss these situations in the third section.

A.2 Nernst's Theorem and its Thermodynamic Consequences

The General Formulation of Nernst's Theorem:

$S(T = 0)/N$ is a finite constant which is independent of parameters X such as V and P (i.e. the degeneracy does not change with X) and $S(T)$ is finite for finite T.

Results of Nernst's theorem for the specific heat and other thermodynamic derivatives:

Let A be the thermodynamic state which is attained on increasing the temperature starting from $T = 0$ at constant X. From $C_X = T \left(\frac{\partial S}{\partial T} \right)_X$, it follows that

$$S(T) - S(T = 0) = \int_0^A dT \frac{C_X(T)}{T} . \tag{A.5}$$

From this, we find furthermore

$$C_X(T) \longrightarrow 0 \quad \text{for} \quad T \longrightarrow 0 \,,$$

since otherwise $S(T) = S(T = 0) + \infty = \infty$. This means that the heat capacity of every substance at absolute zero tends to zero; in particular, we have $C_P \to 0$, $C_V \to 0$, as already found explicitly in Chap. 4 for ideal quantum gases. Thus the specific heat at constant pressure takes on the form

$$C_P = T^x(a + bT + \ldots) \,, \tag{A.6}$$

where x is a positive exponent. For the entropy, (A.5), one obtains from this expression

$$S(T) = S(T = 0) + T^x \left(\frac{a}{x} + \frac{bT}{x+1} + \ldots \right) \,. \tag{A.7}$$

Other thermodynamic derivatives also vanish in the limit $T \to 0$, as one can see by combining (A.7) with various thermodynamic relations.

The thermal expansion coefficient α and its ratio to the isothermal compressibility fulfill the relations

$$\alpha \equiv \frac{1}{V} \left(\frac{\partial V}{\partial T} \right)_P = -\frac{1}{V} \left(\frac{\partial S}{\partial P} \right)_T \to 0 \quad \text{for} \quad T \to 0 \tag{A.8}$$

$$\frac{\alpha}{\kappa_T} = \left(\frac{\partial P}{\partial T} \right)_V = \left(\frac{\partial S}{\partial V} \right)_T \to 0 \quad \text{for} \quad T \to 0 \,. \tag{A.9}$$

The first relation can be seen by taking the derivative of (A.7) with respect to pressure

$$V\alpha = \left(\frac{\partial V}{\partial T} \right)_P = -\left(\frac{\partial S}{\partial P} \right)_T = -T^x \left(\frac{a'}{x} + \frac{b'T}{x+1} + \ldots \right) \,; \tag{A.10}$$

the second relation is found by taking the derivative of (A.7) with respect to V.

From the ratio of (A.10) and (A.6) we obtain

$$\frac{V\alpha}{C_P} = -\frac{a'}{ax} + \ldots \propto T^0 \,.$$

In an adiabatic pressure change, the temperature changes as[3] $dT = \left(\frac{V\alpha}{C_P} \right) T dP$. A finite temperature change requires that dP increase as $\frac{1}{T}$. *Absolute zero therefore cannot be reached by an adiabatic expansion* .

---

[3] $\left(\frac{\partial P}{\partial T} \right)_S = -\frac{\left(\frac{\partial S}{\partial T} \right)_P}{\left(\frac{\partial S}{\partial P} \right)_T} = \frac{T^{-1} C_P}{\left(\frac{\partial V}{\partial T} \right)_P} = \frac{C_P}{TV\alpha}$

To clarify the question of whether absolute zero can be reached at all, we consider the fact that cooling processes always take place between two curves with $X = \text{const.}$, e.g. $P = P_1$, $P = P_2$ ($P_1 > P_2$) (see Fig. A.1). Absolute zero could be reached only after infinitely many steps. An adiabatic change in X leads to cooling. Thereafter, the entropy must be decreased by removing heat; since no still colder heat bath is available, this can be done at best for $T = \text{const.}$ If a substance with a $T - S$ diagram like that shown in Fig. A.2 were to exist, i.e. if in contradiction to the Third Law, $S(T = 0)$ were to depend upon X, then one could reach absolute zero.

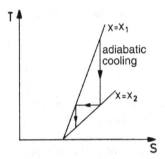
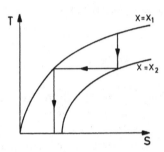

Fig. A.1. The approach to absolute zero by repeated adiabatic changes (e.g. adiabatic expansions)

Fig. A.2. Hypothetical adiabats which would violate the Third Law

A.3 Residual Entropy, Metastability, etc.

In this section, we shall consider systems which exhibit a residual entropy even at very low temperatures, or metastable frozen-in states and other particular qualities which can occur in this connection.

(i) Systems which contain non-coupled spins and are not subject to an external magnetic field have the partition function $Z = (2S + 1)^N Z'$ and the free energy $F = -kTN \log(2S+1) + F'$. The spins then have a finite residual entropy even at $T = 0$:

$$S(T = 0) = Nk \log(2S + 1) .$$

For example: paraffin, $C_{20}H_{42}$; owing to the proton spins of H, the partition function is proportional to $Z \sim 2^{42N}$, from which we find for the residual entropy $S = 42kN \log 2$.

(ii) Metastable states in molecular crystals: the ground state of crystalline carbon monoxide, CO, has a uniformly oriented ordered structure of the linear CO molecules. At higher temperatures, the CO molecules are not ordered. If

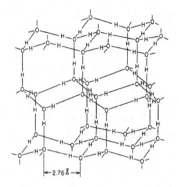

Fig. A.3. The structure of ice[4]

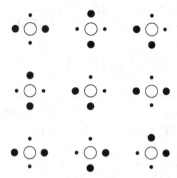

Fig. A.4. Two-dimensional ice: ○ oxygen, ● hydrogen, • other possible positions of H

a sample is cooled to below $T = \frac{\Delta\epsilon}{k}$, where $\Delta\epsilon$ is the very small energy difference between the orientations CO–OC and CO–CO of neighboring molecules, then the molecules undergo a transition into the ordered equilibrium state. Their reorientation time is however very long. The system is in a metastable state in which the residual entropy has the value

$$S(T = 0) = k \log 2^N = Nk \log 2 \,,$$

i.e. $S = 5.76$ J mol^{-1} K^{-1}. The experimental value is somewhat smaller, indicating partial orientation.

(iii) Binary alloys such as β-brass, (CuZn), can undergo a transition from a completely disordered state to an ordered state when they are cooled slowly. This phase transition can also be described by the Ising model, by the way. On the other hand, if the cooling is rapid, i.e. if the alloy is quenched, then the Cu and Zn atoms stay in their disordered positions. At low temperatures, the rate of reordering is so negligibly small that this frozen-in metastable state remains permanent. Such a system has a residual entropy.

(iv) Ice, solid H$_2$O: ice crystallizes in the Wurtzite structure. Each hydrogen atom has four oxygen atoms as neighbors (Fig. A.3). Neighboring oxygen atoms are connected by hydrogen bonds. The hydrogen atom which forms these bonds can assume two different positions between the two oxygen atoms (Fig. A.4). Because of the Coulomb repulsion, it is unfavorable for an oxygen atom to have more or fewer than two hydrogen atoms as neighbors. Thus one restricts the possible configurations of the hydrogen atoms by the ice rule: the protons are distributed in such a manner that two are close and two are more distant from each oxygen atom[5]. For N lattice sites (N oxygen

---

[4] The structure of common (hexagonal) H$_2$O-ice crystals: S. N. Vinogrado, R. H. Linnell, *Hydrogen Bonding*, p. 201, Van Nostrand Reinhold, New York, 1971.

[5] L. Pauling: J. Am. Chem. Soc., **57**, 2680 (1935)

atoms), there are $2N$ hydrogen bonds. The approximate calculation of the partition function[5] at $T = 0$ yields

$$Z_0 = 2^{2N} \left(\frac{6}{16} \right)^N = \left(\frac{3}{2} \right)^N .$$

(The number of unhindered positions of the protons in the hydrogen bonds) times (reduction factor per lattice site, since of 16 vertices, only 6 are allowed). Using $W = \lim_{N \to \infty} Z_0^{1/N} = 1.5$, we find for the entropy per H_2O:

$$\frac{S(T = 0)}{kN} = \log W = \log 1.5 .$$

An exactly soluble two-dimensional model to describe the structure of ice has been given[6] (Fig. A.4). A square lattice of oxygen atoms is bound together by hydrogen bonds. The near-neighbor structure is the same as in three-dimensional ice. The statistical problem of calculating Z_0 can be mapped onto a vertex model (Fig. A.5). The arrows denote the position of the hydrogen bonds. Here, H assumes the position which is closer to the oxygen towards which the arrow points. Since each of the four arrows of a vertex can have two orientations, there are all together 16 vertices. Because of the ice rule, of these 16 vertices only the six shown in Fig. A.5 are allowed.

Fig. A.5. The vertices of the two-dimensional ice model which obey the ice rule (two hydrogen atoms near and two more distant)

The statistical problem now consists in determining the number of possibilities of ordering the 6 vertices in Fig. A.5 on the square lattice. The exact solution[6] of the two-dimensional problem is obtained using the transfer matrix method (Appendix F.).

$$W = \lim_{N \to \infty} Z_0^{1/N} = \left(\frac{4}{3} \right)^{3/2} = 1.5396007 \ldots .$$

The numerical result for three-dimensional ice is[7]:

$$W = 1.50685 \pm 0.00015, \quad S(T = 0) = 0.8154 \pm 0.0002 \text{ cal/K mole}$$

Experiment at 10 K: $\quad S(T = 0) = 0.82 \pm 0.05 \text{ cal/K mole} .$

[6] E. H. Lieb, Phys. Rev. Lett. **18**, 692 (1967); Phys. Rev. **162**, 162 (1967)
[7] Review: E. H. Lieb and F. Y. Wu in: Domb and Green, *Phase Transitions and Critical Phenomena I*, p. 331, Academic Press, New York, 1972.

The approximate formula of Pauling gives a lower limit for the residual entropy.

If the orientations of the hydrogen bonds were allowed to be completely unhindered, the residual entropy per lattice site would be $\log 2^2 = \log 4$. Due to the ice rule (as a result of the Coulomb repulsion), the residual entropy is reduced to $\log 1.5$. If other interactions of the protons were taken into account, there would be still finer energy splittings among the various configurations of the vertex arrangements. Then, on lowering the temperature, only a smaller number would be allowed and presumably at $T \to 0$ no residual entropy would be present. The fact that ice has a residual entropy even at low temperatures indicates that the reorientation becomes very slow under these conditions.

(v) The entropy of a system with low-lying energy levels typically shows the dependence shown in Fig. A.6. Here, the value of the entropy between T_1 and T_2 is not the entropy S_0. In case energy levels of the order of kT_1 are present, these are practically degenerate with the ground state for $T \gg T_1$, and only for $T < T_1$ is the residual entropy (possibly $S_0 = 0$) attained. An example of this is a weakly coupled spin system. The plateau in the temperature interval

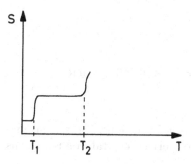

Fig. A.6. The entropy of a system with energy levels of the order of kT_1 and kT_2

$[T_1, T_2]$ could appear as a residual entropy on cooling. In this interval, the specific heat is zero. In the region of T_1, the specific heat again increases with decreasing temperature, then drops below T_1 towards the value zero after the degrees of freedom at the energy kT_1 are frozen out; this could possibly indicate a final decrease of the entropy to its value at $T = 0$.

For degrees of freedom with a discrete excitation spectrum (spins in a field, harmonic oscillators), the excitation energy determines the temperature below which the entropy of these degrees of freedom is practically zero. This is different for translational degrees of freedom, where the energy levels become continuous in the limit $N \to \infty$ and for example the spacing of the first excited state from the ground state is of the order of $\frac{\hbar^2}{mV^{2/3}}$. The corresponding excitation temperature of about $5 \times 10^{-15}\,\mathrm{K}$ is however unimportant for the region of application of the Third Law, which already applies at considerably higher temperatures. The spacing of the energy levels tends towards zero in the thermodynamic limit, and they are characterized by a density of states.

The temperature dependence of the entropy and the specific heat does not depend on the value of individual energy levels, but instead on the form of the density of states. For crystals, the density of states of the phonons is proportional to the square of the energy and therefore it gives $S \propto T^3$ at low temperatures. The density of states of the electrons at the Fermi energy is constant, and thus one obtains $S \propto T$.

(vi) It is also interesting to discuss chemical substances which exhibit allotropy in connection with the Third Law. Two famous examples are carbon, with its crystalline forms diamond and graphite, and tin, which crystallizes as metallic white tin and as semiconducting grey tin. White tin is the high-temperature form and grey tin the low-temperature form. At $T_0 = 292\,\mathrm{K}$, grey tin transforms to white tin with a latent heat Q_L. Upon cooling, the transformation takes place in the reverse direction, so long as the process occurs slowly and crystallization seeds of grey tin are present. On rapid cooling, white tin remains as a metastable structure. For the entropies of white and grey tin, the following relations hold:

$$S_W(T) = S_W(0) + \int_0^T \frac{dT}{T} C_W(T)$$

$$S_G(T) = S_G(0) + \int_0^T \frac{dT}{T} C_G(T) \,.$$

From the general formulation of Nernst's theorem, it follows that

$$S_W(0) = S_G(0) \,,$$

since the two forms are present under identical conditions. (Statistical mechanics predicts in addition for these two perfect crystal configurations $S_W(0) = S_G(0) = 0$.) It thus follows that

$$S_W(T) - S_G(T) = \int_0^T \frac{dT}{T} \left(C_W(T) - C_G(T) \right) \,.$$

From this we find in particular that the latent heat at the transition temperature T_0 is given by

$$Q_L(T_0) \equiv T_0 \left(S_W(T_0) - S_G(T_0) \right) = T_0 \int_0^{T_0} \frac{dT}{T} \left(C_W(T) - C_G(T) \right) \,. \quad \text{(A.11)}$$

The temperature dependence of the specific heat at very low temperatures thus has an influence on the values of the entropy at high temperatures.

(vii) Systems with continuous internal symmetry, such as the Heisenberg model: for both the Heisenberg ferromagnet and the Heisenberg antiferromagnet, owing to the continuous rotational symmetry, the ground state is continuously degenerate. Classically, the degree of degeneracy would not, to be sure, depend on the number of lattice sites, but it would be infinitely large. For N spin-1/2 sites, in quantum mechanics the z-component of the total spin has $N + 1$ possible orientations. The ground state is thus only $(N + 1)$-fold degenerate (see Eq. (A.4)). This degeneracy thus does not lead to a residual entropy at absolute zero.

Reference: J. Wilks, *The Third Law of Thermodynamics*, Oxford University Press, 1961.

B. The Classical Limit and Quantum Corrections

B.1 The Classical Limit

We will now discuss the transition from the quantum-mechanical density matrix to the classical distribution function, beginning with the one-dimensional case. At high temperatures and low densities, the results of quantum statistics merge into those of classical physics (see e.g. Sect. 4.2). The general derivation can be carried out by the following method[8]:

If we enclose the system in a box of linear dimension L, then the position eigenstates $|q\rangle$ and the momentum eigenstates $|p\rangle$ are characterized[9] by[10]

$$\hat{q}\,|q\rangle = q\,|q\rangle \ , \ \langle q|q'\rangle = \delta(q - q') \ , \ \int dq\,|q\rangle\,\langle q| = \mathbb{1} \ , \tag{B.1a}$$

$$\hat{p}\,|p\rangle = p\,|p\rangle \ , \ \langle p|p'\rangle = \delta_{pp'} \ , \ \sum_p |p\rangle\,\langle p| = \mathbb{1} \ ,$$

$$\langle q|p\rangle = \frac{e^{ipq/\hbar}}{\sqrt{L}} \ , \quad \text{with } p = \frac{2\pi\hbar}{L}\,n \ . \tag{B.1b}$$

We associate with each operator $\hat{A}$ a function[9] $A(p, q)$,

$$A(p, q) \equiv \langle p|\,\hat{A}\,|q\rangle\,\langle q|p\rangle\,L \ . \tag{B.2a}$$

These matrix elements are related to the classical quantities which correspond to the operators. For example, an operator of the form $\hat{A} = f(\hat{p})g(\hat{q})$ is associated with the function

$$A(p, q) = \langle p|\,f(\hat{p})\,g(\hat{q})\,|q\rangle\,\langle q|p\rangle\,L = f(p)g(p) \ . \tag{B.2b}$$

---

[8] E. Wigner, Phys. Rev. **40**, 749 (1932); G.E. Uhlenbeck, L. Gropper, Phys. Rev. **41**, 79 (1932); J.G. Kirkwood, Phys. Rev. **44**, 31 (1933) and **45**, 116 (1934).
[9] For clarity, in this section operators are denoted exceptionally by a 'hat'.
[10] QM I, Chap. 8

The Hamiltonian

$$\hat{H} = H(\hat{p}, \hat{q}) = \frac{\hat{p}^2}{2m} + V(\hat{q}) \tag{B.3a}$$

is thus associated with the classical Hamilton function

$$H(p, q) = \frac{p^2}{2m} + V(q) . \tag{B.3b}$$

The commutator of two operators is associated with the function

$$\langle p| [\hat{A}, \hat{B}] |q\rangle \langle q|p\rangle \ L$$

$$= L \int dq' \sum_{p'} \left\{ \langle p| \hat{A} |q'\rangle \langle q'|p'\rangle \langle p'| \hat{B} |q\rangle - \langle p| \hat{B} |q'\rangle \langle q'|p'\rangle \langle p'| \hat{A} |q\rangle \right\}$$

$$\times \langle q|p\rangle$$

$$= L \int dq' \sum_{p'} \left(A(p, q')B(p', q) - B(p, q')A(p', q) \right)$$

$$\times \langle p|q'\rangle \langle p'|q\rangle \langle q'|p'\rangle \langle q|p\rangle$$

$$\tag{B.3c}$$

according to (B.2b), where $\langle p|q'\rangle \langle q'|p\rangle = \frac{1}{L}$ was used. We note at this point that for the limiting case of large L relevant to thermodynamics, the summation

$$\sum_p \leftrightarrow \frac{L}{2\pi\hbar} \int dp \tag{B.3d}$$

can be replaced by an integral and *vice versa*. The expression in round brackets in (B.3c) can be expanded in $(q' - q)$ and $(p' - p)$:

$$A(p, q')B(p', q) - B(p, q)A(p', q) =$$

$$\left(A(p, q) + (q' - q)\frac{\partial A}{\partial q} + \frac{1}{2}(q' - q)^2 \frac{\partial^2 A}{\partial q^2} + \dots \right)$$

$$\times \left(B(p, q) + (p' - p)\frac{\partial B}{\partial p} + \frac{1}{2}(p' - p)^2 \frac{\partial^2 B}{\partial p^2} + \dots \right)$$

$$- \left(B(p, q) + (q' - q)\frac{\partial B}{\partial q} + \frac{1}{2}(q' - q)^2 \frac{\partial^2 B}{\partial q^2} + \dots \right)$$

$$\times \left(A(p, q) + (p' - p)\frac{\partial A}{\partial p} + \frac{1}{2}(p' - p)^2 \frac{\partial^2 A}{\partial p^2} + \dots \right) . \tag{B.3e}$$

The zero-order terms cancel, and pure powers of $(q' - q)$ or $(p' - p)$ yield zero on insertion into (B.3c), since the p'-summation and the q'-integration lead to a δ-function. The remaining terms up to second order are

$$\langle p| [\hat{A}, \hat{B}] |q\rangle \langle q|p\rangle L =$$

$$= L \int dq' \sum_{p'} (q' - q)(p' - p)\frac{\partial(A, B)}{\partial(q, p)} \langle p|q'\rangle \langle p'|q\rangle \langle q'|p'\rangle \langle q|p\rangle$$

$$= L \langle p| (\hat{q} - q)(\hat{p} - p) |q\rangle \frac{\partial(A, B)}{\partial(q, p)} \langle q|p\rangle \tag{B.3f}$$

$$= L \, i\hbar \frac{\partial(A, B)}{\partial(q, p)} | \langle q|p\rangle |^2 = \frac{\hbar}{i} \frac{\partial(A, B)}{\partial(q, p)} \, ,$$

where the scalar product (B.1b) and Eq. (B.1a) have been inserted. For higher powers of $(\hat{q} - q)$ and $(\hat{p} - p)$, double and multiple commutators of $\hat{q}$ and $\hat{p}$ occur, so that, expressed in terms of Poisson brackets (Footnote 4, Sect. 1.3), we finally obtain

$$\langle p| [\hat{A}, \hat{B}] |q\rangle \langle q|p\rangle L = \frac{\hbar}{i}\{A, B\} + \mathcal{O}(\hbar^2) \, . \tag{B.4}$$

Application of the definition (B.2a) and Eq. (B.2b) to the partition function leads to

$$Z = \mathrm{Tr}\, e^{-\beta \hat{H}} = \sum_p \langle p| e^{-\beta H(\hat{p}, \hat{q})} |p\rangle = \sum_p \int dq \, \langle p| e^{-\beta H(\hat{p}, \hat{q})} |q\rangle \langle q|p\rangle$$

$$= \sum_p \int dq \, \langle p| \left(e^{-\beta K(\hat{p})} e^{-\beta V(\hat{q})} + \mathcal{O}(\hbar) \right) |q\rangle \langle q|p\rangle \tag{B.5}$$

$$= \frac{1}{L} \sum_p \int dq \, e^{-\beta H(p, q)} + \mathcal{O}(\hbar) = \int \frac{dp \, dq}{2\pi\hbar} \, e^{-\beta H(p, q)} + \mathcal{O}(\hbar) \, .$$

Z is thus – apart from terms of the order of $\hbar$, which result from commutators between $K(\hat{p})$ und $V(\hat{q})$ – equal to the classical partition integral. In (B.5), $\hat{K} \equiv K(\hat{p})$ is the operator for the kinetic energy.

Starting from the density matrix $\hat{\rho}$, we define the *Wigner function*:

$$\rho(p, q) = \frac{L}{2\pi\hbar} \langle p|q\rangle \langle q| \hat{\rho} |p\rangle \, . \tag{B.6}$$

Given the normalization of the momentum eigenfunctions, the factor $\frac{L}{2\pi\hbar}$ is introduced in order to guarantee that the Wigner function is independent of L for large L.

The meaning of the Wigner function can be seen from its two important properties:

(1)   normalization : $\displaystyle\int dq \int dp \, \rho(p, q) = \int dq \sum_p \langle p|q\rangle \langle q| \hat{\rho} |p\rangle$ $$\tag{B.7}$$

$$= \mathrm{Tr}\, \hat{\rho} = 1 \, .$$

Here, the completeness relation for the position eigenstates, (B.1a), was used.

(2)   mean values : $\displaystyle\int dq \int dp\, \rho(p,q)\, A(p,q)$

$$= \int dq \sum_p \frac{L}{2\pi\hbar} \langle p|q\rangle \langle q|\,\hat{\rho}\,|p\rangle \langle p|\,\hat{A}\,|q\rangle \langle q|p\rangle \qquad \text{(B.8)}$$

$$= \int dq \sum_p \langle q|\,\hat{\rho}\,|p\rangle \langle p|\,\hat{A}\,|q\rangle = \mathrm{Tr}\,(\hat{\rho}\hat{A}) \,.$$

Following the second equals sign, $\langle p|q\rangle \langle q|p\rangle = \frac{1}{L}$ and Eq. (B.3d) were used.
For the canonical ensemble, we find using (B.5)

$$
\begin{aligned}
\rho(p,q) &= \frac{L}{2\pi\hbar} \langle p|q\rangle \langle q|\, \frac{e^{-\beta\hat{H}}}{Z}\,|p\rangle \\
&= \frac{L}{2\pi\hbar} \langle p|q\rangle \langle q|\, \left(e^{-\beta\hat{K}}e^{-\beta V} + \mathcal{O}(\hbar)\right)|p\rangle\, \frac{1}{Z} \qquad \text{(B.9)}\\
&= \frac{L}{2\pi\hbar}\,|\langle p|q\rangle|^2\, \frac{e^{-\beta H(p,q)}}{Z} + \mathcal{O}(\hbar) = \frac{e^{-\beta H(p,q)}}{2\pi\hbar Z} + \mathcal{O}(\hbar)
\end{aligned}
$$

and

$$
\begin{aligned}
\langle \hat{A}\rangle &= \frac{\frac{1}{L}\sum_p \int dq\, e^{-\beta H(p,q)}\, A(p,q)}{\frac{1}{L}\sum_p \int dq\, e^{-\beta H(p,q)}} + \mathcal{O}(\hbar) \\
&= \frac{\int \frac{dp\,dq}{2\pi\hbar}\, e^{-\beta H(p,q)}\, A(p,q)}{\int \frac{dp\,dq}{2\pi\hbar}\, e^{-\beta H(p,q)}} + \mathcal{O}(\hbar) \,.
\end{aligned}
\qquad \text{(B.10)}
$$

The generalization to N particles in three dimensions gives:

$$\hat{H} = \sum_{i=1}^{N} \frac{\hat{\mathbf{p}}_i^2}{2m} + V(\hat{\mathbf{q}}_1,\dots,\hat{\mathbf{q}}_N) \,. \qquad \text{(B.11)}$$

We introduce the following abbreviations for many-body states:

$$|q\rangle \equiv |\mathbf{q}_1\rangle \dots |\mathbf{q}_N\rangle \,, \quad |p\rangle \equiv |\mathbf{p}_1\rangle \dots |\mathbf{p}_N\rangle \,, \qquad \text{(B.12a)}$$

$$\langle p|p'\rangle = \delta_{pp'} \,, \quad \langle q|p\rangle = \frac{e^{ipq/\hbar}}{L^{3N/2}} \,, \quad \sum_p |p\rangle \langle p| = \mathbb{1} \,. \qquad \text{(B.12b)}$$

Applying periodic boundary conditions, the $\mathbf{p}_i$ take on the values

$$\mathbf{p}_i = \frac{L}{2\pi\hbar}(n_1, n_2, n_3)$$

with integer numbers n_i.

The many-body states which occur in Nature are either symmetric
(bosons) or antisymmetric (fermions):

$$|p\rangle_s = \frac{1}{\sqrt{N!}} \sum_P (\pm 1)^P P |p\rangle . \tag{B.13}$$

The index s here stands in general for "symmetrization", and includes symmetrical states (upper sign) and antisymmetrical states (lower sign). This sum includes $N!$ terms. It runs over all the permutations P of N objects. For fermions, $(-1)^P = 1$ for even permutations and $(-1)^P = -1$ for odd permutations, while for bosons, $(+1)^P = 1$ always holds. In the case of fermions, all of the $\mathbf{p}_i$ in (B.13) must therefore be different from one another in agreement with the Pauli principle. In the case of bosons, the same $\mathbf{p}_i$ can occur; therefore, these states are in general not normalized: a normalized state is given by

$$|p\rangle_{sn} = \frac{1}{\sqrt{n_1! \, n_2! \ldots}} |p\rangle_s , \tag{B.14}$$

where n_i is the number of particles with momentum $\mathbf{p}_i$. We have

$$\operatorname{Tr} \hat{A} = \sum_{\mathbf{p}_1,\ldots,\mathbf{p}_N}{}' {}_{sn}\langle p| \hat{A} |p\rangle_{sn} = \sum_{\mathbf{p}_1,\ldots,\mathbf{p}_N} \frac{n_1! \, n_2! \ldots}{N!} {}_{sn}\langle p| \hat{A} |p\rangle_{sn}$$
$$= \sum_{\mathbf{p}_1,\ldots,\mathbf{p}_N} \frac{1}{N!} {}_s\langle p| \hat{A} |p\rangle_s . \tag{B.15}$$

The prime on the sum indicates that it is limited to different states. For example, $\mathbf{p}_1\mathbf{p}_2\ldots$ and $\mathbf{p}_2\mathbf{p}_1\ldots$ would give the same state. Rewriting the partition function in terms of the correspondence (B.2b) yields

$$Z = \operatorname{Tr} e^{-\beta H} = \frac{1}{N!} \sum_{\{\mathbf{p}_i\}} {}_s\langle p| e^{-\beta \hat{H}} |p\rangle_s$$
$$= \frac{1}{N!} \int d^{3N}q \sum_{\{\mathbf{p}_i\}} {}_s\langle p| e^{-\beta \hat{H}} |q\rangle \langle q|p\rangle_s \tag{B.16}$$
$$= \frac{1}{N!} \left(\frac{V}{(2\pi\hbar)^3} \right)^N \int d^{3N}p \int d^{3N}q \, e^{-\beta H(p,q)} |\langle q|p\rangle_s|^2 + \mathcal{O}(\hbar) .$$

The last factor in the integrand has the form $|\langle q|p\rangle_s| = V^{-N}(1 + f(p,q))$, where the first term leads to the partition integral

$$Z = \int \frac{d^{3N}p \, d^{3N}q}{N! \, (2\pi\hbar)^{3N}} \, e^{-\beta H(p,q)} + \mathcal{O}(\hbar) . \tag{B.16'}$$

Remarks:

(i) In (B.16), the rearrangement ${}_s\langle p| e^{-\beta \hat{H}} |p\rangle_s = \int d^{3N}q \, {}_s\langle p| e^{-\beta \hat{K}} e^{-\beta V} |q\rangle$ $\times \langle q|p\rangle_s + \mathcal{O}(\hbar) = \int d^{3N}q \, e^{-\beta H(p,q)} |\langle q|p\rangle_s|^2 + \mathcal{O}(\hbar)$ was employed, where the symmetry of $\hat{H}$ under particle exchange enters.

(ii) The quantity $|\langle q|p\rangle_s|^2 = V^{-N}(1 + f(p,q))$ contains, in addition to the leading term V^{-N} in the classical limit, also p- and q-dependent terms. The corrections due to symmetrization yield contributions of the order of $\hbar^3$. Cf. the ideal gas and Sect. B.2.

(iii) Analogously (to B.16), one can show that the distribution function is

$$\rho(p,q) = \frac{e^{-\beta H(p,q)}}{Z(2\pi\hbar)^{3N} N!}. \tag{B.17}$$

We have thus shown that, neglecting terms of the order of $\hbar$, which result from the non-commutativity of the kinetic and the potential energies and the symmetrization of the wave functions, the *classical partition integral* (B.16') is obtained.

The classical partition integral (B.16') shows some features which indicate the underlying quantum nature: the factors $1/N!$ and $(2\pi\hbar)^{-3N}$. The first of these expresses the fact that states of identical particles which are converted into one another by particle exchange must be counted only once. This factor makes the thermodynamic potentials extensive and eliminates the Gibbs paradox which we discuss following Eq. (2.2.3). The factor $(2\pi\hbar)^{-3N}$ renders the partition integral dimensionless and has the intuitively clear interpretation that in phase space, each volume element $(2\pi\hbar)^{3N}$ corresponds to one state, in agreement with the uncertainty relation.

B.2 Calculation of the Quantum-Mechanical Corrections

We now come to the calculation of the quantum-mechanical corrections to the classical thermodynamic quantities. These arise from two sources:

a) The symmetrization of the wave function

b) the noncommutativity of $\hat{K}$ and V.

We will investigate these effects separately; their combination yields corrections of higher order in $\hbar$.

a) We first calculate the quantity $|\langle q|p_s\rangle|^2$, which occurs in the second line of (B.16), inserting Eq. (B.13):

$$\begin{aligned}
|\langle q|p\rangle_s|^2 &= \frac{1}{N!} \sum_P \sum_{P'} (-1)^P (-1)^{P'} \langle q| P' |p\rangle \langle q| P |p\rangle^* \\
&= \frac{1}{N!} \sum_P \sum_{P'} (-1)^P (-1)^{P'} \langle P'q|p\rangle \langle Pq|p\rangle^* \\
&\stackrel{\wedge}{=} \frac{1}{N!} \sum_P \sum_{P'} (-1)^P (-1)^{P'} \langle q|p\rangle \langle PP'^{-1}q|p\rangle^* \\
&= \sum_P (-1)^P \langle q|p\rangle \langle Pq|p\rangle^* \\
&= \frac{1}{V^N} \sum_P e^{\frac{i}{\hbar}(\mathbf{p_1}\cdot(\mathbf{q_1}-P\mathbf{q_1})+\dots+\mathbf{p_N}\cdot(\mathbf{q_N}-P\mathbf{q_N}))}.
\end{aligned} \tag{B.18}$$

Here, in the second line, we have used the fact that the permutation of the particles in configuration space is equivalent to the permutation of their spatial coordinates. In the third line we have made use of the fact that we can rename the coordinates within the integral $\int d^{3N}q$ which occurs in (B.16), replacing $P'q$ by q. In the next-to-last line, we have used the general property of groups that for any fixed P', the elements PP'^{-1} run through all the elements of the group. Finally, in the last line, the explicit form of the momentum eigenfunctions in their configuration-space representation was inserted.

Inserting the final result of Eq. (B.18) into (B.16), we can express each of the momentum integrals in terms of

$$\int d^3p \, e^{-\frac{\beta p^2}{2m} + i px} = \int d^3p \, e^{-\frac{\beta p^2}{2m}} f(x) , \tag{B.19}$$

with

$$f(x) = e^{-\frac{\pi x^2}{\lambda^2}} , \tag{B.20}$$

where $\lambda = \frac{2\pi\hbar}{\sqrt{2\pi\hbar mkT}}$ [Eq. (2.7.20)] is the thermal wavelength. Then we find for the partition function, without quantum corrections which result from non-commutatitivity,

$$Z = \int \frac{d^{3N}q \, d^{3N}p}{N!(2\pi\hbar)^{3N}} e^{-\beta H(p,q)} \sum_P (-1)^P f(\mathbf{q}_1 - P\mathbf{q}_1) \ldots f(\mathbf{q}_N - P\mathbf{q}_N) . \tag{B.21}$$

The sum over the $N!$ permutations contains the contribution $f(0)^N = 1$ for the unit element $P = 1$; for transpositions (in which only pairs of particles i and j are exchanged), it contains the contribution $(f(\mathbf{q}_i - \mathbf{q}_j))^2$, etc. Arranging the terms according to increasing number of exchanges, we have

$$\sum_P (-1)^P f(\mathbf{q}_1 - P\mathbf{q}_1) \cdots f(\mathbf{q}_N - P\mathbf{q}_N) =$$
$$= 1 \pm \sum_{i<j} \left(f(\mathbf{q}_i - \mathbf{q}_j) \right)^2 + \sum_{ijk} f(\mathbf{q}_i - \mathbf{q}_j) f(\mathbf{q}_j - \mathbf{q}_k) f(\mathbf{q}_k - \mathbf{q}_i) \pm \ldots .$$
$$\tag{B.22}$$

The upper sign refers to bosons, the lower to fermions. For sufficiently high temperatures, so that the average spacing between the particles obeys the inequality (v is the specific volume)

$$v^{1/3} \gg \lambda , \tag{B.23}$$

we find that $f(\mathbf{q}_i - \mathbf{q}_j)$ is vanishingly small for $|\mathbf{q}_i - \mathbf{q}_j| \gg \lambda$, and therefore only the first term in (B.22) is significant; according to the preceding section, it just yields the classical partition integral, (B.16).

The more factors f that are present in (B.22), the stronger the constraints on the spatial integration region in (B.16). The leading quantum correction therefore comes from the second sum in (B.22), which we can rewrite in the following approximate way:

$$1 \pm \sum_{i<j} (f(\mathbf{q}_i - \mathbf{q}_j))^2 \approx \prod_{i<j} \left(1 \pm (f(\mathbf{q}_i - \mathbf{q}_j))^2\right) = e^{-\beta \sum_{i<j} \tilde{v}_i(\mathbf{q}_i - \mathbf{q}_j)} . \quad (B.24)$$

Here, the effective potential

$$\tilde{v}_i(\mathbf{q}_i - \mathbf{q}_j) = -kT \log\left(1 \pm e^{-2\pi |\mathbf{q}_i - \mathbf{q}_j|/\lambda^2}\right) \quad (B.25)$$

is attractive for bosons and repulsive for fermions. This effective potential arises from the symmetry properties of the wave function and not from any microscopic mutual interaction of the particles. It permits us to take the leading quantum correction into account within the classical partition integral. For the ideal gas, these quantum corrections lead to contributions of the order of $\hbar^3$ in the thermodynamic quantities, as we have seen in Sect. 4.2.

b) The exact quantum-mechanical expression for the partition function is given by

$$\begin{aligned}
Z &= \frac{1}{N!} \sum_{\{\mathbf{p}_i\}} {}_s \langle p| e^{-\beta \hat{H}} |p\rangle_s \\
&= \frac{1}{N!} \left(\frac{V}{(2\pi\hbar)^3}\right)^N \int d^{3N}p \int d^{3N}q \; {}_s \langle p| e^{-\beta \hat{H}} |q\rangle \langle q|p\rangle_s .
\end{aligned} \quad (B.26)$$

If we neglect exchange effects (symmetrization of the wave function), we obtain

$$\begin{aligned}
Z &= \frac{1}{N!} \left(\frac{V}{(2\pi\hbar)^3}\right)^N \int d^{3N}p \int d^{3N}q \quad \langle p| e^{-\beta \hat{H}} |q\rangle \langle q|p\rangle \\
&= \frac{1}{N!} \left(\frac{1}{(2\pi\hbar)^3}\right)^N \int d^{3N}p \int d^{3N}q \, I .
\end{aligned} \quad (B.27)$$

To compute the integrands which occur in this expression, we introduce the following relation, initially for a single particle,

$$I = \langle p| e^{-\beta \hat{H}} |q\rangle \langle q|p\rangle V = e^{ipq/\hbar} e^{-\beta \hat{H}} e^{-ipq/\hbar} . \quad (B.28)$$

After the last equals sign and in the following, $\hat{H}$ denotes the Hamiltonian in the coordinate representation. To calculate I, we derive a differential equation for I using the Baker–Hausdorff formula:

$$\frac{\partial I}{\partial \beta} = -e^{ipq/\hbar} \hat{H} e^{-\beta \hat{H}} e^{-ipq/\hbar} = -e^{ipq/\hbar} \hat{H} e^{-ipq/\hbar} I$$

$$= -\left(\hat{H} - i\left[-\frac{pq}{\hbar}, \hat{H}\right] - \frac{1}{2\hbar^2}[pq, [pq, \hat{H}]] + \dots\right) I \qquad (B.29)$$

$$= -\left[\hat{H} - \frac{\hbar^2}{2m}\left(-\frac{2i}{\hbar} p \frac{\partial}{\partial q} - \frac{p^2}{\hbar^2}\right) I\right] .$$

The higher-order commutators (indicated by dots) vanish, so that

$$\frac{\partial I}{\partial \beta} = \left[-H(p,q) + \frac{\hbar^2}{2m}\left(-\frac{2i}{\hbar} p \frac{\partial}{\partial q} + \frac{\partial^2}{\partial q^2}\right)\right] I , \qquad (B.29')$$

where $H(p,q)$ is the classical Hamilton function. To solve this differential equation, we use the ansatz:

$$\chi = e^{\beta H(p,q)} I . \qquad (B.30)$$

We find the following differential equation for χ from (B.29'):

$$\frac{\partial \chi}{\partial \beta} = H(p,q)\chi + e^{\beta H(p,q)} \frac{\partial I}{\partial \beta} = e^{\beta H(p,q)} \frac{\hbar^2}{2m}\left(\frac{2i}{\hbar} p \frac{\partial}{\partial q} + \frac{\partial^2}{\partial q^2}\right) I$$

$$= e^{\beta H(p,q)} \frac{\hbar^2}{2m}\left(\frac{2i}{\hbar} p \frac{\partial}{\partial q} + \frac{\partial^2}{\partial q^2}\right) e^{\beta H(p,q)} \chi$$

$$= \frac{\hbar^2 \beta}{2m}\left[\frac{2ip}{\hbar} \frac{\partial V}{\partial q} - \frac{2ip}{\hbar\beta} \frac{\partial}{\partial q} - \frac{\partial^2 V}{\partial q^2} + \beta \left(\frac{\partial V}{\partial q}\right)^2 \qquad (B.31)\right.$$

$$\left. - 2\frac{\partial V}{\partial q} \frac{\partial}{\partial q} + \beta^{-1} \frac{\partial^2}{\partial q^2}\right] \chi .$$

Transferring to a many-body system with the coordinates and momenta q_i and p_i yields

$$\frac{\partial \chi}{\partial \beta} = \sum_i \frac{\hbar^2 \beta}{2m_i}\left[\frac{2ip_i}{\hbar} \frac{\partial V}{\partial q_i} - \frac{2ip_i}{\hbar\beta} \frac{\partial}{\partial q_i} - \frac{\partial^2 V}{\partial q_i^2}\right.$$

$$\left. + \beta \left(\frac{\partial V}{\partial q_i}\right)^2 - 2\frac{\partial V}{\partial q_i} \frac{\partial}{\partial q_i} + \beta^{-1} \frac{\partial^2}{\partial q_i^2}\right] \chi . \qquad (B.31')$$

The solution of this equation is obtained with the aid of a power series expansion in $\hbar$:

$$\chi = 1 + \hbar \chi_1 + \hbar^2 \chi_2 + \mathcal{O}(\hbar^3) . \qquad (B.32)$$

Because of (B.28) and (B.30), χ must obey the boundary condition $\chi = 1$ for $\beta = 0$. Inserting this ansatz into (B.31'), we obtain

$$\frac{\partial \chi_1}{\partial \beta} = \pm i\beta \sum_i \frac{p_i}{m_i} \frac{\partial V}{\partial q_i} \tag{B.33a}$$

and

$$\frac{\partial \chi_2}{\partial \beta} = \sum_i \frac{1}{2m_i} \left[-2i\beta p_i \frac{\partial V}{\partial q_i} \chi_1 + 2i p_i \frac{\partial \chi_1}{\partial q_i} - \beta \frac{\partial^2 V}{\partial q_i^2} + \beta^2 \left(\frac{\partial V}{\partial q_i}\right)^2 \right] . \tag{B.33b}$$

From this, it follows that

$$\chi_1 = -\frac{i\beta^2}{2} \sum_i \frac{p_i}{m_i} \frac{\partial V}{\partial q_i} \tag{B.34a}$$

$$\chi_2 = \pm \frac{\beta^4}{8} \left(\sum_i \frac{p_i}{m_i} \frac{\partial V}{\partial q_i} \right)^2 + \frac{\beta^3}{6} \sum_i \sum_k \frac{p_i}{m_i} \frac{p_k}{m_k} \frac{\partial^2 V}{\partial q_i \partial q_k}$$

$$+ \frac{\beta^3}{6} \sum_i \frac{1}{m_i} \left(\frac{\partial V}{\partial q_i}\right)^2 - \frac{\beta^2}{4} \sum_i \frac{1}{m_i} \frac{\partial^2 V}{\partial q_i^2} . \tag{B.34b}$$

Inserting (B.30) and (B.27), we finally obtain the partition function

$$Z = \int \frac{d^{3N}q \, d^{3N}p}{(2\pi\hbar)^{3N} N!} e^{-\beta H(p,q)} (1 + \hbar \chi_1 + \hbar^2 \chi_2) . \tag{B.35}$$

The term of order $\mathcal{O}(\hbar)$ vanishes, since χ_1 is an odd function of p_1, so that the remaining expression is

$$Z = \left(1 + \hbar^2 \langle \chi_2 \rangle_{cl}\right) Z_{cl} . \tag{B.36}$$

Here, $\langle\,\rangle_{cl}$ refers to the average value with the classical distribution function, and Z_{cl} is the classical partition function. From it, we thus obtain for the free energy

$$F = -\frac{1}{\beta} \log Z = F_{cl} - \frac{1}{\beta} \log\left(1 + \hbar^2 \langle \chi_2 \rangle_{cl}\right) \approx F_{cl} - \frac{\hbar^2}{\beta} \langle \chi_2 \rangle_{cl} . \tag{B.37}$$

With

$$\langle p_i p_k \rangle_{cl} = \frac{m}{\beta} \delta_{ik} \tag{B.38}$$

and

$$\left\langle \frac{\partial^2 V}{\partial q_i^2} \right\rangle_{cl} = \beta \left\langle \left(\frac{\partial V}{\partial q_i}\right)^2 \right\rangle$$

(proof via partial integration), it follows that

$$F = F_{cl} + \frac{\hbar^2}{24m(kT)^2} \sum_i \left\langle \left(\frac{\partial V}{\partial q_i}\right)^2 \right\rangle_{cl} . \tag{B.39}$$

The classical approximation is therefore best at high T and large m.

Remark: Using the thermal wavelength $\lambda = 2\pi\hbar/\sqrt{2\pi m k T}$ and the length l which characterizes the spatial variation of the potential (range of the interaction potentials), the correction in Eq. (B.39) becomes $\frac{\lambda^2}{l^2}\frac{V^2}{kT}$. This gives as a condition for the validity of the classical approximation

$$\lambda \ll l \qquad \text{(from the non-commutativity of } \hat{K} \text{ and } \hat{V}) \qquad \text{(B.39a)}$$

and, according to Eq. (B.23)

$$\lambda \ll \left(\frac{V}{N}\right)^{1/3} \qquad \text{(from symmetrization of the wave function)}. \qquad \text{(B.39b)}$$

Rearranging Eq. (2.7.20), one gets

$$T[\mathrm{K}] = \frac{5 \times 10^{-38}}{\lambda^2[\mathrm{cm}^2]m[\mathrm{g}]} = \frac{5.56 \times 10^5}{\lambda^2[\mathrm{\mathring{A}}^2]m[m_\mathrm{e}]} .$$

For *electrons* in solids, we have $\left(\frac{V}{N}\right)^{1/3} \approx 1\mathrm{\mathring{A}}$, so that even at a temperature of $T = 5.5 \times 10^5$ K, their behavior remains nonclassical.

For a gas with the mass number A: $m = A \cdot m_\mathrm{p}$, $\left(\frac{V}{N}\right)^{1/3} \approx 10^{-7}$cm, $T \approx \frac{3}{A}$ K

B.3 Quantum Corrections to the Second Virial Coefficient $B(T)$

B.31 Quantum Corrections Due to Exchange Effects

We neglect the interactions; however, the second virial coefficient from Eq. (5.3.7)

$$B(T) = \left(Z_2 - \frac{1}{2}Z_1^2\right)\frac{V}{Z_1^2} \qquad \text{(B.40)}$$

is still nonzero due to exchange effects. A two-particle eigenstate has the form

$$|p_1, p_2\rangle = \frac{1}{\sqrt{2!}}\left(|p_1\rangle\,|p_2\rangle \pm |p_2\rangle\,|p_1\rangle\right) \qquad \text{for } p_1 \neq p_2$$

$$\text{(B.41)}$$

$$|p_1, p_2\rangle = \begin{cases} |p_1\rangle\,|p_1\rangle & \text{bosons} \\ & \text{for } p_1 = p_2 \\ 0 & \text{fermions} . \end{cases}$$

The partition function for two non-interacting particles is

$$Z_2 = \mathrm{Tr}\,e^{-(\hat{p}_1^2 + \hat{p}_2^2)/2mkT} = \frac{1}{2}\sum_{\substack{p_1, p_2 \\ p_1 \neq p_2}} e^{-(p_1^2 + p_2^2)/2mkT} + \begin{cases} \sum_p e^{-p^2/mkT} \\ 0 \end{cases}$$

$$= \frac{1}{2}\sum_{p_1}\sum_{p_2} e^{-(p_1^2 + p_2^2)/2mkT} \pm \frac{1}{2}\sum_p e^{-p^2/mkT}$$

$$= \frac{1}{2}Z_1^2 \pm \frac{1}{2}\sum_p e^{-p^2/mkT} \qquad \text{for } \begin{cases} \text{bosons} \\ \text{fermions} \end{cases}. \qquad \text{(B.42)}$$

From this[11], we find for the second virial coefficient (5.3.7):

$$B(T) = \mp \frac{\lambda^6}{2V} \sum_p e^{-p^2/mkT} = \mp \frac{\lambda^3}{2^{5/2}}$$

$$= \mp \frac{1}{2} \left(\frac{\pi\hbar^2}{mkT} \right)^{3/2} \quad \text{for} \quad \begin{cases} \text{bosons} \\ \text{fermions} \end{cases} \quad . \quad (B.43)$$

B.32 Quantum-Mechanical Corrections to $B(T)$ Due to Interactions

In the semiclassical limit (unsymmetrized wave functions), from Eq. (B.35) we obtain for the partition function of two particles

$$Z_2 = \frac{1}{2} \left(\frac{1}{\lambda^3} \right)^2 \int d^3x_1\, d^3x_2\, e^{-v_{12}(x_1-x_2)/kT} \left(1 + \underbrace{\hbar\chi_1}_{=0} + \hbar^2\chi_2 \right). \quad (B.44)$$

This leads to the following expression for the second virial coefficient ((5.3.7), (B.40)):

$$B = \frac{1}{2} \left(\frac{1}{V} \int d^3x_1\, d^3x_2 \left(e^{-v_{12}(x_1-x_2)/kT} (1 + \hbar^2\chi_2) - 1 \right) \right). \quad (B.45)$$

The quantum correction is therefore given by

$$B_{qm} = \int d^3y\, e^{-v(\mathbf{y})/kT} \frac{1}{kT} \left(\frac{\partial v}{\partial \mathbf{y}} \right)^2 \frac{\hbar^2}{24m(kT)^2}$$

$$= \frac{\hbar^2\pi}{6m(kT)^3} \int_0^\infty dr\, r^2 e^{-v(r)/kT} \left(\frac{\partial v}{\partial r} \right)^2, \quad (B.46)$$

where in the second line we have assumed a central potential. This quantum correction adds to the classical value of B; it is always positive. The exchange corrections (B.43) are of the order $\mathcal{O}(\hbar^3)$. The lowest-order quantum corrections, i.e. (B.46), are of order $\hbar^2$. At low temperatures, these quantum effects (due to non-commuting $\hat{V}$ and $\hat{K}$) become important. The contribution from symmetrization is relatively small.

B.33 The Second Virial Coefficient and the Scattering Phase

One can also represent the second virial coefficient in terms of the phase shift of the interaction potential. The starting point is the formula for the virial

---

[11] $Z_1 \equiv \sum_p e^{-p^2/2mkT} = \frac{V}{\lambda^3}$
We do not take the spin degeneracy factor $g = 2S + 1$ into account here.

coefficient, Eq. (5.3.7)

$$B = -\left(\frac{Z_2}{Z_1^2} - \frac{1}{2}\right) V .$$ (B.47)

The interaction does not appear in the partition function for a single particle

$$Z_1 = \sum_{\mathbf{p}} e^{-\frac{\mathbf{p}^2}{2mkT}} = \frac{V}{(2\pi\hbar)^3} \int d^3p\, e^{-\frac{p^2}{2mkT}} = \frac{V}{\lambda^3} .$$ (B.48)

The Hamiltonian for two particles is given by

$$\hat{H} = \frac{\mathbf{p}_1^2 + \mathbf{p}_2^2}{2m} + V(\mathbf{x}_1 - \mathbf{x}_2)$$ (B.49)

and, introducing coordinates for the center of mass (CM) and the relative position (r):

$$\mathbf{x}_{\mathrm{CM}} = \frac{1}{2}(\mathbf{x}_1 + \mathbf{x}_2) , \qquad \mathbf{x}_r = \mathbf{x}_2 - \mathbf{x}_1 ,$$ (B.50)

it can be written as

$$\hat{H} = \frac{\mathbf{p}_{\mathrm{CM}}^2}{4m} + \frac{\mathbf{p}_r^2}{m} + V(\mathbf{x}_r) .$$ (B.51)

Then the partition function for two particles becomes

$$Z_2 = \mathrm{Tr}_{\mathrm{CM}}\, e^{-\frac{\mathbf{p}_{\mathrm{CM}}^2}{4mkT}}\, \mathrm{Tr}_r\, e^{-\left(\frac{\mathbf{p}_r^2}{m}+V(\mathbf{x}_r)\right)/kT} = 2^{3/2}\frac{V}{\lambda^3} \sum_n e^{-\frac{\varepsilon_n}{kT}} .$$ (B.52)

In this expression, ε_n denotes the energy levels of the two-particle system in relative coordinates taking into account the different symmetries of bosons and fermions. It leads to

$$B = -\left(2^{3/2}\lambda^3 \sum_n e^{-\varepsilon_n/kT} - \frac{V}{2}\right) .$$ (B.53)

We now remind the reader that for non-interacting particles, (B.43) gives for the second virial coefficient

$$B^{(0)} = -\left(2^{3/2}\lambda^3 \sum_n e^{-\varepsilon_n^0/kT} - \frac{V}{2}\right) = \mp 2^{-5/2}\lambda^3 \quad \begin{cases} \text{bosons} \\ \text{fermions} \end{cases} . $$ (B.54)

The change in the second virial coefficient due to the interactions of the particles is thus given by

$$B(T) - B^0(T) = -2^{3/2}\lambda^3 \sum_n \left(e^{-\beta\varepsilon_n} - e^{-\beta\varepsilon_n^{(0)}}\right) .$$ (B.55)

The energy levels of the non-interacting system are

$$\varepsilon_n^{(0)} = \frac{\hbar^2 k^2}{m} , \tag{B.56a}$$

while in the interacting system, along with the continuum states of energy

$$\varepsilon_n = \frac{\hbar^2 k^2}{m} , \tag{B.56b}$$

bonding states of energy ε_B can also occur. The values of k are found from the boundary conditions and are different for the interacting system and for the free system, so that also different densities of states are obtained. The number of energy levels $g(k)dk$ in the interval $[k, k + dk]$ defines the density of states, $g(k)$. We thus find

$$B(T) - B^{(0)}(T)$$
$$= -2^{3/2}\lambda^3 \left[\sum_B e^{-\varepsilon_B/kT} + \int_0^\infty dk \left(g(k) - g^{(0)}(k) \right) e^{-\varepsilon_k/kT} \right] . \tag{B.57}$$

The change in the density of states which occurs here can be related to the derivative of the scattering phase. We assume that the potential $V(r)$ has rotational symmetry and consider the eigenstates of the relative part of the Hamiltonian. Then we can represent the wave functions for the free and the interacting problem in the form[12]

$$\psi_{klm}^{(0)}(\mathbf{x}) = A_{klm}^{(0)} Y_{lm}(\vartheta, \varphi) R_{kl}^{(0)}(r)$$
$$\psi_{klm}(\mathbf{x}) = A_{klm} Y_{lm}(\vartheta, \varphi) R_{kl}(r) . \tag{B.58}$$

The free radial functions are given in terms of spherical Bessel functions. The asymptotic forms for $r \to \infty$ are

$$R_{kl}^{(0)}(r) = \frac{1}{kr} \sin\left(kr + \frac{l\pi}{2}\right)$$
$$R_{kl}(r) = \frac{1}{kr} \sin\left(kr + \frac{l\pi}{2} + \delta_l(k)\right) \tag{B.59}$$

with the phase shifts known from scattering theory, $\delta_l(k)$. The allowed values of k are found from the boundary conditions

$$R_{kl}^{(0)}(R) = R_{kl}(R) = 0 \tag{B.60}$$

at a large radius R (which finally goes to infinity). From this it follows that

$$kR + \frac{l\pi}{2} = \pi n \quad \text{and} \quad kR + \frac{l\pi}{2} + \delta_l(k) = \pi n , \tag{B.61}$$

---

[12] QM I, Chap. 17

where $n = 0, 1, 2, \ldots$. The values of k therefore depend upon l. Neighboring values of k for fixed l differ by

$$\Delta k^{(0)} = \frac{\pi}{R} \quad \text{and} \quad \Delta k = \frac{\pi}{R + \frac{\partial \delta_l(k)}{\partial k}} . \tag{B.62}$$

We still must take into account the fact that every value of l occurs with a multiplicity of $(2l + 1)$. Since each interval Δk or $\Delta k^{(0)}$ contains a value of k, the densities of states are

$$g_l^{(0)}(k) = \frac{2l + 1}{\pi} R \quad \text{and} \quad g_l(k) = \frac{2l + 1}{\pi} \left[R + \frac{\partial \delta_l(k)}{\partial k} \right] . \tag{B.63}$$

From this the second virial coefficient follows:

$$B(T) - B^{(0)}(T)$$

$$= -2^{3/2} \lambda^3 \left\{ \sum_B e^{-\varepsilon_B/kT} + \frac{1}{\pi} \int\limits_0^\infty dk \sum_l{}' f_l (2l + 1) \frac{\partial \delta_l(k)}{\partial k} e^{-\frac{\hbar^2 k^2}{mkT}} \right\} . \tag{B.64}$$

Now we need to determine the values of l which are allowed by the symmetry properties. For bosons, we have $\psi(-\mathbf{x}) = \psi(\mathbf{x})$, and for spin-1/2 fermions, $\psi(-\mathbf{x}) = \pm\psi(\mathbf{x})$, depending on whether a spin singlet or a triplet state is considered. For spin-0 bosons, we thus have $l = 0, 2, 4, \ldots$ and $f_l = 1$. For spin-1/2 fermions,

$$\begin{array}{llll} l = 0, 2, 4, \ldots & f_l = 1 & \text{(singlet)} \\ l = 1, 3, 5, \ldots & f_l = 3 & \text{(triplet)} . \end{array} \tag{B.65}$$

The change in the second virial coefficient is expressed in terms of the binding energies and the phase shifts. An important contribution to the k-integral comes from the resonances. Very sharp resonances have $\frac{\partial \delta_l(k)}{\partial k} = \pi \delta(k - k_0)$, and one obtains a similar contribution to that of the bonding states, however with positive energy. More generally, one can interpret the quantity

$$\frac{1}{\hbar} \frac{\partial \delta_l(k)}{\partial k} = \frac{\partial E}{\partial \hbar k} \frac{\partial \delta_l}{\partial E} = v \frac{\partial \delta_l}{\partial E} ,$$

as velocity times the dwell time[13] in the potential. The shorter the dwell time within the potential, the more nearly ideal is the interacting gas.

Literature:
S. K. Ma, *Statistical Mechanics*, Sect. 14.3, World Scientific, Singapore, 1985
E. Beth and G. E. Uhlenbeck, Physics **4**, 915 (1937)
A. Pais and G. E. Uhlenbeck, Phys. Rev. **116**, 250 (1959)

---

[13] See e.g. QM I, Eq. (3.126)

Problems for Appendix B.:

B.1 Carry out in detail the rearrangements which occur in Eq. (B.3f).
B.2 Carry out the rearrangement in Eq. (B.28).
B.3 Show that (B.29′) follows from (B.29).
B.4 Determine the behavior of the effective potentials $\tilde{v}(\mathbf{x})$ in Eq. (B.25) for small and large distances. Plot $\tilde{v}(\mathbf{x})$ for bosons and fermions.

C. The Perturbation Expansion

For the calculation of susceptibilities and in other problems in which the Hamiltonian $H = H_0 + V$ is composed of an "unperturbed" part and a perturbation V, we require the relation

$$e^{H_0+V} = e^{H_0} + \int_0^1 dt\, e^{tH_0} V e^{(1-t)H_0} + \mathcal{O}(V^2)\,. \tag{C.1}$$

To prove this relation, we introduce the definition

$$A(t) = e^{Ht} e^{-H_0 t}$$

and take its time derivative

$$\dot{A}(t) = e^{Ht}(H - H_0)e^{-H_0 t} = e^{Ht} V e^{-H_0 t}\,.$$

By integrating over time between 0 and 1,

$$A(1) - A(0) = e^{H}e^{-H_0} - 1 = \int_0^1 dt\, e^{Ht} V e^{-H_0 t}\,,$$

we obtain after multiplication by e^{H_0} the exact identity

$$e^{H} = e^{H_0} + \int_0^1 dt\, e^{Ht} V e^{(1-t)H_0}\,. \tag{C.2}$$

Expanding $e^{Ht} = e^{(H_0+V)t}$ in a power series, we obtain the assertion (C.1).

Iteration of the likewise exact identity which follows from (C.2),

$$e^{Ht} = e^{H_0 t} + \int_0^t dt'\, e^{Ht'} V e^{(t-t')H_0}\,, \tag{C.2′}$$

yields

$$e^{H} = e^{H_0} + \int_0^1 dt\, e^{H_0 t} V e^{(1-t)H_0} +$$
$$+ \int_0^1 dt \int_0^t dt'\, e^{H_0 t'} V e^{(t-t')H_0} V e^{(1-t)H_0} + \ldots +$$
$$+ \int_0^1 dt_1 \int_0^{t_1} dt_2 \ldots \int_0^{t_{n-1}} dt_n e^{H_0 t_n} V e^{(t_{n-1}-t_n)H_0} V e^{(t_{n-2}-t_{n-1})H_0} \ldots$$
$$\times V e^{(1-t_1)H_0} + \ldots\,.$$

$$\tag{C.3}$$

With the substitution

$$1 - t_n = u_n, 1 - t_{n-1} = u_{n-1}, \ldots, 1 - t_1 = u_1$$

we obtain

$$e^H = e^{H_0} + \sum_{n=1}^{\infty} \int_0^1 du_1 \int_{u_1}^1 du_2 \cdots \int_{u_{n-1}}^1 du_n \, e^{(1-u_n)H_0} V e^{(u_n - u_{n-1})H_0} \cdots$$
$$\times V e^{(u_2 - u_1)H_0} V e^{u_1 H_0} .$$

$$(C.3')$$

D. The Riemann ζ-Function and the Bernoulli Numbers

In dealing with fermions, the following integrals occur:

$$\frac{1}{\Gamma(\nu)} \int_0^\infty dx \, \frac{x^{\nu-1}}{e^x + 1} = \sum_{k=1}^{\infty} (-1)^{k+1} \frac{1}{k^\nu}$$

$$= \sum_{k=1}^{\infty} \frac{1}{k^\nu} - 2 \sum_{l=1}^{\infty} \frac{1}{(2l)^\nu} = \left(1 - 2^{1-\nu}\right) \zeta(\nu) . \quad (D.1)$$

After the last equals sign, the Riemann ζ-function

$$\zeta(\nu) = \sum_k \frac{1}{k^\nu} \qquad \text{for Re}\,\nu > 1 \tag{D.2}$$

was introduced. The integrals which occur for bosons can also be related directly to it:

$$\frac{1}{\Gamma(\nu)} \int_0^\infty dx \, \frac{x^{\nu-1}}{e^x - 1} = \sum_{k=1}^{\infty} \frac{1}{k^\nu} = \zeta(\nu) . \tag{D.3}$$

According to the theorem of residues, $\zeta(\nu)$ can be written in the following manner:

$$\zeta(\nu) = \frac{1}{4i} \int_C dz \, \frac{\cot \pi z}{z^\nu} = \frac{1}{4i} \int_{C'} dz \, \frac{\cot \pi z}{z^\nu} . \tag{D.4}$$

Definition of the Bernoulli numbers:

$$\frac{1}{2} z \cot \frac{1}{2} z = 1 - \sum_{n=1}^{\infty} B_n \frac{z^{2n}}{(2n)!} , \tag{D.5}$$

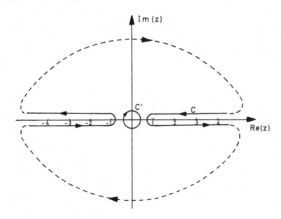

Fig. D.1. The integration path in (D.4)

$$B_1 = \frac{1}{6}, \qquad B_2 = \frac{1}{30}, \qquad B_3 = \frac{1}{42}, \dots$$

$\nu = 2k$:

$$\zeta(2k) = \frac{\pi^{-1}}{4i} \int\limits_{C'} dz \, \frac{1 - \sum_{n=1}^{\infty} B_n \frac{(2z\pi)^{2n}}{(2n)!}}{z^{2k} \pi z} = \frac{(2\pi)^{2k} B_k}{2(2k)!}, \tag{D.6}$$

since only the term $n = k$ makes a nonzero contribution.

$$\frac{1}{\Gamma(2k)} \int dx \, \frac{x^{2k-1}}{e^x + 1} = \frac{\left(2^{2k-1} - 1\right) \pi^{2k} B_k}{(2k)!}$$

$$\int_0^{\infty} dx \, \frac{x^{2k-1}}{e^x + 1} = \frac{\left(2^{2k-1} - 1\right) \pi^{2k} B_k}{2k} \tag{D.7}$$

$$\int_0^{\infty} dx \, \frac{x^{2k-1}}{e^x - 1} = (2k-1)! \frac{(2\pi)^{2k} B_k}{2(2k)!} = \frac{(2\pi)^{2k} B_k}{4k}. \tag{D.8}$$

E. Derivation of the Ginzburg–Landau Functional

For clarity, to carry out the derivation we first consider a system of ferromagnetic Ising spins ($n = 1$), which are described by the Hamiltonian

$$H = -\frac{1}{2} \sum_{l,l'} J(l - l') S_l S_{l'} - h \sum_l S_l, \tag{E.1}$$

where S_l takes on the values $S_l = \pm 1$. We assume a d-dimensional, simple cubic lattice; its lattice constant is taken to be a_0 and the side length of the crystal to be L. This d-dimensional lattice is then divided into cells of volume

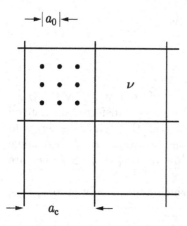

$\rightarrow | a_0 | \leftarrow$

ν

a_c

Fig. E.1. Division of the lattice into cells

$v = a_c^d$, whereby the linear dimensions of the cells a_c are assumed to fulfill the inequality $a_0 \ll a_c \ll L$. The number of cells is $N_c = \left(\frac{L}{a_c}\right)^d = \frac{N}{\tilde{N}}$, and the number of lattice points within a cell is $\tilde{N} = \left(\frac{a_c}{a_0}\right)^d$. Finally, we define the cell spin of cell ν:

$$m_\nu = \frac{1}{\tilde{N}} \sum_{l \in \nu} S_l \,, \tag{E.2}$$

whose range of values lies in the interval $-1 \le m_\nu \le 1$. We now define a new, effective Hamiltonian $\mathcal{F}(\{m_\nu\})$ for the cell spins by carrying out the exact rearrangement

$$Z = \mathrm{Tr}\,e^{-\beta H} \equiv \sum_{\{S_l = \pm 1\}} e^{-\beta H} = \sum_{\{m_\nu\}} \mathrm{Tr}\left(e^{-\beta H} \prod_\nu \delta_{\sum_{l \in \nu} S_l, \tilde{N} m_\nu}\right)$$
$$\equiv \sum_{\{m_\nu\}} e^{-\beta \mathcal{F}(\{m_\nu\})} \,, \tag{E.3}$$

which corresponds to a partial evaluation of the trace, i.e.

$$\mathcal{F}(\{m_\nu\}) = -\frac{1}{\beta} \log \sum_{\{S_l = \pm 1\}} e^{-\beta H} \prod_\nu \delta_{\sum_{l \in \nu} S_l, \tilde{N} m_\nu} \,. \tag{E.4}$$

For sufficiently many spins per cell, m_ν becomes a continuous variable $\left(\Delta m_\nu = \frac{2}{\tilde{N}}\right)$

$$\sum_{m_\nu} \cdots \longrightarrow \frac{\tilde{N}}{2} \int_{-1}^{1} dm_\nu \cdots \tag{E.5}$$

$$\mathcal{F}(\{m_\nu\}) = \tilde{N} f(\{m_\nu\}) \qquad \text{for sufficiently large } \tilde{N} \,.$$

The field term expressed in the new variables is

$$-h\sum_l S_l = -h\sum_\nu \sum_{l\in\nu} S_l = -h\tilde{N}\sum_\nu m_\nu \quad . \tag{E.6}$$

Thus, the factor $e^{-h\sum_l S_l} = e^{-h\tilde{N}\sum_\nu m_\nu}$ is not affected at all by the trace operation after the third equals sign in (E.3), and is transferred unchanged to $\mathcal{F}(\{m_\nu\})$. This has the important effect that all the remaining terms in $\mathcal{F}(\{m_\nu\})$ are independent of h, and due to the invariance of the exchange Hamiltonian (see Chap. 6) under the transformation $\{S_l\} \rightarrow \{-S_l\}$, they are also even functions of the m_ν.

We can decompose $f(\{m_\nu\})$ into terms which depend only on one, two, $\dots m_\nu$:

$$f(\{m_\nu\}) = \sum_{\nu=1}^{N_z} f_1(m_\nu) + \frac{1}{2}\sum_{\mu\neq\nu} f_2^{\nu\mu}(m_\nu, m_\mu) + \dots \quad . \tag{E.7}$$

The Taylor expansion of the functions in (E.7) is given by

$$f_1(m_\nu) = f_1(0) + c_2 m_\nu^2 + c_4 m_\nu^4 + \dots - h m_\nu \tag{E.8a}$$

and

$$f_2(m_\nu, m_\mu) = -\sum_{\mu,\nu} 2K_{\mu\nu} m_\mu m_\nu + \dots \quad . \tag{E.8b}$$

It then follows from (E.3) and (E.5) that

$$Z = \prod_\nu \frac{\tilde{N}}{2} \int_{-1}^{1} dm_\nu e^{-\beta\tilde{N} f(\{m_\nu\})} \tag{E.8c}$$

with

$$f(\{m_\nu\}) = N_c f_1(0) + \sum_\nu \left(a m_\nu^2 + \frac{b}{2} m_\nu^4 + \dots - h m_\nu \right)$$
$$+ \frac{1}{2}\sum_{\mu,\nu} K_{\mu\nu}(m_\mu - m_\nu)^2 + \dots \quad . \tag{E.8d}$$

The coefficients $f_1(0), a, b$ and $K_{\mu\nu}$ are functions of T and $J_{ll'}$. The cells, like the original lattice, form a simple cubic lattice, which is the cell lattice with lattice constants a_z and lattice vectors $\mathbf{a}_\nu$. Let N_i be the number of lattice points (cells) in the direction i, whose product $N_1 N_2 N_3$ must give N_c; we then define the wavevectors with components

$$k_i = \frac{2\pi r_i}{N_i a_c}, \text{where} -\frac{N_i}{2} < r_i \leq \frac{N_i}{2} \quad . \tag{E.9}$$

The reciprocal lattice vectors for the cell lattice are given by

$$\mathbf{g} = \frac{2\pi}{a_z}(n_1, n_2, n_3) \, . \tag{E.10}$$

The Fourier transform of the cell spins is introduced via

$$m_\nu = \frac{1}{\sqrt{N_c}} \sum_{\mathbf{k}} e^{i\mathbf{k}\mathbf{a}_\nu} m_{\mathbf{k}} \tag{E.11a}$$

$$m_{\mathbf{k}} = \frac{1}{\sqrt{N_c}} \sum_{\nu} e^{-i\mathbf{k}\mathbf{a}_\nu} m_\nu \, . \tag{E.11b}$$

The orthogonality and completeness relations of the Fourier coefficients are

$$\sum_{\mathbf{k}} e^{i\mathbf{k}(\mathbf{a}_\nu - \mathbf{a}_{\nu'})} = N_c \delta_{\nu\nu'} \tag{E.12a}$$

$$\sum_{\nu} e^{i(\mathbf{k}-\mathbf{k}')\mathbf{a}_\nu} = N_c \Delta(\mathbf{k} - \mathbf{k}') \equiv N_c \begin{cases} 1 & \text{for } \mathbf{k} - \mathbf{k}' = \mathbf{g} \\ 0 & \text{otherwise} \end{cases} , \tag{E.12b}$$

where $\mathbf{g}$ is an arbitrary vector of the reciprocal lattice.

The transformation of the individual terms of the free energy is given by

$$a \sum_{\nu} m_\nu^2 = a \sum_{\mathbf{k}} m_{\mathbf{k}} m_{-\mathbf{k}} \, ,$$

$$b \sum_{\nu} m_\nu^4 = \frac{b}{N_c^2} \sum_{\nu} \sum_{\mathbf{k}_1,\dots,\mathbf{k}_4} e^{i(\mathbf{k}_1 + \dots + \mathbf{k}_4)\mathbf{a}_\nu} m_{\mathbf{k}_1} \dots m_{\mathbf{k}_4}$$

$$= \frac{b}{N_c} \sum_{\mathbf{k}_1,\dots\mathbf{k}_4} \Delta(\mathbf{k}_1 + \dots + \mathbf{k}_4) m_{\mathbf{k}_1} m_{\mathbf{k}_2} m_{\mathbf{k}_3} m_{\mathbf{k}_4} \, , \tag{E.13a}$$

$$h\tilde{N} \sum_{\nu} m_\nu = h\tilde{N} \sqrt{N_c} m_{\mathbf{k}=0} \quad .$$

Due to translational invariance, the interaction $K_{\mu\nu}$ depends only on the separation,

$$\frac{1}{2} \sum_{\nu,\nu'} K(\nu - \nu')(m_\nu - m_{\nu'})^2$$

$$= \frac{1}{2N_c} \sum_{\nu} \sum_{\delta} \sum_{\mathbf{k}\mathbf{k}'} K(\delta) e^{i\mathbf{k}\mathbf{a}_\nu} \left(1 - e^{i\mathbf{k}\mathbf{a}_\delta}\right) m_{\mathbf{k}} e^{-i\mathbf{k}'\mathbf{a}_\nu} \left(1 - e^{-i\mathbf{k}'\mathbf{a}_\delta}\right) m_{-\mathbf{k}'}$$

$$= \sum_{\mathbf{k}} v(\mathbf{k}) m_{\mathbf{k}} m_{-\mathbf{k}} \, .$$

Here, $\delta \equiv \nu - \nu'$ was introduced, and

$$v(\mathbf{k}) = \sum_\delta K(\delta)(1 - \cos \mathbf{k}\mathbf{a}_\delta) = \sum_\delta K(\delta)\, 2 \sin^2 \frac{\mathbf{k}\mathbf{a}_\delta}{2} \tag{E.13b}$$

was defined. Due to the short range of the interaction coefficients $K(\delta)$, we can expand $\sin^2 \frac{\mathbf{k}\mathbf{a}_\delta}{2}$ for small $\mathbf{k}$ in a Taylor series, and terminate the series after the first term. Taking the cubic symmetry into account, in d dimensions we find

$$v(\mathbf{k}) = \mathbf{k}^2 \frac{1}{2d} \sum_\delta K(\delta)\mathbf{a}_\delta{}^2 + \mathcal{O}(k^4) \quad . \tag{E.13c}$$

Then the partition function in Fourier space is

$$Z = Z_0 \left(\prod_\mathbf{k} \int dm_\mathbf{k} \right) \exp \left\{ - \beta\tilde{N} \left[\sum_\mathbf{k} (a + c\mathbf{k}^2) m_\mathbf{k} m_{-\mathbf{k}} \right.\right.$$

$$\left.\left. + \frac{b}{2} \frac{1}{N_c} \sum_{\mathbf{k}_1 \ldots \mathbf{k}_4} \Delta(\mathbf{k}_1 + \ldots + \mathbf{k}_4)\, m_{\mathbf{k}_1} \ldots m_{\mathbf{k}_4} - h\sqrt{N_c}\, m_{\mathbf{k}=0} + \ldots \right] \right\},$$

$$\tag{E.14}$$

where Z_0 is the part of the partition function which is independent of $m_\mathbf{k}$, as follows from (E.8c).

<u>Definition of $\int dm_\mathbf{k}$:</u>

Due to $m_\mathbf{k}^* = m_{-\mathbf{k}}$, (E.11a) can be written in the form

$$m_\nu = \frac{1}{\sqrt{N_z}} \sum_{\mathbf{k}\in HS} \left(e^{i\mathbf{k}\mathbf{a}_\nu} (\mathrm{Re}\, m_\mathbf{k} + i\mathrm{Im}\, m_\mathbf{k}) \right.$$

$$+ e^{-i\mathbf{k}\mathbf{a}_\nu} (\mathrm{Re}\, m_\mathbf{k} - i\,\mathrm{Im}\, m_\mathbf{k}))$$

$$= \frac{1}{\sqrt{N_z}} \sum_{\mathbf{k}\in HS} \left(\underbrace{\frac{e^{i\mathbf{k}\mathbf{a}_\nu} + e^{-i\mathbf{k}\mathbf{a}_\nu}}{\sqrt{2}} \left(\sqrt{2}\,\mathrm{Re}\, m_\mathbf{k}\right)}_{y_\mathbf{k}} \tag{E.15a}$$

$$+ i\underbrace{\frac{e^{i\mathbf{k}\mathbf{a}_\nu} - e^{-i\mathbf{k}\mathbf{a}_\nu}}{\sqrt{2}} \left(\sqrt{2}\,\mathrm{Im}\, m_\mathbf{k}\right)}_{y_{-\mathbf{k}}} \right),$$

where the sums over $\mathbf{k}$ extend only over half of the $\mathbf{k}$-space (HS). This is an orthogonal transformation

$$\sum_\nu \left(e^{i\mathbf{k}\mathbf{a}_\nu} + e^{-i\mathbf{k}\mathbf{a}_\nu} \right) \left(e^{i\mathbf{k}'\mathbf{a}_\nu} + e^{-i\mathbf{k}'\mathbf{a}_\nu} \right) \frac{1}{2N_c} = \delta_{\mathbf{k},\mathbf{k}'} \; ; \tag{E.15b}$$

and correspondingly for $\sin \mathbf{k}a_\nu$. The cross terms give zero. It follows that

$$\int \prod_\nu dm_\nu \ldots = \int \prod_{\mathbf{k} \in HS} \left(\sqrt{2}d\,\mathrm{Re}\, m_\mathbf{k}\right)\left(\sqrt{2}d\,\mathrm{Im}\, m_\mathbf{k}\right)\ldots = \int \prod_\mathbf{k} dy_\mathbf{k} \ldots \,.$$

(E.15c)

Clearly, from (E.13b),

$$v(\mathbf{k}) = v(-\mathbf{k}) = v(\mathbf{k})^*$$

(E.15d)

and

$$\sum_\mathbf{k} v(\mathbf{k})m_\mathbf{k}m_{-\mathbf{k}} = \sum_{\mathbf{k} \in HS} v(\mathbf{k})\left(\left(\sqrt{2}\,\mathrm{Re}\, m_\mathbf{k}\right)^2 + \left(\sqrt{2}\,\mathrm{Im}\, m_\mathbf{k}\right)^2\right)$$

$$= \sum_\mathbf{k} v(\mathbf{k})y_\mathbf{k}^2 \,. \quad \text{(E.15e)}$$

In the harmonic approximation, it follows from (E.15d), as will be verified in (7.4.47), that

$$\langle m_\mathbf{k}m_{\mathbf{k}'}\rangle = \int \left(\prod_\mathbf{k} dm_\mathbf{k}\right)\frac{e^{-\sum_\mathbf{k} v(\mathbf{k})|m_\mathbf{k}|^2}}{Z}m_\mathbf{k}m_{\mathbf{k}'} = \frac{\delta_{\mathbf{k}',-\mathbf{k}}}{2v(\mathbf{k})} \,.$$

(E.16)

Continuum limit $v = a_c^d \to 0$.

If we consider wavelengths which are large compared to a_z, we take the continuum limit:

$$m(\mathbf{x}_\nu) = \frac{1}{\sqrt{v}}m_\nu$$

(E.17)

$$m(\mathbf{x}) = \frac{1}{\sqrt{N_c a_c^d}}\sum_{\mathbf{k} \in B} e^{i\mathbf{k}\mathbf{x}}m_\mathbf{k} \,.$$

(E.18)

In the strict continuum limit, the Brillouin zone goes to ∞. The terms in the Ginzburg–Landau functional are

$$\sum_\nu m_\nu^2 = \int d^d x\, m(\mathbf{x})^2 \,, \qquad \sum_\nu m_\nu^4 = v \int d^d x\, m(\mathbf{x})^4 \,,$$

$$\sum_\nu h m_\nu = \frac{h}{\sqrt{v}} \int d^d x\, m(\mathbf{x}) \,, \qquad \sum_\mathbf{k} k^2 |m_\mathbf{k}|^2 = \int d^d x \left(\nabla m(\mathbf{x})\right)^2 \,,$$

(E.19)

$$\int \prod_\nu dm_\nu \ldots \to \int \mathcal{D}[m(\mathbf{x})]\ldots \equiv \int \prod_\nu \left(\sqrt{v}dm(\mathbf{x}_\nu)\right) \,.$$

The functional integrals are defined by discretization. Then as our result for the *Ginzburg–Landau functional*, we can write

$$\mathcal{F}[m(\mathbf{x})] = \int d^d x \left[am^2(\mathbf{x}) + \frac{b}{2}m^4(\mathbf{x}) + c(\nabla m)^2 - hm(\mathbf{x}) + \dots \right] \quad \text{(E.20)}$$

which yields the partition function in terms of the following functional integral:

$$Z = Z_0(T) \int \mathcal{D}[m(\mathbf{x})] e^{-\beta \mathcal{F}[m(\mathbf{x})]} . \quad \text{(E.21)}$$

(i) Here, we have redefined the coefficients once again; e.g. $\frac{1}{\sqrt{v}}$ was combined with h. The coefficient $Z_0(T)$ is found from the prefactors defined earlier, but it is not important in what follows.

(ii) Owing to the fact that the trace is only partially evaluated, the coefficients a, b, c and $Z_0(T)$ are "uncritical", i.e. they are not singular in $T, J, \dots$ etc.

(iii) In the following, we extend the integration range for $\int_{-1}^1 dm_\nu \dots =$ $\int_{-1/\sqrt{v}}^{1/\sqrt{v}} dm(\mathbf{x}) \to \int_{-\infty}^\infty dm(\mathbf{x})$, since $m(\mathbf{x})$ is in any case limited by e^{-bm^4}. Its *most probable value is* $m(\mathbf{x}) \sim \sqrt{\frac{-a}{b}}$, and thus $m_\nu \sim \sqrt{v}\sqrt{\frac{a}{b}} \ll 1$.

(iv) General statements about the coefficients in the Ginzburg–Landau functional:

$\alpha)\mathcal{F}[m(\mathbf{x})]$ has the same symmetry as the microscopic spin Hamiltonian; i.e. except for the term with h, $\mathcal{F}[m(\mathbf{x})]$ is an even function of $m(\mathbf{x})$.

$\beta)$ From the preceding rearrangement of the h term, it may be seen that a, b, c are independent of h. In particular, the partial evaluation of the trace produces no higher odd terms.

$\gamma)$ Stability requires $b > 0$. Otherwise, one cannot terminate at m^4. At the tricritical point, $b = 0$ and one must take the term of order m^6 into account, also.

$\delta)$ The ferromagnetic interaction favors parallel spins, i.e. nonuniformity of spin direction costs energy. Thus $c\nabla m\nabla m$ with $c > 0$.

$\epsilon)$ Concerning the temperature dependence of the G.-L. coefficient a, we refer to the main part of the text, Eq. (7.4.8).

(v) In the thermodynamic limit, the linear dimension $L \to \infty$,

$$m(\mathbf{x}) = \frac{1}{L^{d/2}} \sum_{\mathbf{k} \in B} e^{i\mathbf{k}\mathbf{x}} m_{\mathbf{k}} = \frac{1}{L^d} \sum_{\mathbf{k}} e^{i\mathbf{k}\mathbf{x}} m(\mathbf{k})$$

and $m(\mathbf{x}) \overset{L \to \infty}{\Longrightarrow} \int_B \frac{d^d k}{(2\pi)^d} e^{i\mathbf{k}\mathbf{x}} m(\mathbf{k})$, where the integral is extended over the whole Brillouin zone B: $k_i \in \left[-\frac{\pi}{a_c}, \frac{\pi}{a_c} \right]$, and $m(\mathbf{k}) = L^{d/2} m_{\mathbf{k}}$.

Later, the integral over the cubic Brillouin zone will be approximated by an integral over a sphere:

$$m(\mathbf{x}) = \int_{|\mathbf{k}| < \Lambda} \frac{d^d k}{(2\pi)^d} e^{i\mathbf{k}\mathbf{x}} m(\mathbf{k}) .$$

F. The Transfer Matrix Method

The transfer matrix method is an important tool for the exact solution of statistical-mechanical models. It is particularly useful for two-dimensional and one-dimensional models. We introduce the transfer matrix method by solving the *one-dimensional Ising model*. The one-dimensional *Ising model for N spins with interactions between nearest neighbors* is described by the Hamiltonian

$$\mathcal{H} = -J \sum_{j=1}^{N} \sigma_j \sigma_{j+1} - H \sum_{j=1}^{N} \sigma_j , \tag{F.1}$$

where periodic boundary conditions are assumed, $\sigma_{N+1} = \sigma_1$. The partition function ($K \equiv \beta J,\ h \equiv \beta H$) has the form

$$Z_N = \sum_{\{\sigma_i = \pm 1\}} \prod_{j=1}^{N} e^{K\sigma_j \sigma_{j+1} + \frac{h}{2}(\sigma_j + \sigma_{j+1})} = \mathrm{Tr}\left(T^N\right) . \tag{F.2}$$

Following the second equals sign, the *transfer matrix*, defined by

$$T_{\sigma\sigma'} \equiv e^{K\sigma\sigma' + \frac{h}{2}(\sigma + \sigma')} , \tag{F.3}$$

was introduced. Its matrix representation is given by

$$T = \begin{pmatrix} e^{K+h} & e^{-K} \\ e^{-K} & e^{K-h} \end{pmatrix} . \tag{F.4}$$

One readily finds the two eigenvalues of this (2×2) matrix:

$$\lambda_{1,2} = e^{K} \cosh h \pm \left(e^{-2K} + e^{2K} \sinh^2 h\right)^{1/2} . \tag{F.5}$$

The trace in (F.2) is invariant under orthogonal transformations. By transforming to the basis in which T is diagonal, one can verify that

$$Z_N = \lambda_1^N + \lambda_2^N . \tag{F.6}$$

The *free energy* per spin is given by the logarithm of the partition function

$$f(T, H) = -kT \lim_{N \to \infty} \frac{1}{N} \log Z_N . \tag{F.7}$$

In the thermodynamic limit, $N \to \infty$, the largest eigenvalue dominates:

$$f = -kT \log \lambda_1 \quad \text{(due to } \lambda_1 \geq \lambda_2 \text{ for all } T \geq 0\text{)} . \tag{F.8}$$

There is no phase transition in one dimension, since

$$f(T,0) = -kT\bigl[\log 2 + \log(\cosh\beta J)\bigr] \tag{F.9}$$

is a smooth function for $T > 0$. Owing to the short range of the interactions, disordered spin configurations (with high entropy S) are more probable than ordered configurations (with low internal energy E) in equilibrium (where $F(T,0) = E - TS$ is a minimum). The isothermal susceptibility $\chi = \left(\frac{\partial m}{\partial H}\right)_T$ for $H = 0$ is found from (F.8) to be $\chi = \beta e^{2\beta J}$ at low T. There is a pseudo-phase transition at $T = 0$: $m_0^2 = 1$, $\chi_0 = \infty$.

Magnetization: $m = -\partial f/\partial B$ 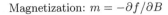   Specific Heat: $C_H = -T\partial^2 f/\partial T^2$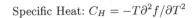

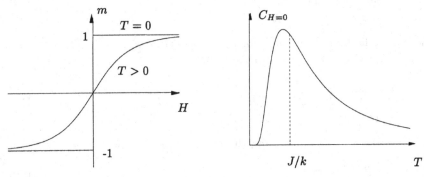

Fig. F.1. The magnetization and specific heat in the one-dimensional Ising model

The *spin correlation function* can also be expressed using the transfer matrix and computed:

$$
\begin{aligned}
\langle\sigma_k\sigma_l\rangle_N &\equiv \frac{1}{Z_N}\sum_{\{\sigma_i=\pm 1\}} e^{-\beta\mathcal{H}}\sigma_k\sigma_l \\
&= \frac{1}{Z_N}\sum_{\{\sigma_i=\pm 1\}} T_{\sigma_1\sigma_2}\ldots T_{\sigma_{k-1}\sigma_k}\sigma_k T_{\sigma_k\sigma_{k+1}}\ldots T_{\sigma_{l-1}\sigma_l} \\
&\qquad\qquad\qquad\times\, \sigma_l T_{\sigma_l\sigma_{l+1}}\ldots T_{\sigma_{N-1}\sigma_1} \\
&= \frac{1}{Z_N}\sum_{\pm}\langle\chi_\pm|\,T^k\tau_z T^{l-k}\tau_z T^{N-l}\,|\chi_\pm\rangle \\
&= Z_N^{-1}\,\mathrm{Tr}\left(\tau_z T^{l-k}\tau_z T^{N-l+k}\right),\quad \tau_z\equiv\begin{pmatrix}1 & 0\\ 0 & -1\end{pmatrix},\quad l\geq k\,.
\end{aligned}
\tag{F.10}
$$

To distinguish them from $\sigma_i = \pm 1$, the Pauli matrices are denoted here by $\tau_{x,y,z}$. The trace in the last line of (F.10) refers to the sum of the two diagonal matrix elements in the Pauli spinor states $\chi_\pm$. Further evaluation is carried out by diagonalizing T:

$$\Gamma\, T\,\Gamma^{-1} = \begin{pmatrix}\lambda_1 & 0\\ 0 & \lambda_2\end{pmatrix} \equiv \Lambda\,,\quad\text{where}\quad \Gamma = \frac{1}{\sqrt{2}}\begin{pmatrix}1 & 1\\ -1 & 1\end{pmatrix}\,.$$

With

$$\Gamma \tau_z \Gamma^{-1} = - \begin{pmatrix} 0 & 1 \\ 1 & 0 \end{pmatrix} \equiv -\tau_x \, ,$$

it follows from (F.10) that

$$\langle \sigma_k \sigma_l \rangle_N = Z_N^{-1} \text{Tr}(\tau_x \Lambda^{l-k} \, \tau_x \Lambda^{N-l+k})$$

and thus for $l - k \ll N$ in the thermodynamic limit $N \to \infty$ we obtain the final result

$$\langle \sigma_k \sigma_l \rangle = \left(\frac{\lambda_2}{\lambda_1} \right)^{l-k} \, . \tag{F.11}$$

For $T > 0$, $\lambda_2 < \lambda_1$, i.e. $\langle \sigma_k \sigma_l \rangle$ decreases with increasing distance $l - k$. For $T \to 0$, $\lambda_1 \to \lambda_2$ (asymptotic degeneracy), so that the correlation length $\xi \to \infty$.

By means of the transfer matrix method, the one-dimensional Ising model is mapped onto a zero-dimensional quantum system (one single spin). The two-dimensional Ising model is mapped onto a one-dimensional quantum system. Since it is possible to diagonalize the Hamiltonian of the latter, the two-dimensional Ising model can in this way be solved exactly.

G. Integrals Containing the Maxwell Distribution

$$f^0(\mathbf{v}) = n \left(\frac{m}{2\pi kT} \right)^{3/2} e^{-\frac{mv^2}{2kT}} \tag{G.1a}$$

$$\int d^3v \, f^0(\mathbf{v}) = n \tag{G.1b}$$

$$\int d^3v \left(\frac{mv^2}{2} \right)^s f^0(\mathbf{v}) = n \left(\frac{m}{2\pi kT} \right)^{3/2} \left(-\frac{\partial}{\partial(1/kT)} \right)^s \underbrace{\int d^3v \, e^{-\frac{mv^2}{2kT}}}_{\left(\frac{\pi}{m/2kT} \right)^{3/2}} \tag{G.1c}$$

$$= n(kT)^s \frac{3}{2} \frac{5}{2} \cdots \frac{1+2s}{2} \qquad s = 1, 2, \ldots$$

$$\int d^3v \, \frac{mv^2}{2} f^0(\mathbf{v}) = \frac{3}{2} nkT \tag{G.1d}$$

$$\int d^3v \left(\frac{mv^2}{2} \right)^2 f^0(\mathbf{v}) = \frac{15}{4} n(kT)^2 \tag{G.1e}$$

$$\int d^3v \left(\frac{mv^2}{2} \right)^3 f^0(\mathbf{v}) = \frac{105}{8} n(kT)^3 \tag{G.1f}$$

$$\int d^3v\, v_k v_i v_j v_l f^0(\mathbf{v}) = \lambda\big(\delta_{ki}\delta_{jl} + \delta_{kj}\delta_{il} + \delta_{kl}\delta_{ij}\big)\,, \quad \lambda = \frac{kT}{m} \qquad \text{(G.1g)}$$

Eq. (G.1g) can be demonstrated by first noting that the result necessarily has the form given and then taking the sum $\sum_{k=i}\sum_{j=l}$: comparison with (G.1e) leads using $\int d^3v\,\left(\mathbf{v}^2\right)^2 f^0(\mathbf{v}) = 15\lambda$ to the result $\lambda = \frac{kT}{m}$.

H. Hydrodynamics

In the appendix, we consider the microscopic derivation of the linear hydrodynamic equations. The hydrodynamic equations determine the behavior of a system at low frequencies or over long times. They are therefore the equations of motion of the conserved quantities and of variables which are related to a broken continuous symmetry. Nonconserved quantities relax quickly to their local equilibrium values determined by the conserved quantities. The conserved quantities (energy, density, magnetization...) can exhibit a time variation only by flowing from one spatial region to another. This means that the equations of motion of conserved quantities $E(\mathbf{x})$ typically have the form $\dot{E}(\mathbf{x}) = -\nabla \mathbf{j}_E(\mathbf{x})$. The gradient which occurs here already indicates that the characteristic rate (frequency, decay rate) for the conserved quantities is proportional to the wavenumber q. Since $\mathbf{j}_E$ can be proportional to conserved quantities or to gradients of conserved quantities, hydrodynamic variables exhibit a characteristic rate $\sim q^\kappa$, i.e. a power of the wavenumber q, where in general $\kappa = 1, 2$. In the case of a broken continuous symmetry there are additional hydrodynamic variables. Thus, in an isotropic antiferromagnet, the alternating (staggered) magnetization $\mathbf{N}$ is not conserved. In the ordered phase, its average value is finite and may be oriented in an arbitrary direction in space. Therefore, it costs no energy to rotate the staggered magnetization. This means that microscopic variables which represent fluctuations transverse to the staggered magnetization likewise belong to the set of hydrodynamic variables, Fig. H.1.

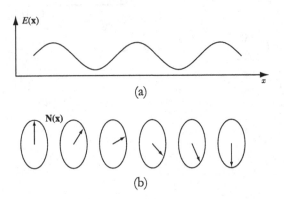

(a)

(b)

Fig. H.1. Conserved quantities and broken-symmetry variables: (a) the energy density $E(\mathbf{x})$ and (b) the alternating (staggered) magnetization $\mathbf{N}(\mathbf{x})$ in an isotropic or planar antiferromagnet

H.1 Hydrodynamic Equations, Phenomenological Discussion

In order to gain some insight into the structure of the hydrodynamic equations, we first want to consider a simple example: the hydrodynamics of a ferromagnet, for which only the magnetization density is conserved. The magnetization density $M(\mathbf{x})$ obeys the equation of continuity

$$\dot{M}(\mathbf{x}) = -\nabla \mathbf{j}_M(\mathbf{x}) \ . \tag{H.1}$$

Here, $\mathbf{j}_M$ is the magnetization current density. This becomes larger the greater the difference between magnetic fields at different positions in the material. From this fact, we obtain the phenomenological relation

$$\mathbf{j}_M(\mathbf{x}) = -\lambda \nabla H(\mathbf{x}) \ , \tag{H.2}$$

where λ is the magnetization conductivity. The local magnetic field depends on the magnetization density via the relation

$$H(\mathbf{x}) = \frac{1}{\chi} M(\mathbf{x}) \ , \tag{H.3}$$

in which the magnetic susceptibility χ enters. Inserting (H.3) into (H.2) and the latter into (H.1), one finds the diffusion equation

$$\dot{M}(\mathbf{x}, t) = D \nabla^2 M(\mathbf{x}, t) \ , \tag{H.4}$$

where the magnetization diffusion constant is defined by

$$D = \frac{\lambda}{\chi} \ .$$

To solve (H.4), it is expedient to apply a spatial Fourier transform; then the diffusion equation (H.4) takes on the form

$$\dot{M}_{\mathbf{q}} = -Dq^2 M_{\mathbf{q}} \ , \tag{H.5}$$

with the obvious result

$$M_{\mathbf{q}}(t) = e^{-Dq^2 t} M_{\mathbf{q}}(0) \ . \tag{H.6}$$

The diffusive relaxation rate Dq^2 decreases as the wavenumber becomes smaller. For several variables $X_{\mathbf{q}}^c$, whose deviations from equilibrium are denoted by $\delta\langle X_{\mathbf{q}}^c \rangle$, the hydrodynamic equations have the general form

$$\frac{\partial}{\partial t} \delta\langle X_{\mathbf{q}}^c \rangle + M^{cc'}(\mathbf{q}) \, \delta\langle X_{\mathbf{q}}^{c'} \rangle = 0 \ . \tag{H.7}$$

Here, $M^{cc'}(\mathbf{q})$ is a matrix which vanishes as $\mathbf{q} \to 0$. For the hydrodynamics of liquids, we recall Eq. (9.4.46a–c).

H.2 The Kubo Relaxation Function

In linear response theory[14], one investigates the effects of an external force $F(t)$ which couples to the operator B. The Hamiltonian then contains the additional term

$$H'(t) = -F(t)B .$$
(H.8)

For the change of the expectation value of an operator A with respect to its equilibrium value, one obtains to first order in $F(t)$

$$\delta\langle A(t)\rangle = \int_{-\infty}^{\infty} dt' \chi_{AB}(t-t')F(t')$$
(H.9)

with the dynamic susceptibility

$$\chi_{AB}(t-t') = \frac{i}{\hbar}\Theta(t-t')\langle[A(t),B(t')]\rangle .$$
(H.10)

Its Fourier transform reads

$$\chi_{AB}(\omega) = \int_{-\infty}^{\infty} dt e^{i\omega t}\chi_{AB}(t) .$$
(H.11)

We now consider a perturbation which is slowly switched on and then again switched off at the time $t = 0$: $F(t) = e^{\epsilon t}\Theta(-t)F$. One then finds from (H.9)

$$\delta\langle A(t)\rangle = \int_{-\infty}^{\infty} dt' \chi_{AB}(t-t')F\Theta(-t')e^{\epsilon t'}$$
$$= \int_{t}^{\infty} du \chi_{AB}(u)F e^{\epsilon(t-u)} ,$$
(H.12)

where the substitution $t - t' = u$ has been employed. The decay of the perturbation for $t > 0$ is thus described by

$$\delta\langle A(t)\rangle = \phi_{AB}(t)F e^{\epsilon t} ,$$
(H.13)

where[15] the *Kubo relaxation function* $\phi_{AB}(t)$ is defined by

$$\phi_{AB}(t) \equiv \frac{i}{\hbar}\int_{t}^{\infty} dt' \langle[A(t'),B(0)]\rangle e^{-\epsilon t'} .$$
(H.14)

Its half-range Fourier transform is given by

$$\phi_{AB}(\omega) \equiv \int_{0}^{\infty} dt\, e^{i\omega t}\phi_{AB}(t) .$$
(H.15)

---

[14] QM II, Sect. 4.3
[15] The factor $e^{\epsilon t}$ is of no importance in (H.13), since $\phi_{AB}(t)$ relaxes faster.

The Kubo relaxation function has the following properties:

$$\phi_{AB}(t=0) = \chi_{AB}(\omega = 0) , \tag{H.16}$$

$$\dot{\phi}_{AB}(t) = -\chi_{AB}(t) \qquad \text{for } t > 0 , \tag{H.17}$$

$$\phi_{AB}(\omega) = \frac{1}{i\omega}\left(\chi_{AB}(\omega) - \chi_{AB}(\omega = 0)\right) . \tag{H.18}$$

Eq. (H.16) follows from the comparison with the Fourier transformed dynamical susceptibility, Eq. (H.11). The second relation is obtained immediately by taking the derivative of (H.14). The third relation can be obtained by half-range Fourier transformation of (H.17)

$$-\int_0^\infty dt\, e^{i\omega t}\chi_{AB}(t) = \int_0^\infty dt\, e^{i\omega t}\dot{\phi}_{AB}(t)$$

$$= e^{i\omega t}\phi_{AB}(t)\Big|_0^\infty - i\omega \int_0^\infty dt\, e^{i\omega t}\phi_{AB}(t) = \phi_{AB}(t=0) - i\omega\phi_{AB}(\omega)$$

and application of $\phi_{AB}(t=\infty) = 0$, (H.16) and

$$\int_0^\infty dt\, e^{i\omega t}\chi_{AB}(t) = \int_{-\infty}^\infty dt\, e^{i\omega t}\chi_{AB}(t) = \chi_{AB}(\omega) .$$

Further, one can show for $t \geq 0$ that

$$\phi_{\dot{A}B}(t) = \int_t^\infty dt'\frac{i}{\hbar}\left\langle[\dot{A}(t'), B(0)]\right\rangle e^{-\epsilon t'} = -\frac{i}{\hbar}\left\langle[A(t), B(0)]\right\rangle = -\chi_{AB}(t) ,$$

i.e.

$$\phi_{\dot{A}B}(\omega) = -\chi_{AB}(\omega) \tag{H.19}$$

and, together with (H.18),

$$\omega\phi_{AB}(\omega) = i\phi_{\dot{A}B}(\omega) + i\chi_{AB}(\omega = 0) . \tag{H.20}$$

Later, we will also require the identity

$$\chi_{\dot{A}B^\dagger}(\omega = 0) = \frac{i}{\hbar}\int_0^\infty dt'\left\langle[\dot{A}(t'), B(0)^\dagger]\right\rangle = -\frac{i}{\hbar}\left\langle[A(0), B(0)^\dagger]\right\rangle , \tag{H.21}$$

which follows from the Fourier transform of (H.10) and the fact that the expectation value $\left\langle[A(\infty), B(0)^\dagger]\right\rangle$ vanishes.

H.3 The Microscopic Derivation of the Hydrodynamic Equations

H.31 Hydrodynamic Equations and Relaxation

We introduce the following notation here:

$$X^i(\mathbf{x}, t) \qquad i = 1, 2, \ldots \qquad\qquad \text{densities (Hermitian)}$$

$$X^i_{\mathbf{q}}(t) = \frac{1}{\sqrt{V}} \int d^3 x \, e^{-i\mathbf{q}\mathbf{x}} X^i(\mathbf{x}, t) \qquad \text{Fourier transforms} \qquad (\text{H.22a})$$

$$X^i(\mathbf{x}, t) = \frac{1}{\sqrt{V}} \sum_{\mathbf{q}} e^{i\mathbf{q}\mathbf{x}} X^i_{\mathbf{q}}(t) \,, \qquad\qquad X^{i\dagger}_{\mathbf{q}} = X^i_{-\mathbf{q}} \qquad (\text{H.22b})$$

$$\chi^{ij}(\mathbf{q}, t) \equiv \chi_{X^i_{\mathbf{q}}, X^j_{-\mathbf{q}}}(t) \qquad\qquad \text{etc.}$$

Conserved densities are denoted by indices $c, c', \ldots$ etc. and nonconserved densities by $n, n', \ldots$.

We now consider a perturbation which acts on the conserved densities. At $t = 0$, it is switched off, so that the perturbation Hamiltonian takes on the form

$$H' = -\int d^3 x X^c(\mathbf{x}, t) K^c(\mathbf{x}) \Theta(-t) e^{\epsilon t} = -\sum_{\mathbf{q}} X^c_{-\mathbf{q}}(t) K^c_{\mathbf{q}} \Theta(-t) e^{\epsilon t}$$

and leads according to Eq. (H.13) to the following changes in the conserved quantities for $t > 0$:

$$\delta\langle X^c_{\mathbf{q}}(t) \rangle = \phi^{cc'}(\mathbf{q}, t) K^{c'}_{\mathbf{q}} e^{\epsilon t} \,. \qquad\qquad (\text{H.23})$$

The decay of the perturbation is determined by the relaxation function.

The situation considered here is, on the other hand, also described by the hydrodynamic equations (H.7)

$$\left\{ \delta^{cc'} \frac{\partial}{\partial t} + M^{cc'}(\mathbf{q}) \right\} \delta\langle X^{c'}_{\mathbf{q}}(t) \rangle = 0 \,. \qquad\qquad (\text{H.24a})$$

If we insert Eq. (H.23) into (H.24a), we obtain

$$\left\{ \delta^{cc'} \frac{\partial}{\partial t} + M^{cc'}(\mathbf{q}) \right\} \phi^{c'c''}(\mathbf{q}, t) K^{c''}_{\mathbf{q}} = 0 \,.$$

Since this equation is valid for arbitrary $K^{c''}_{\mathbf{q}}$, it follows that

$$\left\{ \delta^{cc'} \frac{\partial}{\partial t} + M^{cc'}(\mathbf{q}) \right\} \phi^{c'c''}(\mathbf{q}, t) = 0 \,. \qquad\qquad (\text{H.24b})$$

From this, we find $\phi^{c'c''}(\mathbf{q}, \omega)$ by taking the half-range Fourier transform

$$\int_0^\infty dt\, e^{i\omega t} \left\{ \delta^{cc'} \frac{\partial}{\partial t} + M^{cc'}(\mathbf{q}) \right\} \phi^{c'c''}(\mathbf{q}, t) = 0$$

and carrying out an integration by parts, i.e.

$$-\phi^{cc''}(\mathbf{q}, t = 0) + \left\{ -i\omega \delta^{cc'} + M^{cc'}(\mathbf{q}) \right\} \phi^{c'c''}(\mathbf{q}, \omega) = 0 .$$

Using (H.16), we obtain finally

$$\left\{ -i\omega \delta^{cc'} + M^{cc'}(\mathbf{q}) \right\} \phi^{c'c''}(\mathbf{q}, \omega) = \chi^{cc''}(\mathbf{q}) . \tag{H.24c}$$

Therefore, the relaxation functions $\phi^{cc'}(\mathbf{q}, \omega)$ and thus the dynamic response functions are obtained from hydrodynamics for small $\mathbf{q}$ and ω.

If, conversely, we can determine $\phi^{cc'}(\mathbf{q}, \omega)$ for small $\mathbf{q}$ and ω from a microscopic theory, we can then read off the hydrodynamic equations by comparing with (H.24a) and (H.24c).

We consider an arbitrary many-body system (liquid, ferromagnet, antiferromagnet, etc.) and divide up the complete set of operators $X_{\mathbf{q}}^i$ into conserved quantities $X_{\mathbf{q}}^c$ and nonconserved quantities $X_{\mathbf{q}}^n$. Our strategy is to find equations of motion for the $X_{\mathbf{q}}^c$, where the forces are decomposed into a part which is proportional to the $X_{\mathbf{q}}^c$ and a part which is proportional to the $X_{\mathbf{q}}^n$. The latter fluctuate rapidly and will lead after its elimination to damping terms in the equations of motion, which then contain only the $X_{\mathbf{q}}^c$.

In order to visualize this decomposition in a clear way, it is expedient to choose the operators in such a manner that they are orthogonal. To do this, we must first define the *scalar product* of two operators A and B:

$$\langle A|B \rangle = \chi_{A,B^\dagger}(\omega = 0) . \tag{H.25}$$

Remark: one can readily convince oneself that this definition fulfills the properties of a scalar product:

$$\langle A|B \rangle^* = \langle B|A \rangle$$
$$\langle c_1 A_1 + c_2 A_2 | B \rangle = c_1 \langle A_1 | B \rangle + c_2 \langle A_2 | B \rangle$$
$$\langle A|A \rangle \quad \text{is real and } \langle A|A \rangle \geq 0 \ (0 \text{ only for } A \equiv 0) .$$

We now choose our operators to be orthonormalized:

$$\langle X_{\mathbf{q}}^i | X_{\mathbf{q}}^j \rangle = \chi^{ij}(\mathbf{q}, \omega = 0) = \delta^{ij} . \tag{H.26}$$

To construct these operators, one uses the Schmidt orthonormalization procedure.

The Heisenberg equations of motion $\dot{X}_{\mathbf{q}}^c = \frac{i}{\hbar}[H, X_{\mathbf{q}}^c]$ etc. can now be written in the form

$$\dot{X}_{\mathbf{q}}^c = -iC^{cc'}(\mathbf{q})X_{\mathbf{q}}^{c'} - iC^{cn}(\mathbf{q})X_{\mathbf{q}}^n \tag{H.27a}$$
$$\dot{X}_{\mathbf{q}}^n = -iC^{nc}(\mathbf{q})X_{\mathbf{q}}^c - iD^{nn'}(\mathbf{q})X_{\mathbf{q}}^{n'} . \tag{H.27b}$$

Here, the derivatives $\dot{X}_{\mathbf{q}}^{c}$ and $\dot{X}_{\mathbf{q}}^{n}$ were projected onto $X_{\mathbf{q}}^{c'}$ and $X_{\mathbf{q}}^{n'}$. If we take e.g. the scalar product of $\dot{X}_{\mathbf{q}}^{c}$ with $X_{\mathbf{q}}^{c'}$, we find using Eq. (H.21)

$$\langle \dot{X}_{\mathbf{q}}^{c} | X_{\mathbf{q}}^{c'} \rangle \equiv -iC^{cc'}(\mathbf{q}) = -\frac{i}{\hbar}\langle [X_{\mathbf{q}}^{c}, X_{\mathbf{q}}^{c'\dagger}] \rangle \,.$$

That is:

$$C^{cc'}(\mathbf{q}) = \frac{1}{\hbar}\langle [X_{\mathbf{q}}^{c}, X_{\mathbf{q}}^{c'\dagger}] \rangle \,, \tag{H.28a}$$

and analogously

$$C^{cn}(\mathbf{q}) = \frac{1}{\hbar}\langle [X_{\mathbf{q}}^{c}, X_{\mathbf{q}}^{n\dagger}] \rangle \,, \qquad D^{nn'}(\mathbf{q}) = \frac{1}{\hbar}\langle [X_{\mathbf{q}}^{n}, X_{\mathbf{q}}^{n'\dagger}] \rangle \,. \tag{H.28}$$

These coefficients obey the following symmetry relations:

$$C^{cc'*}(\mathbf{q}) = C^{c'c}(\mathbf{q}) \,, \;\; C^{nc*}(\mathbf{q}) = C^{cn}(\mathbf{q}) \,, \;\; D^{nn'*}(\mathbf{q}) = D^{n'n}(\mathbf{q}) \,. \tag{H.29}$$

It thus follows from (H.20) that

$$\omega\phi^{cc'}(\mathbf{q},\omega) = C^{cc''}(\mathbf{q})\phi^{c''c'}(\mathbf{q},\omega) + C^{cn}(\mathbf{q})\phi^{nc'}(\mathbf{q},\omega) + i\delta^{cc'}$$
$$\omega\phi^{nc}(\mathbf{q},\omega) = C^{nc'}(\mathbf{q})\phi^{c'c}(\mathbf{q},\omega) + D^{nn'}(\mathbf{q})\phi^{n'c}(\mathbf{q},\omega)$$
$$\omega\phi^{nn'}(\mathbf{q},\omega) = C^{nc}(\mathbf{q})\phi^{cn'}(\mathbf{q},\omega) + D^{nn''}(\mathbf{q})\phi^{n''n'}(\mathbf{q},\omega) + i\delta^{nn'} \tag{H.30}$$
$$\omega\phi^{cn}(\mathbf{q},\omega) = C^{cc'}(\mathbf{q})\phi^{c'n}(\mathbf{q},\omega) + C^{cn'}(\mathbf{q})\phi^{n'n}(\mathbf{q},\omega) \,.$$

From (H.30b) we read off the result

$$\phi^{nc}(\mathbf{q},\omega) = (\omega\mathbb{1} - D(\mathbf{q}))_{nn'}^{-1} C^{n'c'}(\mathbf{q})\phi^{c'c}(\mathbf{q},\omega) \,; \tag{H.31}$$

then inserting (H.31) into (H.30a) leads to

$$\left[\omega\delta^{cc'} - C^{cc'}(\mathbf{q}) - C^{cn}(\mathbf{q}) \left(\frac{1}{\omega\mathbb{1} - D(\mathbf{q})} \right)_{nn'} C^{n'c'}(\mathbf{q}) \right] \phi^{c'c''}(\mathbf{q},\omega)$$
$$= i\delta^{cc''} \,. \tag{H.32}$$

For the conserved quantities, the coefficients $C^{cc'}(\mathbf{q})$ and $C^{cn}(\mathbf{q})$ vanish in the limit $\mathbf{q} \to 0$. Therefore, in the limit of small $\mathbf{q}$, we find

$$i\left(C^{cc'}(\mathbf{q})X_{\mathbf{q}}^{c'} + C^{cn}(\mathbf{q})X_{\mathbf{q}}^{n} \right) = iq_{\alpha}j_{\alpha}^{c}(\mathbf{q}) \,. \tag{H.33}$$

We define also the nonconserved part of the current density

$$C^{cn}(\mathbf{q})X_{\mathbf{q}}^{n} = q_{\alpha}\tilde{j}_{\alpha}^{c}(\mathbf{q}) \,. \tag{H.34}$$

In contrast to (H.33) and (H.34), the $D_{nn'}(\mathbf{q})$ remain finite in the limit $\mathbf{q} \to 0$. For the behavior at long wavelengths ($\mathbf{q} \to 0$), we can therefore take $\frac{1}{\omega\mathbb{1} - D}$

in the limit $\omega, \mathbf{q} \to 0$, whereby due to the finiteness of $D(\mathbf{q})$, we can expect that

$$\lim_{\omega \to 0} \lim_{\mathbf{q} \to 0} \frac{1}{\omega \mathbb{1} - D(\mathbf{q})} = \lim_{\mathbf{q} \to 0} \lim_{\omega \to 0} \frac{1}{\omega \mathbb{1} - D(\mathbf{q})} . \tag{H.35}$$

In the limit $\mathbf{q} \to 0$, we can find a relation between $\frac{1}{\omega \mathbb{1} - D}$ from Eq. (H.30c) and a correlation function. Owing to $\lim_{\mathbf{q} \to 0} C^{nc}(\mathbf{q}) = 0$, it follows from (H.30c), with the abbreviation $D \equiv \lim_{\mathbf{q} \to 0} D(\mathbf{q})$, that

$$\left(\omega \mathbb{1} - D \right)^{nn''} \lim_{\mathbf{q} \to 0} \phi^{n''n'}(\mathbf{q}, \omega) = i \delta^{nn'}$$

or

$$\left(\frac{1}{\omega \mathbb{1} - D} \right)_{nn'} = -i \lim_{\mathbf{q} \to 0} \phi^{nn'}(\mathbf{q}, \omega) . \tag{H.36}$$

Inserting this into (H.32) and taking the the double limits, we obtain

$$\left(\omega \delta^{cc'} - C^{cc'}(\mathbf{q}) + i C^{cn}(\mathbf{q}) \left(\lim_{\omega \to 0} \lim_{\mathbf{q} \to 0} \phi^{nn'}(\mathbf{q}, \omega) \right) C^{n'c'}(\mathbf{q}) \right) \phi^{c'c''}(\mathbf{q}, \omega)$$
$$= i \delta^{cc''} ,$$

i.e. finally:

$$\left(\omega \delta^{cc'} - C^{cc'}(\mathbf{q}) + i q_\alpha q_\beta \Gamma_{\alpha\beta}^{cc'} \right) \phi^{c'c''}(\mathbf{q}, \omega) = i \delta^{cc''} , \tag{H.37}$$

with the damping coefficients

$$\Gamma_{\alpha\beta}^{cc'} \equiv \lim_{\omega \to 0} \lim_{q \to 0} \phi_{\tilde{j}_\alpha^c \tilde{j}_\beta^{c'}}(q, \omega) . \tag{H.38}$$

Here, the sums over n and n' were combined into the nonconserved current densities defined in Eq. (H.34).

When the system exhibits symmetry under reflection, rotation, etc., the number of nonvanishing coefficients $\Gamma_{\alpha\beta}^{cc'}$ can be reduced. We assume that for the remaining functions $\phi_{\tilde{j}_\alpha^c \tilde{j}_\beta^{c'}}$, the operators j_c and $j_{c'}$ have the same signature[16] under time reversal: $\epsilon_{j_c} = \epsilon_{j_{c'}}$. Applying (H.18) and the dispersion relations[17], one obtains

$$\phi(\omega) = -\frac{i}{\omega}(\chi'(\omega) - \chi(0)) + \frac{\chi''(\omega)}{\omega}$$
$$= -\frac{i}{\pi} P \int d\omega' \frac{\chi''(\omega')}{(\omega' - \omega)\omega'} + \underbrace{\frac{\chi''(\omega)}{\omega}}_{\frac{1-e^{-\beta\hbar\omega}}{2\hbar\omega} G^>(\omega)} , \tag{H.39}$$

---

[16] QM II, Sect. 4.8.2.2
[17] QM II, Sect. 4.4

which, owing to the fluctuation-dissipation theorem[18] and the antisymmetry of $\chi''(\omega)$, finally leads to $\lim_{\omega \to 0} \phi(\omega) = \lim_{\omega \to 0} \frac{\beta}{2} G^>(\omega)$ and

$$\Gamma_{\alpha\beta}^{cc'} = \frac{1}{2kT} \lim_{\omega \to 0} \lim_{\mathbf{q} \to 0} \int_{-\infty}^{\infty} dt\, e^{i\omega t} \langle \tilde{j}_{\alpha\mathbf{q}}^c(t) \tilde{j}_{\beta-\mathbf{q}}^{c'}(0) \rangle . \tag{H.40}$$

This is the *Kubo formula* for the transport coefficients, expressed in terms of current-current correlation functions. Without taking up their straightforward proofs, we mention the following symmetry properties:

$$\Gamma_{\alpha\beta}^{cc'*} = \Gamma_{\beta\alpha}^{c'c} , \quad \Gamma_{\alpha\alpha}^{cc} > 0 , \quad \Gamma_{\alpha\beta}^{cc'} = \Gamma_{\beta\alpha}^{c'c} \text{ real.} \tag{H.41}$$

In summary, one can read off the following *linear hydrodynamic equations* by comparison with Eqns. (H.24c) and (H.24a):

$$\left[\frac{\partial}{\partial t} \delta^{cc'} + iC^{cc'}(\mathbf{q}) + q_\alpha q_\beta \Gamma_{\alpha\beta}^{cc'} \right] \delta \langle X_\mathbf{q}^{c'}(t) \rangle = 0 , \tag{H.42a}$$

$$C^{cc'}(\mathbf{q}) = \frac{1}{\hbar} \langle [X_\mathbf{q}^c, X_{-\mathbf{q}}^{c'}] \rangle , \tag{H.42b}$$

$$\Gamma^{cc'}(\mathbf{q}) = \frac{1}{4kT} \lim_{\omega \to 0} \lim_{\mathbf{q} \to 0} \int_{-\infty}^{\infty} dt\, e^{i\omega t} \langle \{ j_\mathbf{q}^c(t), j_{-\mathbf{q}}^{c'}(0) \} \rangle . \tag{H.42c}$$

The elements of the *frequency matrix* $C^{cc'}(\mathbf{q}) \sim q$ (or q^2) are functions of expectation values of the conserved quantities and the order parameters and susceptibilities of these quantities. They determine the periodic, reversible behavior of the dynamics. For example, for a ferromagnet, the spinwave frequency (H.42b)follows from $\omega(\mathbf{q}) = \frac{M}{\chi_\mathbf{q}^T} \propto q^2$, where M is the magnetization and $\chi_\mathbf{q}^T \propto q^{-2}$ the transverse susceptibility. The damping terms result from the elimination of the nonconserved degrees of freedom. They can be expressed via Kubo formulas for the current densities. For the derivation it was important that the nonconserved quantities have a much shorter time scale than the conserved quantities, which also permits taking the limit $\lim_{\omega \to 0} \lim_{\mathbf{q} \to 0}$. We note the similarity of this procedure to that used in the case of the linearized Boltzmann equation (Sect. 9.4). The present derivation is more general, since no constraints were placed on the density or the strength of the interactions of the many-body system.

Literature:
H. Mori, Prog. Theor. Phys. (Kyoto) **33**, 423 (1965); **34**, 399 (1965); **28**, 763 (1962)
F. Schwabl and K. H. Michel, Phys. Rev. **B2**, 189 (1970)
K. Kawasaki, Ann. Phys. (N.Y.) **61**, 1 (1970)

---

[18] QM II, Sect. 4.6

I. Units and Tables

In this Appendix we give the definitions of units and constants which are used in connection with thermodynamics. We also refer to the Table on page 562.

Conversion Factors

$$1 \text{ eV} = 1.60219 \times 10^{-19} \text{ J}$$
$$1 \text{ N} = 10^5 \text{ dyn}$$
$$1 \text{ J} = 1 \times 10^7 \text{ erg}$$
$$1 \text{ C} = 2.997925 \times 10^9 \text{ esu} = 2.997925 \times 10^9 \sqrt{\text{dyn cm}^2}$$
$$1 \text{ K} \; \hat{=} \; 0.86171 \times 10^{-4} \text{ eV}$$
$$1 \text{ eV} \; \hat{=} \; 2.4180 \times 10^{14} \text{ Hz} \; \hat{=} \; 1.2399 \times 10^{-4} \text{ cm}$$
$$1 \text{ T} = 10^4 \text{ Gauss (G)}$$
$$1 \text{ Å} = 10^{-8} \text{ cm}$$
$$1 \text{ sec} \equiv 1 \text{ s}$$

Pressure

$$1 \text{ bar} = 10^6 \text{dyn/cm}^2 = 10^5 \text{N/m}^2 = 10^5 \text{Pa}$$
$$1 \text{ Torr} = 1 \text{ mm Hg}$$

Physical Atmosphere:

$$1 \text{ atm} = \text{air pressure at 760 mm Hg} \equiv 760 \text{ Torr} = 1.01325 \text{ bar}$$

This relation between Torr and bar follows from the mass density of mercury $\rho_{\text{Hg}} = 13.5951 \text{g cm}^{-3}$ at $1°$C and the acceleration of gravity $g = 9.80655 \times 10^2 \text{cm s}^{-2}$. Technical Atmosphere:

$$1 \text{ at} = 1 \text{ kp/cm}^2 = 0.980655 \text{ bar}$$

Temperature

The absolute temperature scale was defined in Sect. 3.4 using $T_t = 273.16$ K, the triple point of H_2O.

The zero point of the Celsius scale $0°$C lies at 273.15 K. Thus in this scale, absolute zero is at $-273.15°$C. With this definition, the equilibrium temperature of ice and water saturated with air under a pressure of 760 mm Hg $\equiv$ 1 atm is equal to $0°$C.

Table I.1. Fixed points of the international temperature scale:

$0°$C	ice point of water
$100°$C	equilibrium temperature of water and water vapor
$-182.970°$C	boiling point of oxygen
$444.600°$C	boiling point of sulfur
$960.8°$C	solidification point of silver
$1063.0°$C	solidification point of gold

For a comparative characterization of materials, their properties are quoted at
standard temperatures and pressures. In the physics literature, these are $0°C$ and
1 atm, and in the technical literature, they are $20°C$ and 1 at.

Physical Standard State $\equiv$ standard pressure (1 atm) and standard temperature
$(0°C)$.

Technical Standard State $\equiv$ 1 at and $20°C$.

Density of H_2 at T_t and $P = 1$ atm:

$$\rho = 8.989 \times 10^{-2} g/\text{Liter} = 8.989 \times 10^{-5} g\ cm^{-3}\ .$$

Molar volume under these conditions:

$$V_M = \frac{2.016\,g}{8.989 \times 10^{-2}\,g\,\text{Liter}^{-1}} = 22.414\,\text{Liter}\ \left(\hat{=}\,22.414\frac{\text{Liter}}{\text{mole}}\right)\ .$$

1 mole $\hat{=}$ atomic weight in g (e.g. one mole of H_2 corresponds to a mass of 2.016 g).

$$k = \frac{PV}{NT} = \frac{1\,\text{atm}\,V_M}{L \times 273.16\,K} = 1.38065 \times 10^{-16} erg/K\ .$$

Loschmidt's number $\equiv$ Avogadro's number:

$$L \equiv N_A = \text{number of molecules per mole}$$
$$= \frac{2.016\,g}{\text{mass}\ H_2} = \frac{2.016}{2 \times 1.6734 \times 10^{-24}} = 6.02213 \times 10^{23}\ .$$

Energy

The unit calorie (cal) is defined by

$$1\ \text{cal} = 4.1840 \times 10^7 erg = 4.1840\ \text{Joule}\ .$$

A kilocalorie is denoted by Cal (large calorie). With the previous definition, 1 Cal
up to the fourth place past the decimal point has the meaning

$$1\ \text{Cal} \equiv 1\ \text{kcal} \equiv 1000\ \text{cal}$$

= the quantity of heat which is required to warm 1 kg H_2O at 1 atm from 14.5 to
$15.5°C$[19].

$$1\ J = 1\ Nm = 10^7\ \text{dyn}\ cm = 10^7\ \text{erg}\ .$$

Power

$$1\ W = 1\ VA = 1\ J\ s^{-1} = 10^7 erg\ s^{-1}$$
$$1\ HP = 75\ \text{kp}\ m\ s^{-1} = 75 \times 9.80665 \times 10^5 \text{dyn}\ m\ s^{-1} = 735.498\ W\ .$$

The universal gas constant R is defined via Loschmidt's/Avogadro's number by

$$R = N_A k = 8.3145 \times 10^7 erg\ \text{mol}^{-1}K^{-1}\ .$$

Using the gas constant R, one can write the equation of state of the ideal gas in
the form

$$PV = nRT\ , \tag{I.1}$$

where n is the amount of matter in moles (mole number).

---

[19] Note that the nutritional values of foods are quoted either in kJ or in kilocalories.

We close this section with some numerical values of *thermodynamic quantities*. Table I.2, below, gives values of specific heats (C_P).

As can be seen, the specific heat of water is particularly large. This fact plays an important role in the thermal balance of Nature. Water must take up or release a large quantity of heat in order to change its temperature noticeably. Therefore, the water of the oceans remains cool for a relatively long time in Spring and warm for a relatively long time in Autumn. It therefore acts in coastal regions to reduce the annual temperature fluctuations. This is an essential reason for the typical difference between a coastal climate and a continental climate.

Table I.2. The Specific Heat of Some Materials under Standard Conditions

	Specific heat C [cal K^{-1} g^{-1}]	Molecular weight	Molar heat capacity [cal K^{-1} mole^{-1}]
Aluminum	0.214	27.1	5.80
Iron	0.111	55.84	6.29
Nickel	0.106	58.68	6.22
Copper	0.091	63.57	5.78
Silver	0.055	107.88	5.93
Antimony	0.050	120.2	6.00
Platinum	0.032	195.2	6.25
Gold	0.031	197.2	6.12
Lead	0.031	207.2	6.42
Glass	0.19	—	—
Quartz Glass	0.174	—	—
Diamond	0.12	—	—
Water	1.00	—	—
Ethanol	0.58	—	—
Carbon Disulfide	0.24	—	—

Table I.3. Expansion Coefficients of Some Solid and Liquid Materials in K^{-1}

linear				volume	
Lead	0.0000292	Diamond	0.0000013	Ethanol	0.0011
Iron	120	Graphite	080	Ether	163
Copper	165	Glass	081	Mercury	018
Platinum	090	Quartz Crystal $\perp$ axis	144	Water	018
Invar (^{64}Fe+^{36}Ni)	016	Quartz Crystal $\parallel$ axis	080		
		Quartz Glass	005		

The *linear expansion coefficient* α_l is related to the volume or cubic expansion coefficient in (3.2.4) via

$$\alpha = 3\alpha_l \ .$$

This follows for a rectangular prism from $V + \Delta V = (a + \Delta a)(b + \Delta b)(c + \Delta c) = abc\left(1 + \frac{\Delta a}{a} + \frac{\Delta b}{b} + \frac{\Delta c}{c}\right) + O(\Delta^2)$, thus $\frac{\Delta V}{V} = 3\frac{\Delta a}{a}$ under the assumption of isotropic thermal expansion as found in isotropic materials (liquids, amorphous substances) and cubic crystals.

Table I.4. Some Data for Gases: Boiling Point (at 760 Torr), Critical Temperature, Coefficients in the van der Waals Equation, Inversion Temperature

Gas	Boiling point in K	$T_c[K]$	$a\left[\dfrac{\text{atm cm}^6}{\text{mole}^2}\right]$	$b\left[\dfrac{\text{cm}^3}{\text{mole}}\right]$	$T_{\text{inv}} = \frac{27}{4}T_c[K]$
He	4.22	5.19	0.0335×10^6	23.5	35
H_2	20.4	33.2	0.246×10^6	26.7	224
N_2	77.3	126.0	1.345×10^6	38.6	850
O_2	90.1	154.3	1.36×10^6	31.9	1040
CO_2	194.7	304.1	3.6×10^6	42.7	2050

Table I.5. Pressure Dependence of the Boiling Point of Water

Pressure in Torr	Boiling Point in °C
720	98.49
730	98.89
740	99.26
750	99.63
760	100.00
770	100.37
780	100.73
790	101.09
800	101.44

Table I.6. Heats of Vaporization of Some Materials in $\text{cal} \cdot \text{g}^{-1}$

Ethyl Alcohol	202
Ammonia	321
Ether	80
Chlorine, Cl_2	62
Mercury	68
Oxygen, O_2	51
Nitrogen, N_2	48
Carbon Disulfide	85
Water	539.2
Hydrogen, H_2	110

Table I.7. Heats of Melting of Some Materials in $\text{cal} \cdot \text{g}^{-1}$

Aluminum	94	Silver	26.0
Lead	5.5	Table Salt	124
Gold	15.9	Water (Ice)	79.5
Copper	49		

Table I.8.   Vapor Pressure of Water (Ice) in Torr

$-60°$C	0.007
$-40°$C	0.093
$-20°$C	0.77
$+0°$C	4.6
$+20°$C	17.5
$+40°$C	55.3
$+60°$C	149.4
$+80°$C	355.1
$+100°$C	760.0
$+200°$C	11665,0

Table I.9.   Vapor Pressure of Iodine in Torr

$-48.3°$C	0.00005
$-32.3°$C	0.00052
$-20.9°$C	0.0025
$0°$C	0.029
$15°$C	0.131
$30°$C	0.469
$80°$C	15.9
$114.5°$C	90.0 (melting point)
$185.3°$C	760.0 (boiling point)

Table I.10.   Freezing Mixtures and Other Eutectics

Constituents with Melting Points		Eutectic Temperature in °C	Concentration
NH_4Cl	Ice (0)	-15.4	
NaCl	Ice (0)	-21	29/71 NaCl
Alcohol	Ice (0)	-30	
$CaCl_2 \cdot 6H_2O$	Ice (0)	-55	
Alcohol	CO_2(-56)	-72	
Ether	CO_2(-56)	-77	
Sn (232)	Pb (327)	183	74/26
Au (1063)	Si (1404)	370	69/31
Au (1063)	Tl (850)	131	27/73

Table I.11. Important Constants

Quantity	Symbol or Formula (cgs)	Symbol or Formula (SI)	Numerical Value and Unit (cgs)	Numerical Value and Unit (SI System)
Atomic Quantities:				
Planck's Constant	h	h	6.626075×10^{-27} erg s	6.626075×10^{-34} J s $\triangleq 4.135669 \times 10^{-15}$ eV s
	$\hbar = h/2\pi$	$\hbar = h/2\pi$	1.054572×10^{-27} erg s	1.054572×10^{-34} J s $\triangleq 6.582122 \times 10^{-16}$ eV s
Elementary Charge	e_0	e_0	4.80324×10^{-10} esu	1.602177×10^{-19} C
Speed of Light in Vacuum	c	c	2.997925×10^{10} cm s^{-1}	2.997925×10^{8} m s^{-1}
Atomic Mass Unit (amu)	$1\,\mathrm{u} = 1\,\mathrm{amu} = \frac{1}{12} m_{C^{12}}$		1.660540×10^{-24} g	1.660540×10^{-27} kg $\triangleq 931.5$ MeV
Rest Mass of the Electron	m_e		9.109389×10^{-28} g	9.109389×10^{-31} kg $\triangleq 5.485799 \times 10^{-4}$ amu
–Rest Energy	$m_e c^2$			0.510999 MeV
Rest mass of the Proton	m_p		1.672623×10^{-24} g	1.672623×10^{-27} kg $\triangleq 1.0072764$ amu
–Rest Energy	$m_p c^2$			938.272 MeV
Rest Mass of the Neutron	m_n		1.674928×10^{-24} g	1.674928×10^{-27} kg $\triangleq 1.0086649$ amu
–Rest Energy	$m_n c^2$			939.565 MeV
Mass Ratio Proton : Electron	m_p/m_e			1836.152
Mass Ratio Neutron : Proton	m_n/m_p			1.0013784
Specific Charge of the Electron	e_0/m_e		5.272759×10^{17} esu/g	1.758819×10^{11} C/kg
Classical Electron Radius	r_e — $\dfrac{e_0^2}{m_e c^2}$	$\dfrac{e_0^2}{4\pi\varepsilon_0 m_e c^2}$	2.817940×10^{-13} cm	2.817940×10^{-15} m
Compton Wavelength of the Electron	λ_c — $h/m_e c$	$h/m_e c$	2.426310×10^{-10} cm	2.426310×10^{-12} m
	$\bar{\lambda}_c$ — $\hbar/m_e c$	$\hbar/m_e c$	3.861593×10^{-11} cm	3.861593×10^{-13} m
Sommerfeld's Fine Structure Constant	α — $\dfrac{e_0^2}{\hbar c}$	$\dfrac{e_0^2}{4\pi\varepsilon_0 \hbar c}$		$\dfrac{1}{137.035989}$
Bohr Radius of the Hydrogen Atom's Ground State	a — $\dfrac{\hbar^2}{m_e e_0^2}$	$\dfrac{4\pi\varepsilon_0 \hbar^2}{m_e e_0^2}$	5.291772×10^{-9} cm	5.291772×10^{-11} m
Rydberg Constant (Ground State Energy of Hydrogen)	Ry — $\frac{1}{2} m_e c^2 \alpha^2$	$\frac{1}{2} m_e c^2 \alpha^2$	2.179874×10^{-11} erg	2.179874×10^{-18} J $\triangleq 13.6058$ eV
Bohr Magneton	μ_B — $\dfrac{e_0 \hbar}{2m_e c}$	$\dfrac{e_0 \hbar}{2m_e}$	9.274015×10^{-21} erg G^{-1}	9.274015×10^{-24} J T^{-1}
Nuclear Magneton	μ_K — $\dfrac{e_0 \hbar}{2m_p c}$	$\dfrac{e_0 \hbar}{2m_p}$	5.050786×10^{-24} erg G^{-1}	5.050786×10^{-27} J T^{-1}
Magnetic Moment of the Electron	μ_e		9.28477×10^{-21} erg G^{-1}	9.28477×10^{-24} J T^{-1} $= 1.00115965\,\mu_B$
Magnetic Moment of the Proton	μ_p		1.410607×10^{-23} erg G^{-1}	1.410607×10^{-26} J T^{-1} $= 2.792847\,\mu_K$

Table I.11. (Continuation)

Thermodynamic Quantities:

Quantity	Symbol		
Boltzmann's Constant	k_B	$1.380658 \times 10^{-16}\,\mathrm{erg\,K^{-1}}$	$1.380658 \times 10^{-23}\,\mathrm{J\,K^{-1}}$
Gas Constant	R	$8.314510 \times 10^{7}\,\mathrm{erg\,mole^{-1}K^{-1}}$	$8.314510 \times 10^{3}\,\mathrm{J\,kmole^{-1}K^{-1}}$
Avogadro's Number	L	$6.0221367 \times 10^{23}\,\mathrm{mole^{-1}}$	$6.0221367 \times 10^{26}\,\mathrm{kmole^{-1}}$
Molar Volume (Ideal Gas) under Standard Conditions	V_0	$22.41410 \times 10^{3}\,\mathrm{cm^3 mole^{-1}}$	$22.41410\,\mathrm{m^3 kmole^{-1}}$
Standard Pressure (atmospheric pressure)	P_0	$1.01325 \times 10^{6}\,\mathrm{dyn\,cm^{-2}}$	$1.01325 \times 10^{5}\,\mathrm{Pa}$
Standard Temperature	T_0		$273.15\mathrm{K}(\hat{=}0^{\circ}\mathrm{C})$
Temperature of the Triple Point of Water	T_t		$273.16\mathrm{K}$
Stefan–Boltzmann Radiation Constant	σ	$5.67051 \times 10^{-5}\,\mathrm{erg\,s^{-1}cm^{-2}K^{-4}}$	$5.67051 \times 10^{-8}\,\mathrm{W\,m^{-2}K^{-4}}$
Wien's Displacement Constant	A	$0.2897756\,\mathrm{cm\,K}$	$2.897756 \times 10^{-3}\,\mathrm{m\,K}$
Number of Gas Molecules per $\mathrm{cm^3}$ under Standard Conditions	$n_0 = L/V_M$	$2.686763 \times 10^{19}\,\mathrm{cm^{-3}}$	$2.686763 \times 10^{25}\,\mathrm{m^{-3}}$

Gravitation and Electrodynamics:

Quantity	Symbol		
Gravitational Constant	G	$6.67259 \times 10^{-8}\,\mathrm{dyn\,cm^2 g^{-2}}$	$6.67259 \times 10^{-11}\,\mathrm{N\,m^2 kg^{-2}}$
Standard Acceleration of Gravity	g	$9.80665 \times 10^{2}\,\mathrm{cm\,s^{-2}}$	$9.80665\,\mathrm{m\,s^{-2}}$
Permeability Constant of Vacuum	μ_0		$4\pi \times 10^{-7}\,\mathrm{N\,A^{-2}} = 1.2566 \times 10^{-6}\,\mathrm{N\,A^{-2}}$
Dielectric Constant of Vacuum $\left\{ \begin{array}{l} \epsilon_0 = 1/(\mu_0 c^2) \\ 1/(4\pi\epsilon_0) \end{array} \right.$			$8.85418 \times 10^{-12}\,\mathrm{C^2 m^{-2}N^{-1}}$ $8.98755 \times 10^{9}\,\mathrm{N\,m^2 C^{-2}}$

Table I.12. Ground States of Ions with Partially Filled Shells According to Hund's Rules

d – shell($l = 2$)

n	$l_z = 2$	1	0	-1	-2	S	$L = \lvert\sum l_z\rvert$	J	$^{2S+1}L_J$
1	↑					1/2	2	3/2	$^2D_{3/2}$
2	↑	↑				1	3	2	$^3F_2$
3	↑	↑	↑			3/2	3	3/2	$^4F_{3/2}$
4	↑	↑	↑	↑		2	2	0	$^5D_0$
5	↑	↑	↑	↑	↑	5/2	0	5/2	$^6S_{5/2}$
6	↑↓	↑	↑	↑	↑	2	2	4	$^5D_4$
7	↑↓	↑↓	↑	↑	↑	3/2	3	9/2	$^4F_{9/2}$
8	↑↓	↑↓	↑↓	↑	↑	1	3	4	$^3F_4$
9	↑↓	↑↓	↑↓	↑↓	↑	1/2	2	5/2	$^2D_{5/2}$
10	↑↓	↑↓	↑↓	↑↓	↑↓	0	0	0	$^1S_0$

f – shell($l = 3$)

n	$l_z = 3$	2	1	0	-1	-2	-3	S	$L = \lvert\sum l_z\rvert$	J	$^{2S+1}L_J$
1	↑							1/2	3	5/2	$^2F_{5/2}$
2	↑	↑						1	5	4	$^3H_4$
3	↑	↑	↑					3/2	6	9/2	$^4I_{9/2}$
4	↑	↑	↑	↑				2	6	4	$^5I_4$
5	↑	↑	↑	↑	↑			5/2	5	5/2	$^6H_{5/2}$
6	↑	↑	↑	↑	↑	↑		3	3	0	$^7F_0$
7	↑	↑	↑	↑	↑	↑	↑	7/2	0	7/2	$^8S_{7/2}$
8	↑↓	↑	↑	↑	↑	↑	↑	3	3	6	$^7F_6$
9	↑↓	↑↓	↑	↑	↑	↑	↑	5/2	5	15/2	$^6H_{15/2}$
10	↑↓	↑↓	↑↓	↑	↑	↑	↑	2	6	8	$^5I_8$
11	↑↓	↑↓	↑↓	↑↓	↑	↑	↑	3/2	6	15/2	$^4I_{15/2}$
12	↑↓	↑↓	↑↓	↑↓	↑↓	↑	↑	1	5	6	$^3H_6$
13	↑↓	↑↓	↑↓	↑↓	↑↓	↑↓	↑	1/2	3	7/2	$^2F_{7/2}$
14	↑↓	↑↓	↑↓	↑↓	↑↓	↑↓	↑↓	0	0	0	$^1S_0$

↑= spin $\frac{1}{2}$; ↓= spin $-\frac{1}{2}$.

Subject Index

absolute temperature, 91
absolute zero, 513
acoustic resonances, 466
activation energy, 425
adiabatic change, 515
adiabatic equation, 102
- of the ideal quantum gas, 174
allotropy, 520
ammonia synthesis, 155
amplitude ratios, 372
anharmonic effects, 211
antiferromagnet, 336, 337, 548
antiferromagnetism, 287–288
Arrhenius law, 425
atmosphere, 167
average value, 65
average-potential approximation, 244
Avogadro's number, 92, 558

background radiation, 203, 508
barometric pressure formula, 53, 72, 414
barrier, see reaction rates
BBGKY hierarchy, 442
Bernoulli numbers, 228, 537
Bethe lattice, 393–398
- percolation threshold, 394
Bethe–Peierls approximation, 327
binary alloys, 517
black body, 203
black holes, 188, 508
black-body radiation, 198, 203
Bloch equations, 429
- in the ferromagnetic phase, 435
block-spin transformation, 359, 360
Bohr magneton, 269
Bohr–van Leeuwen theorem, 276
boiling boundary, 133
boiling point
- numerical values for some materials, 560

boiling-point elevation, 263, 265
Boltzmann constant, 36, 92
- experimental determination of, 92, 414
Boltzmann equation, 437–475
- and irreversibility, 443, 445, 505
- derivation of, 438–443
- linearized, 455–468
- symmetry properties of, 440, 444, 476
Boltzmann's entropy, 480, 498–499
- in the urn model, 512
bond percolation, 389
Bose distribution function, 171
Bose–Einstein condensation, 190–197, 223
Bose–Einstein statistics, 170
bosons, 170
- second virial coefficient, 532
Bravais crystal, 208
Brillouin function, 281
Brownian motion, 409–410
- in a force field, 414
- in the limit of strong damping, 414
- microscopic model of, 484–490, 510
- of a sphere in a liquid, 414
bubble point line, 165
Buckingham potential, 240
bulk viscosity, 465, 477

canonical momentum, 269
canonical variables, 80
Carnot cycle, 126, 162
- efficiency of, 127
- inverse, 127
catalyst, 154
cavity radiation, 198, 203
Cayley tree, see Bethe lattice
central limit theorem, 7, 9
characteristic function, 5
chemical constants, 152, 235